POROSITY OF
CERAMICS

MATERIALS ENGINEERING

Additional Volumes in Preparation

POROSITY OF CERAMICS

Roy W. Rice

Consultant
Alexandria, Virginia

MARCEL DEKKER, INC.   NEW YORK · BASEL · HONG KONG

Library of Congress Cataloging-in-Publication Data

Rice, Roy W.
     Porosity of ceramics / Roy W. Rice
        p. cm. -- (Materials engineering ; 12)
     Includes bibliographical references and index.
     ISBN 0-8247-0151-8
     1. Ceramic materials. 2. Porosity. I. Title. II. Series:
Materials engineering (Marcel Dekker. Inc.) ; 12.
TA455.C43R53 1998
620.1'404299--dc21

                                                                             97-52813
                                                                                              CIP

This book is printed on acid-free paper.

Headquarters
Marcel Dekker, Inc.
270 Madison Avenue, New York, NY 10016
tel: 212-696-9000; fax: 212-685-4540

Eastern Hemisphere Distribution
Marcel Dekker AG
Hutgasse 4, Postfach 812, CH-4001 Basel, Switzerland
tel: 44-61-261-8482; fax: 44-61-261-8896

World Wide Web
http://www.dekker.com

The publisher offers discounts on this book when ordered in bulk quantities. For more information, write to Special Sales/Professional Marketing at the headquarters address above.

Current printing (last digit):
10 9 8 7 6 5 4 3 2 1

PRINTED IN THE UNITED STATES OF AMERICA

To Dr. Fennimore N. Bradley—initially a chance acquaintance responsible for my becoming involved in the field of ceramics who then became a longtime friend.

PREFACE

This book is intended to provide a comprehensive view of the porosity dependence of properties, i.e., to be part handbook and part reference book. Both the effects of porosity that are simply the result of available processing technology and the effects of purposely introduced porosity are addressed. The primary focus is on ceramics, especially nominally single (solid) phase ceramics. However, attention is also given to effects of porosity in other materials, including traditional ceramics, ceramic composites, cements and plasters, rocks, and metals, to further show underlying needs and principles, as well as aid understanding of porosity effects in these materials. The major emphasis is on the porosity dependence of mechanical properties, in which the focus is on lower and moderate temperatures where brittle fracture dominates (high temperature creep and stress rupture are briefly discussed). Microcrack effects are also treated, not only because they are important on their own, but also because microcracks are a limiting form of pores. Attention is given to other physical properties, such as surface area, electrical and thermal conductivity, electrical breakdown, dielectric constant, and piezoelectric and magnetic properties. These properties are noteworthy in themselves, but their different dependences on porosity also aid in better understanding and engineering its effects.

The major goal in writing this book was to provide a much broader perspective on the effects, characterization, and modeling of porosity. This is necessary because so many measurements are being done over a narrow range of materials and porosity, often with no characterization of the porosity other than the volume fraction, and because much modeling is being derived or applied without adequate understanding and evaluation of the applicability and limits of the models and the realities of porosity. To provide this perspective, as well as accommodate the many readers primarily interested in porosity effects on one or a few properties, major properties are treated in separate chapters, along with issues such as the self-consistency of the data and models for that property. Other important, almost universally neglected, topics such as the effects of pore orientation and heterogeneity of shape, size, and spatial distribution are addressed to the extent feasible and to point out the need for improved characterization. Since many results, concepts, and models are in conflict, material is presented in a review rather than a textbook format. However, to assist the diversity of research and engineering interests to which this book can be applied, each chapter has a summary and recommendations. Thus, the book is structured to provide detailed insight for a particular need as well as a comprehensive view.

Another significant motivation for writing this book was to aid the many and growing applications for porous materials. Thus, engineering needs are also addressed, for example by noting practical as well as scientific needs for better characterization of porosity, correlation, and differences in the porosity dependence of different properties and true prediction of properties, not just more curve fitting. The final chapter summarizes many applications for porous materials, as well as many of the processing technologies to better tailor porosities for specific applications. This chapter also discusses the important topic of engineering trade-offs that are commonly needed in designing porous materials for a particular application.

Several people contributed in a variety of ways to the development of this book. Dr. Melvin Leitheiser of the 3M Company indirectly planted the idea of this and a companion book in my mind. Dr. Dave Lewis and especially my daughter and son-in-law, Colleen and Philip Montgomery, aided in establishing my computer capability, as well as preparing some figures. Several others assisted by reading drafts of chapters, providing comments and sometimes additional references: Dr. Rasto Brezny (W. R. Grace, Chapter 6), Dr. Aldo R. Boccaccini (Institute of Mechanics and Materials, University of California, San Diego, Chapters 1–10). Dr. L. Eric Cross (Materials Research Laboratory, Pennsylvania State University, Chapter 7), Dr. Dave Lewis (U.S. Naval Research Laboratory, Chapters 1–10), Dr. Robert Ruh (Air Force Materials Laboratory, Chapters 1 and 2), and Dr. Francis J. Young (University of Illinois, Chapter 8). Finally, Drs. Steve Freiman and Sheldon Wiedreharn and Mr. George Quinn of NIST are thanked for making me a visiting scientist there and thus providing me with library access.

Roy W. Rice

CONTENTS

Contents   ix

Symbols and Abbreviations

In covering such a diverse array of properties, it is impossible to avoid some overlap of mathematical symbol designation. The following common definitions, plus any alternate uses, are given in the text in conjunction with their applications.

SYMBOLS

A   parameter in porosity-property relations
B   bulk modulus
b   exponential parameter for porosity dependence (e^{-bP})
C   electrical conductivity
c   flaw size (typically the radius)
D   grain size (diameter)
E   Young's modulus
F   force
G   shear modulus
H   hardness
h   height
K   thermal conductivity
L,l   length
m   Weibull modulus
n   exponent for porosity dependence
P   volume fraction porosity
t   thickness

w   width
X   porosity dependent property
∈   dielectric constant (also strain)
λ   mean free path
ν   Poisson's ratio
ρ   density (subscript t = theoretical)
σ   stress

ABBREVIATIONS

CNB   chevron notch beam test
CVD   chemical vapor deposition
CT   compact tension test
DCB   double cantilever beam test
DT   double torsion test
EA   elastic anisotropy
F   fractography determination of fracture energy/toughness
GS   grain size
HIP   hot isostatic press
IF   indentation fracture test
NB   notch beam test(s)
PCNB   precracked NB test(s)
PG   pyrolitic (CVD) graphite
PSZ   partially stabilized zirconia
PZT   lead zirconate titanate
RSSC   reaction sintered silicon carbide
RSSN   reaction sintered silicon nitride
SSA   specific surface area
TEA   thermal expansion anisotropy
TZP   tetragonal polycrystal
WOF   work of fracture tests

POROSITY OF CERAMICS AND ITS EFFECTS ON PROPERTIES

1

OVERVIEW: POROSITY (AND MICROCRACK) DEPENDENCE OF PROPERTIES

KEY CHAPTER GOALS

1. Introduce the purpose of this book, the subjects covered, and their organization.
2. Identify sources and occurrences of pores and microcracks, and the diversity, mixes and changes of pores.
3. Outline measurement techniques and issues for porous bodies.

1.1 INTRODUCTION

The purpose of this chapter is to introduce and put in perspective the evaluation of primarily the porosity, and secondarily the microcrack, dependence of physical properties of solids. This includes a perspective on the interaction of other microstructural changes accompanying and impacting porosity and microcrack effects on properties. The diverse sources, mixes, changes, and ranges of porosity as well as issues for good measurement of porosity and microcrack effects on properties are discussed. Although there is much less information on microcracking and it is much less pervasive, its effects are also addressed (in a more limited fashion) because they are also important and have parallels with porosity, in part since such cracks represent one limit of pores. These topics will be addressed after a brief outline of the aspects of this book, in particular the properties and materials covered and their organization.

The first of two key aspects of this book is that it is written so it can be used as a reference (i.e., reading one, or a few, sections or chapters for some specific

information; as a survey for a broader perspective; or as a comprehensive treatise.) To help achieve these goals, the final section of each chapter is a summary
of results, needs, and key recommendations and there is considerable cross-ref-
erencing of results within and between various chapters. The second key aspect
is paying attention to scientific as well as practical (i.e., engineering) results and
needs. The main emphasis of this book is on mechanical properties, especially
from ≤ 22 to $\sim 1000°C$, where brittle fracture dominates and the most data and
broadest needs exist; however, both nonmechanical properties and property
dependence at higher temperatures are also covered, although in less detail.

To address the goals of this book, Chapter 2 treats the porosity parameters
impacting property dependences, characterizing porosity, and approaches to
modeling porosity dependences. Chapters 3–5 cover elastic properties
(Young's, bulk, and shear moduli, and Poission's ratio), crack propagation (slow
crack growth, fracture energy and toughness, and R-curve and wake effects),
and tensile strength (mainly uniaxial) with observations on reliability. These are
followed by a chapter on the porosity and microcrack dependence of other
mechanical properties, primarily those involving compressive loading, especially hardness, compressive strength, and wear and erosion. An important task,
one that is seldom done but is undertaken in this book, is assessing the consistency of test results and models that are based on the underlying relations
between various mechanical properties, including the self-consistency of related
properties. This can be done in the most precise fashion for elastic properties,
but is also extended to other mechanical properties via their correlation with
elastic moduli.

Following the above four chapters on mechanical properties, Chapter 7
addresses nonmechanical properties, particularly electrical and thermal conductivity (at and near room temperature). An evaluation is also made of the porosity
dependence of other solid properties, such as the dielectric constant and other
electrical and magnetic properties, since they are important by themselves and
because they provide some guidance for the microstructural dependence of
other properties. In addition to these properties of the solid modified by the
porosity, key aspects of properties of the pores, namely surface area and fluid
flow, are outlined because of their key role in important applications of special
highly porous bodies.

Chapter 8 addresses the porosity and microcrack dependence of the solid
properties of other traditional and designed ceramic composites and other materials (e.g., polymers, metals, cements and plasters, and rocks). These show the
broad similarities of mainly porosity and secondarily microcrack dependence of
properties across many materials as well as some broader variations, both of
which reinforce the underlying principles and effects of porosity and microcracks on properties. Then the more limited data on the temperature effect on
the porosity dependence of properties of bodies is addressed in a Chapter 9 to

further aid in the breadth of perspective and give greater coherence to the temperature effects in view of the much more limited data for any one property. Since much of the occurrence and manifestation of microcracks occurs as a result of temperature changes, they are more extensively treated in this chapter. Chapter 10 is both an overall summary of porosity and microcrack dependence and the need to improve our evaluation and understanding of them, as well as an outline of applications for special porous bodies and the technologies for making such bodies.

1.2 UNDERSTANDING THE MICROSTRUCTURAL DEPENDENCE OF PROPERTIES

1.2.1 Background and Basic Needs

A thorough understanding of the microstructural dependence of ceramic properties is needed since this is the critical link between their forming and processing (i.e., fabrication) and resulting properties and performance. [The terms *fabrication* and *processing* are often interchanged loosely, particularly for powder-based processing of ceramics, which is the dominant mode of manufacturing, and even more dominant in research, on ceramics. Thus, many use the term *processing* to refer mainly to the effects of particle character, green density, and sintering parameters on microstructure, especially density, that are obtained mainly from die pressing of small (e.g., cm scale) pellets. Much less attention is given to issues of forming specific shapes by die pressing or other forming methods (e.g., isopressing, injection molding, extrusion, slip casting) and the effects of binders, mixing, and forming-consolidation-size and shape in making actual uniform components. These highly interactive aspects of making parts from powders are better reflected in the term *fabrication*, which include the important subtopics of forming and processing.] Different fabrication methods and parameters lead to a variety of microstructural ranges which vary in their application and results with different materials; however, microstructure-property relations are basic to all materials and fabrication methods since the resultant microstructural features are a fundamental consequence of the starting materials character and the fabrication process and parameters. Similarly, the resultant properties are a fundamental consequence of the microstructures resulting from fabrication. Thus, microstructure is the critical connecting link between fabrication and properties. Attempting to directly relate properties to fabrication without relating both to microstructure would be analogous to attempting to relate chemical processing to resultant chemical properties without accounting for the intervening chemical reactions. Either would be totally impractical from the standpoint of theoretical understanding as well as effective processing for targeted properties.

The importance of the porosity dependence of ceramic properties, as shown in a previous review (1), is put in perspective by considering both the typical fabrication choices and the scope and magnitude of the impact of microstructure on various properties. Of the microstructural variables, pores, second phases, and grains all have substantial and broad property impacts, but the scope of these impacts generally decreases in the order listed. The magnitudes of the property impacts of porosity are greater than for second phase particles and grain parameters since the properties that pores contribute to a body are usually zero. The property decreases from porosity commonly vary from a few fold to a few orders of magnitude for most important properties such as elastic moduli, fracture toughness, strengths, hardness, wear resistance, and electrical and thermal conductivities. Young's modulus (E), with which most mechanical properties scale, ranges over about an order of magnitude with ceramic composition, e.g., from about 70 GPa for most silicate glasses to 400–500 GPa for most high modulus ceramics (e.g., Al_2O_3, SiC, and B_4C), to 500–700 GPa for a few materials (e.g., WC, along some axes of some stiff ceramic crystals such as the frequent <111> growth axis of SiC whiskers (2), and high stiffness carbon fibers), to about 1100 GPa for diamond. Thus, substantial porosity can reduce most mechanical properties of high modulus materials as much as or more as changing to another material.

The typical fabrication choice for ceramics (and frequently for metals) is to consolidate powder, which then commonly follows one of two paths. One path is to obtain the desired porosity to achieve key application parameters that commonly increase with increasing porosity such as surface area, but these must be balanced against other physical properties that decrease with increasing porosity (see Section 10.3). The other path is where high or maximum levels of physical properties such as strength, optical transmission, thermal conductivity, etc., are sought, which require low to zero levels of porosity. Both paths depend critically on the high impact of porosity on properties, but the second path typically reflects greater importance of other microstructural parameters. Thus, the additional property increases (commonly 50–100%) that are achievable by controlling grain and second-phase microstructural parameters over those achieved by reducing porosity, are often critical for many applications.

Another important need for understanding porosity effects on properties is to combat confusion and conflicts in much of the substantial literature. Thus, expressions for the porosity dependence are often suggested or modified in an ad hoc or empirical fashion, often with implications that they have validity beyond the narrow range of porosity for which they were introduced. Most evaluation of porosity effects is done without any significant pore characterization, and claims of the validity or lack of it for various expressions are often misleading or simply wrong. Further, accurate quantification of the effects of porosity is a complex and challenging task that commonly requires some simplifications or approximations, many of which may be limited in their range of applicability or even their correctness. This raises differences between purists and pragmatists, and confusion because of the diversity of specific approaches and results.

Microcracks are of interest both as an extreme of pores and because they occur in a number of materials and are used to modify properties, e.g., to reduce thermal expansion, and increase toughness and resistance to more extreme thermal shock, however, as with use of porosity for various applications, there are trade-offs with other necessary properties. Such engineering trade-offs again require suitable documentation and understanding of their effects.

Understanding the microstructural dependence of properties is of even greater importance for ceramics for two reasons. First, ceramic microstructures (and hence many properties) are very dependent on fabrication parameters. While this is very true for the dominant ceramic process of sintering, it is also an important factor in all fabrication and processing. Second, brittle fracture, which is dominant in ceramics, is a weak-link process controlled primarily by the statistics of the flaw population(s) causing failure, and secondarily of variations of pertinent material properties, and hence microstructure globally and locally around the failure origin. Such flaws and key microstructural features are commonly: 1) microstructural extremes such as large pores, grains, or second-phase particles, 2) statistical variations of the distributions, e.g., clusters, of less extreme members of these microstructural features, 3) machining flaws or other cracks, or 4) interactions between these flaws. The importance of microstructure in ceramics is compounded by ceramics being much less amenable to postdensification deformation processing, such as forging (as metals are) to further densify them or refine or homogenize their grain structure.

1.2.2 Porosity and Property Changes in Sintering

The above flaw-microstructure statistical effects and the needs for understanding microstructure effects on properties are illustrated by considering sintering, which is the dominant ceramic process and produces a broad range of microstructures. Many properties, e.g., thermal and electrical conductivity and elastic properties, increase at differing rates to varying plateau values for a given material and fabrication with increasing firing conditions. Similarly, other properties especially most mechanical properties, increase at varying rates, but pass through variable maxima, then decrease at varying rates with key processing parameters such as sintering temperature or time (Fig. 1.1). Such differing and variable trends make understanding the underlying microstructural changes which drive and control these essential to understand and control properties. The most basic changes are those of porosity and grain size which are interrelated both to each other during sintering, as well as to starting particle size and

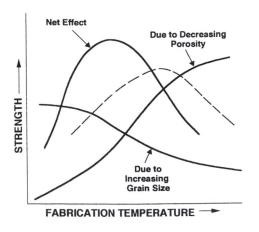

Figure 1.1 Schematic of maxima of properties such as strength that are dependent on both porosity decreases and grain-size increases with increasing firing temperature or time. Opposite effects to these changes occur at differing interrelated rates impacted by grain-pore relations (see Fig. 1.2) and lead to maxima (net effect) that change (dashed curve) with different material and fabrication parameters.

packing. Other parameters such as the size, shape and (local or global) preferred orientations of grains and pores, as well as heterogeneity of their spatial distributions can also be important. All of these may be substantially impacted by other phases present, as well as phase changes.

Property maxima typically result from opposite property trends with porosity and grain size and their respective decreases and increases with processing temperature and time (Fig. 1.1). Thus, most mechanical (and other) properties increase with decreasing porosity, i.e., with increased sintering, but many properties also decrease with increasing grain size, which increases with sintering. This leads to variable maxima of many properties due to the effects of varying starting microstructures, their variable interactions (locally and globally), and the resultant microstructures (3,4). Pores (and second-phase particles) inhibit grain growth; hence, pores limit grain size until sufficiently low levels of porosity or higher driving forces for grain growth are reached. Such levels are dependant on starting particle (grain) sizes and packing (and hence on pore sizes and shapes), additives and impurities, and sintering parameters. The resultant increased grain growth is accompanied by changes in inter- versus intragranular location of pores as well as in shapes and sizes of the pores relative to the grains (Figs. 1.2–1.4).

Properties that are not dependent on grain size rise to various plateaus as porosity decreases with sintering, and approach limits determined by material

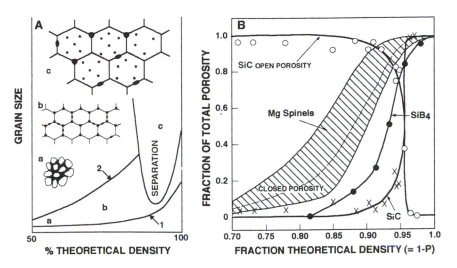

Figure 1.2 Porosity changes in sintering. A) Schematic of the variable interrelations of grain size and density (hence porosity) in sintering. Dark areas in inserts a, b, and c represent pores between particles (a) and grains (b, c) and within grains (c). Progressive microstructural changes with increasing density are indicated by the location of the letters for inserts a–c on the plot. These changes depend on whether the paths followed go below the separation region (e.g., path 1) so the pores remain at grain boundaries (insert b) and hence can be removed, or intersect the separation region (e.g., path 2) where grain boundaries can separate from pores, thus trapping some pores within grains (insert c). B) Fraction of closed porosity as a function of fractional density (= $1 - P$) for various sintered spinels (5) and SiC (6) as well as HIPed SiB_4 (7). Formation of closed porosity (= 1 – open porosity, also shown for the SiC) is a necessary step to pore entrapment in grains from grain growth.

and sintering parameters. Understanding the microstructural dependence of such properties is also needed to control property development since the rate of approaching the plateaus and the levels reached are functions not only of the porosity achieved, but also of its uniformity and character. Such understanding is also important since even properties having little or no dependence on grain size can still be affected by grain-porosity interaction in sintering. This results from the effect on pore size, shape, and location as a result of significant grain growth, which occurs only when grain boundaries are no longer pinned by pores. This results in some pores becoming intragranular, not just, intergranular (Figs. 1.2–1.4), as well as frequent growth of remaining intergranular pores. A precursor to such grain-pore changes is the closure of intergranular pores, which is typically not complete until P ~ 0.05 (8) (Fig. 1.2B). Pore closure and the associated changes occur over significantly varying ranges that reflect changing

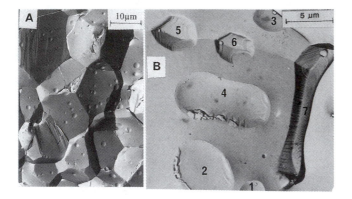

Figure 1.3 Examples of closed intergranular pores (from replica electron micrographs) after considerable grain growth. A) Lower magnification of mainly intergranular fracture in MgO. Note the small, flat, lenticular pores that are approximately circular on grain boundaries (and a few intragranular pores revealed by the transgranular fracture, upper left). B) Higher magnification of an Al_2O_3 intergranular fracture. Note the flat, lenticular (1–4), sometimes elongated (4) pores on grain boundary facets, the more complex cross-sectional shape of pores along triple lines (5,6), and a substantially elongated pore (7).

porosity levels, pore character, grain size, and microstructural heterogeneity. The resultant pore changes can affect properties and give them a correlation with grain size, even if they have little or no intrinsic dependence on grain size. Such effects are of particular importance in crystalline materials where the inherent anisotropies of surface energies result in some to substantial polyhedral character of all pores, especially intragranular ones, and the lenticular or cusp shape of intergranular pores (Figs. 1.3, 1.4).

Whether properties depend on grain size (hence giving property maxima) or are not (hence giving property plateaus with sintering) there are important interrelations between porosity and grain growth. Thus, as outlined in Fig. 1.2, pores between packed powder particles remain at grain boundaries if the starting particle sizes and packings and resultant sintering follow a density-grain size path that remains below the area where grain boundaries can separate from at least some pores (e.g., path 1, Fig. 1.2). Paths (e.g., path 2, Fig. 1.2) intersecting the separation region lead to grain boundaries detaching from associated pores, resulting in entrapment of such pores within the grains. Besides resulting in mixed pore locations, such boundary separation from pores results in different pore shapes and sizes (Figs. 1.3, 1.4) due to differences in surface energies and diffusion along grain boundaries as opposed to within grains. Thus, even for properties not dependent on grain size, grain growth can effect those properties

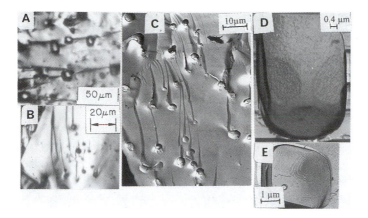

Figure 1.4 Examples of intragranular (hence closed) pores in sintered and hot pressed polycrystalline bodies. A) and B) are optical micrographs of $BaTiO_3$ and $MgAl_2O_4$, respectively. Also shown are replica electron micrographs of pores in C) Al_2O_3, D) MgO showing rounded edges, and E) MgO showing beveled edges. Note the fracture tails emanating from the pores in A–C, the polyhedral nature of the pores in A, D, and E (see also Fig. 2.2) and the approximately spherical pores in B and C.

via effects on pore location, size, and shape. For example, the cusp or lenticular shape of grain boundary pores results in substantially more coverage of the boundary area for a given pore volume than more equiaxed, e.g., spherical or polyhedral, pores such as intragranular ones, thus having greater proportional impact on properties, especially those involving grain boundary fracture. Further, while shrinkage of grain boundary pores can (and often does) occur, such pores also often grow, e.g., via consolidation, with grain growth, giving some correlation between pore size and grain boundary area covered, and grain size.

The changes of properties with sintering temperature (or time), whether with or without applied pressure (e.g., hot pressing), for ceramic composites are often similar to those of monolithic ceramics outlined above. Thus, properties still increase with increasing density (i.e., reduced porosity) and generally decrease with increasing sizes of both the matrix grains and the dispersed phase. Again, pores inhibit growth of matrix grains as well as of dispersed particles, and some pore entrapment in growing grains (and occasionally growing particles) occurs; however, mutual inhibition of grain and particle growth is commonly more important, especially at lower porosity. This grain-particle size limitation is often important in the mechanical properties of composites. More general and pertinent to this book, however, is the impact of there being differing phase interfaces for intergranular pores (see Section 8.3.5).

1.3   SOURCES AND OCCURRENCES OF POROSITY
AND MICROCRACKS

1.3.1   Intrinsic Sources (Other Than Sintering)

The importance of porosity effects on properties which arise from its significant impact as outlined in the previous section, as well as from its pervasive occurrence in most fabrication processes whether it is desired or not, are addressed in this section. Despite the frequent desire for no porosity for many applications, pores pervasively occur due to both intrinsic and extrinsic factors as well as physical and economical limitations. A basic step in understanding the effects of porosity is to be aware of its sources and occurrences. Intrinsic limitations on reducing porosity were outlined for sintering (Fig. 1.2A), where most or all pores start as open, intergranular pores (see Fig. 2.2), and often remain intergranular (Fig. 1.3), i.e., highly cusp-shaped pores. There is substantial diversity in the transition closed, lenticular (still cusp-shaped) pores (e.g., Figs. 1.2B, 1.3), and even more in the transition to intragranular pores; the latter reflects a marked change in shape, i.e., rounded polyhedral to spherical pores (Fig. 1.4). Sources of porosity, especially intrinsic ones, in other fabrication processes are outlined below, then other, generally extrinsic, sources of porosity are discussed, especially for sintering.

Other processes also result in some, often substantial, mainly intrinsic porosity. Most melt-cast ceramics, due to component shape, size, or composition (i.e., not having a single melting point), are not amenable to directional solidification to eliminate intrinsic porosity from the liquid to solid transformation. Thus, melt-cast ceramics typically have 5–20% porosity, i.e., often more than most cast metals since many ceramics have higher liquid-to-solid density increases than metals (e.g., to $\geq 20\%$ for ceramics (9,10) versus 5–10 % for most metals). Much of the resultant porosity occurs as large (e.g., micron to centimeter or greater scale), intergranular, often approximately spherical pores (Fig. 1.5). There can also be some, often heterogeneous, mainly polyhedral, intragranular, solidification pores that reach occasional cm to mm sizes, but scales are usually on the order of microns in dimension (Fig. 1.6), i.e., similar in geometry and size to intragranular pores from sintering (Fig. 1.4). Melt spraying of ceramics, while resulting in approximate directional solidification of the flattened droplets, results in some residual porosity (and often some microcracks), that are frequently laminar in character. Some methods of depositing ceramics, e.g., melt-sprayed ceramics, deposit as lower density, metastable phases, e.g., Al_2O_3 (11) that convert to more stable, denser phases on higher temperature exposure, thus generating intra- or intergranular and hence often nonspherical pores from nano- to micrometer scales.

Both inter- and intragranular pores of similar scale can also be introduced by various processes, e.g., differential diffusion (e.g., Kirkendal), annealing of

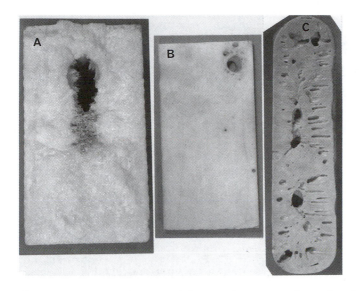

Figure 1.5 Larger (intergranular) pores in cross-sections of melt cast ceramics. A) Refractory Al_2O_3 brick (~ 5 cm thick). There is a large pore near the top (from casting two bricks edge-to-edge for thermal management and production efficiency) above the darker region of more distributed, finer porosity. B) Experimental 99% Al_2O_3 brick; note the larger pores in the upper right (and isolated smaller pores elsewhere). C) Experimental sample (~ 0.5 cm thick); note the denser surface layer.

radiation damage or dislocations (especially entangled ones from deformation), stoichiometric changes, and phase changes due to reaction of constituents or impurities during, or after, sintering (12,13). Creep, via grain boundary sliding, primarily in tension, generates intergranular pores that are typically a fraction of the grain size. Conversion of preceramic polymers to ceramics commonly leaves of the order of 20–30 % porosity (14) resulting from the inherent shrinkage of bonding dimensions and decomposition products produced during polymer to ceramic conversion. Such pores vary in size from nm to mm in size, with these commonly being spherical (15) (Fig. 1.7). Two other sources of porosity are: 1) intrinsic intragranular, crystallographic pores in zeolites and related materials (e.g., some phosphates), and 2) in growth of bones, other natural skeletal, and shell materials (usually with some organic materials).

1.3.2 Extrinsic Sources in Sintering

There are other, often extrinsic aspects of processing that effect the amount and character of pores beyond the more intrinsic sources of porosity noted above, especially in sintering. Thus, more, and possibly somewhat larger pores

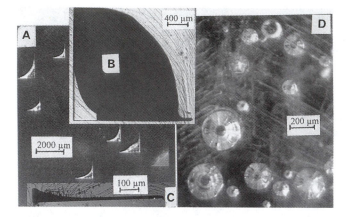

Figure 1.6 Optical micrographs of intragranular pores in large, cleaved grains of melt derived ceramics. A), B), and C) show large pores with some cubic character in large (1 to several cm diameter), usually columnar grains of skull melted MgO. Note the dominant, partial (occasionally nearly fully) cubic {100} character of the pores (similar to intergranular pores from sintering [Fig. 1.4D, E]. D) Smaller (apparently double, i.e., base-to-base) conical pores in cm scale, platey, cleaved grains in fusion cast beta alumina refractory. Radial lines on most pores indicate faceting; cone axis apparently the C axis, i.e., normal to the cleavage (basal) plane.

commonly remain in sintered ceramics not only due to intrinsic effects (Fig. 1.2), but also due to gases left in the originally open pores (e.g., due to firing in air rather than vacuum). Gases not soluble in the surrounding ceramic are entrapped in closed pores since they cannot be diffused out of the pore, e.g., N_2 in oxides, preventing even pores at grain boundaries from being removed. Many

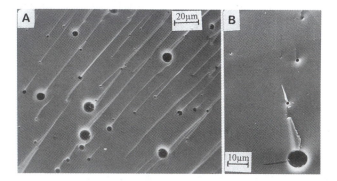

Figure 1.7 Examples of pores in ceramics from a pyrolyzed preceramic polymer. A) Glassy carbon from polyfurfuryl alcohol. B) SiC from a carborane-siloxane. Note the approximately spherical pores in the ~ 1–5 μm dia. in both materials.

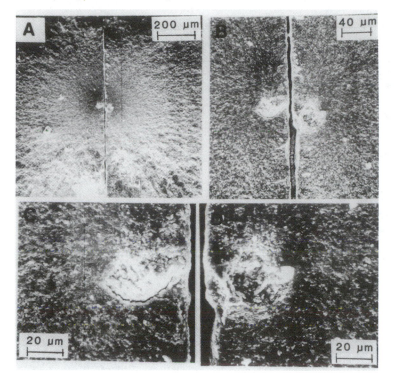

Figure 1.8 Pore around isolated (typically hard) agglomerate at the fracture origin (at ~ 660 MPa) of a pressed and sintered high purity alumina. A) Lower and B) higher magnification SEM of matching fracture halves. Note the peripheral (crescent-shaped pore cross-section) around the bottom of the agglomerate in the left fracture half. Since the fracture typically follows the peripheral pore around such agglomerates, the agglomerates typically remain imbedded in one-half of the fracture and leave an approximately hemispherical pore on the other fracture half.

other extrinsic sources of (commonly intergranular) pores exist, especially for sintered bodies. These other sources of pores, i.e., from the materials and forming operations used, contribute mainly to heterogeneity of the size, shape, and spatial distribution of the porosity.

The powders used for forming bodies often contribute larger, heterogeneous pores from heterogeneous particle sizes, shapes, or spatial distributions (see Fig. 2.2). Larger pores typically occur with agglomerates in the powder (16,17). Isolated agglomerates, which are variable in character and occurrence, often form variable pores around much or all of the agglomerate periphery due to poorer particle packing, differential pressing spring back, sintering, or a combination of those between the agglomerate and the matrix (Fig. 1.8). A similar, broader problem arises from spray-dried agglomerates that are widely used in

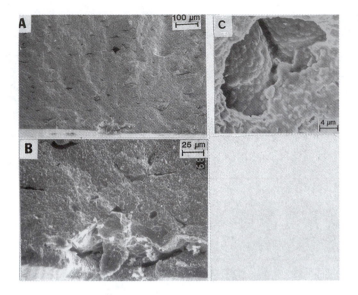

Figure 1.9 Pores from spray-dried agglomerates in commercial TZP. A) A substantial number of laminar pores from incomplete, inhomogeneous deformation of many spray-dried agglomerates due to their interaction with stress inhomogeneities in a different sample; fracture origin, bottom center. B) Pores from a larger, approximately laminar-agglomerate defect. C) Isolated pore between agglomerates with limited deformation in pressing and limited sintering of the agglomerates (which is also representative of pores between lightly sintered individual particles).

manufacturing (especially via die pressing) in order to get the required powder flow. Larger pores result from varying degrees of incomplete deformation/destruction of the agglomerate during pressing (Fig. 1.9). Another important source of extrinsic porosity in powder derived bodies are foreign organic particles (16). These have been shown to include pieces of brush or broom bristles, insect bodies or parts of them, as well as insect feces, even in otherwise high purity powders (18). Many other sources of organic impurities exist, including human or animal (e.g., rodent) dandruff or hair, lint, dust, cigarette ashes, etc. Another important organic source of pores is binder agglomerates or particles (16) (Fig. 1, 10). Larger pores can also be formed in pressure-cast bodies due to gases which dissolve in the fluid media under pressure and revaporize on decompression of the body (19).

The forming aspects of fabricating ceramics, especially via sintering, also have significant impacts on the amount, and especially the character of the pores which are often unique to the process, but vary with process parameters.

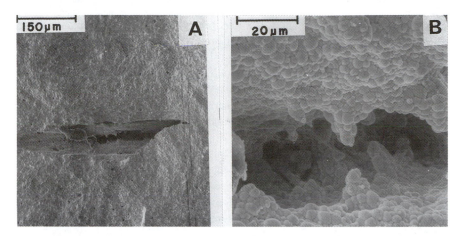

Figure 1.10 Large, laminar pore fracture origin in commercial PZT. A) Intermediate magnification and B) higher magnification showing recesses of the large pore merging into the background porosity at the back of the larger pore channels, indicating it probably originated from binder agglomerate(s). Published with permission of Plenum Press (16).

Thus, for example, electrophoretic deposition can result in electrolysis of the fluid used, especially water, generating bubbles in the green deposit. The amount, size, and spatial distribution of this mostly spherical porosity depends on particle and deposition parameters. (This phenomena has been used to generate special porosity, see Fig. 10.10.)

Many forming operations often result in varying anisotropy of pore shape and degrees of preferred orientation of such pores to give anisotropic properties that, though often neglected, are often important. Extrusion inherently results in approximately aligned binder strands due to the flow necessary to carry out the extrusion (20,21). The extent and degree of the tubular pore character depends on parameters such as the amount of binder and the extrusion pressure, which often depends on component size and number relative to the barrel diameter. Similarly, injection molding involves extrusion of material into a cavity, but the turning and consolidation of plastic strands within the cavity leads to considerable variation and complexity of the orientation of the tubular binder strands. Such binder strands often result in pores of some tubular character. Also, all injection molding and many extrusions involve the knitting together of adjacent masses of material (20,21). When such knitting is incomplete or entails excess binder, it causes complex, locally varying pores that are aligned, at least in extrusion.

Most pressing operations may result in laminar pores, or laminar arrays of equiaxed pores, either generally or locally. While some lamination can occur in other pressing operations such as some isopressing (e.g., on a mandrel) and compression molding, they occur in hot pressing, and particularly in die pressing (20) and tape lamination (e.g., Figs. 1.10, 1.11). Higher die-wall friction, especially at higher pressures and for taller parts, can result in laminar pores, sometimes extensively (such friction also often results in substantial gradients of increasing porosity as a function of increased distance from the die wall, and from top to bottom for single, and top and bottom to center for double-acting presses). The extensive use of spray-dried agglomerates in production die pressing (necessitated by the need for rapid, uniform die fill) commonly results in incomplete deformation of some or many of the agglomerates. This leaves either larger, usually scattered, pores between the agglomerates (Fig. 1.9 C) or laminar pores associated with the agglomerates (Fig. 1.9 A, B) as a result of interactions of pressing tendencies for laminations with incomplete deformation of the agglomerates.

Though little studied, some laminar character is expected in colloidally processed bodies, e.g., as indicated in earlier examination of slip cast sonar transducers (22) and recent evaluation of slip casting SiC (23). The latter showed particle flocculation during casting results in a more porous structure with increased casting time, i.e., a porosity gradient. Also, some gradation of particle size often occurs across the thickness of cast ceramic tape that can result in some laminar porosity between layers. This is consistent with particle packing studies showing more interlayer porosity when larger particles are deposited on finer ones, as is the case with some slip settling in casting (24). Similarly, most if not all rapid prototyping processes involve lamination and thus may retain some resultant laminar character (Fig. 1.11). Other examples of pores (at fracture origins) are described in Section 5.3.3. Also, note again that pores in many bodies or coatings made by deposition processes often reflect aspects of anisotropy of shape, orientation, or spatial distribution.

Another frequently important but widely neglected aspect of pores from ceramic fabrication is gradients of the size, shape, and especially amount of porosity in various bodies. Thus, for example, die-wall friction effects in die pressing commonly result in substantial density gradients. The extent and character of these, which can impact both general and other (especially laminar) pores, depends on the size and shape of the die, whether the press is single or double acting, and the pressure used. Other pressing processes can also exhibit similar gradients, with hot pressing being a particular example, but with less range of variation in the amount of porosity because of the high overall densities typically achieved. Porosity gradients near die walls may also occur due to or be aided by particle packing effects there. Thus, packing of uniform size spheres is disturbed to about five sphere layers in from the wall, two sphere sizes reduces this to about 3 layers, and three sphere sizes reduces it to about 1 layer (25).

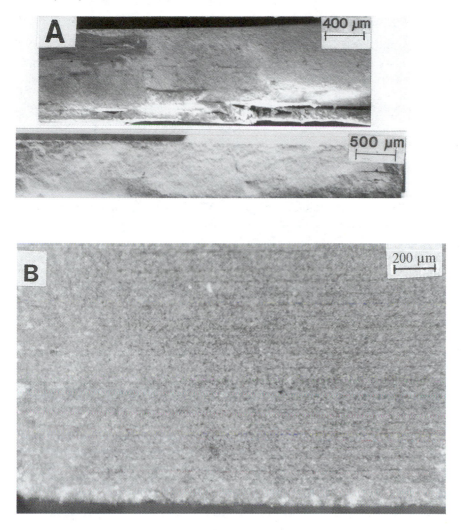

Figure 1.11 Residual laminar character of bodies and pores in bodies laid up layer by layer in experimental photolithiographic curing of each layer for rapid prototyping. A) Two sections of TZP ceramic showing laminar defects. B) Stainless steel (having substantial ductility despite the approximately 5% porosity, much of it associated with a fine, homogeneous laminar character. (Photo courtesy of W. Zimbeck of Ceramic Composites Inc., Annapolis, MD.)

1.3.3 Mainly Extrinsic Sources of Porosity in Other Fabrication Processes

Besides the above material and processing sources of porosity left from sintering, there are a variety of other sources of (often heterogeneous) porosity in bodies made by other fabrication methods. While, hot isostatic pressing (HIP) as well as chemical vapor deposition (CVD), commonly approach theoretical density, i.e., 0 % porosity, there is usually some, e.g., 0.01% to a few tenths of one percent, residual porosity. Thus, pores in CVD and other vapor-deposited bodies can result from: 1) particles, e.g., of foreign material or from gas phase nucleation, falling on the deposition surface, preventing, i.e., shadowing, deposition underneath them, and 2) from inadequate lateral growth of colonies (clusters of similarly oriented grains, typically from a common nucleus), often near their base, being closed off, trapping pores (10). In HIP and especially hot pressing, local heterogeneities of particle packing and especially composition can result in porosity. Thus, having insufficient local densification aid, e.g., from poor mixing, often leaves porous regions after pressing (Fig. 1.12). Also, locally excess additives with volatile components can result in (often large) isolated pores (16). Generally, incorporation of volatile (commonly anion) impurities result in pores, especially in hot pressed, some sintered (26,27), and even some fused bodies (27–29) after subsequent thermal exposure. Most melt-grown ceramic crystals are prepared from previously melted material to minimize the amount of pores trapped in the resultant crystal due to bubbles from gas-producing impurities from the raw materials.

Residual porosity in hot pressed bodies, e.g., from local deficiency of densification aid frequently occurs as laminar pores or porous areas (16,20) (Fig. 1.12) due to the unidirectional nature of the densification. Laminar pores also frequently form in bodies hot pressed to near but imperfect transparency with the entrapped gas producing impurities (26–29), upon subsequent exposure to high temperatures (Fig. 1.13). Since the resultant pores are almost invariably laminar and frequently normal to the pressing direction, but not yet evident in the hot-pressed body, this indicates that the initial laminations can be quite subtle to detect, but pronounced in their effects.

Porosity gradients also occur in other fabrication-processing methods such as melt casting and polymer pyrolysis. In melt casting directional solidification starts from the mold surfaces, and results in gradients of porosity from the intrinsic liquid to solid volume change. The extent and character of the resultant porosity depends on part size, shape, cooling rate, and composition (Fig. 1.5). Gradients also result from diffusion of gases dissolved in the melt; they diffuse out of the material near its surface, but are entrapped as bubbles as the depth from the surface increases. Similarly, denser surface regions result from diffusion of gases from polymer pyrolysis out of the ceramic near the surface, which clearly depends on various parameters, such as the diffusivity of the gas in the

Figure 1.12 Example of a laminar porous region at the fracture origin in a dense hot-pressed Si_3N_4. A) Lower magnification SEM photo showing fracture and origin (white region, bottom center). B) Higher magnification of typical dense region. C) Higher magnification of the porous origin.

material as well as the time and temperature. This results in part from polymer pyrolysis frequently having increasing porosity as component size, especially cross-sectional thickness, increases (10).

The diversity of the pore character thus reflects the diversity of the forming and processing methods and parameters of fabrication. The porosity of a body usually consists of one dominant type of porosity with one or more secondary types. The latter, which may or may not be related to the mainstream porosity of the body, are often a minor component of the total porosity, but can be a major component. Further, the nature of the porosity mix changes with the size, shape, and forming of a component, and with process parameters and resultant density and grain size, e.g., changes from inter- to intragranular porosity. Thus, comprehensive porosity modeling must be able to address both mixes of porosity of different character and changes in pore character as the amount of porosity changes.

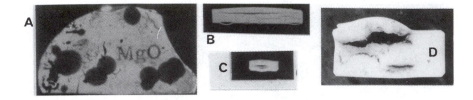

Figure 1.13 Laminar pores from gases released on annealing of partially transparent hot-pressed MgO at A) ~3.8 cm dia., B) ~3.8 cm dia., and C) ~1.8 cm wide, and D) dense Al_2O_3 (~2.5 mm thick). Note the laminar plane perpendicular (A-C) and parallel (D) to the hot pressing directions. (Reproduced with permission of Am. Inst. of Chemical Engrs. (27). Copyright © 1990 AIChE. All rights reserved.)

1.3.4 Uses, Fabrication, and Design of Special Porous Bodies

A variety of applications such as insulation, filtration, burners and catalyst supports (see Section 10.3) often require control of the character and amount of porosity, with the latter often being near or beyond the limits of normal use of typical fabrication methods. Maximum volume fraction porosity (P) levels in typical powder compacts (whether for sintering or hot pressing) are P ~ 0.4–0.6. Other fabrication methods typically have lower limiting porosity levels; however, many applications require higher porosity levels as well as different pore characteristics than are normally produced by typical fabrication, and thus require modification of existing fabrication or alternate methods. There are a variety of ways to purposely (or unintentionally) introduce pores. The five most common are: 1) incomplete sintering of dense, porous particles (including balloons), or mixtures of these with various degrees of packing, 2) introducing bubbles, e.g., for foaming of green bodies, 3) introduction of fugitive material to be removed in the early stages of sintering, 4) forming operations such as extrusion, injection molding, and (now emerging) rapid prototyping methods that form pores, e.g., tubular ones, and 5) use of fiber systems, e.g., felts. Other, newer techniques for making special porous bodies are summarized in Section 10.3. Some of these can be used in conjunction with conventional fabrication, e.g., most or all closed pores in the struts and walls of open cell foams can be eliminated by HIPing foams (30).

A substantial diversity in specialized techniques is needed to achieve the variety of pore structures desired. Extensive engineering is also typically needed to balance the property trade-offs required since the benefits of porosity must almost always be balanced with other needed properties that are substantially reduced by porosity, e.g., mechanical properties. The engineering to make bodies with good balances of porosity dependent properties is both more complex and more fruitful because the balances can be shifted substantially by

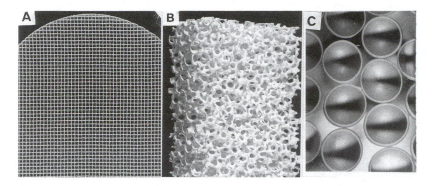

Figure 1.14 Highly porous ceramic bodies. A) Honeycomb extrusion for autocatalyst support; cell walls ~0.15 mm thick. B) Replicating a polymer reticulated foam with ceramic, e.g., for filtration. C) Uniform, thin-wall ceramic balloons (~2.5 mm dia.) that can be readily self-bonded to form very uniform closed-cell foams. These all reflect multiple pore populations, i.e., in addition to the obvious pores there are pores in the ceramic walls or struts. (Photo courtesy of J. Cochran, Ceramic Fillers, Inc., Atlanta, GA.)

changing pore character by fabrication changes. Typical pores between packed particles or bubbles, or from fugitive additives, extrusion, and foaming are well known (Fig. 1.14).

The challenge of understanding and characterizing porosity is illustrated by three factors that affect both the occurrence and character of porosity in processes based on powder consolidation and sintering. First, several porosity sources can be operative in a given sample, as discussed above. Second, some sources can be interactive, e.g., volatile species from impurities or decomposition products, in expanding laminar defects from various forming processes and heterogeneity of starting porosity and the forming of laminations. Third, the operation and effect of separate or interactive sources of pores can be quite variably dependant on not only the type of forming, but also on aspects of a given forming method, e.g., the size, shape, and number of components made in a given die or mold. Another complication (but also a potential tool) is that some, e.g., elastic and conductive, properties are less and some, e.g., hardness, wear, and strengths, are more sensitive to heterogeneities and gradients of porosity. Also, as discussed above and substantially in Chapters 3–9, pores may often have anisotropic shape, be oriented or arrayed in anisotropic patterns, or both, thus giving anisotropic properties. Unfortunately these issues of gradients, heterogeneity, and anisotropy have been almost universally neglected. Another complication, addressed to the limited extent of data, is the effect of micro-cracks whose formation and impact often depends on pores, grains, and particles, or combinations of these.

1.3.5 Sources and Occurrence of Microcracks

The first of three basic sources of microcracks on specimen surfaces is from transient, local impact on the specimen surface, e.g., from rain drops, ice, dust, particles of sufficient sizes, velocities, and impact angles. The second, most prevalent and important source of such surface microcracks is from abrasive machining, which introduces two populations of cracks, one normal to the abrasive particle motion and the other parallel with it (31–36). The dimensions of such machining cracks are determined primarily by the machining parameters and secondarily by material parameters; there is a limited effect of microstructure. Both sets of cracks are of similar depths and are generally planar, with those formed normal to the abrasive motion being roughly semicircular. The cracks formed parallel to the abrasive motion are more elongated when they are larger or smaller than the grains around or through which they form, but their lengths are constrained by grain dimensions as the crack size approaches the size of single grains in or along which they form. Such machining cracks are commonly the flaws causing mechanical failure and hence determine the strengths of the material.

Such cracks are not addressed in detail in this book since they are a large, separate topic. They are noted here for completeness as well as a source of the second type of microcracks, namely from thermal stress and shock, whose porosity dependence is discussed in Sections 9.2.2 and 9.3.2. Surface cracks from machining or surface impact are common sources of larger cracks from significant transient or steady thermal stresses (along with other sources of cracks such as pores and grain mismatch strains). The second and more important reason for briefly discussing such cracks is their interaction and competition with pores as sources of failure (31–34). Thus, as discussed in Chapter 5, machining flaws frequently are associated with individual pores (e.g., Fig. 5.14) or clusters of them, especially larger, isolated ones, and particularly in tests with higher surface stresses, e.g., flexure. This can affect the impact of such pores on strength; however, there can also be a transition from pore to flaw determined failure as porosity decreases impacting the extrapolation of strengths to P = 0 (e.g., Fig. 5.6). For finer, more homogeneous size and spatial distributions of porosity, machining flaws still commonly dominate failure until at least P ~ 0.3, as directly identified by fractography as well as effects of machining direction relative to the stress axis on strength (35,36) (e.g., Fig. 5.10); however, there must be some transition from the dominance of such machining flaws on strength as porosity significantly increases, e.g., in foams. While fractography is likely to be less definitive in such situations, the anisotropy of strength as a function of the stress direction to the machining direction (and hence which machining flaw population is stressed) can be a major tool in defining this transition. (This strength anisotropy can also be a tool to indicate inhomogeneities of porosity distributions, e.g., Fig. 5.10.)

The third source of microcracks, the one most pertinent to specific topics in this book, is from mismatched microstructural strains, especially from phase transformations (including crystallization of glasses) or thermal expansion of grains or particles in a body (37–46). The latter arise from anisotropic thermal expansion of individual grains (that occurs to varying extent in all noncubic materials) or from differential expansion between grains, particles, inclusions, etc., of different phases, hence also different thermal expansions (and elastic properties). Phase transformations that result in locally incompatible strains of grains or particles, whether the result of temperature or pressure changes or reactions (e.g., in making cementatious materials) are also an important microcrack source. Mismatches of elastic properties on the microstructural scale can also contribute to the resultant cracking by accentuating local or global stresses (43). Thus, while compressive stresses can close some microcracks, as discussed later, microcracks are also commonly formed at high compressive stresses as a precursor to failure.

While the stresses from such mismatch strains are independent of the scale of the grain or particle strain source, they cause cracking only once a threshold grain or particle size is reached that depends on the local properties (39–45), e.g., Eq. 10.1. Such cracking is thus typically on the scale of the grains or particles, but obviously varies in size and character with the statistics of the local sizes, shapes, clustering etc., of the grains or particles. It is also commonly intergranular, consistent with the stresses being across the grain boundaries, but there can also be reasonable levels of transgranular fracture (40), presumably due to the complex, varying multiaxial character of the resulting stresses and fracture character of the grains or particles. Resulting microcracks are typically distributed throughout the volume of the body, with their density depending on the statistics of the spatial, size, shape, and orientation distributions of the grains or particles and resultant variations in their mismatch strains. Since many phase transformation and all thermal expansion mismatch stresses depend on temperature changes involved in forming the cracks, crack densities and the resultant property effects typically have significantly temperature dependences (see Chapters 9, 10). While a number of observations clearly show the presence and effects of such cracks, detailed quantification of microcrack populations, e.g., in their spatial distributions, sizes, etc., are challenging (see Section 2.3).

1.4 MEASUREMENT OF PROPERTIES TO BE EVALUATED

1.4.1 Wave Velocities and Elastic Moduli

It is useful to briefly consider issues in measurements used to obtain the properties of interest since different techniques can have differing utilities, especially for porous bodies. The need is often not for the ultimate in precision, since scatter for most samples means that techniques of the highest precision are often not necessary and may be counterproductive if they restrict the number and versatility of measurements made. Two often critical but almost universally neglected issues are effects of heterogeneity and anisotropy in the specimens. Both are far more prevalent than commonly assumed, especially in porous bodies, as will be discussed further in subsequent chapters; however, note the frequent anisotropy of individual pores (Figs. 1.9, 1.10, 1.13, and 1.14A) or arrays of fine pores (Figs. 1.11 and 1.12) that affects properties, including making them anisotropic with a preferred orientation of such pores. Even in the absence of a preferred orientation of anisotropically shaped pores, they still typically have property effects different from isotropically shaped pores. While microstructural characterization is an important component in evaluation of these issues, measurements to detect, or preferably quantify, them are important.

Consider first elastic property measurements, which are an important example of this need to consider measurement methods. There are basically two ways to measure elastic properties, of which Young's, bulk, and shear moduli are most common. Preferably at least two moduli are measured so the third one and Poission's ratio can then be calculated (per eqns. 3.1–3.3). The first way is to measure strain in response to some quasi statically applied stress, commonly in conjunction with strength testing. There are several limitations to this, an important one is that often only Young's modulus is obtained. Another is that since most ceramics have high moduli and hence low strains, using the easiest measurement of strains, i.e., from test machine head travel, is generally grossly inaccurate due to substantial parasitic deflections of the loading train being of similar or greater magnitude as the specimen strains. For example, common medium to high technology ceramics have true tensile strengths of 300–400 MPa and Young's moduli of 300–400 GPa giving failure strains of $\sim 10^{-3}$, e.g., only $\sim 10\mu m$ per cm of gauge length. While flexure amplifies specimen deflections (but of course not strains) by several fold, and reduces system deflections (due to reduced loads), measurements of head travel are still generally inadequate. Increased microcracking and especially porosity increases strains and reduces loads, thus reducing these errors but often with increasing errors due to crushing at contact points (a broad problem for bodies of intermediate and especially high porosity).

Direct measurement of specimen strains is needed. Several methods are available, but not without their limitations, e.g., clip-on gauges may present problems of stress concentrations or damage at the contacts. The use of strain gauges is often very effective at modest temperatures, but can present problems in measurements on porous bodies due to the gauge bonding adhesive penetrating into the specimen pores and thus modifying the properties. This can be a particular problem for more porous bodies in flexure versus true tensile or compressive testing since flexure measurements are more affected by the behavior of the specimen surface region which is most affected by the gauge adhesive. Another problem of strain measurement at high temperatures is that creep may occur (the amount obviously affected by the stain rate), so measurements of elastic strains are often compromised by limitations in separating elastic and plastic strains for an accurate measure of elastic behavior. Also, as noted later (e.g., Fig. 3.9A) there can be differences between tensile and compressive loading in some porous and microcracked bodies.

The second and generally preferred method of measuring elastic properties is but one of two sets of wave motion measurements (47–51). These are versatile, often being applicable on components, and generally offer good accuracy and other advantages. One basic way of doing this is by transmission of ultrasonic waves, or transmission and reflection of pulses (i.e., pulse echo). Thus, elastic properties can be calculated from the material density (ρ) and measurable velocities of longitudinal and shear waves, respectively v_l and v_s, from the following equations:

$$\rho v_l^2 = \frac{G(4G - E)}{(3G - E)} = E(1 - v)[(1 + v)(1 - 2v)]^{-1} \tag{1.1}$$

$$\rho v_s^2 = G = E[2(1 + v)]^{-1} \tag{1.2}$$

$$v = \left[1 - (\tfrac{1}{2})\left(\frac{v_l^2}{v_s^2}\right)\right]\left[(1 - \left(\frac{v_l^2}{v_s^2}\right)\right]^{-1} \tag{1.3}$$

as well as other equations interrelating elastic properties (Eqs. 3.1–3.3). Another basic way of measuring elastic properties by wave methods is by resonance vibration of specimens. This can be done with spheres (possibly as small as 100 μm diameter), but is more commonly done with bars, rods, or plates. In the latter cases, the resonances can be longitudinal, torsional, or transverse. For the former two resonances, the number of half-wavelengths along the length (L) of the specimen is simply 2L. This is not exactly the case for transverse flexural resonance since the node points and wave lengths are effected by Poission's ratio. Representative equations are:

$$G = \left(\frac{4}{n}\right)(L \cdot f_G)^2 \rho S_t \tag{1.4}$$

where n = 1 for the fundamental mode and 2, 3, etc., for subsequent overtones, f_G = the corresponding torsional resonant frequency, and S_t = torsional shape factor (1 for a cylinder), and

$$E = 4\rho(\pi L \cdot f_E)^2 S_E \tag{1.5}$$

where S_E is a collection of correction factors depending on sample geometry, vibration mode, and Poission's ratio (48,49). While E can be determined by either longitudinal or flexural resonance, the latter is much easier to excite, especially in thinner specimens, and hence is used much more often. Note that for either wave method, the determination of E requires ν, so at least two measurements have to be made, which is also important for comparison of results as discussed later. Resonance methods with bars or rods are most amenable for higher temperature tests (47,51), but some ultrasonic and pulse echo tests can also be done at elevated temperatures. One caution in the use of wave techniques is relaxation that can occur at lower frequencies (e.g., at < 10 kHz), especially in materials with weaker grain boundaries such as in graphite (52) or at elevated temperatures. Such lower frequency boundary relaxation effects have been observed to initiate at modest P levels and increase with increasing porosity levels (52).

Both sets of wave techniques can be used to detect and potentially to analyze the complexities of anisotropic samples. This can be important for anisotropy from preferred orientation of anisotropic shaped pores (and grain, particle, platelet, and whisker). Most of the wave techniques are also of value since they use specimens that are precursors to or the actual specimens used for other subsequent tests of other properties. An important demonstrated (50) advantage of flexural resonance is investigating the heterogeneity of the spatial distribution of the porosity (and hence also the distribution of second phases in composites). Thus, resonance of bars, rods, or plates allows measurement at various harmonics in one specimen and test setup. Since each harmonic particularly reflects effects at the antinodes, and these shift position in the sample with the order of the harmonic, this provides a partial sampling of different regions of the same sample. This provides a ready check and partial indication of the homogeneity in the same specimen by simply modestly increasing the limited test time; however, measurements can also readily be made on samples covering a range of sizes from the same starting billets with results compared with the specific porosity of each specimen versus that of the piece from which it was cut. This should also be done with the direct stress-strain method, but wave techniques also often readily allow measurements of various areas of larger samples, billets, or billet sections. Pulse echo techniques can be particularly useful due to the smaller sample sizes needed, but the sphere resonance method is poorly suited for this due to the complex effects of anisotropy on resonance modes. It is also important to note that most wave techniques also can be used to measure internal specimen damping, which can be valuable in indicating cracking and crack bridging as well as some aspects of the degree of their heterogeneous distribution in the sample. Another factor favoring wave techniques, especially resonance ones, are measurements at high temperatures where measurable creep occurs which makes such strains unsuitable for calculating elastic moduli; however, wave techniques with frequencies high enough so significant relaxation cannot occur during measurement are useful. Measurement of internal friction can also be valuable for assessing the extent and character of creep, as well as microcracking.

Use of all three measurements of elastic properites for intercomparison is advantageous. Static measurements can be particularly valuable for revealing the occurrence of stress-induced microcracking. While ultrasonic measurements showed no change of E in Tomaszewski's (53) study of Al_2O_3, his static measurements showed E decreasing with increasing grain size indicating progressive microcrack formation confirmed by direct microscopic observations. Moduli measurements via ultrasonic velocity allow for an addition check of self consistency of results via the relations between velocities and the moduli per eqns. 1.1–1.3 (see also p. 117 regarding intrinsic relations between the P dependence of sonic velocities and elastic moduli).

1.4.2 Crack Propagation, Fracture Energy, and Toughness

There are several techniques that have been developed for studying crack propagation and measuring fracture energy and toughness (54). Advantages, disadvantages, and uncertainties of common or representative methods are outlined in Table 1.1. These techniques can be classified as: 1) plate, i.e., double

Table 1.1 Comparison of Representative Fracture Toughness Tests[a]

Test method	Advantages	Disadvantages/uncertainties
Double torsion (DT)	Fairly well established; high temperature use.	Large specimen and crack scale; effects of curved crack front and grove up or down?
Double-cantilever beam (DCB)	Well established; some versions are constant K	Generally not practical at high temperatures. Larger specimen and crack scale.
Single-edge notch beam (NB)	Smaller (flexure) specimen; good for high temperatures.	Uncertain flaw character (without fractography verification) in some versions.
Indentation fracture (IF)	Smaller (flexure) specimen; high temperature use (within limits of fracture oxidation).	Residual stresses; often unsuitable cracks with many coarser microstructures.
Indentation (I)	Smallest sample; lowest cost.	Often unsuitable cracks with coarser microstructures; residual stresses; not practical at high temperatures.
Fractography (F)	Smaller (flexure) specimen; high temperature use (within limits of fracture oxidation)	Adequate determination of failure initiating flaws.

[a]Listed in approximate order of decreasing specimen size/crack propagation scale.

cantilever (DCB), double torsion (DT), and compact tension (CT), 2) beam, i.e., notch (NB), chevron notch (CNB), indent fracture (IF), and fractography (F, i.e., calculated from the strength and subsequently observed failure initiating flaw size, location and geometry), and 3) indentation (also including IF noted in 2) techniques. There are often several variations of some of these, especially of the DCB and NB techniques. While some tests are used more than others, none is dominant, so there is a diversity of data. Measuring the related phenomena of crack propagation, particularly wake effects such as crack bridging, and fracture energy and toughness is complex not only because of the diversity of the test methods also but because of their variable, incompletely understood results, including some, possibly significant, dependence on crack shape (55). The importance of measurement issues is clearly illustrated by the differences between different tests, as well as within a given test method, e.g., between and within different notch beam techniques. This importance is accentuated by the significant discrepancies that commonly occur between the microstructural dependence of these measurements on the one hand and those of Young's modulus and tensile strength indicated above, and later in Chapters 3–5, and 8.

The neglected issues of specimen homogeneity and anisotropy play a role in the above test differences; however, more important is the issue of the scale of the crack. The absolute crack scale can be important, with toughness values generally increasing as the general scale of the crack increases, roughly in the order: F, I, IF, NB, CNB, DCB, and DT (e.g., Tables 1.1, 4.2); however, this trend is only approximate for two reasons. First, the extent of crack propagation in these tests may vary, changing the relative crack size ranking, especially for tests with less difference in crack sizes between them. Second, the scale of the crack to the toughness and strength affecting microstructure (e.g., the pore, grain, and particle sizes as well as their heterogeneities), and to the strength controlling flaw sizes is more important. Thus a key need is to directly address this crack scale issue which can be a particular issue in foam materials, especially with larger cells. Fractography (56–60), which is usually more (but not exclusively) applicable to bodies of zero through moderate porosity, is a critical factor in this for two reasons. First it is an important, often essential, means of determining, or verifying, crack sizes in strength as well as propagation and fracture energy and toughness tests. This can be particularly important for notch beam tests, where the issue of the presence and character of a sharp crack is important (61), but can also reveal sources, including microstructural ones, of other test variations, e.g., in double cantilever tests (62). Second, fracture energy and toughness values obtained from identifying failure causing flaws in strength tests, reflect values for the smallest crack sizes which are clearly consistent with the crack sizes controlling strength. Where they can be obtained, which is in a reasonable fraction of materials and specimens, they are more consistent with the microstructural dependence of Young's modulus and tensile strength with fracture energy and toughness.

While crushing, e.g., at contact points can be an issue, it is less critical in these versus other tests, except for indentation techniques. Finally, three comments about high temperature testing. First. as temperature increases, grain boundary weakening is likely to increase crack bridging, so attention to this and its effects are needed. Second, such boundary weakening or other sources of plasticity lead to significant strain rate dependant effects that must be addressed. Third, the transition to crack tip plasticity can exacerbate differences between tests, e.g., Hirsch and Roberts (63) showed that the ductile-brittle transition in single crystals of Si varied by up to 250°C between DCB and indentation fracture tests. They showed the presence of dislocations at or near the crack tip due to the indent was an important factor in lowering the ductile-brittle transition temperature, e.g., polishing off most of the indent raised the transition temperature 50°C and abrading the area lowered the temperature 40°C.

1.4.3 Tensile Strength and Thermal Shock

While there are several tensile strength tests, they fall in logical groups based on stress state, with substantial focus on a few tests, generally with fewer basic variations of specimen configurations and test parameters within a given test type. Further, variations of results within and between tests can often be understood by successful fractography and by statistical aspects of the scaling of brittle fracture. Two main groupings of tensile tests are multiaxial (mainly biaxial) and uniaxial tension (e.g., Figs. 1.15–1.17). Triaxial tests are challenging, costly, and seldom used. Biaxial flexure tests are simpler and used fairly commonly, but have challenges associated with the specimen contacts. Contact problems are accentuated for the ball-on-ring and cylinder-on-three ball support

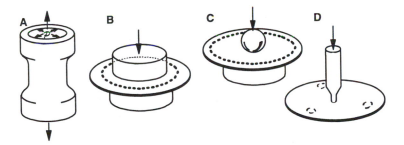

Figure 1.15 Schematic of multiaxial tensile tests. A) A triaxial test using a tensile-loaded, pressurized tube. B) Ring-on-ring biaxial flexure test (probably more commonly used). C) Ball-on-ring biaxial flexure test. D) Biaxial flexure test using a small central cylinder to apply the load on the top and three-ball support on the bottom of (primarily for thin) disc specimens.

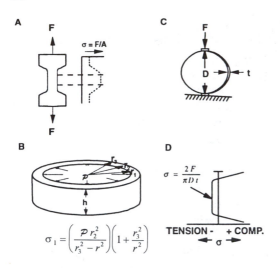

A

F

$\sigma = F/A$

F

C

F

D — — t

B

p

r_3

r_1

h

$$\sigma_1 = \left(\frac{p\, r_2^2}{r_3^2 - r^2} \right) \left(1 + \frac{r_3^2}{r^2} \right)$$

D

$\sigma = \dfrac{2F}{\pi D t}$

TENSION -   + COMP.

$\blacktriangleleft \; \sigma \; \blacktriangleright$

Figure 1.16 Schematic of tensile tests. A) True tensile "dog bone" or "dumbbell" specimen (which may have rectangular or round cross-sections. B) Hoop tensile ring tests, which (approach true tension for thinner walls) are very useful for cylinder sections (e.g., used for sonar transducer rings). Compression of rubber plugs prema-chined or in place (silicone rubber) are very practical for conducting such tests at modest temperatures, and should also be suitable for fairly porous bodies. C) Diametral compression of discs can also approximate true tension (and can be done on fairly small specimens), but presents problems of loading contact stresses (accentuated in many porous materials) and correlation with other tests.

tests with small contact areas, being more prone to crushing under the loading members in more porous materials, thereby limiting their use.

For uniaxial tension tests, true tension tests are desired for design data, but are demanding in terms of avoiding stress concentrations from gripping and misalignment, thus making them expensive. Substantial progress has been made in making these tests more practical, so they are being used more. Dog bone specimens (Fig. 1.16A) offer the advantage of reducing possible crushing prob-lems in very porous specimens. Hoop tension tests using uniaxial compression of a rubber plug to generate hoop tension in ring specimens (Fig. 1.16B) may be advantageous for avoiding crushing problems where such specimens are practi-cal. The simple diametral test is used sometimes, but presents problems of controlling loading contact effects, which may rule it out for higher porosity, e.g., foam materials; however, it can be quite useful for characterizing some aspects of tensile behavior of sections of rods. Similarly, diametral loading can be valuable for characterizing tensile failure of beads and balls (see Fig. 6.13). Hiramatsu and Oka (64) report that, very similar to the equation in Fig. 1.16C, the tensile stress at the failing load (F) of a diametrely loaded sphere is:

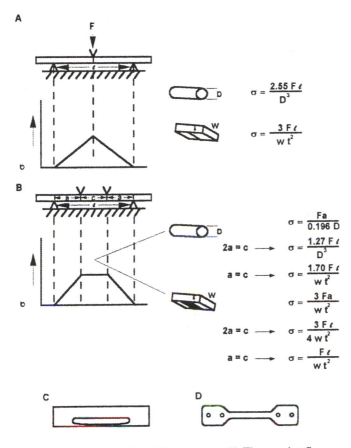

Figure 1.17   Schematic of flexure tests. A) Three-point flexure with equations shown for both cylindrical and rectangular specimens. B) Four-point flexure with equations for the two-specimen geometries and the two common span ratios. C) and D) Examples of other flexure specimens previously considered (the latter held by pins).

$$\sigma_t = 0.9F(D)^{-2} \qquad\qquad (1.6)$$

where D is the sphere diameter. They note that this expression is often valid even for rather irregular bodies (with D then being the separation of the two loading points) since the stress over about the central half of D is uniform tension for substantial variation on geometry.

Flexure tests are widely used, especially with 4-point loading of rectangular test bars; however, 3-point flexure, which is very valuable for early material development, can have advantages for microstructural studies. Four-point flexure (and especially true tension) tests, by stressing larger specimen surface areas and volumes evaluate more of the microstructural and other variations of the specimens (which is why they are better for design purposes). They are thus also more useful for reflecting effects of microstructural extremes on strength, rather than of the average microstructure. Three-point flexure, by using smaller samples and stressing smaller specimen areas and volumes, can often thus give a better picture of the effects of a more uniform and hence typically a broader range of microstructures. Such flexure is also often more amenable to fractography (again, generally more successful, but not exclusively so, at lower porosity) for evaluating overall mechanical behavior as well as anisotropy via differences in overall fracture character, fracture origins, or both. The primary need for all tensile tests is to do more fractography, including obtaining fracture toughness values as noted above; however, testing of different size specimens in true tension or flexure for comparison is valuable not only in assessing specimen homogeneity, but also in aiding fractography and verifying test values, i.e., whether tests reflect representative or atypical flaws. In all of these tests, specimen sizes need to be large relative to the pore sizes. In very porous bodies crushing at loading points may become a problem, for which alternate specimen configurations might be considered (Fig. 1.17 C, D).

Thermal stress and shock resistance are extensive, more specialized subjects for which there is less standardization of test methods, hence a somewhat more detailed and referenced summary is given here. The focus here and in most testing is on thermal shock effects, i.e., of rapid changes of temperature from quenching, for example, on the mechanical integrity of test specimens or components since this reflects a basic and pervasive limit on the rate of temperature change they can tolerate. Reader are referred to other reviews (65–68) on the subject for additional background. While thermal shock clearly involves temperature differences, it commonly involves fracture at or near room temperature since ceramics commonly fail from tensile loading of flaws at or near their surfaces and thermal stresses are typically tensile in the surface region and compressive in the interior on cooling of bodies (and the reverse on heating). There are tests for "up quenching" (i.e., rapidly heating all or part of samples), e.g., by dropping them into heated baths (including molten metals) or exposing them to a high temperature heat source such as torches (including plasma torches) and lasers (e.g., for tests of coatings (69) to address those situations where such temperature effects are most pertinent to specialized performance. A very important case of such up quenching are refractory applications for such applications as handling molten steel. There such refractories (commonly composites of ceramics such as Al_2O_3, MgO, or SiC with graphite) thermal shock tests are conducted using highly vigorous, exothermic, thermite reactions of mixed iron oxide and aluminum metal powders that rapidly react to produce molten iron (70).

The primary emphasis is on thermal shock testing by down quenching. The most common technique for such testing is to heat test bars in a furnace to a set temperature, and then to quench them into a liquid bath, typically water, usually at room temperature, but occasionally boiling (71). For such tests to be valid for comparison of materials, drop factors (72) and heat transfer factors in both the sample (including size and configuration) and quench media must be appropriate (73). Quenching by cold air blasts (74) and quenching into fluidized beds (75) have also been investigated. Crack initiation and extension, which can be initiated in some ceramics at temperatures differences of ~ 100°C or less, is detected by various nondestructive and destructive means. The former include dye penetrant observations (68), acoustic emission (76), or changes in ultrasonic or flexural resonance and damping (47); however, changes in tensile (flexure) strength by subsequent testing a series of sample sets, each at different quench temperatures is particularly common. This typically shows a critical quench temperature differential (ΔT_C) is reached where strengths precipitously drop, followed by further progressively decrease with increasing ΔT if the initial drop is not to zero strength. The onset and nature of this decrease has been related to aspects of the thermal stress and material parameters (65–69) allowing reasonable comparison of different materials or bodies. Although there can be microstructural issues, e.g., the effects of open porosity on some of these tests and property anisotropy on all tests, general trends are revealed as discussed in Sections 9.2.2 and 9.3.2. While thermal shock results can often be related to uniaxial tensile (flexure) strength, it should be noted that thermal stresses are typically more complex, usually at least biaxial in nature.

1.4.4 Other Properties

Turning to other measurements, hardness has well-known differences with different indenter geometry (e.g., Knoop and Vickers) and generally decreases with increasing indent load, but also depends on other parameters such as surface finishing, e.g., due to variable (e.g., microstructural dependent) surface work hardening (77). Another key issue can be the effect of microstructural heterogeneities. Thus, areas of microstructural heterogeneity, especially of higher porosity or microcrack content, and larger pores or microcracks (or grains) often give irregular or less distinct indents, especially when the microstructural features or their heterogeneities are on a scale similar to that of the indent (which is of course load dependant). Indents reflecting such effects are often neglected or discarded, thus often biasing results toward lower porosity (or finer grain sizes). Such indents should, at the minimum, be taken as an indication of heterogeneities that require further investigation. Similarly, indent cracks (e.g., for toughness measurements) may interact with pores and thus be affected. Biswas (78) noted that strengths of Al_2O_3 samples with up to 8% larger added pores showed much less or no degradation of strength as a function of indent loads compared to the normal decrease of strength of dense bodies with increased indent load. Since hardness requires only simple, small samples, checking for anisotropy is in principle relatively easy (but may be challenging due to limited sensitivity and scatter); however, it is almost never checked.

Key issues in obtaining valid, reliable compressive strengths are alignment (as in true tensile testing) and differential interfacial, lateral expansion between the loading platens and the specimen ends. While use of loading platens of the same material as the specimen is important in giving compatible elastic strains between the specimen ends and the loading platens, this does not eliminate other end-effect problems with specimens of uniform cross-section. Such end effects can be a particular problem in measuring microstructural effects on compressive strengths, especially with porous specimens. Use of strain matching platen materials is less feasible for bodies with substantial porosity. Further, general or local machining damage associated with heterogeneities of the microstructure, especially porosity can result in serious, parasitic end failures. Use of reduced gauge section, i.e., "dumbbell" specimens can be of significant aid by moving end damage away from the region of maximum stress. In either case suitable thin, compliant, e.g., polymer, sheets on the loading platens can be important (79). The higher levels of compressive strength indicate that it should be sensitive to anisotropic specimen character; however, the size and (frequently) shape of test specimens and their interaction with fabrication may complicate checking for anisotropy. Considerable variations in compressive strength of porous bodies indicates that the higher strength levels in compression may accentuate sensitivity to heterogeneity more than it is ameliorated by compressive failure generally being a cumulative damage process. Again, large specimen to pore size is needed, e.g., to minimize edge effects (80). For testing specimens with microcracks it should be noted that the cracks will often close when properly oriented in compressive loading.

There are a variety of wear tests, which present problems due to their diversity and incomplete understanding. Some tests use pins of various sizes and geometries wearing against a rotating disk (usually about an axis parallel to the pin axes) and others use a sample disk rotating under an abrasive wheel (with their axes of rotation perpendicular), tumbling or ball-milling specimens in grit, or grit blasting. In many of these, weight loss is a major measurement, but observations of the surfaces are also often important. Although all encompass more area than hardness tests, the test areas and volumes are still generally modest so the issue of specimen homogeneity is also critical for them. Thus, since wear can be significantly exacerbated by the presence of more porous or microcracked areas and especially larger pores of cracks or clusters (and larger grains) due to these enhancing penetration and plowing effects, the issue of homogeneity is very important from both a measurement as well as an application standpoint. Such effects are also of concern in erosion measurements.

Measurement of the porosity dependence of other properties also raises similar issues. Thus, heterogeneous and anisotropic porosity affect the dielectric constant and thermal and electrical conductivities. Such effects are greater for those many important properties that are nonlinearly dependant on porosity; however, even properties that are linearly dependant on microstructure, e.g., the dependence of dielectric constant on porosity, can show substantial variation with heterogeneity due to porosity differences in the complete specimen versus the area of measurement. There are also issues specific to particular measurements. Thus, laser flash measurement of thermal conductivity on porous materials can present problems due to penetration of the beam or the absorptive coating into surface pores, especially with larger pores and thinner samples (often needed for measurement of porous specimens). Measurement of electrical conductivity may be affected by absorption of conductive contaminants in the pores.

1.5 KEY ASPECTS OF THE POROSITY DEPENDENCE OF PROPERTIES

Before proceeding in the subsequent chapters to addressing in detail the porosity dependence of properties, it is of value to highlight key aspects of behavior, especially those that are not known or widely addressed. Thus, it is generally recognized that many properties decrease, often substantially, as porosity increases, and that the rate of decrease often increases with increasing porosity; however, the fact that there is a hierarchy of the porosity dependence of properties ranging from those not dependant on porosity, those dependant on only the amount of porosity, to various classes of dependence on both the amount and character of the porosity has not been addressed. This is addressed in Section 2.1, along with the important, but also commonly neglected, issues of the effects of pore character on many key properties, an issue emphasized throughout this book. Similarly, the fact that there is an upper limit to the volume fraction porosity (P) where the percolation limit of the solid is reached so properties go to zero since there is no longer a continuous solid structure, which frequently occurs at volume fractions of porosity (designated P_C) < 1, often substantially so, is also emphasized. These varying effects of pore character and amount of porosity are key to designing porous bodies for various engineering functions of porous materials. They are also key reasons why a single, "universal" model is not feasible; rather, at least one family or more of models is needed.

Self-consistency of the porosity dependence of properties, which entails two aspects, is also emphasized. First is the consistency of results for a given property over some P range with other and overlapping porosity ranges. Key aspects of this are the porosity dependence at low P, especially as it approaches zero, and at higher than normal porosity levels, especially where properties decrease much more rapidly toward zero with increasing porosity, often at P_C < 1. The second critical aspect of self-consistency is the underlying relations between various properties that provide perspective and proscriptions on their P dependence. This is most extensively and definitively applicable for elastic properties, where underlying interrelations set clear bounds on the P dependences of related properties. Other mechanical properties are generally related, often explicitly, to elastic moduli, commonly Young's modulus, providing further comparison of consistency.

The comparison of the P dependence of mechanical properties with that of elastic moduli shows the overall similarity expected as well as deviations. Definite and possible deviations provide some insight to as well as motivation for further evaluation of the causes of these, with detailed microstructural characterization, including fractography, being key factors. Significant deviations and their probable or established causes are outlined as follows. First, some fracture toughness (and especially fracture energy) data temporarily shows no decrease with increasing porosity or increasing to maxima. These deviations are attributed to varying combinations of measuring parameters, especially with larger cracks allowing more complex crack-microstructural interactions, porosity heterogeneity, grain-pore relations, and grain size effects. Second, tensile strength sometimes shows lower P dependence at low P, and often higher overall P dependence. The former often reflect effects of increased grain size with decreased P, while the latter commonly results from heterogeneity of the porosity, including failure from isolated larger pores or pore clusters. Third, mechanical properties such as hardness, compressive strength, wear, and erosion also commonly show deviations toward greater P dependence than elastic moduli, which again is attributed in part to effects of porosity heterogeneity. The size of individual pores or pore clusters can be important, e.g., for hardness, wear, and erosion. At intermediated to higher P levels, collapse of the pore structure can also become important.

These effects of grain size arise due to either the effects of grain growth on pore location and character (Fig. 1.2), direct effects of grain size on properties themselves, or combinations of these. While some variations of elastic properties probably occur with changes in inter- versus intragranular pore locations due to the associated pore character changes, greater effects can occur for properties involving fracture since intergranular pores often aid fracture more than intragranular pores. The direct contributions of grain size to these deviations arise since it has no effect on elastic properties (unless grain size correlates with grain orientation, or leads to microcracking, 39–41), while other mechanical properties have substantial dependence on grain size, usually substantially decreasing with increasing grain size (1). Wider variations of the porosity dependence of fracture toughness and especially energy with grain size itself arise due to the substantial variations they commonly show due to grain size effects that are often test, especially crack size, dependent. Frequent substantial variations of other mechanical properties due to grain size arise from effects of intergranular pores on fracture paths and their general decrease with increasing grain size.

1.6 SUMMARY

The effects of porosity (and, secondarily, of microcracks) on mechanical properties (with inclusion of other properties such as electrical and thermal conductivities) was outlined along with the scope and sequence of presentation of the book. Porosity and microcrack effects are very important because of their large impacts on many properties, and while limiting mechanical and other properties, they, especially porosity, are also essential or important for many important functions. The latter include thermal insulation, thermal shock resistance, filtration and high surface areas for sensors and catalysis for porosity and fracture toughness-thermal shock resistance for microcracking. Achieving such beneficial effects requires important design compromises between the favorable functions sought versus those, e.g., strength and stiffness, that are needed but are degraded by porosity or microcracking. Porosity is particularly important since its effects are more pervasive and variable, but also more tailorable due to the impact of pore character. The trade-offs and tailorability in turn require important engineering attention, especially fabrication-porosity interactions, to obtain the best balance of enhanced and limited properties central to a particular application.

Beyond the above background, property measurement methods and issues were outlined, noting for example advantages of multiple tests and samples involving moderate test areas or volumes to detect and minimize scatter effects of various heterogeneities and anisotropies of porosities. Finally, the reader is alerted to watch for issues of heterogeneity of porosity, orientation of anisotropic-shaped porosity and resultant property anisotropy, self-consistency of results, and impacts of other microstructural variables on porosity dependence of properties, important factors treated in this book, but generally neglected elsewhere.

More specific porosity results of this chapter are as follows. While there are important patterns in the pore amount and character from various forming-processing methods and parameters of fabrication, the amount and character of pores can vary widely. There are usually two or more pore populations, one of which is often but not always dominant, and varying gradients and heterogeneity of pore populations occur. The amount and character of pores typically shifts somewhat, sometimes substantially, with the size and shape of the component for a given fabrication method. While secondary pore populations that are modest in amount, often have a secondary effect on many, e.g., elastic and conductive, properties, they may play a more significant role in some, e.g., tensile strength. Some widely used forming methods, especially extrusion and die or hot pressing, can result in anisotropicaly shaped, oriented pores. Such pores, which may be the primary or a secondary component of the body porosity, result in varying degrees of anisotropic properties that are widely neglected. Both basic processing factors as well as subsequent use can change the mix of porosity, especially the balance of inter- versus intragranular porosity.

Microcracks are an extreme of some pores and have similar effects to those of pores on some properties, and hence are also considered. Such cracks commonly form spontaneously from microstructural mismatch strains in bodies with grains of anisotropic thermal expansions due to their noncubic crystal structures or dispersed particles of a second phase, or from phase transformations. While the local microstructural stress are independent of the microstructural scale, these cracks form above critical grain or particle sizes that are dependent material properties. Microcracks may also form due to the superposition of such local and applied, i.e., global, stresses. Such cracks, which form in the bulk, are more challenging to quantitatively characterize than those on the surface (and pores), and are discussed in Section 2.3.

REFERENCES

1. R. W. Rice, "Microstructural Dependance of Mechanical Behavior of Ceramics," Treatise on Materials Science and Technology, **11**, Properties and Microstructure (R. K. MacCrone, Ed.), Academic Press, New York, pp. 191–381, 1977.
2. J. J. Petrovic, J. W. Milewski, D. L. Rohr, and F. D. Gac, "Tensile Mechanical Properties of Silicon Carbide Whiskers," J. Mat. Sci., **20**, pp. 1167–77, 1985.
3. M. A. Spears and A. G. Evans, "Microstructure Development During Final/Intermediate Stage Sintering-Grain and Pore Coarsening," Acta Met., **30**, pp. 1281–89, 1982.
4. M. Sakarcan, C. H. Hsueh, and A. G. Evans, "Experimental Assessment of Pore Breakaway During Sintering," J. Am. Cer. Soc. **66**(6), pp. 456–61, 1983.
5. D. F. Porter, J. S. Reed, and D. Lewis, III, "Elastic Moduli of Refractory Spinels," J. Am. Cer. Soc., **60**(7–8), 345–49, 1977.
6. S. Prochazka, C. A. Johnson, and R. A. Giddings, "Investigation of Ceramics for High Temperature Turbine Components," GE report for NADC contract N62269–75–C-0122, 12/1975.
7. R. Tremblay and R. Angers, "Mechanical Characterization of Dense Silicon Tetraboride (SiB$_4$)," Ceramics Intl., **18**, pp. 113–17, 1992.
8. C. P. Cameron and R. Raj, "Grain-Growth Transition During Sintering of Colloidally Prepared Alumina Powder Compacts," J. Am. Cer. Soc., **71**(12), 1031–35, 1988.
9. J. J. Rasmussen, "Surface Tension, Density, and Volume Change on Melting of Al$_2$O$_3$ systems, Cr$_2$O$_3$, and Sm$_2$O$_3$," J. Am. Cer. Soc., **55**(6), pp. 326, 1972.
10. R. W. Rice, "Advanced Ceramic Materials and Processes" Design of New Materials (D. L. Cocke, and A. Clearfield, Eds.), Plenum Press, New York, pp. 169–94, 1984.
11. R. McPherson, "On the Formation of Thermally Sprayed Alumina Coatings," J. Mat. Sci., **15**, pp. 3141–49, 1980.

12. R. W. Rice and W. J. McDonough, "Intrinsic Volume Change of Self-Propagating Synthesis," J. Am. Cer. Soc., **68**(5), pp. C-122–23, 1985.

13. R. W. Rice, "Processing of Ceramic Composites," Advanced Ceramic Processing and Technology, **1** (J. G. P. Binner, Ed.), Noyes Pub., Park Ridge, NJ, pp. 123–213 (1990).

14. R. W. Rice, "Ceramics from Polymer Pyrolysis, Opportunities and Needs-A Materials Perspective," Am. Cer. Soc. Bul. **62**(8), pp. 889–92, 1983.

15. W. S. Rothwell, "Small-Angle X-Ray Scattering from Glassy Carbon," J. Appl. Phy., **39**(3), pp. 1840–45, 1968.

16. R. W. Rice, "Processing Induced Sources of Failure," Processing of Crystalline Ceramics (H. Palmour III, R. F. Davis, and T. M. Hare, Eds.) Plenum Press, New York, pp. 303–19, 1978. Materials Science Research, vol. 11.

17. R. W. Rice, "Fracture Identification of Strength-Controlling Flaws and Microstructure," Frac. Mech. Cer. (R. C. Bradt, D. P. H. Hasselman, and F. F. Lange, Eds.) **1**, Plenum Press, New York, pp. 323–343, 1974.

18. W. H. Rhodes, P. L. Berneburg, R. M. Cannon, and W. C, Steele, "Microstructure Studies of Polycrystalline Refractory Oxides," Avco Corp. Lowell, MA, Summary Report for Naval Air Systems Contract NOOO19–72–C-0298, 4/1973.

19. P. H. Rieth, J. S. Reed, and A. W. Naumann, "Fabrication and Flexural Strength of Ultrafine-Grained Yttria-Stabalized Ziriconia," Am. Cer. Soc. Bul., **55**(6), pp. 717–27, 1976.

20. J. S. Reed, "Introduction to the Principles of Ceramic Processing," John Wiley & Sons, New York, 1988.

21. D. W. Richerson, "Modern Ceramic Engineering," Marcel Dekker, Inc., New York, 1992.

22. B. K. Molnar and R. W. Rice, "Strength Anisotropy in Lead Zirconate Titanate Transducer Rings," Am. Cer. Soc. Bull., **52**(6), pp. 505–9, 1973.

23. A. G. Haerle and R. A. Haber, "Experimental Evaluation of a New Floculation-Filtration Model for Ceramic Shape Forming Processes," J. Am. Cer. Soc., **79**(9), pp. 2385–96, 1996.

24. M. Propster and J. Szekely, "The Porosity of Systems Containing Layers of Different Particles," Pwd. Tech., **17**, pp. 123–38, 1977.

25. J. S. Goodling, R. I. Vachon, W. S. Stelpflug, and S. J. Ying, "Radial Porosity Distribution in Cylindrical Beds Packed with Spheres," Powd. Tech., **35**, pp. 23–29, 1983.

26. R. W. Rice, "The Effect of Gaseous Impurities on the Hot Pressing and Behavior of MgO, CaO, and Al_2O_3," Proc. Brit. Cer. Soc., **12**, pp. 99–123, Mar. 1969.

27. R. W. Rice, "Ceramic Processing: An Overview," AICHE J., **36**(4), pp. 481–510 (1990).

28. A. Briggs, "Hydroxyl Impurity and the Formation and Distribution of Cavities in Melt-Grown MgO Crystals," J. Mat. Sci., **10**, pp. 729–36, 1975.

29. A. Briggs, "The Formation of Hydrogen-Filled Cavities in MgO Crystals Annealed in Reducing Atmospheres," J. Mat. Sci., **10**, pp. 737–46, 1975.

30. A. Takata and K. Ishizaki, "Mechanical Properties of Hiped Porous Copper," Porous Materials, Ceramic Trans., vol. 31 (K. Ishizaki, L. Shepard, S. Okada, T. Hamasaki, and B. Huybrechts, Eds.), Am. Cer. Soc., Westerville, OH, pp. 233–42, 1993.

31. R. W. Rice, "Machining of Ceramics," Ceramics for High Performance Applications, (J. J. Burke, A. E. Goroum, and R. N. Katz, Eds.), Metals & Ceramic Info. Center, Columbus, OH, pp. 287–343, 1974.

32. J. J. Mecholsky, Jr., S. W. Freiman, and R. W. Rice, "Effect of Grinding on Flaw Geometry and Fracture of Glass," J. Am. Cer. Soc., **60**(3–4), pp. 114–17,1977.

33. R. W. Rice and J. J. Mecholsky, Jr., "The Nature of Strength Controlling Machining Flaws in Ceramics," NBS Special Publication 562, The Science of Ceramic Machining and Surface Finishing II, (B. J. Hockey and R. W. Rice, Eds.), U.S. Govt. Printing Office, Washington, D.C., pp. 351–78, 1979.

34. R. W. Rice, J. J. Mecholsky, Jr., and P. F. Becher, "The Effect of Grinding Direction on Flaw Character and Strength of Single Crystal and Polycrystalline Ceramics," J. Mat. Sci., **16**, pp. 853–862, 1981.

35. R. W. Rice, "Porosity Effects on Machining Direction-Strength Anisotropy and Failure Mechanisms," J. Am. Cer. Soc. **77**(8), pp. 2232–36, 1994.

36. R. W. Rice, "Effects of Ceramic Microstructural Character on the Machining Direction-Strength Anisotropy," Machining of Advanced Materials, Proc. Intl. Conf. Machining of Advanced Materials 7/20–22/93 (S. Johanmir, Ed.), NIST Spec. Pub. 847, U.S. Govt. Printing Office, Washington, D.C., pp. 185–204, 6/1993.

37. W. R. Bussem, "Internal Ruptures and Recombinations in Anisotropic Ceramic Materials," Mechanical Properties of Engineering Ceramics (W. W. Kriegel and H. Palmour III, Eds.), Interscience Pubs., New York, pp. 127–148, 1961.

38. W. R. Bussem and F. F. Lange, "Residual Stresses in Anisotropic Ceramics," Interceram, **15**(3), pp. 229–31, 1966.

39. A. G. Evans, "Microfracture from Thermal Expansion Anisotropy-I. Single Phase Systems," Acta Met., **26**, pp. 1845–53, 1978.

40. R. W. Rice and R. C. Pohanka, "Grain-Size Dependence of Spontaneous Cracking in Ceramics," J. Am. Cer. Soc. **62**(11–12), pp. 559–63, 1979.

41. R. W. Davidge, "Cracking at Grain Boundaries in Polycrystalline Brittle Materials," Acta Met., **29**, pp. 1695–1702, 1981.

42. Y. M. Ito, M. Rosenblatt, L. Y. Cheng, F. F. Lange, and A. G. Evans, "Cracking in Particulate Composites due to Thermalmechanical Stress," Int. J. Fract., **17**, pp. 483–91,1981.

43. V. Tvergaard and J. W. Hutchinson, "Microcracking in Ceramics Induced by Thermal Expansion or Elastic Anisotropy," J. Am. Cer. Soc., **71**(3), pp. 157–66, 1988.

44. F. Ghahremani, J. W. Hitchinson, and V. Tveraard, "Three-Dimensional Effects in Microcrack Nucleation in Brittle Polycrystalline Materials," J. Am. Cer. Soc., **73**(6), pp. 1548–54, 1990.

45. E. D. Case, J. R. Smyth, and O. Hunter, "Grain-Size Dependence of Microcrack Initiation in Brittle Materials," J. Mat. Sci., **15**, pp. 149–53, 1980

46. J. Lankford, "Mechanisms Responsible for Strain-Rate-Dependent Compressive Strength in Ceramic Materials," J. Mat. Sci., **64**(2), pp. C-33–34, 1981.

47. E. Schreiber, O. L. Anderson, and N. Soga, "Elastic Constants and Their Measurement," McGraw-Hill, New York, 1973.

48. J. B. Wachtman, Jr., "Determination of Elastic Constants Required for Application of Fracture Ceramics to Ceramics," Fracture Mechanics of Ceramics, **1** (R. C. Bradt, D. P. H. Hasselman, and F. F. Lange, Eds.), Plenum Press, New York, pp. 49–68, 1974.

49. S. Spinner and W. E. Tefft, "A Method for Determining Mechanical Resonance Frequencies and for Calculating Elastic Moduli from These Frequencies," Proc. ASTM, **61**, pp. 1221–38, 1961.

50. S. M. Lang, "Properties of High-Temperature Ceramics and Cermets, Elasticity and Density at Room Temperature," National Bureau of Standards Monograph 6, U.S. Govt. Printing Office, Washington, D.C., 3/1960.

51. D. R. Larson, L. R. Johnson, J. W. Adams, A. P. S. Teotia, and L. G. Hill, "Ceramic Materials for Advanced Heat Engines, Technical and Economic Evaluations," Noyes Pub., Park Ridge, NJ, 1985.

52. R. Pampuch and J. Piekarczyl, "The Texture and Elastic Properties of Porous Graphites," Sci. Cer., (P. Popper, Ed.), Br. Cer. Soc., **8**, pp. 257–68, 1976.

53. H. Tomaszewski, "Influence of Microstructure on the Thermomechanical Properties of Alumina Ceramics," Cer. Intl., **18**, pp. 51–55, 1992.

54. A. G. Evans, "Fracture Mechanics Determinations," Fracture Mechanics of Ceramics, vol. 1 (R. C. Bradt, D. P. H. Hasselman, and F. F. Lange, Eds.), Plenum Press, New York, pp. 17–48, 1974.

55. R. W. Rice, "Crack-Shape-Wake-Area Effects on Ceramic Fracture Toughness and Strength," J. Am. Cer. Soc., **77**(9), pp. 2479–80, 1994.

56. R. W. Rice, "Fracture Topography of Ceramics" Surfaces and Interfaces of Glass and Ceramics, (V. D. Frechette, W.C. LaCourse, and V. L. Burdick, Eds.), Plenum Press, New York, pp. 439–472, 1974. Materials Science Research, **7**.

57. R. W. Rice, "Ceramic Fracture Features, Observations, Mechanisms, and Uses" Fractography of Ceramic and Metal Failures ASTM STP 827 (J. J. Mecholsky Jr. and S. R. Powell Jr., Eds.), ASTM pp. 5–103, 1984.

58. R. W. Rice, "Perspective on Fractography," Fractography of Glasses and Ceramics (J. R. Varner and V. D. Frechette, Eds.), Am. Cer. Soc., Westerville, OH, pp. 3–56, 1988. Advances in Ceramics, **22**.

59. V. D. Frechette "Fracture Analysis of Brittle Materials," Am. Cer. Soc., Westerville, OH, 1990.

60. G. D. Quinn, "Fractographic Analysis and the Army Flexure Test Method," Fractography of Glasses and Ceramics (J. R. Varner and V. D. Frechette, Eds.), Am. Cer. Soc., Westerville, OH, pp. 319–49, 1988. Advances in Ceramics, **22**.

61. K. R. McKinney and R. W. Rice, "Specimen Size Effects in Fracture Toughness Testing of Heterogeneous Ceramics by the Notch Beam Method," Fracture Mechanics Methods for Ceramics, Rocks, and Concrete (S. W. Freiman and E. R. Fuller Jr., Eds.), ASTM, pp. 118–26, 1981, ASTM STP 745.

62. R. W. Rice, S. W. Freiman, and J. J. Mecholsky Jr., "The Dependance of Strength-Controlling Fracture Energy on the Flaw-Size to Grain Size Ratio," J. Am. Cer. Soc., **63**(3–4), pp. 129–36,1980.

63. P. B. Hirsch and S. G. Roberts, "The Brittle-Ductile Transition in Silicon," Phil. Mag. A, **64**(1), pp. 55–60, 1991.

64. Y. Hiramatsu and Y. Oka, "Determination of the Tensile strength of Rock by a Compression Test of an Irregular Test Piece," Int. J. Rock Mech. Min. Sci., **3**, pp. 89–99, 1966.

65. W. D. Kingery, "Factors Affecting Thermal Shock resistance of Ceramic Materials," J. Am. Cer. Soc., **38**(1), pp. 3–15, 1955.

66. D. P. H. Hasselman, "Unified Theory of Thermal Shock Fracture Initiation and Crack Propagation in Brittle Ceramics," J. Am. Cer. Soc., **52**(11), pp. 600–04, 1969.

67. D. P. H. Hasselman, "Figures-of-Merit for the Thermal Stress Resistance of High-Temperature Brittle Materials: a Review," Ceramurgia Intl., **4**(4), pp. 147–50, 1978.

68. R. W. Davidge and G. Tappin, "Thermal Shock and Fracture in Ceramics," Trans. Brit. Cer. Soc., **66**(8), pp. 405–22, 1967.

69. L. Loh, C. Rossington, and A. G. Evans, "Laser Technique for Evaluating Spall Resistance of Brittle Coatings," J. Am. Cer. Soc., **69**(2), pp. 139–42, 1986.

70. C. F. Cooper, "Graphite Containing Refractories," Refractories J., **6**, pp. 11–21, 11/12, 1980.

71. P. F. Becher, "Effect of Water Temperature on the Thermal Shock of Al_2O_3," J. Am. Cer. Soc., **64**(3), pp. C-17–18, 1981.

72. J. P. Singh and D. P. H. Hasselman, "Effect of Drop Height on Critical Temperature Difference ($-T_C$) for Brittle Ceramics Subjected to Thermal Shock by Quenching into Water," J. Am. Cer. Soc., **66**(10), pp. C-194–95, 1983.

73. P. F. Becher, D. Lewis III, K. R. Carman, and A. C. Gonzalez, "Thermal Shock Resistance of Ceramics: Size and Geometry Effects in Quench Tests," Am. Cer. Soc. Bul., **59**(5), pp. 542–48, 1980.

74. K. T. Faber, M. D. Huang, and A. G. Evans, "Quantative Studies of Thermal Shock in Ceramics Based on a Novel Test Technique," J. Am. Cer. Soc., **64**(5), pp. 296–301, 1981.

75. K. Niihara, J. P. Singh, and D. P. H. Hasselman, "Observations on the Characteristics of a Fluidized Bed for the Thermal Shock Resistance Testing of Brittle Ceramics," J. Mat. Sci., **17**, pp. 2553–59, 1982.

76. K. J. Konsztowicz, "Crack Growth and Acoustic Emission in Ceramics During Thermal Shock," J. Am. Cer. Soc., **57**(2), pp. 107–108, 1974.

77. P. F. Becher, "Surface Hardening of Sapphire and Rutile Associated with Machining," J. Am. Cer. Soc., **73**(3), pp. 502–8, 1990.

78. D. R. Biswas, "Crack-Void Interaction in Polycrystalline Alumina," J. Mat. Sci., **16**, pp. 2434–38, 1981.

79. C. O. Dam, R. Brezny, and D. J. Green, "Compressive Behavior and Deformation-Mode Map of an Open Cell Alumina," J. Mat. Res., **5**(1), pp. 163–71, 1990.

80. R. Brezny and D. J. Green, "Characterization of Edge Effects in Cellular Materials," J. Mat. Sci., **25**, pp. 4571–78, 1990.

2

EVALUATION OF THE POROSITY DEPENDENCE OF PROPERTIES

KEY CHAPTER GOALS

1. Introduce the hierarchy of microcrack and especially porosity dependence of properties.
2. Present idealizations, key parameters, and characterization of pores and microcracks for modeling property dependences.
3. Outline the development of, and present, the MSA model family.
4. Discuss generic character and issues of mechanistic models.

2.1 HIERARCHY OF POROSITY IMPACTS ON PROPERTIES

The purpose of this chapter is to provide the necessary background primarily for evaluating the porosity, and secondarily the microcrack, dependence of properties. First, the hierarchies of the porosity and microcrack dependences of properties are considered. Then idealizations of pores and microcracks for modeling are presented, followed by an evaluation of what parameters are needed and effective for modeling and discussion of actual pore and microcrack characterization. Then modeling approaches and resultant models having broad applicability are presented and evaluated in terms of their functional character and applicability to various porosities. These and discussion of critical needs provide perspective for detailed evaluations of specific properties in subsequent chapters.

Table 2.1 Hierarchy of the Categories of the Porosity Dependence of Properties

Category/level of porosity dependence	Examples of properties/uses
I) No dependence on porosity	Lattice parameter, unit cell volume, thermal expansion, emissivity (but not emmittance)
II) Dependence only on the amount of porosity	Density, much dielectric constant data, heat capacity per unit volume
III) Dependence on both the amount and character of porosity	
A. Flux or stress dominant in the solid phase	Mechanical properties. Electrical and thermal conductivity at low to moderate temperature and porosity (finer).
B. All flux in the pore phase and filtration	Surface area and tortuosity, e.g., for catalysts
C. Flux in both pore and solid phases	Thermal conductivity, with larger and more open pores at higher temperature

A key example of the limited and often confused discussion of porosity effects is the almost total neglect of even identifying the generic levels, or categories, i.e., hierarchy, of porosity dependence of properties and the parameters needed to characterize porosity effects. Both have recently been considered (1–3); the hierarchy is considered here (Table 2.1) and the parameters in the following section. Basic but simple considerations show that there are three broad categories of porosity dependence that increase in their complexity of dependence on porosity in a hierarchial fashion. At the lowest level, there are properties that have no dependence on porosity; typically, these are those that depend only on the local atomic bonding and are not affected by long-range disruption of this by pores. Thus, properties such as lattice parameter and derivative properties such as unit cell volume fall in this first category, as do melting and boiling temperatures, ablation energy, and emissivity. An important extension of these atom bonding determined properties to macroscopic behavior that does not depend on porosity is thermal expansion.

The second hierarchial category consists of those properties dependant only on the amount, and not the character, of porosity. This category consists of properties dependant only on the mass in a given volume and not the specifics, i.e., the microstructure, of its distribution within that volume. This includes density, heat capacity (per volume), and much dielectric constant data (see Chapter 7). Such properties follow a rule of mixtures relation for the volume fractions of the pore phase (P) and the solid phase (1 − P):

$$X = X_s (1 - P) + X_p P \qquad (2.1)$$

where X = the property of the body, X_s = the solid property (i.e., at P = 0), and X_p = the property of the pore phase (often zero, but not always, e.g., typically about 1 for the refractive index and dielectric constant).

The third, largest, most complex and important category in the hierarchy of the porosity dependence consists of those properties dependant on both the amount as well as the character of the porosity. Three important subsets of this are (following the designations in Table 2.1): A) properties determined only by flux or stress transmission through the solid but not through the pore, phase, B) properties determined only by flux through the pore phase (e.g., filtration), and C) properties dependant on transmission through both the solid and pore phases. This categorization of properties assumes that the pores are empty or contain a fluid whose properties are not significantly different from that of a vacuum. Bodies with fluids, e.g., water or solutions, not meeting this criteria have to be treated as composites for properties they significantly impact. Only the first subset, i.e., with flux or stress transmission in the solid phase, is considered in detail in this book. The other two subsets, especially B, and the effect of different pore fluids are important and large topics, detailed treatments of which are beyond the scope of this book; however, surface area and tortuosity are briefly treated, and the extensive treatment of III A porosity effects in this book provides a basis for, and complements, treatment of these other topics. Comparison of the property dependence within and between the three subcategories can be valuable in characterization of porosity and its effects on properties as noted throughout this book.

It is important to note that some properties can change categories with the nature of the porosity or the conditions. Thus, many properties, including most or all mechanical and conductive properties, of bodies with tubular pores aligned parallel with the stress or flux axis follow a $1 - P$ dependence, which is that of Category II properties (and the upper limit of porosity dependence of properties). Mechanical and conductive properties of other pore structures all fall below this $1 - P$ dependence, i.e., have greater rates of property decreases, typically in a nonlinear fashion over reasonable P ranges. The dielectric constant of bodies with substantial fine porosity, high dielectric constant material, or both, deviate below the rule of mixtures relation (apparently when surface charges on the pores becomes significant) moving it from Category II to Category III. Similarly, pores commonly have negligible effect on optical absorption, but substantial amounts of pores near or at the optimum size for optical scattering can increase net optical absorption due to increased optical path length from scattering. At lower temperatures and finer, closed porosity where convective and radiative heat transfer across and between pores are insignificant, thermal conductivity falls in Category III A; however, as temperature and the size, amount, and openness of pores increases, thermal conductivity transitions to Category III C, and at high temperatures and porosity can approach Category II B due to thermal transport by radiation. Also closely related properties may be in quite different categories, e.g., ablation energy falls in category I, but the rate of ablation for fixed thermal parameters can fall in categories II or III depending on the extent of contribution of surface connected pores to ablation. Similarly, while emissivity is a category I property, the energy emitted per unit area dependence on porosity varies with the extent to which pores open to the surface contribute, e.g., the extent to which they act like black body emitters; however, some properties are difficult to categorize, e.g., dielectric breakdown where the stress is in the solid phase, but much of the actual breakdown may occur via the pore phase.

There is also a hierarchy of property dependences for microcracks, which are of interest as an extreme of pores for their occurrence in a number of materials, and for their use to modify properties, e.g., to increase toughness and resistance to more extreme thermal shock. The hierarchy for microcracks has overall similarity to, but also some differences from, that for porosity. There are the same three basic categories, with many similarities in which properties belong in which category. Thus, Category I properties, i.e., not dependent on microcracks, includes many properties determined only by the local atomic bonding such as unit cell dimensions and theoretical density; however, an important difference between microcrack and porosity dependence of properties is that thermal expansion is a Category II (and in some cases Category III) property for microcracks (see Fig. 7.10), but Category I for porosity (in the absence of microcracks). For both hierarchies, actual density is a Category II property (though with much less decrease with increasing microcrack versus pore content). Category III properties present major similarities and differences for microcrack and porosity dependences of properties. Thus, mechanical and conductive properties both show significant dependences on both the amount of porosity or microcracking as well as their character, with the latter involving the same basic geometrical parameters of shape and stacking, i.e., orientation and spatial distribution for both pores and microcracks. The most significant differences between the two hierarchies is that the extents of Category III B and C type properties for microcracks are far more proscribed because of the small volume of microcracks, and especially their common discrete, i.e., noninterconnected, nature, hence commonly limited or no percolation. Some of these differences are well defined and others poorly defined since, while in some respects microcracks have a more proscribed range of character and amount than pores, detailed understanding of their character and effects is both more incomplete and complex. As with porosity, some properties may also have different microcrack dependences, i.e., change category, depending on the nature of the microcracks, material, or environment.

The complexities of microcracks arises basically from their more challenging, hence less developed, characterization, due mainly to their typically very small, and probably variable, opening. This is compounded by observing their intersection with a free surface being a common, and in principle, easier way of observing them; however, their opening, and possibly their extent of, or very, occurrence may be different at free surfaces versus in the bulk, and probably vary with the nature of the surface on which they are observed. This basic characterization issue is further compounded by the inherent anisotropic shape of microcracks leading to local, and sometimes global property anisotropy, and their occurrence and character may depend on test and microstructural factors. Thus, microcracks can form, change, or disappear during test due to interaction with the test parameters, mainly stress and temperature, and their effects are not always independent of pores as commonly assumed. Their effects are included in this book since they have similarities to those of pores, e.g., being an extreme of pores and thus give greater insight into porosity effects and because their effects are important and may be impacted by interaction with pores. They are also included to aid more research on microcrack character and effects.

2.2 IDEALIZED PORES AND MICROCRACKS AND MODELING PARAMETERS

2.2.1 Model Microcracks and Pores

First, consider model (i.e., idealized) microcracks and their parameters since they are simpler in actual form and in modeling relative to pores, and provide some perspective on pore modeling. Microcracks are idealized as planar cracks of zero thickness, i.e., a mathematical discontinuity of the material properties at the crack, with all cracks in a given model being of the same simple geometrical shape and size, and spatial distribution. Shapes assumed are circular, elliptical, or slit (i.e., long rectangular) cracks. They are typically assumed to be in regular arrays or randomly distributed with random, or more commonly fixed orientations between them. Modest densities of cracks are assumed such that there is limited or no interaction between adjacent cracks. The key parameters for their modeling are thus those of size and shape, (i.e., a radius, eccentricity, or length and width), spacing, and orientation, (i.e., how they are arrayed). While such idealizations are quite useful for a variety of modeling, there is possible need for treating more open, e.g., "inflated" microcracks, i.e., a transition between cracks and pores. As outlined in Chapter 3, solutions to such problems as well as the important problem of mixes of pores and microcracks are becoming available.

Consider now idealized pores and parameters to be used in modeling the porosity dependence of properties, which is a more complex subject. Idealized pores and ideal or real pore parameters and characterization (covered in the listed order in this, and the subsequent subsection and section) should be closely

Table 2.2 Porosity Parameters for Stacked Spherical Particles

Stacking	Coord. No. (C_n) of: Stacked spheres	IBS[a]	Resultant dense grain shape[b]	Volume fraction porosity Maximum[c]	At pore closure
Simple Cubic	6	6	Cube	0.524	0.035
Orthorhombic	8	5	Hexagonal prism	0.395	0.165
Tetragonal	10	—	—	0.30	—
Rhombohedral[d]	12	4	Rhombic dodecahedron	0.26	0.036
Body-Centered-	8		Tetrakai-decahedron	0.32	
Cubic[e]	14			0.006	0.006

Source: ref. (1); published with permission of J. Am. Cer. Soc.

[a] IBS= Interstices (i.e., pores) between (solid) spheres.

[b] See Fig. 2.1.

[c] Maximum porosity, i.e., P_C value, is where solid spheres just touch so further separation (hence greater porosity) means the spheres no longer touch and therefore no longer form a solid body. This important porosity level is discussed further later.

[d] Encompass both FCC and HCP packings.

[e] Densification of body centered stacked spheres proceeds in two stages: the critical transition coming when six spheres originally not in contact (i.e., not nearest neighbors) come in contact to increase the C_n to 14 at P~0.04, but porosity remains open till P~0.006.

interrelated in a practical fashion, but most current concepts and methods fall far short of this.

Identifying key parameter (s) that determine pore character effects on category III A porosity dependence is critical to addressing these important properties, which are the focus of this book. Many characteristics of pores have been considered to correlate with properties along with the amount of porosity. These include pore size, the degree of open versus closed porosity, the modality of the pores (i.e., the number of different types of pore populations present), the stacking relation of the pores (i.e., the coordination number, C_n, of the particles or webs defining the pores, the pores themselves, or both; Table 2.2), and especially pore shape; however, until recently there has not been any systematic, comprehensive attempt to evaluate which parameters are suitable and practical. This analysis (1) is summarized here as a basis for subsequent evaluation of porosity behavior falling in category III A.

As a precursor for both the evaluation of porosity parameters as well as subsequent modeling and evaluation of porosity behavior, it is necessary to first consider geometrical aspects of modeling pore structures. A basic, widely used, approach is that of assuming pores of idealized, simple, specific geometrical shapes, very commonly spheres, or of pores formed between packed spherical particles. It is further common to consider various stackings of such idealized particles or pores, especially regular stackings of uniform size spheres (Fig. 2.1). Such idealized pores reflect the two most common and extensive methods of

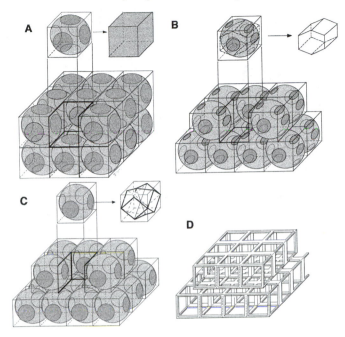

Figure 2.1 Schematic of regular stackings of uniform spherical particles (or pores). A) Simple cubic, B) orthorhombic, and C) rhombic stacking. Note the unit cell shapes are the resultant grain shapes at P = 0 (see also Table 2.2). The minimum solid area (MSA) for: 1) stacked particles is the area of the typical bond between two particles normal to the reference (e.g., stress) direction, and 2) stacked pores is the minimum area of the solid web area per pore between two pores, i.e., at the pore equator normal to the reference direction (see Fig. 2.10). D) Schematic of an idealized foam based on orthorhombic stacking of (originally spherical, then cylindrical) pores. The MSA is the cell strut cross-section area normal to the reference direction. Typical strut tapering to a minimum at their center is, for simplicity, not shown here. Published with permission of J. Mat. Sci. (2).

creating pores in ceramics and other materials. Thus, regular stacking of uniform spherical particles reflects pores formed in bodies made by typical powder processes such as by chemical bonding, e.g., cementatious processes, and especially by sintering particles. Such pores are inherently intergranular and smaller than the particles/grains that define and surround them. The progress of densification in sintering such ideal systems can be followed in numerical models by recognizing that each sphere is centered in a conceptual cell whose geometry reflects, and is defined by, the stacking, and is thus also the shape of (but of larger initial size than) the final resultant idealized grain at full density, i.e., zero porosity (Table 2.2, Fig. 2.1). Thus, as sintering progresses, the

spherical particles intersect, and hence truncate, each other at the cell boundaries. Since mass must be conserved in sintering, such truncation reflects
shrinkage of the cell dimensions so the mass removed in truncation of each
sphere is returned to increase the diameter of the truncated sphere. The volume
fraction porosity at any degree of sintering is simply 1 minus the ratio of the
volume of the truncated sphere divided by the total volume of the cell. (Some
modification of this numerical treatment is needed for chemical bonding process
such as cementatious ones since these commonly change, usually increasing, the
mass of the initial powder compact; see Section 8.2.2.) Pores that result from
condensation of lattice vacancies from radiation, differential diffusion, and
reaction at or near grain boundaries also fit, at least approximately, such models
when the pores form, or end up, on grain boundaries. Pores formed by imperfections in stacking particles (Figs. 2.1, 2.2) cover a range of pore sizes relative
to the particle sizes, varying from less than to greater than or equal to the size of
the spherical particles. Pores from particle stacking imperfections represent a
transition from the above pore structure to that which immediately follows.

The other related, idealized model is that in which uniform spherical (or
other shaped) pores are similarly stacked (Fig. 2.1). This reflects forming
processes where the pores can vary from much smaller to much larger than the
dimensions of the solid struts defining and surrounding the pores. Such pores
are inherently larger than any grain structure, and are thus intergranular. Such
stacking of pores simply reflects the interchanging of the pore and solid phases
from that of stacking spherical particles, i.e., the spherical particles become the
pores, and the solid phase becomes the portion of the cell surrounding the
(truncated) spheres. The same basic numerical procedure for spherical particles
can be applied to following increases or decreases of density for spherical pores;
however, at high levels of porosity, most foam structures are better approximated as the result of the intersection of short (hemispherically capped) cylindrical pores since the intersection of these would lead more closely to actual
foam cells than spherical pores. Practical processes that result in such types of
pores include forming of bodies, e.g., refractories, with a fugitive phase, incorporation of bubbles (as in foaming operations), and bonding of balloons or
particles with large internal pores. Some pores from processes such as condensation of lattice vacancies from radiation damage, differential diffusion, and
reactions are also approximated by such models, especially when they are, or
become, intragranular, i.e., within the grains. Also, as noted above, particle
stacking imperfections due to particle bridging also are a source of such pores
(Fig. 2.2).

Such idealized pore structure models clearly define key aspects of their pore
character. These include key parameters such as the coordination number (C_n)
of the solid and pore structures, when the pores become open or closed, and the

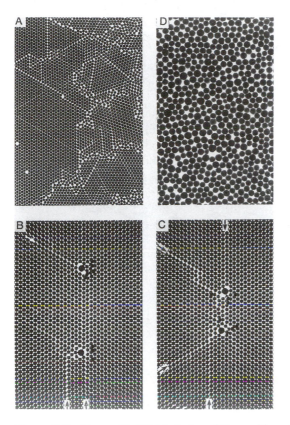

Figure 2.2 Transmitted light photos of the packing of single layers of spherical particles. A) All particles of the same size. B) and C) All particles of the same size except two being somewhat larger, but with different spacings between them. D) All particles with limited deviations from perfect sphericity and uniform diameter. Note individual stacking defects due to particle bridging in A–D and linear stacking defects in A–C (e.g., arrows in B and C). Published with permission of Plenum Press (4).

maximum allowable volume fraction porosity (P, i.e., P_C) where the solid no longer transmits stress or flux. All of these parameters cover a substantial range, depend on pore/particle character (Table 2.2, Figs. 2.11, 3.1), have some interrelationship, and can also depend on stress and orientation factors in even ideal pore systems. In real pore systems they also depend on the distributions of different particle-pore factors, e.g., an open-closed porosity transition occurs over a substantial porosity range (e.g., 0.05–0.30, Fig. 1.2B). These parameters, their interrelationship, stress-orientation dependence, and variations are discussed further in Section 2.4.4.

2.2.2   Porosity Parameters for Modeling

Microcrack parameters were briefly considered in the previous subsection in conjunction with their idealization. The long-standing issue of what other porosity parameters besides the volume fraction porosity (P) are appropriate to Category III A of the hierarchy of porosity dependence (Table 2.1), though only recently systematically addressed, is a larger subject. Rice (1) addressed this by plotting percent or fractional theoretical density $(1 - P)$ versus the ratio of the radius of the pore (r_p) to that of the particle (or solid element, r_s) defining the porosity resulted in an S-shaped curve for stackings of uniform spherical particles or pores (Fig. 2.3). This was done for each regular stacking, and is carried through its full range of feasible porosity, i.e., $P < \sim 52\%$ and 100%, respectively, for stacked spherical particles or pores. Thus, as intersections of the spherical pores are reduced and the degree of intersection (i.e., bonding) of the

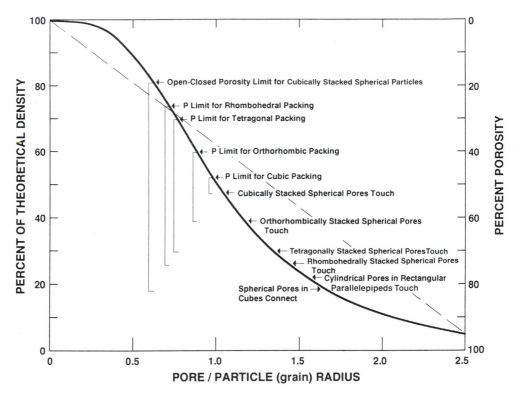

Figure 2.3   Plot of percent theoretical density or porosity versus r_p/r_s showing the highly restricted porosity opportunities. Note the dashed straight line representing a rule of mixtures combination of different porosities. Published with permission of the American Ceramic Society (1).

spherical particles are increased, they all followed the S-shaped curve back to the left from the maximum P value allowed for each stacking. Before turning to evaluation of this curve, note some factors about it. First, the r_p/r_s ratio goes to infinity at $P = 1$, but ratios of up to only 2–3 encompasses P to ~ 0.94, thus readily covering the normal useful P range. The inverse of this ratio might have been used, but has the disadvantage of going to infinity at $P = 0$. Second, to make handling of the size of the nonspherical solid element or pore, respectively, between the spherical pores or particles more manageable, its radius was taken as that of the sphere of the same volume. This results in changes in dimensions of about 25% or less, but gives a single, consistent number to compare with that of the spherical particle or pore.

Pore parameters to be considered include the pore modality, the dimensions and coordination number of the pores or solid elements defining the pores, whether the pores are open or closed, pore size and shape as noted earlier, as well as the solid bond area between pores or particles. For these initial, idealized cases, the modality of the porosity is one, and there is no distribution of it since there is only one type of porosity (excluding body centered stacking because of the complication of C_n increasing as second and first nearest neighbors come in contact with shrinkage, Table 2.2). The S-shaped curve of Fig. 2.3 provides insight into the merit of other pore parameters besides P to characterize the porosity and correlate with properties of interest as follows. First, it readily shows that the ratio r_p/r_s provides no new information over P, since there is one and only one value for this ratio for each value of P. This does not rule out either, or both, r_p and r_s being important separately; however, these by themselves are typically not a factor, unless either, especially r_p, is on the scale of the critical element determining the property of interest, especially the crack or flaw size in crack propagation or strength tests, or the indent or asperity size in respectively hardness or wear tests (see Chapter 6). Second, the curve also readily shows that the open–closed pore transition provides no new information since, for each idealized pore structure, this is a single discrete point at a specific r_p/r_s and a corresponding P value for a given stacking. That the open–closed porosity transition itself is not a fundamental factor is also shown by other arguments later.

The coordination number, C_n, of the pores, particles, or both (Table 2.2), contains new information beyond that provided by P or r_p/r_s (Fig. 2.3). First, while C_n is fixed for a given particle stacking, as different stackings of spherical particles of pores undergo different degrees of bonding or intersection, there are common values of P and r_p/r_s for different stackings. Further, experimental data of Fischmeister et al. (5) (Fig. 2.4) showed that while iso- and die-pressing of spherical metal particles give similar $P - C_n$ trends, they are not identical. Isopressing gave lower C_n values, but lower P levels than die pressing due to the former giving larger bond areas than the latter. Thus, different porosities are not fully characterized, i.e., differentiated, unless C_n is accounted for directly or

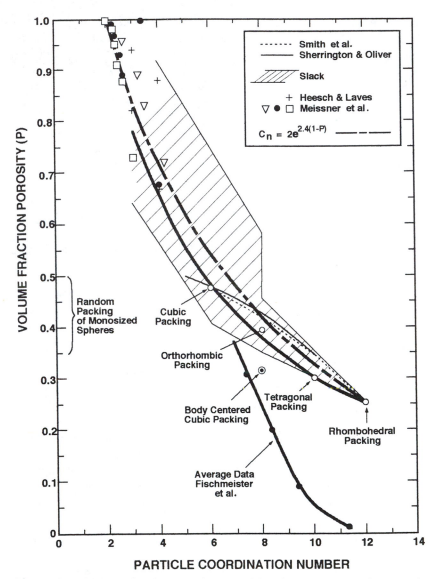

Figure 2.4 Volume fraction porosity (P) arising from the spaces between (mostly uniform spherical) particles versus the coordination number, C_n, of the particles. The lower curve gives average experimental data for uniaxial and isopressed spherical metal particles, after Fischmeister et al. (5). Note the exponential dependence of C_n on P of Meissner et al. (6), which is another source of the exponential dependence of properties on P. Most data is for stacked (unbonded) particles from a review by Rice (1); published with permission of J. Am. Cer. Soc.

indirectly; however, C_n is included in the other two candidate parameters: pore shape and minimum solid area (MSA; see Figs. 2.1 and 2.10 for the definition of MSA in ideal systems) since it is clear that any change of coordination number in such ideal stackings makes unique changes in pore shape and MSA, which account for effects of C_n. Pore shape and MSA clearly contain information beyond that of C_n alone or in combination with P and r_p/r_s. Thus, use of C_n is inadequate by itself, since it is redundant with pore shape or MSA, both of which each contain other needed information.

This leaves the choice between pore shape and MSA as the only additional parameter besides P needed to characterize porosity for effects on properties of interest (i.e., Category III A, Table 2.1). Pore shape is a frequently cited, but never specifically defined, factor except for limited idealized, e.g., spheroidal pores, and thus lacks a specific quantification for modeling or measurement of many idealized and all real pore systems. A key example of such pore shape definition problems is for pores between particles (Fig. 2.5). It is more complex for real systems where pore shapes vary within a given body. In fact even defining where one pore and a connecting one each begin and end is challenging. Further, pore shape may contain extra, extraneous information having little or no effect on properties of interest here, e.g., "hills" or "hollows" of the solid particles or webs defining the pores, that impact pore shape but are well away from the minimal solid cross-sections that commonly dominate properties are an example. Also, pore shape becomes even more complex as the percolation limit of the solid is approached, but is not particularly effective in defining this limit.

On the other hand, MSA reflects the pertinent pore shape and C_n information and is readily calculated for a variety of models. Also, though there are measurement issues (discussed later) MSA should also be calculable for other, including real, porosities with modern computational techniques. Further, MSA avoids the extraneous information that accompanies and complicates pore shape, and it very clearly defines the solid percolation limit, which may not be clearly delineated by pore shape. This MSA approach, though not referred to by the MSA term until recently (1–3), has been widely used as a basis of modeling porosity effects of interest here, as discussed later. Thus, MSA is also a realistic, key, practical parameter for characterizing actual pore structures.

The first of two obvious questions is what the curve in Fig. 2.3 would be for other ideal shaped pores, e.g., cylindrical or cubic pores (as modeled later). Both normalization of r_p and r_s respectively for the pore or the solid phase segments defining the pores as those for spheres of the same volume, and the evaluation of other pore structures below show that the same curve results. The second and broader question is how well this analysis pointing to the MSA as the key other parameter besides P to characterize porosity effects on properties of interest apply to more complex and real pore systems. Rice (1) argued that real pore systems would start along the S-shaped curve from either end, and broaden into

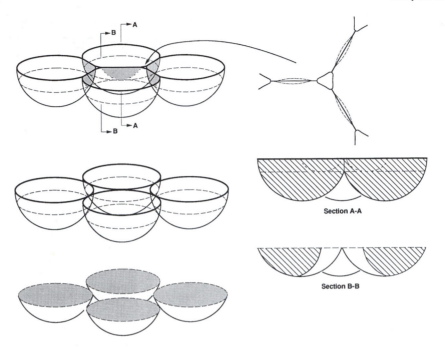

Figure 2.5 Illustrations indicating the complexity of pore shape factors (e.g., radii of curvature), even in an idealized body of partially sintered, uniform spheres. The changing character of a triple-point pore is shown by: top section, through grain centers, with blow-up of pore to the right (arrow) showing closed section of pore; the middle section, where the two center grains just touch, showing almost total open nature of the pore there, and the lowest section showing total connection of that portion of the pore with adjacent pores. Sections A–A and B–B show two views of a pore perpendicular to the aforementioned sections. Note closed portion of pore within bounds of sintering (above dashed line), and open below, while in section B–B the pore is totally open. Published with permission of the American Ceramic Society (1).

a modest band about the curve. This conclusion is predicated on the concept that more complex porous bodies can be broken into smaller units, each approximating relative uniform ideal pore structures, and that the P and r_p/r_s for the body can be calculated by averaging those units into which the body was divided. If the S-shape curve were instead a straight line, then the line for ideal systems and that obtained by averaging values for various cells in a body would be the same for such a rule-of-mixtures approach for a linear system; however, while the curve is S-shaped, it does not differ greatly from a straight line, e.g., the maximum difference between it and a straight line between its termini is about 15% of theoretical density (Fig. 2.3). Further, because of the modest S-shape to the curve, a straight line average through the curve crosses it. There-fore, since averaging different units assumes the straight-line relationship, deviations from the S curve can be either above or below the S curve, e.g., by the amounts of the difference between the S curve and the average line. Thus, since the deviations from the S curve are both commonly positive and negative, the band formed about the S curve is modest.

Having argued that more complex ideal and real pore systems follow the S curve as a modest band about the S curve for ideal systems, again consider the arguments for other pore parameters, as done for ideal pore structures. For more complex ideal and real pore systems the modality may no longer be only one, and hence must be considered, but is reflected in other parameters. One of these is the coordination number of pores or particles, C_n, which is no longer neces-sarily a unique value for a given body (Fig. 2.4), i.e., varies within the body. It thus reflects more information about the porosity, and can therefore be an important parameter for characterization, e.g., indicating the range of particle stacking encountered. The mean particle coordination number should indicate the proportions of the main ideal particle stackings found in a body, and hence a guide to which models to combine to approximate the body's behavior (e.g., Fig. 2.11); however, C_n is still reflected in pore shape and MSA. Also, the tran-sition between open and closed porosity is no longer at a discrete P for a given pore character, but now occurs over a (variable, often substantial) range of porosity. While detailed studies are limited, these (Fig. 1.2B) and more general observations on sintering powder compacts both show porosity not becoming totally closed until P is ≤ 0.06, which is consistent with the fact that C_n for grains in nearly dense bodies is greater than or equal to 12 (Table 2.2, consistent data from with Fischmeister et al. (4); Fig. 2.4). Going to a distribution of pore sizes or shapes can make only limited decreases in the values of P at which all pores become closed, but significantly extends the porosity range over which the transaction to open porosity occurs; however. the impacts of these changes are all more modest than, and contained within either pore shape or MSA. Thus, the choice again is between the latter two, and again MSA is seen as the only truly defined, and practical parameter for any porosity from both a computational and a measurement standpoint. That MSA is a unique parameter for a given pore type over the entire pertinent porosity range is shown by the separate curves for different idealized pore systems in Figs. 2.11 and 3.1.

2.3 POROSITY AND MICROCRACK CHARACTERIZATION

In principle characterizing porosity is fairly simple for most properties since this requires only the volume fraction porosity (P), or this with the MSA or pore shape information, or both; however, there are complications, limits, and uncertainties with all of these in simple pore systems, which are outlined below. These are compounded in real porous bodies, e.g., due to mixes, heterogeneities, anisotropies, and gradients of porosity, all of which change with fabrication methods and parameters and commonly with P, even for a fixed fabrication method and parameters. Thus, a variety of observations and measurements, including partially or completely duplicative ones are needed, the degree and extent of these necessarily being constrained by various practical limitations, as outlined below.

Clearly the most basic and broadest porosity measurement needed is P. This is in principle, and often in practice, simple, since it is just 1 minus the ratio of the actual density (ρ) to the theoretical density ($\rho_t = \rho_0$ at P = 0), i.e., P = 1 $-$ ρ/ρ_t; however, there are important cases where the theoretical density is not known accurately, or at all, or changes as a function of processing and thus with density. Such uncertainties and changes occur in many carbons, cements and related materials, porcelains (see Fig. 8.13) and other partially or fully crystallized glasses and microcrystalline or glassy materials, or composites involving these materials. In many cases pores are clearly observable by microscopy, so P can be obtained by stereological measurements, since the area fraction occupied by pores on a random cross-section of the sample is also the volume fraction porosity. This is typically done on a polished surface, which may raise issues of distinguishing pores from pullouts and of pores partially or fully obscured by debris. The area fraction observation is also reasonably, if not fully, accurate on fracture surfaces, which avoids the above issues with many polished surfaces, and has far lower preparation cost. Such fracture observations may be biased towards the more extremes of pores in terms of numbers, sizes, or both; however, if this is so, this difference is the type of information needed for properties dependant on fracture, especially on a larger scale. Thus, serious study of the comparison of porosity measurements from both polished and fracture surfaces is recommended. Information on very fine porosity, e.g., those at or beyond practical microscopies, can be obtained by x-ray (7,8) or neutron scattering (9). Other than density measurements, the above measurements also give information on pore sizes, especially stereological measurements (which can also give pore shape, anisitropy, and orientation information, but comparison of measurements at differing magnifications can be important). The volume fraction of open porosity and its size distribution can be measured by mercury porosimetry, gas absorption (giving surface area) and electrical conductivity via conductive fluids in the pores (10,11). However, results of these involve some assumptions of pore character, and mercury porosimitry can also begin to damage pore structures at higher material and pore dependent pressures, especially in softer or weaker materials (12–14), so comparison of various methods can be important. Ultrasonic attenuation can also give some information of pore sizes (15).

The MSA is defined, and in principle measurable (as opposed to pore shape which is poorly defined and thus of uncertain measurability), but presents complications. One complication is differences for differing pore types. For pores between MSA of grains is essentially the projection of the contiguity of the grains normal to the reference stress or flux direction (s) (Figs. 2.1, 2.10), e.g., of the necks between grains in earlier stages of sintering. This should be the same as determining the size and orientation distributions of microcracks, for which theoretical analysis has recently be presented to guide stereological measurements (16); however, experimental tests of this, e.g., of its practicality, reproducibility, and accuracy, are not available. For intragranular pores or pores larger than the grain structure, e.g., from burnout of material or foaming, the MSA is the minimum wall or strut area normal to the reference stress or flux direction(s). This should also be stereologically measurable, but again demonstration of the practicality of this is not available. Other complications are determining what average MSA, especially with mixed porosity, to use and the ratio of this to the representative cell (Figs. 2.1, 2.10) or body area. Averages for a single pore type based on normal area rather than on linear dimensions appears appropriate. Given the uncertainties involved, recourse to ascertaining the character of the pores in a body to determine which MSA model or combinations of models are most applicable, appears to commonly be necessary until direct measurement techniques are established. For partially sintered powder bodies the particle coordination, C_n (Fig. 2.4) is important for this (as well as giving more insight into the pore character, e.g., heterogeneity via its variation). Green density is of fundamental importance in determining the P_C values and starting C_n. Observations on the relation of C_n to starting density as well as changes with densification (5,6,17) can also be important.

A very important aspect of porosity characterization, as will be extensively shown in subsequent chapters, is determining its heterogeneity in the bodies used, i.e., its degree of systematic or random variations. This can often be indicated, or partially done for P via tomography techniques, and determined by density measurements of sectioned bodies; however, a direct, critical, but seldom used method, is to simply make a sufficient number of suitably accurate density or stereological measurements for porosity that allow a statistical evaluation of the resultant values. As discussed later in more detail (Table 10.6), it is recommended that porosity measurements be evaluated using the Weibull modulus (m) for direct comparison with that for mechanical and other physical properties. Note that a reasonable calculation of m can be obtained from the average value (A, i.e., of P in this case) and its standard deviation (S) (18):

$$m = 1.21 \left(\frac{A}{S} \right) - 0.47 \qquad (2.2)$$

Determining the Weibull modulus for aspects of pore character is also important in indicating heterogeneity (as well as statistical variation of the character). Thus, another, way of indicating the variation of P is by determining the global and local average coordination numbers and their variations for particles, pores, or both. As shown in Table 2.2 and Figs 2.11 and 3.1, these, clearly impact both porosity level and character, especially for stacked particles. More broadly stereological techniques should again be of substantial value, e.g., using the technique of Stoyan and Schnabel (19) for determining the spatial distribution of larger grains applied to pores. They used a pair correlation approach, characterizing the frequencies of interpoint distances (preferably between grain centers, which could thus instead be pore centers). There are also important property measurement methods of indicating heterogeneity or gradients of porosity as noted in Section 1.4 and later.

In addition to direct porosity measurement, there are four other aspects of characterization of bodies that are important to documenting and understanding the effects of porosity. The first is raw material characterization, which is especially pertinent to making samples via powder consolidation-densification routes. Thus pycnometric density measurements of the powder can indicate whether the particles are dense or have measurable internal porosity. Particle size, shape, and distribution can be important in indicating particle packing, and hence of resultant porosity, as can the similar characterization of any hard or purposely formed (e.g., spray dried) agglomerates, and their degree of deformation or destruction in the consolidation process. Second is characterization of the consolidation process and product. The forming method and parameters, e.g., specimen sizes and shapes, consolidation pressures, deposition rates, mold sizes, and injection sprue relations to part geometry can be important indicators of pore character. For powder consolidation techniques, the green density is a crucial parameter since it is ~ the P_C value. Examination of the green and sintered body to estimate the average coordination number of the particles (and of agglomerates and their deformation or destruction) is also valuable. Third is other microstructural characterization. A key factor of this is grain size, shape and distribution, since these often significantly impact the amount, shape (s), and location (s) of pores, e.g., the amount of inter- versus intragranular pores (Figs. 1.2–1.4). In composites similar determination of pores at the matrix-dispersed interfaces versus in the second phase can be quite important (Chapter 8).

Fourth, and equally important, is the conduct of property measurements. As noted earlier and later, measurement and comparison of more properties, such as at least two elastic moduli, strength, hardness, and conductivity on different samples sizes and orientations can be extremely valuable in identifying and ultimately characterizing, porosity, e.g., its spatial, shape, and orientation variations. Thus, it is expected that by iterative comparison of more detailed sample microstructure characterization and more comprehensive property measurements, the latter become a larger, and often cheaper and more effective tool to characterize many of the needed aspects of porosity that are difficult to directly obtain in a detailed quantitative fashion.

There are obviously a number of tradeoffs that must be made in determining the extent of characterization, e.g., time, cost, amount of material, and tools available; however, it should again be emphasized that even qualitative observations and a few extra property measurements can give much more information than the commonly very limited extent of porosity (and other) characterization in so many studies. Thus, microstructural photos of starting powders, green bodies, and bodies of varying densities can be valuable for giving at the minimum a qualitative, and often at least a semi-quantitative, indication of key porosity factors and trends. Again, photographs of fractures are often of particular practicality and use, especially to identify extremes of porosity and their effects (see Chapter 5). Table 2.3 outlines minimum and desired levels of porosity characterization to accompany suitable (preferably broader, more comprehensive) property measurements.

Microcrack characterization for needed size, spatial, and orientation, which is theoretically feasible (16), is also often challenging in practice, in part due to their frequent small openings (often nanometers) and their often being either

Table 2.3 Porosity Characterization

Factors	Minimum characterization	Desired characterization
Porosity source	Raw material summary, especially for powder derived bodies: average particle size and shape. Consolidation method and basic parameters, and green density	Raw material (particle and agglomerate density, size and shape distributions), consolidation and densification specifics
Amount, P	Average value, specimen sizes used relative to those for testing and from processing	Statistical variation of P. Spatial distribution, i.e., gradients and heterogeneity, RFB&TSD[a]
Size	Average and approximate range	Size distribution(s) and RFB&TSD[a,b]
Geometry	Description of: shape and range, open and closed and inter- versus intragranular pores, and orientation	Shape, spacing, and orientation distributions; pore and especially grain contiguity, MSA, related stereological data RFB&TSD[a,b]
Related micro-structure	Average size of grains (and dispersed phase(s) for composites), and approx. distribution of pores between and within grains (and dispersed phase)	Size, shape (e.g., aspect ratio), and orientation distribution(s) of grains and dispersed phase(s), RF&TSD[a,b]

[a] RFB&TSD= relative to fabricated body and test specimen dimensions.
[b] Data on extremes—mainly from fractography.

obscured or altered by their intersecting free surfaces. Their presence can be directly confirmed in some cases by transmission (20), and in some cases by scanning electron microscopy (the latter mainly on fracture surfaces). however, such observations are at best quantitatively uncertain in whether all cracks are observed even in the small areas examined, and the extent to which the presence of the surface examined has affected the presence or scale of the cracks observed. Microcracks (and small pores) can be detected and their volume fraction and spatial distribution at least estimated by X-ray and especially neutron diffraction (21). More generally both the presence and some characterization of the microcracks is obtained from their effects on the properties, both those of direct interest and others. It is particularly important to use correlations of combinations of these various measurements over a range of pertinent test and microstructural parameters. Covering a range of temperatures and grain size can be particularly valuable because of their impacts on the occurrence, extent, and scale of microcracks.

Three important techniques of detecting (and partially characterizing) microcracks are via thermal expansion (22), acoustic emission (23–25), and internal friction (26–28), the latter is readily done directly but measuring elastic properties by wave techniques. Other techniques for detection and some characterization include local probing methods such as thermal wave techniques and instrumented indentation (29) and (for electronic or ionic) conductors via AC impedance spectroscopy (30). However, characterization of microcracks, especially those typically on a fairly fine microstructural scale, via iterative evaluation of them and their effects on a variety of properties is particularly important. For larger scale cracks, e.g., in some cementatious bodies, stereological techniques can be applied (31).

2.4 MODELING OF POROSITY AND MICROCRACK EFFECTS ON PHYSICAL PROPERTIES

2.4.1 Introduction and Background

Modeling of microcrack effects on properties is a more limited subject for several reasons. A basic one is the much less extensive occurrence of microcracks in terms of the materials in which they occur and extent of their occurrence in a given material. Another is the much more proscribed character range of microcracks versus pores. The difficulty of characterizing microcracks and the complexity of the microcrack population changes with temperature and around some propagating cracks have also been factors. Thus, modeling, which is more recent and mechanics-based, is limited to a few properties and is included in the treatment of those properties.

Modeling of porosity effects on physical properties is an extensive and complex subject because of the greater diversity of pore character. While there are a variety of modeling approaches for which there is no perfect categorization, most models can be seen as being predominately in one of three categories: 1) totally empirical, 2) mechanistic, and 3) geometric, especially those based on MSA. Mechanistic models focus more on rigorous evaluation of the specific mechanism (s), e.g., mechanics, for the porosity effects on the specific property, and are thus generally specific to the property and hence are covered in detail in conjunction with discussion of the particular property addressed. Such models, in order to be more rigorous in their physics, can typically only treat simple idealized pore shapes (mainly spheres) in a very narrow fashion, e.g., only noninteracting pores, hence limited P levels (neglecting pore stacking), and can treat only a very limited number of cases in comparison to those encountered in practice. Also, for mechanical properties, mechanistic models are rigorous for only bulk and sometimes shear moduli. Geometry-based models also typically focus individually on a simple, narrow pore geometry, but directly address pore stacking, cover the full range of feasible porosity, and are more general and simple in the physics involved; however, since these models are much more tractable and versatile in the pore geometries they can address, collectively they are much more comprehensive in their coverage of pore character.

The goal of modeling is either, preferably both, understanding and prediction of reasonable property values given only the amount and character of the porosity, but the relative balance of these can vary substantially. Thus, empirical models may be useful for some prediction within the bounds of reasonable extrapolation, but very limited in providing understanding of the mechanisms of pores affecting properties. On the other hand, mechanistic models may give more insight to mechanisms, but they may be quite restricted in their predictive capabilities because of very narrow idealization of the pore character, and often amount of porosity, required to be rigorous in the physics of the analysis. Clearly properties dependent on both the amount and character of the porosity present substantially more challenges and complexity than those dependent only on the amount of porosity, i.e., Category III A versus Category II (Table 2.1), though the latter are not trivial. Since Category III A properties are more complex and represent the majority of properties in this book, they receive greater attention in this and subsequent chapters.

The first of three sets of challenges in evaluating porosity models is their number and diversity. Another is the uncertainties and variations in the suitability of the models since many, especially mechanistic, models are modified or applied in an ad hoc fashion to porosities having little or no relation to that of their derivation. The third challenge is that the quality of the data against which the models must be evaluated is often not as good as it might appear, e.g., based on its scatter, due to systematic but neglected errors (mainly from heterogeneity and anisotropy). Thus it is essential that evaluations be based on comprehensive comparisons of data with individual models, whose fits are in turn compared. Unfortunately evaluations have typically not been comprehensive, hence fostering comprehensive evaluations is an important goal for this book.

The remainder of this section first addresses aspects of model and data plotting and bounding, followed by key underlying phenomena and issues for modeling mechanical properties, and models of broad applicability. Subsequent specific topics addressed are first effects of stress concentrations of pores, specifically questions and limitations showing that there are serious limitations on their effects in contrast to frequent intimation of their playing a major role in mechanical effects of pores. Next, the derivation of the major family of geometry (MSA) models, which are the broadest and thus far the most successful and only truly predictive models over a range of porosities, including at least some mixes and changes of pores, is presented. Then, general trends and character of various other models and their comparison are addressed. Models focused on a specific property are dealt with in the evaluation of the porosity dependence of that property in subsequent chapters; however, before turning to these subjects, empirical and bounding approaches to modeling are addressed.

Initially the evaluation of the porosity dependence of elastic properties was primarily empirical, with substantial work on materials other than ceramics, especially metals. Plotting properties versus density was a common procedure to gain insight into the effects of porosity, and for extrapolation to other porosity regimes. Using relative density, i.e., the percent or fraction of theoretical density, or the percent or volume fraction porosity (P) are preferable since all plots are normalized to a 0 to 1 scale. Similarly, plotting the percent or fraction of the property relative to its value at $P = 0$ (if known with sufficient accuracy) is also useful.

Various plots have been used empirically. Linear plots have been used since data over the low porosity range often varies, at least approximately, linearly with P; however, most data over greater P ranges, especially at higher P levels, shows substantial nonlinear property decreases with increasing porosity. Semilog plots of the log of the absolute or relative property values versus P on a linear scale give a straight line to higher P levels than linear plots, often the complete P range commonly investigated, e.g., to $P < 0.5$. The use of such plots and the resultant exponential relation, e^{-bP}, where b is the semilog slope of the property dependence over the initial range of porosity for linearity on a semilog plot was first shown by Duckworth (32) in discussion of Ryskewitch's (33) paper on compressive strength of porous alumina and zirconia. Such plots were later extensively used by Spriggs and colleagues for both elastic properties and flexure strengths of oxides, so some have referred to the exponential dependence as the "Spriggs' Equation".

Log-log plots of properties versus porosity have also commonly been used, again initially based on a purely empirical basis since they also allow ready presentation of a broad range of data. Such plots are also commonly linear (for the common range of P), leading to relations for the porosity dependence of the form $(1 - AP)^n$, which is also the common form for various mechanics and other models (see Table 3.1). Usually, A was taken as 1, but more recently has often been taken as $1/P_C$. There are three issues with the use of log-log plots: 1) accurate extrapolation to P = 0, or at P or P/P_C = 1 (depending on whether the abscissa is P or P/P_C, or $1 - P$ or $1 - P/P_C$) since such plots do not go to zero, 2) they seriously obscure the transition to, and the actual, P_C values (in part because of issue 1), and 3) they are of doubtful use in interpolating between various models (in part because log-log plots of properties over substantial P ranges are not linear, see Fig. 2.14).

Both semilog and log-log plots are useful in relating elastic properties and sound velocity results versus porosity. The latter, e.g., from pulse echo tests, commonly also are linear on linear, semilog, and log-log plots, e.g., from P = 0 to P = 0.15–0.3 or more. Since E and G vary as the square of the pertinent sound velocities (Eqs. 1.1–1.3) the slope of their porosity dependence over the range where velocities are linear on such plots is theoretically twice that for the corresponding velocity plus 1 (due to the porosity dependence of density, plus a correction for E due to effects of Poission's ratio; see Section 3.2.2). Also both the resultant exponential and power relations for semi- and log-log plots reduce to the linear relation at low P, i.e., $1 - bP$ and $1 - AP$ respectively (the latter with n = 1).

Evaluation of the property bounds between which the models should fall can be useful in their development and evaluation. For many properties an upper bound is for a composite with planer slabs of the two phases parallel with the reference, e.g., stress, direction. This results in a rule of mixtures relation, i.e., Eq. 2.1. The lower property bound is (34) (e.g., for planar slabs perpendicular to the reference, e.g., stress, direction):

$$X^{-1} = (X_S)^{-1} (1 - P) + (X_P)^{-1} (P) \qquad (2.3)$$

where $X_S = X_0$ is the matrix property at P = 0, and X_P for the dispersed phase, pores in this case; however, X_P is zero for pores for most properties, including mechanical properties. This gives X = 0 as the lower bound and

$$X = X_S (1 - P) \qquad (2.4)$$

for the upper bound. Hashin and Shtrickman (35) derived bounds that are much narrower for composites than the bounds noted above, but reduce to the above bounds for porosity as the second phase (with properties = 0). The utility of these bounds arises from several properties coinciding with the upper bound for bodies with tubular pores aligned parallel to the stress or flux direction. Other pore structures result in greater property decreases with increasing porosity, but well above zero property values for $P < P_C$, so the lower bound of zero properties is of quite limited use. This illustrates one of the limitations of using composite models for effects of porosity by setting the properties of the second phase = 0. Another problem of starting from composite models is that they commonly obscure P_C values, since in composites the matrix phase can become discontinuous, but this cannot occur in a porous body since it then is no longer a coherent solid. Another aspect of bounds is to check the models at the bounds of P, i.e., at P = 0 or 1. While this is of some utility, its value is limited to models that should be applicable at these values. This is not always true, especially at P = 1 since many models are not applicable there by derivation (the exponential approximation to MSA models discussed below is a major example), or more generally because properties often go to zero before P < 1 (Section 2.2).

2.4.2 Stress Concentration Versus Load-Bearing Area

Having established that MSA is the second parameter needed, besides P, for characterizing pore structures and their effects on Category III A properties, it is important to examine in an overall fashion how this parameter fits in with models for pore effects on properties of interest. Since the effect of pores on mechanical properties is a dominant consideration of this book, this raises questions of stress concentrations in porous materials. While mechanics models may vary depending on the properties (i.e., elastic, crack propagation, strength), considering the role of stress concentrations, which directly enter some of these models, is informative. Clearly both stress concentration and load-bearing areas play a role in mechanical properties, and load-bearing area also relates to electrical or thermal conduction through the solid phase. The role of stress concentrations for mechanical properties, while widely cited, is both variable for different properties and generally not examined in detail. Since this is not a mechanics text, the focus will be on summarizing an overall, more qualitative, view, from recent evaluations (36,37). Important questions about the impact of stress concentrations on various specific properties are considered in later chapters on those properties. These evaluations show that stress concentration effects are much more restricted than commonly assumed, and in fact lend support to analysis based on MSA (discussed further later).

Consider now the broad aspects of stress concentrations, first of individual pores by themselves, i.e., in isolation, then some aspects of interaction of cylindrical pores, for which stress concentrations are frequently defined. The first of two questions for such isolated pores is what stress concentration: the maximum

Table 2.4 Maximum Stresses[a] on Isolated Spherical or Cylindrical Holes (Pores)[b]

Applied stress	Spherical hole[c]	Cylindrical hole[b,c]
Uniaxial tension	2 (e),	3 (e), 1 (p)
Uniaxial bending		2
Uniaxial compression	1 (e), −1 (p)	1 (e), −1 (p)
Biaxial tension		2 (p), 2 (e)
Biaxial tension/compression		4 (p), 4 (e)
Triaxial tension	1.5	

Source: ref. (36); published with permission of J. Mat. Sci.

[a] Stresses are shown as multiples of the applied stress; a factor of 1 means the peak stress is the applied stress and factors >1 mean a stress concentration. A negative sign means a reversal of the stress, i.e., from compressive to tensile or vice versa. Values typically depend somewhat on Poission's ratio.

[b] For cylindrical pores oriented perpendicular to the stress axis (or axes). There is no stress concentration for such pores parallel to the stress axis.

[c] Locations of the maximum stresses are p for polar and e for equatorial for spherical pores where these are defined by the coincidence of the stress and sphere axis. The same designations are used for circular cross-section of a cylindrical pore except each of the two poles are now each a line and not a point and the "equator" is two lines rather than a circle.

or the average. Stress concentrations are complex, even for simple isolated pores, consisting of multiaxial components of both tensile and compressive stresses, so spatial averages around a pore are much lower than the maximum. Second and more fundamental is that the maximum stress concentration, which is the only one typically cited, varies with the stress state for a given pore shape (Table 2.4; 36). Thus, note that there is no stress concentration for spherical or cylindrical pores under uniaxial compression (only a local reversal of the compressive to a tensile stress) in contrast to stress concentrations of 2 and 3 respectively under uniaxial tensile stress. Correlation of properties with maximum stress concentration would thus imply less effect of pores on compressive versus tensile properties, which is not true. In fact if there is a difference, it is in the opposite direction, i.e., greater porosity dependence for compressive than for tensile stresses (see Chapter 6). Also, note the difference in maximum stress concentration between an internal (spherical) and a surface (hemispherical) pore (i.e., a surface pit), e.g., 2 versus 2.5 in flexure, but no indications of broad basic differences of effects of surface versus bulk pores. Similarly, aligned cylindrical pores stressed normal to their axes have higher stress concentration than spherical pores, but both pore geometries give similar effects in models (see Figs. 3.1, 3.2A) and data. Further, there is no stress concentration (or stress reversal) for aligned tubular pores stressed parallel to their length. Bodies with such pores clearly show loss of mechanical (and conductive) properties despite this absence, further questioning a determining role of stress concentrations.

Finally, note that: 1) these stress concentrations are much less than the maxima for the cusp-shaped pores between grains (Fig. 1.3), and 2) the general decrease in the level of the maximum stress concentration as the complexity of the stress state increases, which is consistent with interactive effects of stress concentrations discussed below. Thus, maximum stress concentrations of individual pores do not play a major role in porosity effects.

Consideration of isolated pores is reasonably valid only over a limited range of porosity because of increasing interactions of the stress fields around the pores. Such interactions generally become significant by the time the level of porosity has risen so that the surface to surface separation of adjacent pores is equal to their diameter (D, the center-to-center spacings of twice the pore diameter). For spherical pores in simple cubic stacking this is

$$\Lambda = L - 2R = L - D,$$

(see Fig. 2.1), so this point is reached at

$$P = \frac{\pi}{48} \sim 0.065$$

For such stacking of aligned cylindrical pores it is at $P \sim 0.2$; however, comparable interaction for cylindrical pores to that for spherical pores occurs at lower concentration since the stress concentration for cylindrical pores is higher, i.e., 3 (and still higher for other tubular pore geometries) for uniaxial stress normal to the axis of cylindrical pores versus 2 for spherical pores. Also, all tubular pores that are not aligned interact at lower porosities, i.e., have lower percolation limits than spherical pores.

Use of Fullman's equation for the mean free spacing (λ) between randomly dispersed features (uniform pores in this case):

$$\lambda = \left(\frac{2}{3}\right) D \left(\frac{1-P}{P}\right) \tag{2.5}$$

where D = the pore diameter; these might be considered for evaluating the P levels where interactions between pores become significant. This gives much higher P levels for such interaction since $\lambda = D$ for spherical pores occurs at $P = 0.4$, but there are problems with use of this equation for this purpose. First, since this is the mean spacing, at any value of λ, half of the pores are closer and half are further apart than the mean. The former clearly have much more interaction, which, since much porosity dependence is nonlinear, means that the closer pores have more effect on properties. Thus, significant interaction is expected at $\lambda = D$, and hence at P substantially < 0.4 for both spherical and general pore

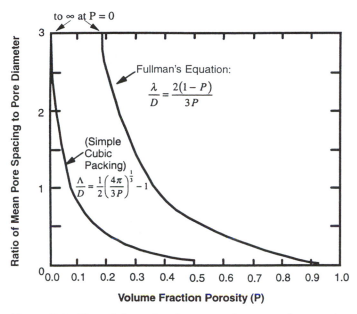

Figure 2.6 Plot of the ratio of mean surface to surface spacing of spherical pores versus volume fraction porosity (P) for simple cubic packing (Λ/D) and Fullman's equation (Eq. 2.5, λ/D). Note that in the former case the Λ/D values apply to all pores, but in the latter case for random spacing of the pores the λ/D values are mean values, so some pores are much closer and some much further apart.

systems. Further, the use of this equation is uncertain since it predicts very different behavior from that for simple cubic packing of spherical pores and real porosity, e.g., recall that the former gives P ~ 0.065 and Eq. 2.5 gives 0.4 for Λ and λ = 1. More generally note the marked difference for the mean free separation to pore diameter ratios from these two pore structures, e.g., λ/D is still 6 versus Λ/D ~ 0.7 at P= 0.1, and pores become open, i.e., at λ/D and Λ/D = 0 at respectively 1.0 and ~ 0.52 (Fig. 2.6). The Λ/D values are much more consistent with spherical pores than those for λ/D.

There are as or more serious issues for using Eq. 2.5 for pores between packed particles, e.g., uncertainties in defining a pore diameter for such cases that is pertinent to eqn. 2.5. More fundamentally, in such cases the size, shape, and spatial distribution of such pores between particles are all determined by the particle size, shape and packing, which will typically be incompatible with a random spatial distribution of the pores on which Eq. 2.5 is based. Finally, use of Eq. 2.5 for pores between particles and an estimate of P ~ 0.4 before significant pore interaction is inconsistent with the solid percolation limits since these

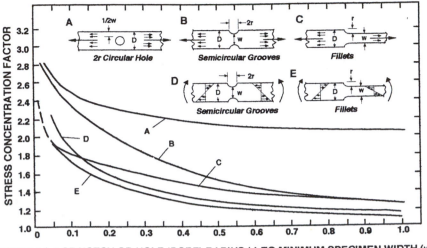

Figure 2.7 Stress concentration versus the ratio of the radius (r) of a single hole, groove, or surface step (i.e., intersections of pores and the surface) and the resultant minimum bar thickness (t) for both tensile and flexure stresses. Note that: 1) there is generally less stress concentration in flexure versus tension, 2) w is essentially the MSA (Figs. 2.1, 2.9), and 3) there is a decrease in stress concentration as the MSA decreases. Published with permission of J. Mat. Sci. (36).

typically occur at P ~ 0.24–0.5 for such pores (Fig. 2.10, Table 2.2). Thus, prediction of no significant pore interaction until P ~ 0.4 neglects the fact that some pore structures no longer exist at P ~ 0.5 this value, and those that do have most or all pores intersecting each other.

Analytical solutions for the stress interactions between adjacent pores are clearly difficult, existing primarily for simple arrays of cylindrical pores. Solutions also exist for surface notches and grooves, many of which are models for tubular pores exposed on the surface (Fig. 2.7) and for simple arrays of cylindrical holes (pores, Fig. 2.8). These typically show stress concentrations decreasing with decreasing separation between pores, which is consistent with the decrease in stress concentrations as the complexity, e.g., the multiaxial character, of the applied stress increases (Table 2.3). This is logical and reinforces the above trends since the presence of pores locally converts uniaxial applied stresses to multiaxial stresses. More generally, increasing the number and closeness of surface notches reduces their effects, since little stress is carried by the webs of material between them and nominally normal to the stress (Fig. 2.9). This means that the load, and hence stress, is increasingly carried by the material webs,

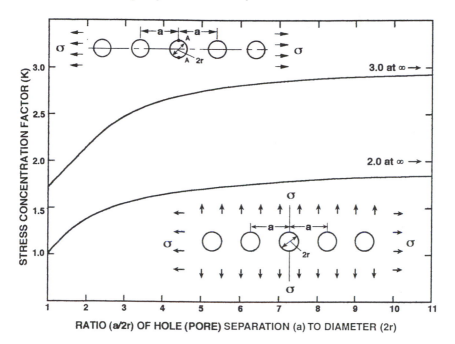

Figure 2.8   Stress concentrations in a plate with a linear array of holes (cylindrical pores) as a function of the ratio of the hole spacing (a) to hole diameter (2r) for a uniaxial and a biaxial stress. Note: 1) the reduced stress concentrations with decreased spacings of cylindrical holes and with bi- versus uniaxial stress and 2) the stress situation is clearly more complex with other, more discrete, e.g., spherical, pore shapes; however, the resulting interaction stresses between such nearby pores are much more limited in spatial extent relative to that for aligned cylindrical pores or notches normal to the stress axis. Published with permission of J. Mat. Sci. (36).

which are the minimum solid area as defined in Figs. 2.1, and 2.10. Thus, at low P with large spacings between pores MSA is not as accurate, but the difference between it and the average area is limited; at higher P, however, where the differences between average and minimum solid areas are substantial, MSA is more appropriate. Also note from Fig. 2.9 that the schematic of the stress contours is also the schematic for flux contours for thermal and electrical conductivities. This results from similar physical reasons, i.e., as more pores occur, there is less conduction directly between them, i.e., most conduction occurs in the MSA. Thus such conductivities will follow the same, or very similar, porosity dependence as stress-based, e.g., elastic, properties.

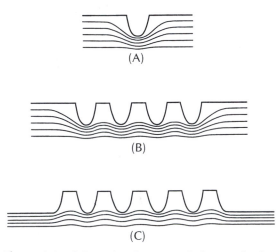

(A)

(B)

(C)

Figure 2.9 Schematic of stress and also conductive flux contours as the number of cylindrical notches (hence also pores) increases A–C, showing how the stress (flux) is increasingly carried by the remaining cross-sectional (i.e., minimum solid) area (Fig. 2.10).

2.4.3 Basic Models Based on Minimum Solid Area

Models such as those of Fig. 2.1 provide both a conceptual basis via idealization of pore structures as well as a quantitative basis for modeling in two aspects. The first and most extensive of these is for actual modeling based on load bearing area, especially MSA. Such modeling, which competes with or compliments

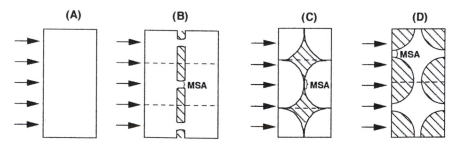

Figure 2.10 Schematic of the minimum solid area (MSA) model concept (cross-hatched areas = pores). For a homogeneous solid (A) the minimum and average solid cross sections are the same. However, it is clear that flux or stress transmission in (B) must be dominated by the minimum solid area (MSA) normal to the direction of the stress or flux (dashed lines delineate cell structure). Corresponding definitions of the MSA for pores between spherical particles (C) and spherical pores (D). Published with permission of J. Mat. Sci. (2).

other modeling approaches, constitutes the first and major portion of this section; however, the identification of the P levels for the open-closed porosity transition and P_C values, which are a direct factor and outcome of MSA modeling, are of importance themselves, and hence a second quantitative aspect of such idealized models. The open-closed porosity transition is not particularly important for mechanical and conductive properties, but is critical for applications based on surface area, flow, and permeability. P_C values are of broad importance for all applications. Thus, since MSA approaches are the primary method of obtaining these values, they are discussed further in the next section.

Turning to property modeling, load bearing area, i.e., the cross-sectional area normal to the stress or flux, is a logical basis for modeling much mechanical and conductive behavior of materials, as noted earlier. It is particularly effective for modeling effects of pores, but not effective for microcracks (see Section 3.2.3), so only the former is addressed. Because of their breadth of application to pores, the basic development of MSA models is presented here, versus other mechanistic models, which are typically more specific to the particular property of interest. The underlying concepts, calculation procedures, and resultant established MSA models are given first; then specifics of the calculation procedure for one model is given followed by discussion of ramifications and extensions of this family of models.

Choices for the load bearing area used clearly range from the average to the minimum cross-sectional area. The two areas are the same at P = 0 for any porosity and for any aligned tubular pores parallel to their axis for any P. In all other cases differences between the two areas increases to very substantial levels, e.g., two orders of magnitude, at high P, so a selection must be made. Of these two areas, the latter, i.e., MSA, is generally most logical for three reasons. First, the average solid area is simply 1 − P, which depends only on P and not pore character, and hence is pertinent primarily to Category II properties. Except for also being the upper bound for all porosity dependence, which coincides with the P dependence of aligned tubular pores with the stress or flux parallel with their axes, all Category III properties of primary interest here show greater decreases with increasing P. Second, the minimum area is intuitively much more important, e.g., as indicated in Figs. 2.9 and 2.10, as strongly suggested by stress concentration considerations in the previous section. Third, MSA was shown to be a suitable and practical parameter to characterize porosity for category III properties. It is conceivable that some intermediate (possibly changing) fraction of the bond or contact area might be used, e.g., an electrical conductivity model (see Chapter 7) assumed the pertinent area to be the end area of a cylinder having the same volume as the spherical particles. This gives a contact area that is two-thirds of the maximum (i.e., at the spherical particle equator), and hence greater than the MSA for all but low P; however, there is no clear physical reason for such an intermediate value, and MSA yields good results as shown in (see Chapters 3–10).

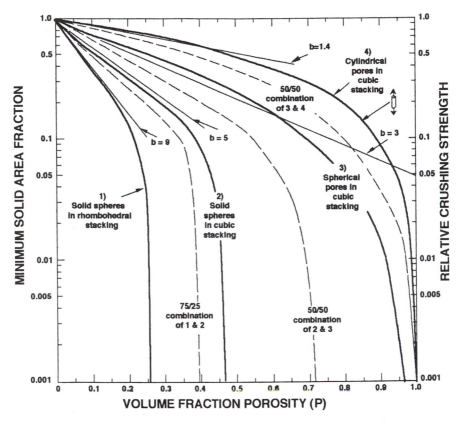

Figure 2.11 Examples of some important MSA models plotted on a semilog plot of the calculated ratios of minimum solid (bond) area per cell to the cell cross-section for the least and most dense sphere stackings per Knudsen (43), and for cubic stacking of spherical and cylindrical pores (39–41). Note their similar plotted forms (despite the differing functional forms, Table 2.4) and simple rule of mixtures combinations, dashed lines, for (a) 75% of rhombohedrally and 25% of cubically stacked spherical particles, (b) 50:50 mixture of cubically stacked particles and bubbles (i.e., of hollow spheres), and (c) 50:50 mixture of cubic stacked spherical and cylindrical voids. For reference, slopes (i.e., b values of e^{-bP}) are shown as mostly integer values from 1.2–9.

Both its intuitive appeal, the relatively simple mathematics, and reasonable results made the MSA concept a common approach. The methodology, illustrated below, is to calculate the minimum solid area per unit cell (Figs. 2.1, 2.10 and 2.11), and assume that the relative property of interest equals the ratio of the average MSA between two pores or particles to their cell area normal to the reference stress or flux direction. There are two model versions, both following the same basic procedures. The first version was for stacking of pores where there is either no grain structure or it is negligible relative to the pores. Such pore structures, which can typically extend to P = 1, lead to foam (Fig 2.1) or honeycomb structures at higher P, and hence will be referred to as foam/honeycomb MSA models.

Foam-type models were initially derived for spherical pores (38,39), e.g., for bubbles in melt or slip cast materials (or foams), then for cubic (34) pores (initially with their sides oriented parallel and normal to the reference axis and later for other orientations; 39). Models were also later derived for aligned cylindrical pores oriented either normal or parallel with the stress (or flux) axis (39–41), then for other shapes of tubular pores, i.e., for honeycomb structures. With rare exception these models have been derived only for simple cubic stacking of the pores due to computational complexity with other stackings (handleable by computers). These models, the typically simple equations they follow until pore intersection occurs (which are a major complication in these model calculations, especially for stackings other than simple cubic), along with pertinent parameters of the closed-open porosity transition and P_C values are given in Table 2.5. Most representative models are plotted in Figs. 2.11, 3.1, and 3.3A, often to or beyond the point of pore or particle intersection.

The second basic version of MSA models is for bodies consisting of packed uniform spherical particles and the resultant pores between the particles. The initial and major development of such models was Knudsen's (42,43), which entailed three aspects. The first was derivation of such models for three different particle stackings (simple cubic, rhombic, and orthorhombic; Fig. 2.1 A, B, C), showing that stacking of the particles, and hence of the pores, is significant (Figs. 2.11. 3.1). Second, he demonstrated that plotting of the resultant models on semilog plots resulted in the characteristic form, i.e., about linear at low to intermediate P values, then a rollover to rapid decreases of properties toward their going to zero at P_C values of $\leq \sim 0.5$, clearly dependant on stacking. The initial, approximately linear semilog portion thus provided an analytical justification for the (up to then purely empirical) use of the exponential dependence on P, relating the substantial range of b values (~5–9) to particle stacking, hence pore geometry. Third, he demonstrated that data from studies he made of a few partially sintered powder bodies was generally consistent with the models.

Consider now the specific calculation methods for MSA models, using the easier calculations for simple cubic stacking of uniform spherical particles. This specific model is selected for illustration not only because it is easier, and hence a good illustration, but also because it is important since this model is also close to random packing (3,44,45; Table 2.2; Fig. 2.4). It is also potentially important for further development of MSA models. The basic approach has been to first recognize the unit cell that is characteristic for the uniform stackings typically addressed. In the case of simple cubic stacking selected for demonstration, it is a cubic cell (Figs. 2.1 and 2.12), which just encloses each particle. At the

Table 2.5   MSA Porosity Models and Parameters

A) Cubic, spherical, and ellipsoidal pores[a]

Pore shape, orientation	$X/X_0 =$[b]	P: open-closed[c]	P_C[d]
Cube, <100>	$1 - P^{2/3}$	1	1
Cube, <110>	$1 - (2)^{1/2}P^{2/3}$	$2^{-3/2} \sim 0.354$	1
Cube, <111>	$1 - (3)^{1/2}P^{2/3}$	$3^{-3/2} \sim 0.193$	~ 1
Sphere	$1 - \pi/4\,(6/\pi P)^{2/3}$	$\pi/6 = 0.524$	~ 0.964
Ellipsoid[e]	$1 - (\pi/4)\varsigma_2\,(6P)^{2/3}\,(\pi\varsigma_1\varsigma_2)^{-2/3}$	$\pi/6 = 0.524$	~ 1

B) Aligned, mainly tubular pores[a]

Pore geometry	Stress or flux parallel to pore axes $X/X_0^b =$	P:o-c[c]	P_C[d]	Stress or flux perpendicular to pore axes $X/X_0^b =$	P:o-c[c]	P_C[d]
Oblate S[e]	$1 - 1/4(\pi\varsigma_1)^{1/3}(6P)^{2/3}$	0.524	>0.52	$1 - 1/4(\pi)^{1/3}(\varsigma_2)^{-2/3}(6P)^{2/3}$	0.524	>0.52
Cylinder	$1 - P$	0.785	≥0.78	$1 - [(4/\pi)P]^{1/2}$ (f)	0.785	≥0.78
Square	$1 - P$	1	1	$0.5(1-P),\ 1.5(1-P)^3$ (G)	1	1
Hexagon	$3/2(1-P)^3$	1	1	$3/2(1-P)^3$	1	1
Triangle	$1.8(1-P)$	1	1	$0.6(1-P)$	1	1

[a] All in simple cubic stacking, except for the close packing of the aligned hexagonal and triangular tubes. See also Table 3.3 for more details

[b] X/X_0 is the ratio of the property, X, at some P to that at P=0. Equations become more complex once pores become open; see Figs. 2.10 and 3.1 for plots of the relations.

[c] P: open-closed is the porosity level where the transition between open and closed porosity occurs (designated P:o-c in B portion of this table).

[d] The value of P where the pore structure disrupts the solid structure so properties go to zero. Note that the P_C values of cylindrical tubes differs depending on their orientation since the opening of the pores disrupts the continuity of the solid in a direction normal to the pore axis, which reduces properties in that direction to zero, but not in the perpendicular direction (i.e., parallel to their axes and the stress or flux direction) so properties do not go to zero.

[e] Oblate S = (aligned) oblate spheroids (a = long axes, ς_1 = a/c and ς_2 = b/c; the equation for ellipsoids reduces to that of a sphere when both ratios are zero).

[f] Note the original publication (41) and a subsequent review (42) of the MSA model for properties normal to the axes of aligned cylindrical pores was seriously in error giving a P_C value about one-half, and a corresponding b value about twice, the correct values (see Figs. 3.1 and 3.2A).

[g] Equation depends on orientation; see Table 3.3.

percolation limit of the solid particles, the imaginary cubic cell is tangent to the enclosed spherical particle at the two poles plus at the four intersections of two normal equatorial diameters. Subsequent bonding of the particles to one another by sintering is simulated by progressively shrinking the enclosing cubic cell dimensions. Since sintering conserves mass and hence solid volume (assuming there is no solid transformations or volume correction for them), the six spherical segments outside the shrinking cubic cell are removed and their volume is combined back into the truncated sphere. Thus, at each stage of sintering evalu-

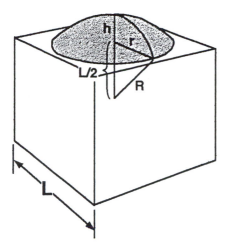

Figure 2.12 Schematic of the unit cell and particle structure and associated notation and relations for the MSA model for uniform spherical particles in simple cubic packing (i.e., $C_n = 6$) and their simulated sintering (for clarity, only one spherical segment is shown, on the top). See also Fig. 2.1A.

ated, there are two truncated particle radii. The first is $R_i = 1$, the initial particle radii, which then increases to a new value, R, with addition of the volume from the six spherical segments removed from outside the cell walls back into the truncated particle such that it conserves the original particle volume within the cubic cell wall positions reflecting that stage of sintering.

Putting the above in equation form using the notation of Fig. 2.12, first consider the volume fraction porosity (P). Since mass, hence normally volume, is conserved, the solid volume (V_s) in each cell is always that of the initial spherical particle, i.e., $(4\pi/3)(R_i)^3$, where R_i = the initial, i.e., untruncated, particle radius. Thus

$$P = 1 - \frac{V_s}{L^3}$$

where L – the cell edge length. Note that the initial length, $L_i = 2 R_i$ and that since all cell (and hence model) calculations are only dependent on the relative values of the parameters within a cell, they can all be normalized, e.g., by setting $R_i = 1$ (hence $L_i = 2$). Thus, :

$$P = 1 - \left(\frac{4\pi}{3}\right)(L)^3 \tag{2.6}$$

and initially

$$P_c = 1 - \pi/6 \sim 1 - 0.52 = 0.48$$

Next, since the volume of a spherical segment is $(\pi/3)h^2(3R - h)$ and the particle volume is conserved, i.e., the initial, complete, spherical particle volume equals the volume of the truncated sphere,

$$R^3 = 1 + \frac{3}{2}\left[\left(R - \frac{L}{2}\right)^2\left(2R + \frac{L}{2}\right)\right] \tag{2.7}$$

Thus, note that values of P can be calculated from Eq. 2.6 for various values of L, or versa, then these can be used to calculate values of R from Eq. 2.7 (though this is complicated by its cubic nature), then r (h, or both) can be calculated from expressions relating them in Fig. 2.12. Results of such calculations (and related considerations) are summarized in Table 2.5.

Additional insight can be gained by further considering geometrical aspects of the model. First, the above calculations neglect an additional complication at intermediate P once the spherical segments begin to overlap. This begins to occur when $r = L/2$ (i.e., at $P < 0.1$, hence an obviously invalid value of r at $P = 0$ in Table 2.6 since it is greater than 2L, and larger increases in values of R,

Table 2.6 Summary of MSA Model and Related Calculations for Simple Cubic Stacked Uniform Spherical Particles[a]

P[b]	L[c]	R[d]	h[e]	r[f]	r/R	MSA[g]	R'[h]	R_L[i]	E/E_0[j]	$1 - P/P_C$
0.48	2	1	0	0	0	0	1	1	0	0
0.4	1.91	1.005	0.05	0.313	0.311	0.08	1.05	1.08	.09	0.17
0.3	1.82	1.017	0.11	0.454	0.446	0.20	1.10	1.16	0.21	0.375
0.2	1.74	1.05	0.18	0.585	0.558	0.36	1.15	1.24	0.37	0.583
0.1	1.67	1.09	0.25	0.701	0.643	0.55	1.20	1.30	0.56	0.792
0	1.61	1.23	0.43	0.910	0.74	1	1.24	1.39	0.82	1

[a] Much of the notation is given in Fig. 2.12.
[b] Volume fraction porosity (Eq. 2.6).
[c] Length of cubic cell edge (Eq. 2.7).
[d] Radius of truncated particle of volume equal to the original spherical particle (Eq. 2.7).
[e] Height of missing spherical segment, $h = R - L/2$ (Fig. 2.12).
[f] Radius of bond area between particles, per $r = (R^2 - L^2/4)^{1/2}$.
[g] MSA $= \pi (r/L)^2$.
[h] From $(1 - P)^{1/3} (1 - P_C)^{-1/3}$ per Fischmeister and Artz (50).
[i] Assuming a linear change between the two geometrical extremes of 1 and 1.39, as discussed in the text.
[j] From Arato et al. (49), see Table 2.7.

r, and h which are effected by such effects); however, porosity effects are small there and can be extrapolated to P = 0. Further, the final cell dimension, L_f, can be calculated since at

$$P = 0 \ (L_f)^3 = 4\pi/3$$

$$L_f = 1.61$$

Similarly, the final truncated particle radius, R_f, must be one-half the final cell body diagonal

$$R_f = \tfrac{1}{2}(3)^{\frac{1}{2}} L_f = 1.39$$

$$R = L/2 + h$$

$$R^2 = r^2 + L^2/4$$

The calculation results (Table 2.6) show that the radius, R, of the truncated solid particle increases slowly as P decreases, e.g., less than a linear interpolation between the two bounds, except at the extreme of P = 0. (As noted earlier, there are complexities due to overlap of the spherical segments near P = 0, and hence in their volume calculation, which have been neglected in these sample calculations.) On the other hand, the radius of the bond area, r, increases rapidly, as does r/R, In reality, surface tension effects result in a sintered neck that will have a minimum dimensions somewhat below that calculated. This is likely to have the greatest relative effect at high P, hence small r, but the model should still be a reasonable approximation even there. Other comparisons of results in Table 2.6 are discussed below in conjunction with discussion of generalizing such MSA models to random stackings.

2.4.4 Character and Utility of the MSA Models

MSA models were developed individually or as groups of a few related models. They were applied individually, not only to pores closely reflecting the pore geometry for which they were derived, but generally rather arbitrarily to other more diverse pore geometries. Subsequently Rice and Freiman (46), Rice (1,2), and Brown et al. (39) pointed out directly or indirectly what seemed obvious but not generally recognized or practiced. This was that the MSA model most closely reflecting the actual pore geometry being dealt with in a given body should be the one selected for application from the array or family of MSA models for different pore geometries. This selection was commonly feasible since this family of models spanned the range of most practical porosities. Thus, as discussed earlier, this family clearly reflects qualitatively, as well as in a reasonable quantitative, fashion, the character and range of porosity expected in practice. This ranges from pores between various particle stackings to nominally spherical pores from bubbles, particle bridging, or burnout of beads, and angular (e.g., square) pores from burnout of angular material (e.g., sawdust) and some inter- and intragranular porosity, as well as tubular pores from such forming operations such as extrusion, laminar pores from die pressing (commonly at higher pressure and die wall friction). This diversity arises from the individual models being derived for such specific pore geometries, i.e., shape, stacking, and orientation, which have varying effects. Thus, while there is little or no differences for some cubic and spherical pores (again questioning pore shape as a key parameter), the combination of their shape, orientation, and stacking, hence MSA, has pronounced effects on properties as shown here (Figs. 2.11 and 3.1) and extensively in subsequent chapters.

Much of the utility of these models derives from both their distinct numerical factors for a given model pore geometry as well as their similarities of functional shape. Thus, MSA models for both sintering particles and for foams/honeycombs have similar forms on a semilog plot where they collectively, systematically cover the full range of the semilog property-volume fraction porosity space from the upper bound of $1 - P$ to nearly the lower bound, 0. This and other characteristics are born out by data summarized in Fig. 2.13 and subsequent chapters. The first, and most distinct, of three key characteristics is their approximately linear character to about $1/3$ to $1/2$ of their effective P range (Figs. 2.11 and 3.1; 1,2,46). This linearity provides an analytical basis for the exponential relation, i.e., where the ratio of the property at some P to that at P = 0 is equal to e^{-bP}. This approximate linearity was noted by Knudsen for his stacked particle models and was later shown to be true for MSA foam/honeycomb models as well (46). Second, the models clearly and directly predict the percolation limit, i.e., P_C value, for a given pore model and the semilog plot clearly displays them. While there can be some variation in these values (as discussed later), these models and the calculated P_C values provide insight needed for these values. Third is the transition or rollover from the approximately linear behavior to the precipitous property decrease to P_C values, which is also impacted some by variations of P_C values.

Consider the above three characterizing factors further. Individual b values for specific MSA pore model geometries range from about 1.5 to 9 and vary from somewhat to substantially distinctly different for a given pore geometry. P_C values also vary substantially, i.e., from about 0.25 to 1 and thus can be quite different for some different, but not all, porosities (Table 2.4). The rollover ranges also vary substantially for many of the model porosities, but are less specific numerically and less distinct to a given porosity. There is qualitatively an inverse relation between the slopes and the rollover, and especially the percolation limits, i.e., a higher b means lower P_C values and vice versa. Thus, individual pore geometry models are distinctly characterized mainly by the

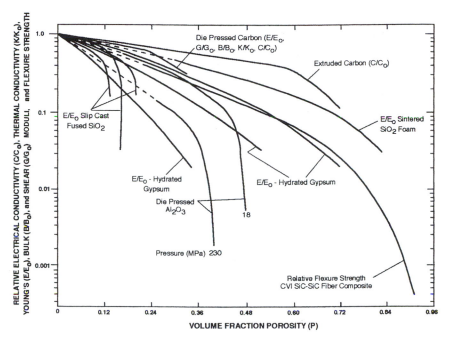

Figure 2.13 Examples of various properties and materials on a semilog plot versus P. While these will be discussed in subsequent chapters, note here the agreement of different properties for a given material (carbon) and increasing b and decreasing P_C values in proceeding from properties parallel to the extrusion axis through various degrees of increasing particle packing via die pressing and slip casting. Published with permission of Trans. Tech. Pub. (3).

lower P slopes, i.e., b values, themselves, secondarily the P_C values (despite some variations of these as discussed below), and overall by the combination of all three factors.

This common characterization of models by the b values may lead some to think of them as a single model, which is also encouraged by the common empirical use of e^{-bP} for porosity dependence; however, it must be recalled that the different b values are reasonable approximations for individually derived specific pore geometry models. Further, the approximately linear portion is true only for one-half or less of the complete pertinent P range. Attempts to fit complete MSA curves to a single simple function with one or two pore parameters have been unsuccessful [although e^{-bP} and $1-e^{-b'(1-P)}$, Table 3.1, have been suggested as a possible useful combination for respectively the lower and higher portions of the pertinent P range (1) where $1 - P$ is replaced by $1 - P/P_C$ (2)]. Also note that while MSA models fill the range from the upper bound to nearly the lower bound, the bulk of encountered porosities lie in the middle of this range. Thus, 6 of 9 b values are in the 2–6 range with one value (~1.2) below and two (~7 and 9) above this range, the latter being for less frequently encountered porosities, respectively aligned tubular pores and pores between particles of high packing densities (Figs. 2.11 and 3.1). Similarly, 4 of 9 P_C values are in the range of 0.45–0.8 for more common porosities and two lower (~ 0.25–0.4) and three higher (~1). The tendency for a mainstream of porosity toward the middle of the ranges of these values is expected to be increased by development of other models for other stackings of pores (e.g., Fig. 3.2A). It is substantially increased by most real porosities being combinations of various ideal porosities, which are characterized, at least approximately, as averages of the individual b and P_C values for each pore type involved as discussed below.

Recognition of the similarity of shape for each MSA model curve noted above lead to the logical suggestion to interpolate between model curves to reflect combinations of different porosities or porosity changes (2). Examples of fixed mixtures of porosity are shown in Fig. 2.11. Changing combinations, which can readily be computed on a point by point basis are considered in Chapters 3–8. The above interpolation between and combination of MSA models readily gives the complete combined behavior, including specific b and P_C values, which thus reflect weighted averages of the values of the individual porosities in the body based on their volume fractions. Thus, these values are no longer unique to that pore combination since other combinations can also give the same or very similar average b and P_C values. Such combination of porosities is supported by results discussed later, as well as by the known theoretical and experimental approximate equivalence of random and simple cubic packing of spheres (1,3,44,45), which is the least dense of the common regular particle packings. This similarity results from random packing being a mixture of denser sphere packing than simple cubic packing as well as less dense packing, i.e., larger pores than the normal interstices between close packed particles, due to particle bridging (Fig. 2.2). While results of such combinations of different porosities are encouraging, further evaluation and development is recommended as discussed further below.

Consider now variations in P_C values, which are commonly the P values beyond which the solid percolation limits are exceeded. Solid percolation limits for various ideal pore structures, range from about 1/4 to 1; however, P_C values can vary not only with even modest changes in pore structure, but with the basic pore structure and the nature of the stress or flux in the body, particularly pore structure orientation-stress conditions. Thus, for example, the solid percolation limits are at P = π/4 is about 0.785, 1, and 0.5, respectively, for tubular pores of circular and square cross-sections, the latter two respectively with tube walls and diagonals parallel with the sides of the imaginary surrounding square cell walls; however, the intersection of two of the three pore geometries at P < 1 simply divides the bodies into a series of parallel filaments which may still sustain some conductive fluxes or tensile loads (i.e., as in a rope, but probably not compression) until their cross sectional area vanishes at $P = 1$. On the other hand, the above solid percolation limits are commonly also the P_C values for stressing or a conductive flux normal to the pore axes, since the separation into filaments eliminates continuity of stress and flux paths in this orientation. On possible exception is compressive loading, especially biaxial loading normal to the tubular pore axes since such loading may maintain some solid contiguity. The same variations, though at different values, hold for bodies of cylindrical-prismatic particles, e.g., aligned elongated grains, whiskers of fiber.

Based on the above and earlier (Table 2.4) observations, consider three important aspects of the variation of P_C values. First, they often vary, sometimes significantly, with different particle-pore geometry and stacking. Second, they can be quite sensitive to stacking imperfections, depending on not only their extent, but also their scale relative to the specimen scale and the type of test (Fig. 3.2A); however, there are some trends to these variations, namely that high P_C are more likely to decrease than increase (e.g., they cannot exceed 1), while lower P_C values are more likely to increase. Thus, packing of smaller particles between densely packed larger particles can increase P_C values, but only by limited amounts, while bridging of the larger particles commonly makes substantially greater decreases in P_C values. (Also for the intermediate, unstable, orthorhombic stacking, i.e., Fig. 2.1B, P_C as well as the open–closed porosity transition (P:o-c) values will vary depending on the choice of the hexagonal cell dimensions since the cell is inherently not completely equiaxed.) Third, P_C variations correspondingly shift MSA curves, at least to, or into, or beyond the rollover regime since the latter is the transition to the solid percolation limit, hence the P_C value. The general inverse trend between b and P_C values suggests the latter.

P_C values are also related to the P values for the open-closed porosity transition (P:o-c). This transition is not directly particularly important for mechanical or conductive properties (as noted earlier and below, contrary to some earlier thinking); however, P:o-c values are useful because of their relation to P_C values, their possible relation to changes from inter- to intragranular pores (Figs. 1.2–1.4), and very important themselves for surface area and flow and permeability applications. Besides P_C values being greater than or equal to P:o-c values, there are other more specific relations which are important since P:o-c values vary more widely with pore stacking and orientation than P_C values. For example, 1) for spherical pores in simple cubic and orthorhombic stackings, P:o-c values are respectively 0.476 and 0.604 to 0.781, and 2) for simple cubically stacked cubic pores of <100>, <110>, and <111> orientations give respective P:o-c values of 1, 0.354, and 0.192, while P_C values for all are, about 1. For cylindrical and spherical particles or pores, P:o-c and P_C values occur when the maximum pore or particle circular cross-section is respectively just inscribed in, or just circumscribed around, the unit cell cross-section. For cylindrical pores and particles the P:o-c and P_C values for particles are simply 1- the corresponding value for the pores and vice versa. Thus, for example, P:o-c values for dense packed cylindrical particles are 0.093 and 0.907 for such pores. P:o-c values for spherical pores are $1 - P_C$ values of spherical particles and P_C values for such pores are $1 - $ P:o-c values for such particles.

Consider four additional aspects of MSA models and data over a substantial range of applicable porosity levels (P, Fig. 2.13). First, is that they readily address both open and closed porosity and the transition from one to the other since this must be accounted for in the model calculations; however, while the onset of open porosity is a factor in, it is not the sole cause of, the property rollover and rapid decrease to zero at P_C. There is a partial correlation of the degree of pore opening and the rollover for spherical, cylindrical, and <110> and <111> cubic pores, but there is no correlation for <100> cubic pores or pores between spherical particles. Second, oriented, anisotropically shaped pores such as tubular ones result in substantial property anisotropy clearly reflected in MSA (and mechanistic) models. Third, MSA (and mechanistic) models for stress (or flux) parallel to the axes of aligned cylindrical or rectangular tubular pores are identical with the upper property bound (Eq. 2.3.). Fourth, again note that the P bounds can be substantially < 1 (Fig. 2.13). Thus, properties going to zero at P = 1, which is often cited as indicating validity of models, is of uncertain value, e.g., the rollover of MSA models and data beyond $P = 1/3\text{-}\frac{1}{2} P_C$ obviate the (misguided) criticism that the exponential relation does not satisfy conditions at P = 1 (47).

2.4.5 Improvement and Extensions of MSA Models

The above developments and discussion along with extensive corroboration in subsequent chapters shows the MSA models to be very useful; however, a variety of further validations, developments, and improvements are needed as outlined as follows, starting with two aspects of the anisotropy of properties need further evaluation. First, MSA models for aligned tubular pores clearly show significant anisotropy of such pore structures for the extremes of stressing parallel or perpendicular to the axes of aligned pores. Modeling of varying degrees of mixed orientations is needed. While, this can be approximated by interpolation between the models for the two orientation extremes, direct analytical evaluation of this should be of value to both give more accurate values as well as additional insight into the accuracy of combining different porosities.

Second, MSA models for spherical pores, and especially particles, have been developed and used for ideal stackings with stress or flux along the <100> axes, and have received only limited considerations for other orientations and stack-

Table 2.7 Qualitative Summary of Angular Variations for Various Spherical Particle Stackings

Particle stacking	Number of particle bonds perpendicular ($\perp$), at intermediate angles (IA), or parallel ($\parallel$) to the reference axis:								
	<100>			<110>			<111>		
	$\perp$	IA	$\parallel$	$\perp$	IA	$\parallel$	$\perp$	IA	$\parallel$
Cubic ($C_n = 6$)	2	0	4	0	6	0	0	4	2
Orthorhombic ($C_n = 8$)	0	4	4	0	4	4	0	8	0
Rhombic ($C_n = 12$)	0	8	4	0	8	4	0	12	0

Source: ref. (2); published with permission of the J. Mat. Sci.

ing imperfections. Models should not introduce any anisotropy not inherent in the structures being modeled, but should reveal such inherent anisotropy. Rice (2) noted that all the models are isotropic at P = 0, and that simple cubic stacking of spherical particles is isotropic in MSA for <100>,<110>, and <111> directions. He also indicated qualitative isotropy for other key particle stackings based on observations of the number of particle-particle bonds perpendicular, at intermediate angles, or parallel to the stress or flux direction and that these represent respectively maximum, intermediate, and minimum contributions to properties (Table 2.7). Thus, qualitatively, there is an approximate balance between opposite changes in the number of particle bond areas having components perpendicular to the reference axis versus the net amount of such area for each bond as the angle of the reference (stress or flux) axis relative to the pore or particle stacking changes (since bonds parallel with the stress or flux axis have limited effect). Another related issue is assuring isotropy of stackings which result in other than regular polyhedral grains, e.g., orthorhombic stacking (Fig. 2.1B) where assumptions on cell dimensions can vary results some as noted earlier.

Two broader aspects of varying or different pore (hence also often particle) stackings or combinations should be noted. First, as discussed earlier and further later, the similar shapes of the MSA curves on semilog plots allows ready interpolation between models to estimate effects of various porosities combined in a given body, e.g., via a rule of mixtures, giving useful estimates; however, this neglects specifics of interaction of combined pore populations. Thus, refinement of combining effects of different porosities in a body is needed, either directly, or by providing better guidelines for interpolation between various model curves. Second, only a few pore systems have been evaluated for differing stackings, i.e, spherical particles and aligned cylindrical pores (Figs. 2.11, 3.1, and 3.2A; Section 3. 2.2). Broader evaluation of effects of different stackings (much of this probably by computer) will thus be of value. An important aspect of this is to account for changes in, and variations of, particle or pore stacking. This raises issues of the unit cell geometry, e.g., whether it is space filling, changes, etc., or whether alternative approaches not using unit cells are feasible. Wang (48) proposed corrections to MSA models for simple cubic stacking of particles to account for stacking imperfections. While his assumptions and approximations are generally individually reasonable, their combination and the resultant recommendation of substituting $b_1P+b_2P^2+\ldots$ for bP in the exponential relation seems both ad hoc and cumbersome. Two other developments may aid in this. Further development and verification of models discussed above for randomly stacked particle may be of considerable value to the above issues by allowing evaluation of different stackings, i.e., different C_n values, and comparison of these with models of fixed stacking yielding the same mean C_n values.

Arato et al. (49) proposed an MSA model not using unit cells based on a model for sintering of randomly packed uniform spherical particles developed by Fischmeister and Artz (50). The latter used methods very similar to those used by Knudsen, i.e., simulating sintering by truncating the spherical particles to form sintered contacts and adding the truncated volumes back into the particles to increase their dimensions to conserve their solid volumes and masses. The primary difference between this model and Knudsen's is the lack of an identified unit cell structure and its aid in calculation. Instead, they utilized the particle coordination number, C_n, as a measure of the mean stacking, which they could then increase to account for denser packing with consolidation and sintering. They did this by using empirical and analytical considerations, giving:

$$C_n = C_{nI} + 15.5 \left(\frac{R'}{R-1}\right) - 0.47 \qquad (2.8)$$

where C_{nI} = the initial or starting C_n and R'/R (= R' when normalized by taking R =1 as done earlier) is the ratio of the initial particle radius to that of the truncated particle. This equation agrees with data of Fischmeister et al. (5) in Fig. 2.4 till P $\simeq$ 0.9, extrapolating to $C_n \simeq 10$ at P = 0 instead of about 12.

Arato et al. (49) derived models for several mechanical properties (Table 2.8) based on the above developments assuming that the critical factor was the MSA, i.e. the load carrying capacity of the resultant particle-particle bonds normal to the stress axis. Two factors indicate the need for further evaluation of these models. First is the underlying greater complexity of handling a random distribution of particles and the question of whether procedures used to do this are sufficiently accurate. Second is comparison of the basic model with those of Knudsen (43) for a fixed C_n of one of the particle stackings evaluated by Knudsen. Simple cubic stacking is the easiest and most appropriate since its density is very similar to that commonly encountered for random packing. Direct comparison of the two models via data is not feasible, but an approximate comparison

Table 2.8 Alternate MSA Models of Arato et al. (49) and Green and Colleagues (51–53) for Random or Unspecified Particle Stackings

Relative property	Arato et al. (49)[a]	Green and colleagues (51–53)[b]
Bulk modulus, B/B_0		$C_B C_n (r/R)$
Young's modulus, E/E_0	$(C_n/4)(1-P)(r^2/R^2)$	$C_E C_n (r/R)$
Fracture toughness, K_{IC}/K_{IC0}	$(\sigma_n R^{1/2})(C_n/4)(1-P)(r^2/R^2) = \sigma_n R^{1/2}(E/E_0)$	$C_K \sigma_n R^{1/2}(r/R)^2 \propto \sigma_n R^{1/2}C_n^{-2}(E/E_0)^2$
Tensile strength, σ	$\sigma_n (E/E_0) R^{1/2}g(c)^{-1/2}$	
Hardness, H	$3\sigma_n (E/E_0)$	

[a] Strength of sintered neck between particles = σ_n, g = flaw geometry factor, c = flaw size.
[b] C_B, C_E, and C_K are unspecified geometrical constants.

can be made using values in Table 2.6 calculated assuming a cubic cell structure for both models. This comparison shows that R' of the Fischmeister and Artz model, which should be comparable to R of Knudsen's model, increases substantially faster than R. The differences appear to be greater than the uncertainty in calculating R, and raises questions of why Fischmeister and Artz gave (without derivation), $R' = (1 - P)^{1/3}(1 - P_C)^{-1/3}$. (The fact that maximum values of R and R' are nearly identical, may be coincidence since neither is accurate at $P = 0$, as noted earlier.) On the other hand, however, two factors should be noted. First, the Arato et al.' models are all self-consistent with the overall scaling of mechanical properties with elastic moduli, especially Young's modulus as discussed extensively in this book. Second, while their MSA is different from that of Knudsen, i.e., $(r/R)^2$ instead of $\pi(r/L)^2$ since there is no defined unit cell and hence no L value, both give nearly identical values for E/E_0 (compare the MSA and E/E_0 columns of Table 2.6 respectively for values from the Knudsen simple cubic and Arato et al. random models); however, such agreement may in part reflect partial use of a (cubic) unit cell for the random packing model.

This model family (Table 2.8), though derived for pores between randomly stacked particles, offers the potential of modeling pores between particles as a function of the mean C_n. Thus, if substantiated by further evaluation, it offers an alternative to not only both the three individual particle stacking models of Knudsen, but also combinations of them. Such combinations would be particularly valuable, especially if they can reasonably represent effects of particle size and shape variations via their impact on mean C_n values (and also possibly on the variation of such mean values).

Green and colleagues (51-53) recently derived models for some mechanical properties based on the response of a single bond area formed by sintering between two spherical particles. Since they are based on only a single bond, the resultant equations all depend on an unspecified constant of proportionality

(Table 2.8). Their equations are inconsistent with an overall scaling of other mechanical properties with E, and show a much faster rise in E with decreasing P than the Knudsen and Arato et al. models (Table 2.6; compare the r/R column to that of the MSA and E/E_0 columns respectively); however, their analysis may be of use in further development of MSA models of variable stacking.

Other significant refinements or developments of MSA models are needed. Important models derived specifically for mechanical properties of foam and honeycomb materials by Gibson and Ashby (54,55) (Eq. 4, Table 3.1) and others (56) are basically load bearing models. They are based on the mechanical properties of the foam cell struts (for open cells, e.g., Figs. 2.1D, and 3.3) and walls (for closed cells), while neglecting the junctions of the struts and walls and the stress concentrations associated with these junctions. They assumed cell struts and walls of uniform dimensions (e.g., Fig. 2.1D), while in reality these will often be tapered toward and thinner at their center points (e.g., due to surface tension effects in forming the original foam, Fig. 1. 14B). Correction for such tapering is one approach for improving these models and would make them even more explicitly MSA models. Cell stacking appears to be important, but has received very limited attention.

Finally, MSA models were derived for mass conserving bonding of particles such as sintering, but should also be an approximation for at least some other bodies, e.g., of chemically bonded particles such as cement or gypsum; however, they can be improved for bodies bonded by non-mass conserving processes. Models derived by Rumpf (57) for packed particles and modified by Onada (58) to account for effects of binders (in his case organic binders) show some possible applicability to inorganic binders, e.g., in cements (Fig. 8.2); however, more development in this area is needed seeking models reflecting different specific non-mass-conserving processes. Further, a newer computer modeling technique that essentially generates microstructures (59) and allows their analysis, may be a valuable combination with the above and earlier MSA models for properties.

2.4.6 Mechanics-Based Models and Comparison of Broadly Applicable Models

There are a variety of mechanistic based model for pores and some for microcracks. They are all specific to the more limited properties for which they have been derived, mainly one or two elastic moduli (and for ultrasonic wave transmission, i.e., starting from Eqs. 1.1–1.3, as discussed in Chapter 3), as well as for thermal conductivity; however, a brief outline and summary of the more broadly used and applicable resultant model forms for porosity effects is given here for broader comparison of the form and character of such models not restricted to a single property. (The limited models for microcracks are similar to those for similar porosity levels see Chapter 3).

A significant family of mechanics models has been developed for elastic properties, giving a rigorous bulk and sometimes shear modulus, and other elastic properties by bounding or other approximations. Much of this development is based on analysis of Hashin (60) for bodies consisting of either isolated (i.e., dilute concentration of) spheroidal (mainly spherical) inclusions or entirely of these. In the latter case it is assumed that each inclusion experiences the same uniform pressure applied externally to the body, hence neglecting interactions of the inclusions with one another similar to the dilute concentration approach. Such models commonly give the porosity dependence of the relative properties as the power relation: $(1 - AP)^n$, usually with $A = 1$ (61,62). ($A = 1/P_C$ has more recently been introduced in an entirely ad hoc fashion.) There is also a more recent family of models of Waugh and colleagues (63) for various properties based on considering a dense body made up of uniform aligned rectangular blocks. Pores are then introduced by randomly replacing individual blocks with cylinders of the same length and orientation, but with varying diameters less than the thickness of the blocks. This also yields the above power relation.

There are other forms for the porosity dependence of the relative property, i.e., the ratio of its value at some P to the value at $P = 0$, but three basic and widely used forms are e^{-bP}, $1 - A_1 P^n$, and $(1 - A_2 P)^n$ (Tables 2.5 and 3.1–3.3). The latter two forms are identical when $n = 1$ and all three are approximately equal at low P if $A_1 \simeq A_2 \simeq b$. The exponential form results from empirical semilog plotting of data and as a good approximation for all MSA models up to $1/3$–$1/2$ of P/P_C (Figs. 2.11 and 2.13); however, the exponential equation has also recently been obtained for elastic properties by Anderson (64) from strain analysis of isolated ellipsoidal pores. The second equation is a common explicit form of some MSA models (Table 2.5), but has also been obtained by Boccaccini et al. (65) using mechanics analysis for spheroidal pores for elastic properties, but his mechanics analysis for the same pore structure for strength (66) gave the third, i.e., power, form. The latter power form arises from empirically plotting data on log-log plots (or linear plots for $n = 1$), and for a number of mechanics models (Tables 3.1, 3.2). Thus, while the frequency of occurrence varies with the nature of the derivation, all three equation forms result from both mechanistic and geometrical modeling approaches. The selection of appropriate models should always depend substantially on their fitting and predicting data trends, with their rigorousness and indication of mechanisms also being factors. However, because 1) there are so many models and derivations with equations of the same and differing forms from differing mechanistic model approaches and 2) many, especially more rigorous mechanistic models being modified in an ad hoc fashion to obtain better data fits, there is little or no rational for selecting mechanistic models based on the rigor of their original derivation. Thus, as should be the case, agreement with substantial data as a sound rationale for the model, and it being reasonable to use are important factors in accepting models.

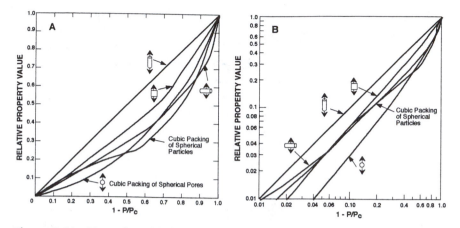

Figure 2.14 Plots of models for basic porosity types (i.e, geometries) as A) a linear function of $1 - P/P_C$ and B) a log-log plot. Note that: 1) only one pore type gives a linear plot in A, so data following this linear behavior must have that pore character or some fortuitous combination of porosity showing similar behavior, and 2) n is typically not a constant.

Basic evaluations show greater limitations of the form $(1 - AP)^n$ than is commonly recognized. Thus, the common assumption that n is constant for a given pore character, and hence allows accurate extrapolation to other ranges of porosity (e.g., via log-log plots) and that n can be used to characterize the porosity are not valid over a broad range. This can be seen in general, since dilute solutions theoretically and experimentally commonly give n = 1 and higher P levels, e.g., near the percolation limit give n > 1 (as well as often different A values) as well as two more specific examples. The recent suggestion (67) that the simplified P dependence, $1 - P/P_C$, is broadly applicable, i.e., with n = 1 so it is also the upper bound when $P_C = 1$ (but not for cylindrical pores aligned parallel to the stress) and is linear over the entire applicable P range may be a partial exception. While also used in conjunction with die pressed alumina (51) and other studies of Green and colleagues (52,53), the introduction of $A = 1/P_C$, though logical, is totally ad hoc and is only an approximation of uncertain validity (Section 3.4.3). More broadly, since MSA models (Figs. 2.11 and 3.1) agree with data as outlined in Fig. 2.13 and as shown in subsequent chapters, they also serve as a means of evaluating other models. This shows that only the model for cylindrical pores stressed parallel to their length (or other combinations of porosity having b ≃ 1.5) is linear on such a plot (Fig. 2.14). All other models for basic pore geometries are significantly nonlinear on such plots, as well as on log-log plots (Fig. 2.14B) showing that n is not constant over a larger P range.

2.5 COMPARISONS OF MSA AND MECHANISTIC MODELS

First consider implications of MSA models for other, e.g., mechanistic models. MSA models, while being derived for very specific pore character, were initially often arbitrarily applied to any or unknown porosity. This reflected the common tendency to neglect pore character and assume the broad, often universal, application of individual MSA models, providing, at best, limited utility for them. Their utility has been substantially enhanced by recognizing and using: 1) individual models for the applicability of the specific pore character on which each is based, and 2) the fact that collectively, i.e., as a family of models, they cover the entire applicable range of both the amount and character of porosities encountered. Thus, they cover the practical ranges needed with reasonable to good self-consistency (Figs. 2.11 and 2.13). On the other hand, mechanistic models are often more rigorous for the primarily limited (unspecified) quantities of limited pore shapes (e.g., cylindrical or spherical pores of unspecified stacking) for which they have been derived. While some models are consistent and some inconsistent with each other and data, this has been obscured by their being widely modified and applied in an ad hoc fashion. Neglecting the important role of pore character, in part to minimize the significant limitations of the narrow pore character thus far addressed by these models reflects the common explicit or implicit assumption that one or more nearly or fully universal models exist.

Two basic factors of porosity evaluation are that 1) the character, not just the amount, of porosity is important for all Category III properties (Table 2.1), and 2) there is no universal model for even a single given Catagory III property, therefore a family of models is needed. This problem, which must be watched closely in reading the literature, is aided by the fact that there can be significant overlap in porosity effects for a considerable range of pore character. Thus, as noted above and extensively elsewhere in this book (e.g., Figs 2.11, 2.13, 3.1, and 3.2A), a number of different pore characters overlap to form a main stream of pore behavior, e.g., having an average b value of about 3–5. This includes pores of approximately spherical or cubic character, tubular pores aligned normal to the stress or flux axis, and pores between randomly packed particles. Thus, apparent agreement between data and models is frequently claimed based on evaluating them against a modest set of literature data, with only a few significant disagreements since statistically much data (e.g., about 60%) lies in this mainstream range; however, such apparent success is misleading since it does not identify the source of variations within this b ~ 3–5 range, which is substantial, let alone the important cases outside of the mainstream. Further and more fundamentally, such accidental fitting provides no insight to combinations of porosity, which can yield the same or similar porosity dependence.

While MSA models are thus far generally more useful than mechanistic models, both need improvement or replacement. The greatest need is for better mechanistic models, with four needs being most important:

1. Development for other porosities other than just spheroidal (mainly spherical and cylindrical) pores, with porosities between particles (e.g., spheroidal or at least spherical ones) being paramount.
2. Addressing pore (and particle) stacking.
3. Addressing broader porosity ranges via one or more models and the P bounds for existing and new models.
4. Addressing combinations of different porosities.

Other needs for all models are addressed below. Encouraging progress is being made in mechanics models which improves their use and may allow further advances by combining their results with those from MSA models (see Chapter 3).

2.6   DATA EVALUATION NEEDS AND SELF-CONSISTENCY

There are basic needs in evaluating models and data for the porosity dependence of properties that can be summarized as demonstrating the self-consistency of the assumptions, parameters, relations and values. Thus, it is important to determine whether the porosity assumed in a model or in evaluating data is valid for the body or bodies being evaluated. This includes the amount of the porosity, since many, mainly mechanistic models are derived only for dilute concentration of porosity and an explicit pore shape (but with unspecified pore stacking). A particularly important factor is the handling of mixed porosity, whether of mixed spatial distribution (including stacking) and pore type since, while this is very common and important, it is not addressed by any single model and rarely addressed in studies of porosity effects. Variations in stacking of both uniform spherical particles results (Figs. 2.2, 2.11, and 3.1) and of aligned tubular pores (Fig. 3.2A) clearly effect properties (and b and P_C values). A few larger (or irregular) particles further broaden the range of pore character, especially as the separation between them decreases (Fig. 2.2B, C), and even limited variations in particle size and shape results in an approximately bimodal pore structure (Fig. 2.2D). More practical is the fact that random and simple cubic packing of particles give similar porosity, since random packing entails both denser particle packing as well as less dense packing due to particle bridging (e.g., Fig. 2.2). From a practical standpoint, handling the important case of mixed pores within and between spray dried agglomerates (Fig. 1.9) or other porous particles as well as special situations such as pores in and between ceramic balloons is important. Handling mixed porosity can also be of importance due to changes in porosity as densification proceeds, e.g., pores within spray dried agglomerates will typically be removed much faster and more extensively by sintering than pores between agglomerates. The issue of handling mixed porosity is an important aspect of the treatment in this and subsequent chapters. Another important need is to have some models that cover a broad, preferably the whole, porosity range to give a broader understanding of porosity effects and provide transitions between models for differing, more restricted, porosity ranges.

Another major need that is also seldom addressed is determining the consistency of the assumed and actual porosity. Key factors are porosity gradients and heterogeneity and anisotropic-shaped, especially oriented, pores and the resultant variable or anisotropic properties that are frequently a consequence of various fabrication methods, as discussed earlier. Inhomogeneous and anisotropic pores clearly have important affects on measured elastic properties (Sections 1.3,1.4) and discussed further later for elasticity and other properties (Chapters 3–7). An aspect of self-consistency of the models are their values at the bounds of P to which they are applicable. Thus, they should give the matrix property value at $P = 0$, unless the model is valid only at higher P. Such evaluation is commonly applied to models, but much less to actual data. Thus, properties at $P = 0$ are often obtainable directly from measurements on other bodies, especially single crystals, but this is not widely used. Evaluation of models at $P = 1$ is also often cited as a requirement; however, this is often not of significance since properties commonly go to zero at $P < 1$, often substantially so (Fig. 2.11). Thus, a model giving zero values for properties at $P = 1$, while often cited as indicating consistency of the model, instead indicates a problem if $P_C < 1$. It is essential that percolation limits on the effects of porosity on properties be defined and a rational aspect of models and analysis. This is a key plus for MSA and a key problem for mechanistic models.

There are fundamental relations of either a precise analytical or general character that can be used as a guide in evaluating the microstructural dependence of properties. While the former are of greatest use, the latter are also of value. These have not been used in most studies, which has contributed substantially to inconsistency of some results. These relations are used in this book to improve and validate results. First are analytical relations among elastic properties, and similar relations between Young's modulus, fracture energy and toughness, and tensile strength. Then there are more qualitative relations between Young's modulus and hardness, and between both of these and compressive strength, wear and related properties.

2.7   SUMMARY

The first chapter showed that effects of microcracks and pores are important because of their impacts, both positive and negative, on properties, with the scope of porosity effects being much broader, hence particularly important. Thus, the function of this chapter was to introduce modeling of their effects, which entails idealization of pores and microcracks, the parameters needed, and their actual characterization. One key aspect of this is recognition of the hierarchy of microcrack and especially porosity dependence of properties, e.g., those properties independent of porosity, dependent on only the amount of porosity, or depend on both the amount and the character of the porosity (Table 2.1). Category III A properties, i.e., those dominated by stress or flux transmission through the solid (as well as through the pore phase) are dependent on both the amount and character of the porosity, the latter being an often neglected factor. Thus pore character (and its relation to fabrication) are critical factors emphasized in this book. The minimum solid area (MSA) was shown to be both theoretically and practically the only other parameter for pore characterization needed for properties of primary focus in this book. While stress concentrations have often been seen as dominant for mechanical properties, it is argued that they often play a secondary role and support MSA typically being the dominant factor for mechanical properties (and imply this for electrical and thermal conductivity). Finally, the self-consistency of the porosity dependence of a property with inherently interrelated properties was stressed since this is important to check on models and data trends and is almost totally neglected in the literature. This entails the self-consistency of the property models proposed with each other and the basic physics of the properties. Characterization of microcracks and especially pores is a critical and complex step in applying models. Substantial further development is needed, e.g., to establish practical, reliable measurement of MSA. In the interim, a range of qualitative and quantative characterization is needed with close coupling to a range of microcrack and especially porosity dependent property measurements. It is expected that with such iterative correlation that many such property measurements will be increasingly used to characterize microcracks and especially porosities.

Empirical models have often been based mainly on semilog or log-log plots of the absolute or relative property versus P or $1 - P$ ($= \rho/\rho_t$) since these often give straight line relations for various properties over the common P range (e.g., to < 0.5). The resulting relations are respectively the exponential, e^{-bP}, and power, e.g., $(1 - AP)^n$, relations. Linear plots have also sometimes been used for the same reason, but often over lower P. The linear relations of $1 - AP$ or $1 - bP$ are low porosity approximations for the preceding expressions. A third common relation is $1 - Ap^n$, which along with the exponential relation are associated more with MSA models and the power relation with mechanistic models, each form has been obtained from both approaches, as well as other approaches. Thus, there is little basis for selecting one form over another based on its derivation, especially in view of more rigorously derived models being extensively modified in an ad hoc fashion because of their very narrow but often ill-defined range of application.

While both semilog and log-log plots allow ready presentation of data over most or all of the P range, semilog plots are advantageous for several reasons. The most important is that it results in curves clearly related to the pore character (via MSA models, Table 2.3 and 3.1–3.3), including explicitly showing P_C values. MSA models, which are based on simple, but rational, load bearing concepts, are currently the only ones encompassing the entire P range and, collectively, anywhere near the range of pore character encountered in practice. There is extensive data agreement with such models as shown in summary form in Figs. 2.11 and 2.13. These models also readily show that pore shape alone is inadequate to describe properties, e.g., cubic and spherical pores can give very nearly the same behavior. Instead it is the nature or character of the pores, i.e., the combined effects of shape, orientation, stacking, and intersection, all of which are reflected in MSA, that lead to the diversity of behavior, including varying anisotropy of properties with anisotropic shaped pores of varying degrees of orientation. These models also allow for combining effects of various types of porosity. Thus, MSA models are currently the only truly predictive models, but improvements are needed and suggested.

Mechanistic models, e.g., for elastic properties, have been more restricted in the pore character addressable, i.e., for noninteracting, closed (isolated) spheroidal, primarily spherical and cylindrical, pores, with no specification of the packing or stacking of the pores. Thus, they have been modified, mainly in an ad hoc fashion, e.g., to account for P_C values, and used, generally indiscriminately, for any (usually unspecified) pore character and over a much broader range of P than the narrow pore character and P range on which they are generally based. Such use has typically really been more of a curve fitting exercise, for which these models have no advantage over purely empirical curve fitting methods; however, a broader range of mechanics models reflecting a broader range of pore character are emerging that often agree with and compliment MSA models for the same character and amount of porosity (see Chapter 3). Much more comparison of their agreement with data as a function of pore character is needed since so little of this has been done. Seeking models of broad, especially universal, applicability that are based on one single, idealized, pore type is of little value, as is applying models to bodies of little or no porosity characterization.

REFERENCES

1. R. W. Rice, "Evaluating Porosity Parameters for Porosity-Relations," J. Am. Cer. Soc., **76**(7), pp. 1801–808, 1993.

2. R. W. Rice, "Evaluation and Extension of Physical Property-Porosity Models Based on Minimum Solid Area," J. Mat. Sci., **31**, pp. 102–18, 1996.
3. R. W. Rice, "The Porosity Dependance of Physical Properties of Materials-A Summary Review," Porous Ceramic Materials, Fabrication, Characterization, Applications (D. M. Liu, Ed.), Trans Tech Publications, Ltd. Zurich, Switzerland, pp. 1–19, 1995.
4. R. W. Rice, "Advanced Ceramic Materials and Processes" Design of New Materials (D. L. Cocke and A. Clearfield, Eds.), Plenum Press, New York, pp. 169–94, 1984.
5. H. F. Fischmeister, E. Arzt, and L. R. Olson, "Particle Deformation and Sliding During Compaction of Spherical Powders: A Study by Quantitative Microscopy," Pwd. Met., **21**(4), pp. 179–87, 1978.
6. H. P. Meissner, A. S. Michaels and R. Kaiser, "Crushing Strength of Zinc Oxide Agglomerates," Ind. Eng. Chem. Process. Des. Div., **3**(3), pp. 202–5 (1964).
7. J. Lipowitz, J. A. Rabe, L. K. Frevel, and R. L. Miller, "Characterization of Nanoporosity in Polymer-Derived Fibers by X-Ray Scattering Techniques," J. Mat. Sci., **25**, pp. 2118–24, 1990.
8. W. S. Rothwell, "Small-Angle X-Ray Scattering from Glassy Carbon," J. Appl. Phy., **39**(3), pp. 1840–45, 1968.
9. K. Hardman-Rhyne, N. F. Berk, and E. R. Fuller, "Microstructural Characterization of Ceramic Materials by Small Angle Neutron Scattering Techniques," J. Research, **89**(1), pp. 17–34, 1984.
10. N. F. Astbury and J. Vyse, "A New Method for the Study of Pore Size Distribution," Brit. Cer. Soc. Trans. & J., **70**, pp. 77–85, 1971.
11. D. G. Beech, "A New Method for the Study of Pore Size Distribution: Some Notes on Theory," Brit. Cer. Soc. Trans. & J., **70**, pp. 87–96, 1971.
12. J. M. Dickinson and J. W. Shore, "Observations Concerning the Determination of Porosities in Graphite," Carbon, **6**, pp. 937–41, 1968.
13. D. J. Baker and J. B. Morris, "Structural Damage in Graphite Occurring During Pore Size Measurements by High Pressure Mercury," **9**, pp. 687–90, 1971.
14. R. F. Feldman, "Pore Structure Damage in Blended Cements Caused by Mercury Intrusion," J. Am. Cer. Soc., **67**(1), pp. 30–33, 1984.
15. A. G. Evans, B. R. Tittmann, L. Ahlberg, B. T. Khuri-Yakub, and G. S. Kino, "Ultrasonic Attenuation in Ceramics," J. Appl. Phy., **49**(5), pp. 2669–79, 1978.
16. A. M. Gokhale, "Estimation of Bivariate Size and Orientation Distribution of Microcracks," Acta. Mater., **44**(2), pp. 475–85, 1996.
17. H. F. Fishmeister, "Characterization of Porous Structures by Stereological Measurements," Pwd. Met. Intl., **7**(4), pp. 178–87, 1975.
18. J. Neil, "Calculating Weibull Modulus from Average and Standard Deviation Values," GTE Tech. Memo. Report TM-0135–07–89–066, 7/1989.
19. D. Stoyan, and H. D. Schnabel, "Description of Relations Between Spatial Variability of Microstructure and Mechanical Strength of Alumina Ceramics," Ceramics Intl., **16**, 11–18, 1990.
20. D. R. Clarke, "Observation of Microcracks and Thin Intergranular Films in Ceramics by Transmission Electron Microscopy," J. Am. Cer. Soc., **63**(1–2), pp. 104–5, 1980.
21. E. D. Case and C. J. Glinka, "Characterization of Microcracks in $YCrO_3$ Using Small-Angle Neutron Scattering and Elasticity Measurements," J. Mat. Sci., **19**, pp. 2962–68, 1984.

22. Y. Ohya and Z. Nakagawa, "Measurement of Crack Volume Due to Thermal Expansion Anisotropy in Aluminum Titanate Ceramic," J. Mat. Sci., **31**, pp. 1555–59, 1996.

23. R. E. Wright, "Acoustic Emission of Aluminum Titanate," J. Am. Cer. Soc., **55**(1), p. 54, 1972.

24. G. Kirchhoff, W. Pompe, and H. A. Bahr, "Structural Dependence of Thermally Induced Microcracking in Porcelain Studied by Acoustic Emission," J. Mat. Sci., **17**, pp. 2809–16, 1982.

25. N. Sato and T. Kurauchi, "Microcracking During a Thermal Cycle in Whisker-Reinforced Ceramics Composites Detected by Acoustic Emission Measurements," J. Mat. Sci. Let., **11**, pp. 590–91, 1992.

26. C. J. Malarky, "Effects of Microstructure on the Elastic Properties of Selected Ta2O5–Eu2O3 Composites," M.S. Thesis, Iowa State U., Ames, Iowa, 6/1977.

27. K. Matsushita, S. Kuratani, T. Okamoto, M. Shimada, "Young's Modulus and Internal Friction in Alumina Subjected to Thermal Shock," J. Mat. Sci. Let., **3**, pp. 345–48, 1984.

28. D. R. Larson, L. R. Johnson, J. W. Adams, A. P. S. Teotia, and L. G. Hill, "Ceramic Materials for Advanced Heat Engines, Technical and Economic Evaluations," Noyes Pub., Park Ridge, NJ, 1985.

29. D. T. Smith and L. Wei, "Quantifying Local Microcrack Densities in Ceramics: A Comparison of Instrumented Indentation and Thermal Wave Techniques," J. Am. Cer. Soc., **78**(5), pp. 301–04, 1995.

30. C. Leach, "Microcrack Obcervations Using A. C. Impedance Spectroscopy," J. Mat. Sci. Let., **11**, pp. 306–07, 1992.

31. P. Stroeven, "Geometric Probability Approach to the Examination of Microcracking in Plain Concrete," J. Mat. Sci., **14**, pp. 1141–51, 1979.

32. W. Duckworth, "Discussion of Paper," J. Am. Cer. Soc., **36**(2), p. 68, 1953.

33. E. Ryskwitch, "Compression Strength of Porous Alumina and Zirconia—9th Communication to Cermography," J. Am. Cer. Soc., **36**(2), pp. 65–68, 1953.

34. B. Paul, "Prediction of Elastic Constants of Multiphase Materials," Trans. Metal. Soc. of AIME, **218**, pp. 36–41, 1960.

35. Z. Hashin and S. Shtrickman, "A Variational Approach to the Theory of the Mechanical Behavior of Multiphase Materials," J. Mech. Phys. Solids, **11**, pp. 127–40, 1963.

36. R. W. Rice, "Limitations of Pore-Stress Concentrations on Mechanical Properties of Porous Materials," J. Mat. Sci., **32**, pp. 4731–4736, 1997.

37. R. W. Rice, "Comparison of Stress Concentration Versus Minimum Solid Area Based Mechanical Property-Porosity Relations," J. Mat. Sci., **28**, pp. 2187–90, 1993.

38. M. Eudier, "The Mechanical Properties of Sintered Low-Alloy Steels," Pwd. Met., **9**, pp. 278–290, 1962.

39. S. D. Brown, R. B. Biddulph, and P. D. Wilcox, "A Strength-Porosity Relation Involving Pore Geometry and Orientation," J. Am. Cer. Soc., **45**(9), pp. 435–38, 1964.

40. D. P. H. Hasselman and R. M. Fulrath, "Effect of Cylindrical Porosity on Young's Moduls of Polycrystalline Brittle Materials," J. Am. Cer. Soc., **48**(10). p. 545, 1965.

41. R. W. Rice, "Extension of the Exponential Porosity Dependence of Strength and Elastic Moduli," J. Am. Cer. Soc., **59**(11–12), pp. 536–537, 1976.

42. R. W. Rice, "Microstructural Dependence of Mechanical Behavior of Ceramics," Treatise on Materials Science and Technology **11**, (R. McCrone, Ed.), Academic Press, pp. 199–381, 1977.

43. F. P. Kundsen, "Dependence of Mechanical Strength of Brittle Polycrystalline Specimens on Porosity and Grain Size," J. Am. Cer. Soc., **42** [8] 376–88, 1959.

44. P. J. Sherrington and R. Oliver, "Particle Size Enlargement," Granulation, Heyden & Son, Ltd. (London), 1981.

45. D. J. Cumberland and R. J. Crawford, "The Packing of Particles," Handbook of Powder Technology **6** (J. C. Williams and T. Allen, Eds.), Elsevier, New York, 1987.

46. R. W. Rice and S. W. Freiman, "The Porosity Dependance of Fracture Energies" Ceramic Microstructures '76: With Emphasis on Energy Related Applications (R. M. Fulrath and J. A. Pask, Eds.), Westview Press, Bolder, CO, pp. 800–23, 1977.

47. S. C. Nanjangud, R. Brezny, and D. J. Green, "Strength and Young's Modulus Behavior of a Partially Sintered Porous Alumina," J. Am. Cer. Soc., **78**(1), 266–8, 1995.

48. J. C. Wang, "Young's Modulus of Porous Materials," J. Mat. Sci., **19**, pp. 801–8, 1984.

49. P. Arato, E. Besenyei, A. Kele, and F. Weber, "Mechanical Properties in the Initial Stage of Sintering," J. Mat. Sci., **30**, pp. 1863–71, 1995.

50. H. F. Fischmeister and E. Artz, "Densification of Powders by Particle Deformation," Pwd. Met., **46**(2), pp. 82–88, 1983.

51. D. J. Green, R. Brezny, and C. Nader, "The Elastic Behavior of Partially-Sintered Materials," Mat. Res. Soc. Symp. Proc., **119**, pp. 43–48, 1988.

52. S. C. Nanjangud and D. J. Green, "Mechanical Behavior of Porous Glasses Produced by Sintering of Spherical Particles," J. Eu. Cer. Soc., **15**, pp. 655–60, 1995.

53. D. J. Green and D. Hardy, "Fracture Toughness of Partially-Sintered Brittle Materials," J. Mat. Sci. Let., **15**, pp. 1167–68, 1996.

54. L. J. Gibson and M. F. Ashby, "Cellular Solids, Structure & Properties," Pergamon Press, New York, 1988.

55. L. J. Gibson and M. F. Ashby, "The Mechanics of Three-Dimensional Cellular Materials," Proc. Roy. Soc. Lond. A **382**, pp. 43–59, 1982.

56. A. N. Gent and A. G. Thomas, "The Deformation of Foamed Elastic," Rubber Chem. Tech., **36**, pp. 597–610, 1963.

57. H. Rumpf, "The Strength of Granules and Agglomerates" Agglomeration (W. A. Knepper, Ed.), Intrescience Pub., Inc., New York, pp. 379–418, 1962.

58. G. Y. Onada, Jr., "Theoretical Strength of Dried Green Bodies with Organic Binders," J. Am. Cer. Soc., **59**(5–6), pp. 236–39, 1976.

59. A. P. Roberts and M. A. Knackstedt, "Mechanical and Transport Properties of Model Foamed Solids," J. Mat. Sci. Let., **14**, pp. 1357–59, 1995.

60. Z. Hashin, "Elasticity of Ceramic Systems," Ceramic Microstructures '76: With Emphasis on Energy Related Applications (R. M. Fulrath and J. A. Pask, Eds.), Westview Press, Bolder, CO, pp. 313–41, 1977.

61. R. M. Christensen, "A Critical Evaluation for a Class of Micro-Mechanics Models," J. Mech. Phys. Solids, **38**(3), pp. 379–404, 1990.

62. N. Ramakrishnan and V. S. Arunachalam, "Effective Elastic Moduli of Porous Ceramic Materials," J. Am. Cer. Soc., **76**(11), pp. 2745–52, 1993.

63. A. S. Waugh, R. B. Poeppel, and J. P. Singh, "Open Pore Description of Mechanical Properties of Ceramics," J. Mat. Sci., **26**, pp. 3862–68, 1991.

64. C. A. Anderson, "Derivation of the Exponential Relation for the Effect of Ellipsoidal Porosity on Elastic Modulus," J. Am. Cer. Soc., **79**(8), pp. 2181, 1996.

65. A. R. Boccaccini, G. Ondracek, P. Mazilu, and D. Windelberg, "On the Effective Young's Modulus of Elasticity for Porous Materials: Microstructure Modeling and Comparison Between Calculated and Experimental Values," J. Mech. Behav. Mat., **4**(2), pp. 119–28, 1993.

66. A. R. Boccaccini and G. Ondracek, "On the Porosity Dependence of the Fracture Strength of Ceramics," Presented at the 3rd European Ceramic Soc. Conf., Madrid, 1993.

67. D.C. Lam, F. F. Lange, and A. G. Evans, "Mechanical Properties of Partially Dense Alumina Produced from Powder Compacts," J. Am. Cer. Soc., **77**(8), pp. 2113–17, 1994.

3

POROSITY AND MICROCRACK DEPENDENCE OF ELASTIC PROPERTIES AT LOW TO MODERATE TEMPERATURES

KEY CHAPTER GOALS

1. Present MSA models for foams and honeycombs as well as mechanistic models for pores and microcracks.
2. Compare models with data and examine their consistency, including showing similarities of porosity and microcrack models and behavior.
3. Demonstrate that a universal model is not feasible or suitable; families of models such as the MSA family are needed, as well as more attention to pore character, anisitropy, and heterogeneity.

3.1 POROSITY DEPENDENCE BACKGROUND

Three of four basic elastic properties are Young's (E), shear (G), and bulk (B) moduli; these reflect the ratios of an applied stress to the resultant elastic (i.e., reversible) strain, respectively, from a uniaxial tensile or compressive stress, a shear stress, and a hydrostatic stress. The fourth basic elastic property, Poisson's ratio (ν), is the ratio of the transverse strain to the axial strain in response to a uniaxial tensile or compressive stress. These four properties are interrelated by the following basic relations for isotropic materials:

$$\nu = \frac{E}{2G} - 1 = \frac{9B}{6B + 2G} - 1 \tag{3.1}$$

$$E = \frac{9BG}{3B + G} \tag{3.2}$$

$$B = \frac{EG}{9G - 3E} \tag{3.3}$$

Only two of the four properties need to be measured, since the other two can be calculated from the above relations. Since ν is a small number, commonly in the range of 0.2–0.4, it is the most difficult to measure, and is thus a particular candidate for calculation from two of the moduli. This also makes it sensitive to small variations in moduli values used to calculate it, so ν values can vary substantially, making accurate determination of ν's porosity dependence challenging. However, this sensitivity of ν also makes it an indicator of variations in the data, e.g., due to heterogeneities of porosity. It also makes it of value to measure the three moduli, then compare the values calculated for ν per Eq. 3.1.

The above moduli (as well as related combined moduli) all decrease continuously with increasing amounts of porosity, even for small amounts of porosity. Typically the decrease is substantial, e.g., the percentage of modulus decrease is at least proportional to the percentage of porosity present, i.e., at or below the upper bound (Eqs. 2.3, 3.4). Poisson's ratio also commonly decreases with increasing porosity, but whether this is universal, and what the rates and limits of decrease are has been uncertain. Some proposals have been made for ν to decrease or increase to a fixed value. Quantification and discussion of $\nu - P$ trends will be presented in this chapter.

As discussed in Chapter 2, initial evaluations of the porosity dependence of properties were primarily empirical, typically plotting absolute or relative properties (i.e., those at some volume fraction porosity, P, to those at $P = 0$) versus P, density, or preferably relative density, $\rho/\rho_t = 1 - P$. Since most properties over greater P ranges, especially at higher P levels, show substantial nonlinearity (i.e., greater rates of property decreases with increasing porosity) either semilog plots of the log of the absolute or relative property value versus P on a linear scale, or log-log plots of the property versus $1 - P$, were used. The common linearity of semilog plots to 0.33–0.5 of P_C gives the porosity dependence as e^{-bP}, where b is the slope of the property dependence over the linear range (Table 3.1; Figs. 2.9, 2.10, and 3.1). Such semilog plots allow the extensive range of property changes to be displayed, the porosity for the solid percolation limits, i.e., P_C values, to be clearly identified, and interpolation between, or combinations of, different porosity models. Log-log plots also commonly give (for reasonable, but not necessarily extended, P ranges) linear plots, leading to porosity dependence relations of the form $(1 - AP)^n$ (Table 3.1). Both semilog and log-log plots are also useful for displaying sound velocities versus porosity, e.g., from pulse echo tests, as discussed in Section 3.2.2; however, log-log plots seriously obscure the percolation limits, i.e., P_C values, have at best limited demonstrated ability to relate porosity parameters to pore character (i.e., have demonstrated no predictive ability, and may have intrinsic problems for predictions, Fig. 2.14B). Data over a limited porosity range, especially at lower porosity, often varies, at least approximately, linearly with P, so linear plots are also used, i.e., for n = 1; however, recent proposals for use of linear plots of properties versus $1 - P/P_C$ (1) are valid for only a limited range of pore character (Fig. 2.13A), as discussed further in Section 3.6.

Since elastic properties of the pore phase are zero, the upper bound for Young's modulus (Eq. 2.1), reduces to

$$E = E_0 (1 - P) \qquad (3.4)$$

and the lower bound is zero. The resulting broad range of these bounds limits their utility in developing and evaluating models, and reflects one of the limitations of adapting composite models for porosity; however, as noted earlier and in Tables 3.2, 3.3, tubular pores aligned parallel with the stress axis follow the $1 - P$ dependence.

3.2   MODELS

3.2.1   Models Based on Load-Bearing Area, Especially Minimum Solid Area

Load-bearing areas ranging from the average to the minimum solid cross-sectional area (MSA) normal to the stress or flux are a logical basis for modeling much mechanical (and conductive) behavior of materials, as noted in Section 2.3. These two areas are the same at P = 0, have modest differences at modest porosity, but large differences, e.g., greater than or equal to two orders of magnitude, at higher P, so a selection must be made. MSA is the most logical for three reasons. First, the average solid area is simply $1 - P$, which depends only on P and not pore character, and hence is pertinent to Category II properties, not the Category III properties (Table 2.1) that are of primary interest here, except at the upper limit (for tubular pores aligned parallel with the stress axis; Tables 3.2, 3.3). Second, MSA is intuitively much more pertinent, as indicated by Figs. 2.1, 2.7–2.9 and interactive stress concentration considerations (Section 2.3.2). Third, MSA was shown to be a suitable and practical parameter to characterize porosity for Category IIIA properties (2,3), Sections 2.2.2, Fig. 2.3. Both the intuitive appeal and the diversity of porosity addressable with relatively simple mathematics made the MSA concept a common approach and allows models covering various idealized pores that collectively reflect the range of porosity encountered in practice (Tables 2.5, 3.1–3.3; Figs. 2.11, 2.13, 3.1).

The most extensive development of MSA models has been for fixed ideal (commonly simple cubic) stackings of uniform, often spherical, particles or

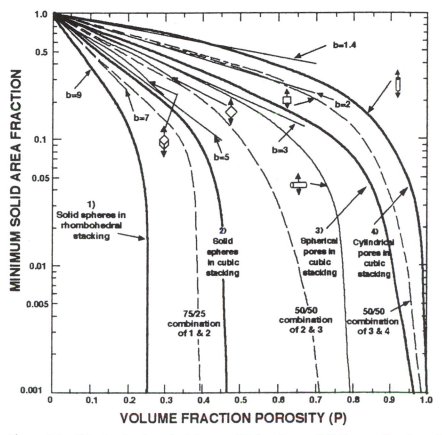

Figure 3.1 Calculated ratios of minimum solid (bond) area (MSA) per cell to the cell cross-section normal to the stress or flux direction, and hence of the relative property, versus volume fraction porosity (P) for two basic sphere stackings [per Knudsen (4)], and for cubic stacking of spherical and cylindrical pores (9). Note: 1) that curves for two orientations of cubical pores have not been calculated beyond the point of pore intersection [for reference, slopes (i.e., b values of e^{-bP}) are shown as mostly integer values of 1–9], and 2) in the original publication (9) and a subsequent review (33) of the MSA model for Young's modulus normal to the axes of aligned cylindrical pores there was serious error giving a P_C value approximately one-half, and a corresponding b value about twice the correct values. See also Fig. 3.2 for more details of such models and their anisotropy.

pores. The methodology is to calculate the minimum solid area per unit cell (Figs. 2.1, 2.10) and assume that the relative property of interest is equal to the ratio of the MSA to the cell cross-sectional area. The resultant simple model equations are valid until the pores intersect, i.e., to substantial values of P, sometimes the full range (Tables 2.5, 3.1 (Eqs. 1–6), 3.2 (Eqs. 1–2), 3.3; Figs.

2.11, 3.1. Beyond the point of pore intersection, the relations become more complex, but are still reasonable closed form solutions. Typically resulting models for a given pore type were applied without checking whether the porosity of the body was even approximated by the pore geometry of the model; however, it was subsequently recognized that (3,5–7): 1) these models collectively reflect the range of pore geometry from typical, practical forming processes, 2) model parameters (mainly b and P_c) are readily associated with the dominant (or average) pore character, and 3) the similar shape of the MSA model curves on a semilog plot readily allows interpolation between various curves to account for mixed and changing porosity characters. An important aspect of the reality of these idealized porosities is their clear inclusion and handling of anisotropic-shaped pores and the resultant property anisotropy from their frequent alinement from the processing creating them (e.g., tubular pores from extrusion and laminar pores from die pressing at higher pressure). Thus, note the anisotropies shown in Tables 3.2, 3.3, and Figs. 3.1, 3.2A for stress parallel or perpendicular to the axes of aligned tubular pores. Note also anisotropy perpendicular to the aligned tubular pore axes ranging from very little to a great deal as a function of tubular pore geometry and stacking (Fig. 3.2B) (29).

Various improvements have been sought for MSA models (Section 2.4.5), mainly for low-intermediate porosity levels, which have been studied most. Helmuth and Turk (30) presented a MSA model (for cement, see Section 8.2.2) based on random stacking of aligned cubic pores for which they argue that probabilities of adjacent cubic pores resulted in a P dependence of $(1 - P)^3$, instead of $1 - P$ for a regular array of cubic pores, thus again showing the importance of pore stacking. Wang (31) considered variations of stacking in Knudsen-type particle models, but his substitution of $b_1P + b_2P$ for bP was very empirical. Sudduth (32) applied a model he developed for composites to pore structures, including the variations Wang considered, reporting some differences from Wang's 'porosity interaction coefficients', but claiming justification for Wang's resultant equation form; however, modeling approaches based on

---

Figure 3.2 Plots showing anisotropy of elastic (and other) properties for some tubular pores. A) Semilog plot of relative property versus P for cylindrical pore MSA mechanics (Table 3.2) and from computer models (34,35). Note the general agreement between the different models, effect of stacking of pores on results, and differences in P_C values from computer versus other modeling (this is attributed to the small specimen to pore size in the computer model, e.g., three pores; see also Fig. 3.1). B) Comparison of elastic anisotropy as a function of the angle relative to the bse of the cell shape in a plane normal to the axes of aligned triangular or square cross-section pores (29). Note the variation from very little to very large anisotropy as a function of pore geometry (i.e., shape and the probable effects of stacking).

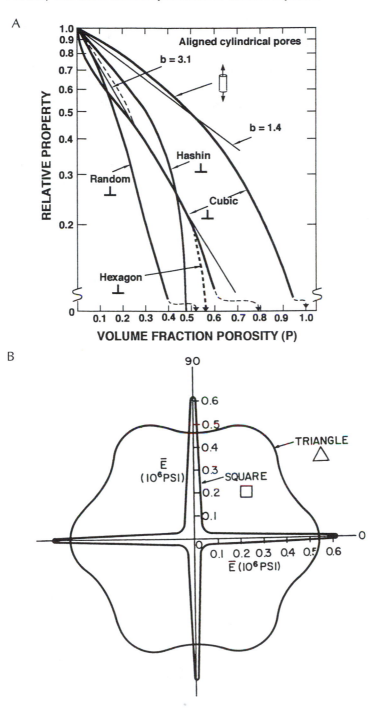

random stacking of uniform spherical particles and using the particle coordination number as a key characterization parameter, appears to be a more promising approach to handling pores from variable particle stacking. Such models
were discussed in Section 2.4.3 (Tables 2.5, 2.6) with that of Arato et al. (36)
appearing to be promising.

Consider now important models derived specifically for foam materials, i.e.,
those of Gibson and Ashby (15,16) (Eq. 4, Table 3.1) and others (39) which are
basically load-bearing models. Thus they are based on the mechanical
properties of the foam cell struts (for open cells) and walls (for closed cells),
totally neglecting the junctions of the struts and walls and the stress
concentrations associated with these junctions. For simplicity, they assumed
struts and walls of uniform dimensions (e.g., Figs. 2.1D, 3.3), while in reality
these will often be tapered toward, i.e., thinner at, their center points (e.g., due
to surface tension effects in forming the foam; Fig. 1.14B). Correction for such
tapering is one approach for improving these models and would make them
even more explicitly MSA models. The first of two other important modifications partly addressed for foam models is foam cell stacking, e.g., for $E/E_0 \propto$
$(1 - P)^n$, n = 1 for an approximately simple cubic stacking (Fig. 3.2A) and n = 2
(Eq. 4, Table 3.1) for a staggered stacking (Fig. 3.2B), but more is needed.
Second, anisotropy often occurs in foams, e.g., elongation of foam cells in the
direction of the foaming (i.e., the rise) direction. Huber and Gibson (37) have
shown that if l and d are the cell dimensions normal and parallel to the rise
direction, respectively properties can be expressed as functions of the ratio d/l =
R, for example, the ratio of Young's modulus in the rise direction (E_3) to that in
the transverse direction (E_1), is:

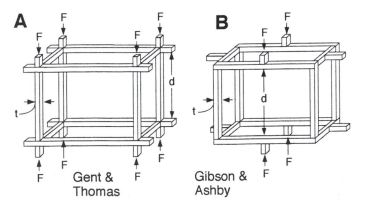

Figure 3.3 Schematic of two open, cubic, cell foam structures modeled. A) Simple
cubic stacking of Gent and Thomas (39) gives E/E_0 infinity $(1 - P)^n$, with n = 1. B)
Staggered cell stacking of Gibson and Ashby (15,16) gives n = 2. Note the definition of
the aspect ratio R for cases of foams elongated in the rise (vertical) direction.

$$\frac{E_3}{E_1} = 2R^2 \left[1 + \left(\frac{1}{R^3}\right)\right] \qquad (3.5)$$

Another important type of high porosity body consists of mainly fibers, e.g., felts rigidized by bonding fibers at their contact points, as for the glass fiber felt tiles used for the Space Shuttle. Green and Lange (38), again using load-bearing concepts, derived a model giving $E/E_0 = 1 - P$. They also addressed anisotropy of E in such rigidized glass felts due to some preferred orientation of fibers from consolidation by assuming that E in the felt plane was proportional to the fraction of fibers aligned normal to the stress directions of interest.

Table 3.1 Typical Equations for the Porosity Dependence of Elastic Properties, Primarily for Equiaxial Pores

Equation	Definition of terms[a]	Source
1) $E/E_0 \sim G/G_0$ $\sim B/B_0 \sim e^{-bP}$ $v/v_0 \sim e^{-bP}$	e: naperian log base, b and $b = (b_v)$ = slopes of initial approximately linear portions of semilog plots; b_v substantially < b (3).	Originally empirical, subsequently related to semilog plots of MSA models (3–7). Recently derived by Anderson (8) from strain analysis.
2) $E/E_0 \sim G/G_0 \sim$ $B/B_0 \sim 1 - e^{-b'(1-P)}$	b' related to b, for $P =$ $(P/P_C) > \sim 0.5$.	Interchanging solid and pore phases (and using P/P_C for P) (9).
3) $E/E_0 = 1 - AP^n$ for nonintersecting, i.e., closed pores	A commonly varies from 1 to $(3)^{1/2}$, and n is typically 2/3; see Table 2.4.	MSA analysis mainly for spherical or cubical pores (10–13) and mechanical analysis of spheroidal pores (14).
4) $E/E_0 \sim (1-P)^2$ $G/G_0 \sim 0.4 (1 - P)^2$		Dimensional analysis of foams by Gibson and Ashby (15,16)
5) $E/E_0 = (1 - P^{2/3})$ $[1 - P^{2/3} (1 - P^{1/3})]^{-1}$		Paul's mechanical analysis for cubic inclusions/pores (13).
6) $E/E_0 \sim G/G_0 \sim B/B_0$ $\sim (1 - AP)^n$ $v/v_0 \sim (1 - A'P)^{n'}$	A and A' are empirical constants, n and n' are slopes on log-log plots	Empirical, based on log-log plots, for unspecified pore character.
7) $G/G_0 = 1 - A_G P$ $B/B_0 = 1 - A_B P$	$A_G = 15 (1 - v_0) (7 - 5v_0)^{-1}$ $A_B = 1.5 (1 - v_0) (1 - 2v_0)^{-1}$	Mechanical analysis by Dewey (17), Hashin (18) (limited, spherical P), and Budiansky (with v_0 replacing v) (19)
8) $G/G_0 = A_G^{-1} (1 - P)$ $B/B_0 = A_B^{-1} (1 - P)$	A_G and A_B as defined above	Mechanical analysis (for spherical pores at high P) by Hashin (18).

Table 3.1 Continued

Equation	Definition of terms[a]	Source
9) $E/E_0 \sim G/G_0 \sim B/B_0$ $\sim 1 - A'P[1 + (A' - 1)]^{-1}$	A' defined in terms of v_0, similar to above, but with differing values for E, G, and B	Hasselman's (20) generalization of Hashin's (18) mechanical analysis of any concentration spherical pores
10) $E/E_0 = (1 - P)^2$ $[1 + (1/\xi - 1)]^{-1}$	ξ, a P-dependent factor reflecting pore shape and interaction; see Table 3.4	Nielsen's (21,22) analysis starting from a composite sphere model (23) derived from Hashin's approach
11) $B/B_0 \sim E/E_0 \sim G/G_0$ $= (1 - P)$ $(1 + P\psi/\phi)^{-1}$	$P_C = 1 \quad 0.5 \quad 0.25$ $\psi = 1 \quad 1+2P \quad 1+4P$ ϕ = fn. pore geometry, $\sim 0.5-10$	Nielsen's (24) modification of Kerner's and others generalized composite Eq. for pore factors
12) $B/B_0 = 1 - P\{1 - (1 - P)[1 + (4/3) (G_0/B_0)]^{-1}\}^{-1}$ $G/G_0 \sim \alpha(1 - P)$, $G/G_0 \sim 1.7(1 - P)^{-3}$	$v \quad 0 \quad 0.2 \quad 1/3 \quad 0.5$ $\alpha \; 2.05 \; 2.38 \quad 3 \quad \infty$ for $\alpha = 0.5$, at high P	Mechanical analysis of Christensen (25) (Generalized Self-Consistent Model)

[a] For terms of the various equations that are not defined elsewhere.

3.2.2 Mechanistic Models for Pores

Mechanistic modeling of mechanical properties is most extensive for elastic properties since these are most amenable to, but are still challenging, for rigorous mechanical analysis; however, to do this and to address the porosity amount and character is a complex problem, requiring considerable simplification using one of two basic approaches and related sets of simplifications (Section 2.4.6). Load-bearing, especially minimum solid area (MSA), approaches summarized in the last section are an alternate (and preferably, as discussed later, a complimentary) approach that is more versatile in its treatment of the amount and character of the porosity, but involves simpler treatment of the mechanical response of the bodies. The focus of this section is the second basic approach of a generally more rigorous analysis of the mechanical response of bodies (commonly composites, setting the properties of the second phase equal to zero), which has been more restricted in the amount and character range of porosity addressable. Primarily nonintersecting and noninteracting, i.e., limited, quantities of pores or second-phase particles, of ideal spheroidal, mainly spherical, geometries have been treated, e.g., as an isolated pore or spherical shell imbedded in a matrix.

Table 3.2 Equations for the Porosity Dependence of Elastic Properties with (Nonintersecting) Cylindrical Pores[a]

Equation	Orientation	Source
1) $G/G_0 \sim B/B_0 \sim E/E_0 = 1 - P$	Stress parallel to aligned pore axis[b,c] and others	MSA model of Hasselman and Fulrath (26), Rice (9)
2) $G/G_0 \sim B/B_0 \sim E/E_0 = 1 - (2/\sqrt{\pi})\sqrt{P}$, $E/E_0 \sim 1 - 1.3\sqrt{P}$	Stress perpindicular to aligned pore axis[d]	Same
3) $E = E_0(1 - P)$, $v = v_0$ $G = G_0(1 - P)(1 + P)^{-1}$ $B = E_0 G_0(1 - P)$ $[9G_0 - 3E_0(1 + P)]^{-1}$	Stress parallel to aligned pore axis	Mechanistic model of Hashin and Rosen (27)

[a] See Fig.3.2A for plots of these and related models.
[b] This equation holds exactly (up to the percolation limit of the porosity) regardless of the cross-sectional shape or stacking of the tubular pores, i.e., the coefficient for P remains 1.
[c] Property decreases continue beyond the interconnection of these pores, but calculations must account for the overlap of the pores, e.g., as for spherical pores or particles (3–7).
[d] This equation is only for tubular pores of circular cross-section in simple cubic packing; for other pore cross-sections the coefficient of P changes, e.g., it is 1 for tubular pores of square cross-section. In either case, the equation is only valid up to the percolation limit of the porosity, which in this case is where the pores overlap and thus become open.

Table 3.3 Elastic Properties of Honeycomb Ceramics[a]

Cell cross-section	Axial E/E_W[b]	Normal E, G, or v[c]
Rectangle	$t(1 + w)(1 \cdot w)^{-1} \sim 1 - P$	
Square	$2t/1 \sim 1 - P$	$E^{0,90}/E_W = 1/2\,(1 - P)$, $E^{45}/E_W \sim 2(t/L)^3 = 1/4(1 - P)^3$, $v^{0,90} \sim 0$, $v^{45} \sim 1$
Equilateral triangle	$\sim 3.5 t/L \sim 1.8\,(1 - P)$	$E/E_W \sim 1.1^{d} t/1 \sim 0.6\,(1 - P)$, $v^{0,90} \sim 0.3$
Regular hexagon	$\sim 2.3\,(t/L)^3 = 1.5\,(1 - P)^3$	$E/E_W = 4\,(3)^{-\frac{1}{2}}\,(t/1)^3 = 3/2\,(1 - P)^3$, G/E_W $\sim 0.6(t/1)^3 \sim 3/8\,(3)^{\frac{1}{2}}\,(1 - P)^3$ $\sim 0.65(1 - P)^3$, $v^{0,90} \sim 1$

[a] Typically produced by extrusion. Data mainly from refs. (11), (12), (28), and (29).
[b] Properties parallel to aligned axes of the tubular pores (channels), t = wall thickness, l and w are the lengths of the cell walls normal to the cell axis, E_W = Young's modulus of (assumed) isotropic wall material (which is often porous, in which case E_W is for that porosity). Note that many of the values shown are valid only when t is small relative to l and w, and that some variation can occur depending on whether l and w are measured at the inside of the cell wall or along its center.
[c] Properties in a plane perpendicular to the cell (pore) axes; 0, 45, and 90 refer to directions respectively along aligned cell walls and 45 and 90 degrees to this direction (i.e., 0 and 90 are the same for rectangular cells which are in simple cubic stacking).
[d] The factor 1.1 is an average, actual values vary from ~ 1.2 at 0 to ~ 1 at 90.

A detailed review of these models and their development here is not feasible or appropriate since this is an extensive and complex subject. This increasingly involves more sophisticated analytical mathematics, computer analysis, or both (e.g., since some models are not in closed form). Instead, an overview is presented, primarily from a users stand point, focusing on not only the advances but also remaining gaps and needs from the still restricted treatment of porosity. Thus, it is also important to consider how these results can be combined with others, especially from MSA models, to provide a more comprehensive view for those needing guidance in addressing real porosity effects. First, representative earlier models still commonly noted or used in the literature are reviewed. These were typically derived or used with an explicit or implicit assumption that they might be applicable to porosity levels and character well beyond that of the specific model derivation, i.e., of broad or universal applicability. Then newer developments and their ramifications are summarized; this is followed by comments on the consistency and combination with MSA results.

Despite the more limited P regime addressable, a diversity of mechanistic models have been derived. Typical equations from such models are given in Table 3.1 for mostly spherical pores and in Table 3.2 for aligned cylindrical pores. Some of these equations also result from other derivations discussed later. Note that MSA, mechanistic, and computer models (discussed below) all generally agree for cylindrical pores (Fig. 3.2A) with two exceptions: 1) that the mechanistic models do not account for stacking, and 2) the computer models show lower P_C values, which is attributed to the small specimen-to-pore size ratio, e.g., three pores across a specimen.

Several issues arise for such models (Tables 3.1, 3.2), as well as for a number of more complex ones not shown (often not in closed form, requiring iterative solution); two key ones arise from analysis of a composite and then setting the second-phase properties = 0 for pores. First, the mechanical analysis may be less valid for the more extreme property differences between the pore and solid phase, versus those of a matrix and a dispersed solid phase. Second, and potentially more serious in a composite, the second phase can become the only continuous phase, but the pore phase cannot be the only continuous phase and still have a solid body. Thus, composite models generally may not be valid near, and cannot reveal, P_C values less than 1. More generally, mechanistic models typically rigorously give only bulk (B) and (sometimes) shear (G) moduli, so Young's modulus, which is often of greater interest, must be estimated by bounding techniques using B and G (which may also be estimated).

Of broader concern is the narrow basis of these models versus that of real porosities, and hence the usefulness of the resultant model. This is compounded by the diversity of these models in their forms and values for a given amount and type of porosity, thus illustrating both the complexity and uncertainty of this approach. These models mainly address limited (undefined) quantities of spheroidal pores, and do not address pore stacking or P_C values. Parameters for differing pore character have commonly added in an entirely ad hoc manner (rather than in a rigorous fashion) thus questioning the net rigorousness of the resultant model. Attempts to use the shape or orientation (or preferably both) of spheroidal pores as measurable parameters are laudable. This clearly provides a basis for testing such models for pertinent pores such as those formed by (limited) foaming or burn-out of fugitive organic particles reasonably modeled as spheroids; however, application of these models to pores between particles is clearly a large deviation in pore geometry, e.g., exchanging pore and solid phases (Fig. 2.1A–C), which, at the minimum, seriously questions the rigor of the resultant evaluations. Thus, two key questions with these and other models are: 1) how seriously do the simplifications restrict the validity of the model and how might such restrictions might be minimized, and 2) can the models be truly predictive, i.e., can model parameter values reliably be explicitly associated with a given type or mix of porosity. Thus, the validity of these and other models must be judged by extensive evaluation against sound data (not by the sophistication of their mechanics, as implied by some). The diversity of mechanistic models for such a narrow range of porosity adds to both the need and challenge of this task.

Consider now comments on some of the models in Tables 3.1, 3.2, and related models. Models of Dewey (17) and Kerner (40), both derived for composites, are identical and linear in P, with both coefficients A_B and A_G for the P dependence = 2 at $v_0 = 0.2$, and A_G decreasing to 1.8 and A_B increasing to 4.5 at $v_0 = 0.4$ (most of these changes occur between $v_0 = 0.3$ and 0.4, which is commonly 0.2 to 0.4). Budiansky's model (19), also derived for composites but of different microstructure, is identical except for the substitution of v for v_0—i.e., making v, and hence the coefficients of P, dependant on P (which couples the equations since v depends on both G and B). Hashin's equations for any concentration of spherical pores actually differ some in the porosity dependence of E, G, and B (18). The original coefficients of P (which are different from A') for each of these three moduli are typically ~ 1, and exactly so at $v = 0.2$, with those for the first two moduli slightly decreasing respectively by 8 and 10%; however, the third coefficient (i.e., for B) increases to 3.5 as v goes to 0.4 (with only slightly more than half of the change occurring for v between 0.3 and 0.4). Thus, Hasselman's (20) reformulation and simplification of Hashin's equations represents some degree of approximation.

An earlier derivation of Mackenzie (41), based on dilute quantities of spherical pores of statistically uniform size and spatial distributions, gives G and B in terms of both the P = 0 values of these moduli, but include terms of P^2 or P^3. (While the mechanical analysis and the resulting equations of Hashin are different, Hashin's physical model based on spherical shells is a derivative of the spherical shell structure modeled by Mackenzie.) Wagh et al. (42) analyzed porosity effects based on a body made of uniform, aligned rectangular blocks. Pores are introduced by randomly replacing individual blocks with cylinders of the same length and orientation as the blocks, but of less volume due to differing radii. Considering bonding between the cylinders being analogous to springs, they proposed $E/E_0 = (1 - P)^n$, where n is a factor presumably reflecting the tortuosity of the pore structure. They could not analytically determine n, but instead used it as a parameter to fit fairly well 12 of 13 data sets (of unspecified pore character). They concluded that n ~ 2 for bodies that were not fabricated by hot pressing or with sintering aids, use of either of which increases n due to apparent accelerated densification (without any pore character evidence).

There have been several other analysis, often based on spherical shell models of Hashin (18). Rossi (43), noting that the coefficients (involving ν for the P term in Hashin's equation for low concentration of spherical pores (Eq. 7 of Table 3.1) partially reflected the stress concentration of such pores, generalized the equations for spheroidal pores (Fig. 3.14). Bert (44) followed Rossi's stress concentration hypothesis, but concluded that Rossi's equation could not be generalized, proposing instead an equation of the familiar form: $E/E_0 = (1 - P/P_C)^n$, where $n = \alpha P_C$ and α = the maximum stress concentration. Based on very limited testing of this equation with data for spherical or cylindrical pores, it was claimed that it provided a good fit to P = 0.2 and it predicted lower values than the data. Nielsen's model for the porosity dependence of Young's modulus (21–23) is also an extension of his model for composites (23), which in turn is an extension of Hashin's composite sphere model (18); however Nielsen, recognizing the diversity of composite, and hence pore-solid microstructures, attempted to model versatile microstructures. He did this, as some other investigators have, by introducing a pore shape factor, ξ, (Eq. 9, Table 3.1), but which changes per:

$$\frac{\xi}{\xi_0} = \left(\frac{1-P}{P_C}\right)^q \tag{3.6}$$

where q is greater than or equal to 0; however, his definition of the pore shape parameter ξ_0 is only semi-quantitative, thus reducing the rigorousness of his model (Table 3.4). It has been used as a data-fitting parameter, which is of limited or no value for predictive purposes until values of ξ_0 and q can be at least approximately specified based on known or expected pore character with reasonable reliability.

Nielsen (24) modified generalized equations of Kerner and others for composites to account for such factors as the shape and stacking limitations of the second phase (via ϕ and ψ respectively of Table 3.1), then these were applied for porosity. The values of ψ in Table 3.1 are based on P for maximum stacking P_C. Note the similarities and differences between the theories of the two Nielsens (two different investigators, one with initials L. F. and the other L. E.), and that the latter model was also derived and used for electrical and thermal conductivities (see Chapter 7).

Table 3.4   Nielsen's Pore Shape Factors

Predominant pore geometry at low P	Pore shape factor, ξ_0	Comments
Enveloping network tending to subdivide solid phase into particles	Low (0–0.4)	Shell-like pore network and compact solid particles decrease magnitude of ξ_0
Dendrites, ribbons	Medium (0.3–0.7)	Coarser and more compact pore geometry increases magnitude of ξ_0
"Pockets" defined by an enveloping network of the solid phase	High (0.6–1.0)	Shell-like solid network and pore "pockets" of compact shapes increasing ξ_0 magnitude

Source: ref. (21).

Spheroidal pore models, which are generally derivatives of Hashin's models, and of more recent work of Mori and Tanaka (45) and Kuster and Toksoz (46) for composite materials, have been extended to the limit of prolate spheroids: cylindrical shells (or inclusions) (25,47). Useful reviews of these models for (composites and) pores are by Christensen (25), Ramakrishnan and Arunachalam (48,49), Berryman and Berge (50), and Kachanov et al. (51). Again these models calculate B and G, with the latter often being obtained from bounds rather than an exact calculation, thus making obtaining E from bounds more uncertain. These models often have the form: $(1 - P)^n (1 + AP)^q$, where n = 1 or 2, q = 0, −1, or −2, and A is often a function of ν, giving some equations the same or similar to those for one, but not necessarily both, of the moduli of one of the models in Table 3.1. The porosity form of the model that Christensen found best fit for composite data is Eq. 10 of Table 3.1; however, as noted earlier and discussed by Ramakrishnan and Arunachalam (48,49), there are often problems in applying composite models to porosity, especially at P > 0.5. Ramakrishnan and Arunachalam modified models based on Hashin's spherical shell approach by substituting ν for ν_0, i.e., allowing for the P-dependence of ν to increase the P-dependence of the moduli by making the coefficients of P dependant on P. This makes the equations more difficult to work with, but they are readily handled by computer. Further, they claimed based on a two-dimensional analysis, that ν was either: 1) constant at an intermediate value of about 0.25, or 2) depending on whether $\nu_0 < 0.25$ or > 0.25 increases or decreases to about 0.25, at either P = 0.5 or P = 1. Some or all of this ν behavior is supported by only limited data (49) and a few analyses (48,49), and is contrary to much data (52) and analysis (53) (both discussed later).

Much of the mechanics development can be put in perspective by considering more recent reviews, especially that of Kachanov et al. (51). Limitations and areas of validity of earlier newer approaches have become better recognized and substantial advances are being made. Thus, the self-consistent scheme tends to overestimate effective compliances, especially as pores approached microcracks (the latter requiring special limiting procedures), e.g., at low P. Mori-Tanak's scheme was found to be exact as P → 1 (i.e., valid for foams). While analysis of three-dimensional (3D) pores is still generally restricted to ellipsoidal pores, handling of mixtures of pore orientations and geometries, of pore-crack combinations, and pore-crack transitions has substantially improved, as has treatment of microcracks. A broad improvement is the clearer recognition that another pore-geometry parameter is needed besides P to specify the porosity dependence. Application of models for spherical pores to other pores can substantially overestimate the porosity present or underestimate the decrease in moduli for a given P.

Specific advances for effects of pore geometry, primarily for two-dimensional, i.e., tubular, pores loaded normal to the tube axes, i.e., for transverse moduli show:

1. More elongated elliptical cross-sections have higher compressibility (α, the ellipse perimeter, is about 4.22 $(a^2 + b^2)^{1/2}$ to < 5.5%), so circular cross-section tubular pores are the stiffest (also indicated for ellipsoidal or spheroidal, 3D, pores).
2. Tubular pores of rectangular or elliptical cross-sections with the same cross-sectional areas and similar aspect ratios yield very similar transverse moduli.
3. Tubular pores of polyhedral cross-sections that have smooth, especially rounded, edges are stiffer (i.e., have less modules decrease, hence reduced P dependence) than such pores with sharp, especially cusped, edges.

Other results regarding transverse anisitropy with tubular pores and cracks, as well as crack-pore interactions are:

1. Tubular pores with square cross-sections have high transverse anisitropy for nonrandom orientations about their axes, while all other tubes of regular polygonal cross-sections are approximately isotropic for any orientational distribution (Fig. 3.2 B).
2. While the moduli for bodies with slightly deformed spherical pores are similar to those for spherical pores, the former result in anisotropic moduli.
3. In bodies having both combinations of separate cracks and pores, or of more circular and more elongated or flattened pores, the pores have more effect on the cracks and the more circular pores have more effect on the elongated pores than vice versa, e.g., the more symmetrical pores reduce the body anisotropy.

These summarized results support, extend, and compliment those from MSA models, which are corroborated by data in subsequent sections. Thus, from a broad perspective both clearly show the critical need for a second pore parameter besides P, i.e., one reflecting pore geometry. More specific agreements, e.g., Figs. 2.11, 3.1, include: 1) the general efficiency of pores of circular cross-sections in limiting decreases in moduli, and 2) limited differences for pores of similar character, e.g., of elliptical and rectangular tubular pores and spherical and cubic pores, and greater effects of pores with cusps (e.g., those between particles versus spherical, cubic, or cylindrical pores). It is thus clear that mechanistic and MSA models can be very valuable in various combinations, which requires greater understanding by the materials community about both types of models and by the mechanistic community of MSA models and results (e.g., Kachanov et al. in their review were unaware of correlating different b values with pore character). Thus, MSA models frequently can give trends over a broader range of the amount, character, or combinations of the porosity, and mechanistic models can provide reinforcement, refinement, or both for MSA models. On the other hand mechanistic models clearly provide more insight into the pore–microcrack transition and combinations of pores and microcracks.

There are other model developments beyond those outlined above that are briefly noted, e.g., as examples of other approaches. Boccaccini et al. (14) derived a model for Young's modules based on stress-strain analysis around spheroidal pores of various orientations, obtaining $E/E_0 = 1 - AP^{2/3}$, where A is an explicit function of the aspect ratio of the spheroids and their average orientation relative the stress direction. They transformed this to $E/E_0 = (1 - P^{2/3})^A$ using the fact that the first equation can be seen as the first term of a series expansion of the second equation, presumably to more closely (but obviously not exactly) agree with the more common mechanics result of the form: $(1 - P)^n$, which they also obtained for tensile strength (54).

Anderson (8) has derived the exponential relation from strain analysis of ellipsoidal pores, where $b = 1 + 4\varsigma_1(1 - v^2)\pi^{-1}$ and (ς_1 = the pore aspect ratio, a/c. Thus b is primarily dependent on pore shape (hence also MSA), and is 1 for cylindrical pores aligned parallel to the stress axis and 2.2 for spherical pores (with $v = 0.25$), both in fairly good agreement with, but somewhat lower than, MSA models (Figs. 2.9, 3.1). This difference may reflect the lack of accounting for pore stacking, or simply differences in initial versus average slopes. Anderson also noted that a weighted average can be used to account for different pore types present (i.e., as proposed for MSA models), but since ς_1 is a ratio, larger values of it dominate. His derivation thus provides symmetry between mechanistic and MSA models, since each approach thus produces all three basic forms of the common P relations for properties, i.e., e^{-bP}, $1 - AP^n$, and $(1 - AP)^n$.

Boccaccini and Fan (55) adapted a model for the conductivity of a two-phase composite (56) to porosity effects. The composite model takes a different microstructural approach, mapping a mixtures of α phase grains and β-phase grains into three slabs parallel with the flux (stress) direction. The slabs consist of: 1) all α grains, 2) all β phase grains, and 3) mixtures of α and β grains such that slabs 1 and 2 have only like-phase grain boundaries and slab 3 has only α–β grain boundaries. Letting one phase be pores so its modules is zero gives the porosity dependence of Young's modulus as;

$$\frac{E}{E_0} = R\,(1-P)^2\,[P+(1-P)R]^{-1} \qquad (3.7)$$

where R is a parameter derivable from quantitative microstructural analysis. They showed this equation fit a sampling of data using R as a fitting parameter since data was not available to determine it, again illustrating the need for characterization and for parameters that can be at least estimated with limited data (such as the correlation of b values with basic pore character). This model again raises questions of applying composite models to porosity in general, and, in this specific case, whether the mapping of part of the mixed microstructure body into one of all solid, all porosity, and solid-pore grain boundaries has physical validity. Thus questions of whether this mapping is valid for both inter- as well as intragranular pores and more fundamentally whether the mixed pore-solid structure exists at the level of porosity required for such a structure, e.g., since P_C for pores between particles is often < 0.5, which would often appear to be the P value for the mixed pore-solid structure of the mapped microstructure.

There are also models based on percolation concepts for random pores (mainly from the physics community). These models give the porosity dependence of the relative property (e.g., Young's modulus, strength, electrical conductivity, and permeability) of the form:

$$(P_C - P)^n \qquad (3.8)$$

i.e., similar to some of the models in Tables 3.1 and 3.2. Although investigators of these types of model originally thought that the exponent n would be the same value for both two-dimensional and three-dimensional pores of a common cross-sectional geometry (e.g., tubular pores, with their axes perpendicular to the measuring direction, and spherical pores), this has been found to not be the case (57–60). Thus, n for properties of two-dimensional pores is often lower by a factor of 0.5–1, e.g., 5 versus 6 for two-versus three-dimensional pores for Young's modulus. That the exponents would be different for two and three-dimensional models is clearly expected from comparing such previously developed models (Tables 3.1–3.3; e.g., tubular versus spherical pore models), but not necessarily the level of differences as $P \to P_C$.

Some models for the porosity dependence of elastic properties have been based on the porosity dependence of sound velocities, e.g., from pulse echo measurements. The most prevalent use of velocity-porosity relations for elastic-porosity dependences is empirical, based on the former also typically being linear on log-log or semi log plots, e.g., to $P = 0.15$ to 0.3. Thus, as per Eq. 1.2, if the P dependence of shear wave velocity (v_s) is $(1 - P)^q$, then since $\rho = \rho_0 (1 - P)$, the exponent for the power relation for the P-dependence of G, and the log-log slope will be $n = 2q + 1$ [or alternatively $q = 1/2(n - 1)$] (61,62). Similarly, since $1 - P \sim e^{-P}$ for modest P, the semilog slope and exponential factor b for G has the same relation to the corresponding factor for shear wave velocity as n and q. The slope for E is also close to these slope-exponent relations for both plots and functions (varying some due to P dependant of v). Thus, for modest P and v (e.g., < 0.5), the numerator and denominator terms of Eq. 1.1 involving v are approximately equal, so letting f_P be the P dependence of v the terms $(1 - v)$ and $[(1 + v)(1 - 2v)]$ become respectively $f_P (f_P^{-1} - v_0)$ and $[f_P (f_P^{-1} + v_0) (1 - 2v_0 f_P)$. However, the latter becomes $f_P (f_P^{-1} - v_0 + 2v_0^2 f_P)$, and since $2v_0^2 f_P < v_0 < f_P^{-1}$, the numerator and denominator for the v terms for E for longitudinal wave velocity measurements approximately cancel, so the slopes and exponents for longitudinal wave velocity and Young's modules are typically very close to those for shear waves and modules, respectively. The above exponential relation will vary if the P-dependence of sonic velocity differs from that of ρ, i.e., from $1 - P$, e.g., if it is $(1 - AP)^q$, with $A \neq 1$.

Kupova (63) derived an expression for the porosity dependence of Young's modules based on more basic aspects of wave propagation in solids. He starts from the dispersion equation for phonons at small wave vectors, accounting for perturbations due to the removal of material by introducing pores and the remaining matrix response. Then using assumptions such as pores being isolated and small relative to the wave length he obtains:

$$E = E_0 (1 + aP + bP)(1 + cP)^{-1} \qquad (3.9)$$

However, there is again no clear relation of a, b, and c to pore character and there has been no significant evaluation of this equation, other than a preliminary favorable comparison to Wang's modified exponential equation (31).

The power and versatility of modern personal computers provides significant capabilities for readily handling coupled equations, iterative solutions, and a variety of other (often not fully utilized) opportunities, as illustrated here and later. An important computer application is via finite element analysis, but it is important to distinguish three (64) versus two dimensional (34,35,48,49) analysis, as discussed later and shown in Tables 3.2, 3.3, and Fig. 3.2A. Note the reasonable agreement between MSA, mechanistic, and computer-derived models for tubular pores (Fig. 3.2A), but there are effects of pore stacking and pore-specimen not addressed by current mechanistic models. More recently another computer modeling approach has been under development (65). This essentially generates microstructures, and apparently allows some analysis of them, which may have considerable promise.

3.2.3 Microcrack Models

Models have also been derived for effects of microcracks (i.e., cracks on the microstructural, often the grain, scale), e.g., as reviewed by Case (66). Simple, planar crack geometries (e.g., the limits of oblate spheroids) in random or simple ordered arrays with limited or no crack stress field interactions, i.e., moderate to "dilute" concentrations, are assumed (66–68). Most resultant models give the ratio of E/E_0, and in some cases G/G_0, in terms of δ (the crack density parameter) as $1 - A\delta$, except when the microcracks are all oriented parallel with the stress (in which case they have no effect on elastic properties due to the assumption of zero cross-section normal to their plane). A is determined by functions of Poisson's ratio of the uncracked material, ν_0, having the form $A_1 - A_2{}^{\nu n}$, where A_1 and A_2 are integers and $n = 1$ or 2, times $\delta = 2N/\pi(\alpha^2/p)$, where N = the number of cracks per unit volume, α and p are respectively the area and perimeter of a crack, and () represents the average of the bracketed quantity. Such earlier models did not address the important question of the effects of pores, e.g., pore-microcrack interactions, which are now being addressed, as noted above.

A representative set of models for dilute microcracks, which are linear in the decrease of elastic properties with increasing crack content are given in Table 3.5 (with their A values). Anderson (8) derived an exponential relation for cracks that is directly analogous to that for porosity. The exponential terms are very similar to those of Table 3.5, consistent with the linear and exponential expressions agreeing at low P.

Models allowing at least some crack interaction, hence some increase in crack densities, typically give nonlinear decreases in elastic properties with increasing crack density (66,69), e.g., very similar to comparable models for thermal conductivity (69) (Fig. 7.2). Such models commonly have the form $(1 - A\delta)^{-1}$. Hasselman and Singh (69) showed that since the ν dependence of E/E_0 is typically limited, it can be taken as $\sim (1 + 1.78\delta)^{-1}$. Again, the A values for the microcrack dependence of Poisson's ratio are 0.33 those for E, as is true for the dilute, linear models. Nielsen modified his pore model (21) for simple cubic arrays of three orthogonally intersecting sets of penny cracks (diameter = 2d) with spacings of 2h, again giving a nonlinear relation:

Table 3.5   Elastic Moduli for Dilute Concentrations of Various Microcrack Configurations

Crack configuration	Elastic moduli equation[a]
1. Random penny cracks	$E/E_0 \sim 1 - 1.76\,\delta$, $G/G_0 \sim 1 - 1.30\,\delta$
2. Aligned penny cracks (Stressed $\perp$ to crack plane)	$E/E_0 = 1 - 16/3\,(1 - v^2)\,\delta \sim 1 - 1.76\,\delta$
3. Slit cracks randomly oriented in all dimensions	$E/E_0 \sim 1 - 1.64\,\delta$, $G/G_0 \sim 1 - 1.36\,\delta$
4. Slit cracks randomly oriented in a plane $\perp$ to stress	$E/E_0 = 1 - \pi^2/4\,(1 - v^2)\,\delta \sim 1 - 2.31\,\delta$
5. Aligned slit cracks (stressed $\perp$ to crack plane)	$E/E_0 = 1 - \pi^2/2\,(1 - v^2)\,\delta \sim 1 - 4.62\,\delta$

[a] δ = crack density parameter, as defined in text.

$$\frac{E}{E_0} = \left[1 + W \ln \left(1 - \left(\frac{d}{h}\right)^3 \right)^{-1} \right]^{-1} \tag{3.10}$$

where $W \sim 0.67$ for cracks emanating from points and $\sim \pi$ when emanating from lines.

The crack density parameter, δ, is at best difficult to directly determine since microcracks are difficult to accurately measure for their size and especially density due to their small size and especially opening. Comparison of models and data indicate that δ is commonly in the range of 0–5 (69); however, Case (66) has shown that this parameter can often be eliminated by solving for one elastic property in terms of another, e.g., E in terms of v or G via their mutual microcrack dependence. Case used this evaluation to show that most microcrack models give very similar mutual property trends and reasonable model-data correlations (Section 3.3.4).

Though it is difficult to directly and numerically relate the effects of micro-cracks and pores on elastic properties, four useful comparisons are evident. First, there is a direct parallel between these equations and those for comparable (i.e., dilute) porosity solutions, e.g., their linearity for low P and δ levels (generally with low A values) and their nonlinear property decreases for higher P and δ levels. Second, the crack models typically show less decrease for v than elastic moduli, e.g., 1/3 as much a decrease as noted earlier, which is consistent with a smaller decrease of v with P. Third, the microcrack models show clear differences due to the shape, orientation, and stacking of the cracks, similar to that shown for pores. Fourth, the moduli do not go to zero with the presence of a number, let alone a few, cracks, e.g., as indicated by Rossi's model (43,70).

Note that the application of the MSA concept would at first appear to be applicable to aligned arrays of microcracks, e.g., Eqs. 2 and 5 of Table 3.5;

however, closer examination shows the MSA concept does not give good results, thus illustrating a limitation of it. Such MSA models predict the same form of dependence as in Table 3.5, but the numerical factor for the δ term and especially the form of these terms expressed in microcrack dimensions and separations, are different. They appear to not be correct since they do not depend on the spacing of aligned cracks normal to their plane due to their mathematically zero (or practical very small) thickness.

3.3 COMPARISON OF MODELS AND DATA

3.3.1 Data for Idealized Pores

A key task is model–model and especially model–data comparison. First, MSA models are compared with pores approximating those idealized in the various models, then their applicability to more general porosity is evaluated. While the number and extent of studies of bodies whose porosity reasonably approximates the ideal pore structures modeled are limited, especially for any one property, they are sufficient to test the model concept. Further, trends found for other properties in subsequent chapters show similar agreement, thus substantially reinforcing the results shown here.

Consider first results for bodies with spherical pores. Hasselman and Fulrath (71) showed that Young's and shear moduli of cast glass plates respectively decreased from 79.5 and 33.3 GPa at P = 0 to 75.5 and to 30.5 GPa at P = 0.025 from bubbles. MSA models for cubic stacking of uniform spherical bubbles (which is very similar to random packing) predict respective decreases to ~ 72 and ~ 30 GPa, which is in reasonable agreement with but a somewhat greater decrease than that observed. The two give slopes (b values) respectively of 2.1 and 3.5, both of which, and especially the average, agree with the MSA model for cubic stacking of uniform spherical pores. Data covering greater ranges of porosity are shown for PZT-sintered with latex spheres, leaving spherical pores (72,73) (Fig.3.4) and giving b values for Young's modules averaging in the range of 2.5–3, which is in good agreement with the MSA model for cubic (hence approximately random) stacking of spherical pores. Probable reasons for the limited differences between the MSA model and the experimental data are variations in pore distribution, the limited porosity range (hence limited data, i.e., note that the greatest variations are for the lowest P ranges), and residual porosity from sintering. (See also Fig. 8.3 showing similar agreement for two cast polymer resins with varying levels of porosity from bubbles.)

Consider next the relative elastic moduli (mainly Young's modules) data for ceramic foams, e.g., with approximately spherical pores and intermediate to high porosity levels; first, we will address MSA models spanning the entire porosity range. Most foam glass data from three sources (74–76) (hence three

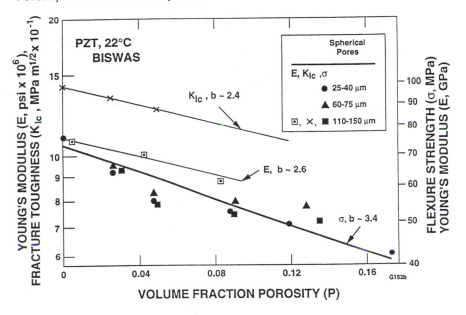

Figure 3.4 PZT (lead zirconate titanate) data of Biswas (72,73) with mostly approximately spherical pores (from burn out of organic beads) for Young's modules (E), fracture toughness (K_{IC}), and flexure strength (σ) versus P at 22°C.

different glass compositions, processing conditions, and pore variations, etc.) agree fairly well with the MSA model for simple cubic packing of uniform spherical pores (Figs. 3.1, 3.5). The primary deviation is at P ~ 0.9, where the data generally follows the MSA model for simple cubic packing of cylindrical pores aligned with the stress direction. This transition from agreement with spherical to cylindrical pore models is consistent with a transition in pore shapes that is often expected, as noted earlier. Thus, as foam cells become quite open at high porosity the cell structure may become better approximated by cubical instead of spherical pores (Fig. 2.1, or approximately rectangular pores if there is anisotropy). Such cubical cells more logically result from the intersection of (hemispherically capped) cylindrical pores (e.g., of aspect ratios of ~ 1, and may or may not be in various orientations, and not result in anisotropy). The data of Hagiwara and Green (77) for three open-cell Al_2O_3 foams approximately follows the trends for MSA models for cubic stacking of spherical or cylindrical pores (aligned with the stress axis), but falls somewhat below them, i.e., relative E values of 0.1 to 0.0025, respectively at P ~ 0.82 and 0.92. At least some of this difference is probably due to the limited reduction in P_C values reflecting statistical variations of cell stacking and strut or wall character, which can thus be addressed in general and specific foam MSA models; however, at such extreme

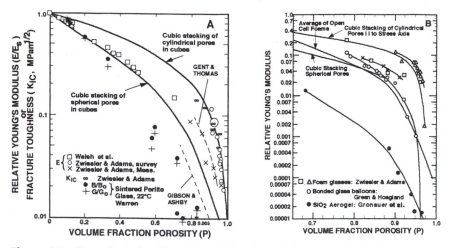

Figure 3.5 Plot of relative Young's modules (i.e., of the porous body, E, divided by that of the dense body, E_S) and of fracture toughness versus volume fraction porosity (P) for sintered glass with bubbles (71,72) and for foamed glasses (76). A) Full scale semilog plot of relative Young's modulus (and fracture toughness) and models. B) Plot of some of this modulus data at high porosity on an expanded porosity scale along with data for bonded glass balloons and a silica aerogel. See also Fig. 3.6.

P values, specific foam models may be more useful, with general MSA models being a guide to overall P dependence, i.e., connecting different models for specific pore structures and P ranges.

With regard to evaluation of specific foam models, Green (78) showed the glass foam data of Zwissler and Adams (76) was much closer to the model of Gent and Thomas (39) than of Gibson and Ashby (15,16) in both absolute value and the exponent of the 1 − P term, i.e., 1.2 versus 1 and 2, respectively (Fig. 3.6). Subsequently, Hagiwara and Green (77) evaluated E and G for three commercial open cell Al_2O_3 foams with nominal cell sizes of 0.25–1 mm and P ~ 0.75–0.92, showing these foams were much more consistent with the overall trend of Gibson and Ashby's model (Fig. 3.3B). Thus, the exponential for the term 1 − P (the slope on a log-log plot of E or G vs 1 − P) was 1.93 for E and 1.99 for G versus 2 for the model. However, the coefficients multiplying the $(1 − P)^2$ term were not ~ 1 and 0.4, respectively, (Eq. 4 of Table 3.1), but instead ~ 0.3 and 0.14, i.e., approximately a factor of 3 lower. They noted that neither the hollow nature of the cell struts nor some anisotropy of the degree of cell openness were the primary cause of this approximately threefold variation. Instead, they suggested it may be due to the distributed porosity in the struts, again showing the need for combined models. They also noted that: 1) the hollow struts made the foams more efficient on a stiffness/mass ratio basis, 2) Poisson's ratio was ~ 0.21 and nearly independent of P as predicted by Gibson

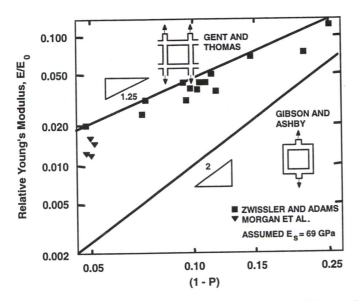

Figure 3.6 Comparison of models of Gent and Thomas (39) and of Gibson and Ashby (15,16) for the Young's modulus dependence of volume fraction porosity (P) for foamed glass [after Green (78)]. Note the much closer agreement with the former model in both slope (i.e., ~ 1.2 versus predictions of 1 and 2, respectively, i.e., closer to simple cubic stacking of the cells) and absolute value. See also Figs. 3.3 and 3.5. Published with permission of the J. Am. Cer. Soc.

and Ashby's model, but 3) there was a modest increase in the modulus with increasing cell size contrary to this model. Green and Lange (38) did not directly corroborate their model for E of rigidized glass felts (i.e., space shuttle tiles), but indirectly did so by showing that it predicted data trends for K_{IC} and tensile strength (Fig. 4.6).

Studies have been carried out on bodies of spherical glass beads poured into a container (i.e., randomly packed) and sintered. The results of Coronel et al. for Young's modulus, using beads nominally 100 ± 50 μm diameter (79), agree fairly well with the MSA model for simple cubic stacking of uniform spherical particles, i.e., with the range of data scatter (Fig. 3.7A). Subsequently Berge et al. (80) measured sonic velocities and B versus P for similar bodies sintered from larger (mainly 230 ± 20 μm diameter), more uniformly sized glass beads. Their data (Fig. 3.7B), which has less scatter, shows similar trends, i.e., on a semilog plot an initial nearly linear decreases, but with lower b values and then a rollover to faster decreases at P ~ 0.2–0.3. The lower b values for sonic velocities are approximately consistent with expectations, i.e., the resultant b value for the corresponding moduli being about twice plus 1 for that of the velocity. However, their b values are generally low, e.g., relative to those of Coronel et

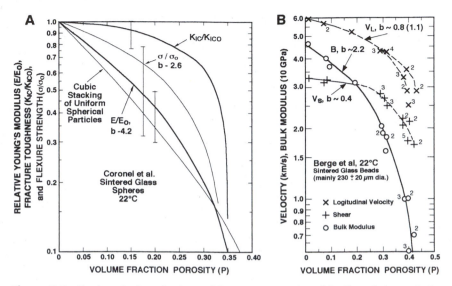

Figure 3.7 Sonic velocity, elastic, and fracture properties of bodies of sintered glass beads versus volume fraction porosity (P). A) Relative Young's modulus (E/E_0), fracture toughness (K_{IC}/K_{ICO}), and flexure strength (σ/σ_0) data of Coronel et al. (79). Note the greatest deviation of K_{IC} (but moderate range bar), and intermediate deviation of σ (but greatest range bar) from the trend for E. B) Data of Berge et al. (80) for sonic velocities and B versus P.

al. The differences between these two studies is probably due to lower density particle packing (i.e., lower C_n) in the latter case since Coronel et al. used finer beads with a broader size range and tapped the mold to increase the bead packing.

Studies have also been conducted on sintered silica gels (81–85), which typically entail very fine, fairly uniform spherical particles (~ 10 nm diameter) (81,82). Particularly note Ashkin et al.'s data (81,82) covering a more extensive range of porosity shows Young's modules following the MSA model for simple cubic (hence also random) packing of uniform spherical particles to P ~ 0.4, then transitions to close agreement with the MSA model for simple cubic stacking of uniform pores by P ~ 0.7 (Fig. 3.8). The good agreement with the spherical particle model at lower porosity is expected since there the bodies mainly reflect partial sintering of fine packed spherical particles. The transition to the model for cubically (hence also randomly) stacked spherical pores is also reasonable since as the porosity increases the pores become defined more and more by longer and longer chains of small spherical particles. Then a transition to a cylindrical pore model is indicated, as expected and noted earlier. This data is an important example of the need to address changes of pore character with P and thus the need for models for mixed porosity.

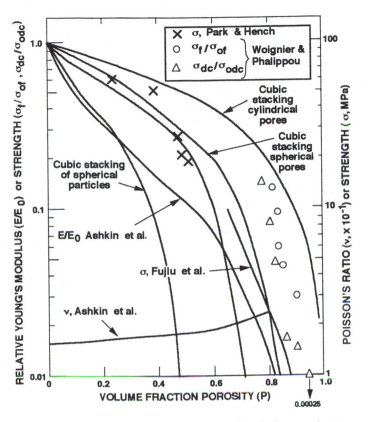

Figure 3.8   Plot of absolute or relative mechanical properties versus porosity (P) for bodies made from partially sintered SiO_2 gels (81–85). These bodies reflect transitions from stacked sintering particles to arrays of bubble-type pores as shown by the various models.

Krasulin et al. (86) investigated elastic (and other mechanical) properties of bodies of ZrO_2 beads of different average diameters (D) partially sintered at contact points (giving P ~ 0.3–0.4; Fig. 3.9A). Variations of bead dimensions and stacking prevent relating average bead size and specific P levels and thus properties with MSA models; however, note the higher levels but much greater decreases of E (and strength) in compression versus in tension as D increases (along with some increase in P). This is another example of significant differences that can occur between compressive and tensile loading along with well-established behavior of rocks with microcracks due to their closure under compression (see Section 3.3.4). It is possible that a similar effect is occurring with the beads, i.e., that the smaller gaps between smaller balloons just past the sintered necks can be closed more by compressive stresses, but more research is

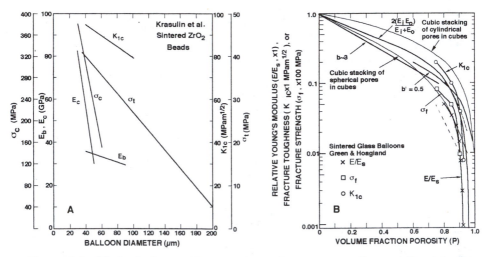

Figure 3.9 Mechanical properties versus porosity parameters for partially sintered bodies of ceramic beads or glass balloons. A) Properties for bodies from ZrO$_2$ (plasma spherodized) beads versus average balloon diameter after Krasulin et al. (86). B) Properties for bodies from glass balloons versus P after Green (87) and comparison to models for stacked bubbles. E = Young's modulus at any P (subscripts s = solid; I and o, respectfully, are values due to porosity inside and outside of the balloons; curve marked 2 $(E_I E_o)(E_I + E_o)^{-1}$ shows combining these two porosities in parallel, that marked b'= 0.5 refers to Rice's Eq. (9) for porosity dependence of $1 = e^{-b'(1-P)}$.

needed on such stress effects. Note that fracture of both bead-based bodies was through the sintered necks consistent with MSA model concepts.

Another clear example of the need to handle mixed porosity is in bodies of partially sintered balloons, where the porosity consists of the spherical pores in the balloons as well as the pores between the packed balloons. The balance of these two pore populations varies with the balloon size and wall thickness, and the degree of their sintering (and the amount and sintering of any filler powder between the balloons). Green's (87) data for Young's modulus for partially sintered glass balloons, where much of the (0–38%) porosity at higher P was due to the pores within the balloons (average outside diameter and wall thickness were respectively 36 ± 11 and ~ 1 μm) agrees reasonably well with the MSA model for cubically stacked spherical pores (Fig. 3.9B). Attempts to account for the porosity between the balloons with that in the balloons by combining models for cubically stacked pores and particle were considered with encouraging results, but detailed analysis was hindered by the variation in balloon dimensions and breakage (6). Similar levels of agreements were obtained with two foam cell models (6,15,16). The investigators noted that the data at higher P was due to the properties being dominated by the small sintered contacts (i.e., their minimum solid area), and derived a mechanistic-based model for such partially bonded balloons (88).

3.3.2 Data for General Porosity

A key test of the scope of utility of any model is its applicability to the range of porosity encountered in typical bodies rather than just those with more specialized or idealized pores considered above. Unfortunately, almost no characterization of porosity, other than P, is given in most studies of porosity effects; however, Rice (7) proposed that the generally expected trends of pore character with different fabrication (i.e., consolidation-forming) processes could be used as a guide (as outlined in Section 1.3 and Table 3.6), which gives the corresponding trends in b and P_C values. While most processes produce a range of porosity, one or two particular pore shapes are often dominant over part or much of the range of fabrication parameters. Broader particle size or agglomerate distributions, and more or harder agglomerates will commonly reduce b values and increase P_C values due to particle and agglomerate bridging and more pores around and between agglomerates (which can also be a source of laminar pores; Fig. 1.9). P_C values will commonly be ~ 1 minus the relative green density for most of the above bodies when the anisotropy of pore structures is limited, i.e., when laminar or tubular pores are not a significant fraction of P. Tighter

Table 3.6 Summary of Pore Shape Trends with Forming Consolidation Processes[a]

Pore character	Forming-consolidation process[b]	Representative ranges[c] b	P_C values
Tubular, mixed orientation	Injection molding	1–3	0.4–0.8
Tubular (or platelet), oriented	Extrusion (tape lamination, die pressing)[d]	1–2 (∥) ~3 (⊥)	~1 0.5–0.8
Pores between particles of varying packing	Colloidal processing; die, iso, and hot pressing, HIPing[e]	4–9	0.3–0.5

[a] See Section 1.3.

[b] All processes yield a mixture of these, other pores, or both; processes shown here commonly produce measurable to substantial pores of the character listed over some (e.g., as listed below) or substantial ranges of their parameters.

[c] Values shown are for elastic moduli, most or all the pores being of the listed character, and for the orientation of the stress or flux relative to the pore axis or plane for tubular and laminar pores. Note: higher P_C values correspond to lower b values; b values for wave velocities will be approximately those shown −1, then divided by 2 (61,62); and ∥ = stress and pore axes aligned, ⊥ = stress and pore axes perpendicular to each other.

[d] Laminar pores occur primarily at higher pressure and greater ram travel and die-wall friction for die pressing, and can occur in other pressing and colloidal consolidation (Fig. 1.12).

[e] Though there is variability in the position and overlap of the processes, the degree of consolidation, i.e., average particle coordination number, C_n, and hence b values increase in the order listed (and P_C values decrease).

Table 3.7 Summary of Ceramic Property Porosity Dependence for Different Processing[a]

Property	Extruded	Cold (die) pressed	"b" Values Iso- pressed	Colloidal	Hot- pressed
Young's modulus (E)	2.3 (1)	3.6 ± 0.8 (25)	$4 \pm 0 \pm 1.2$ (14)	3.9 ± 1.2 (4)	4.5 ± 0.7 (31)
Shear modulus (G)	2.2 (1)	3.2 ± 0.9 (8)	3.9 ± 1.3 (12)		4.3 ± 0.6 (17)
Bulk modulus		4.0 (1)			5.1 ± 0.8 (8)
Poisson's ratio		0.5 ± 0.5 (3)	0.5 ± 0.3 (3)	0 (1)	0.4 ± 0.6 (2)
Tensile strength (σ)	5.1 ± 2.1 (3)	4.0 ± 1.1 (3)	7 (1)	4 (1)	4.5 ± 1.2 (15)
Hardness (H)		4.1 ± 1.1 (5)			5.8 ± 1.4 (8)
Compressive strength					5.8 ± 1.5 (3)
Thermal conductivity	2.5 ± 1 (6)				4.4 ± 0.3 (3)
Electrical conductivity	1.9 ± 1 (4)	2.9 (1)			

Source: ref. (7); published with permission of J. Mat. Sci.

[a] Values shown are averages and ± 1 standard deviation where more than one value was available. Values in ( ) = number of studies averaged.

packing of particles, e.g., via higher pressure, increases b and decreases P_C, though the former increase can be limited or reversed by forming laminar pores (which also introduce anisotropic properties).

Results of the most extensive survey of the porosity dependence of ceramic properties (Table 3.7) (7) corroborate the expected trends from the models and as a function of fabrication–pore character as outlined above. Consider the elastic properties here. First, the average b value from about 85 different studies was ~ 4, as would be expected from averaging values for models of different pore character and orientation (Tables 3.1, 3.2; Figs. 2.11, 3.1). Second, trends for different densification-forming processes are consistent with the trends expected as a function of pore character. Thus extrusion clearly gives the lowest b values for moduli measured parallel with the extrusion axis, and hot pressing gives the highest b values, with die and isopressing and colloidal processing having intermediate values as expected. Further, all values for bodies whose pores are primarily those between packed powder particles have higher b values than for bodies with larger pores, e.g., from bubbles (Figs. 3.1, 3.3–3.5), except for extruded bodies. All of these results are as predicted by MSA models.

These results are strongly reinforced and extended by examination of pertinent individual studies. Thus, a compilation of relative Young's modules versus porosity curves for colloidally processed (89,90) and sintered (89) SiO_2, Al_2O_3 die-pressed at various pressures (91,92), and die pressed, as well as extruded, carbon bodies (94–96) and $Yba_2Cu_3O_{7-x}$ superconductors (97–104) (Figs. 3.10, 3.11) very clearly support the above b trends for different processing. Thus, note the progression from higher b and lower P_C values to lower b and

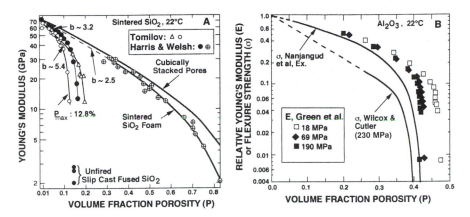

Figure 3.10 Young's modules versus volume fraction porosity (P) for: A) sintered fused quartz at 22°C [data of Harris and Welsh (89) and Tomilov (90)] made via slip casting and by prior foaming (and crushing). B) Die-pressed and sintered Al_2O_3 [data of Green et al. (91)] for differing (listed) compaction pressures, with P_C values decreasing with increasing pressure as expected. Note the similar, consistent trend of Al_2O_3 flexure strength data of Wilcox and Cutler (92).

higher P_C values on progressing from colloidally processed SiO_2 to die-pressed Al_2O_3 and carbon to sintered foam SiO_2, and finally to extruded carbon and superconductor (measured parallel to the extrusion axis). These studies also illustrate effects of changing parameters within a given process, i.e., the shifts for different colloidal processing and die-pressing pressures (i.e., increasing b and decreasing P_C values as pressing pressures, and hence particle packing, increases, as expected). The expected progression from die- to isopressing indicated in Table 3.8 is further supported by the fact that the b values for bodies that were only isopressed were higher than for those that were first die-then isopressed. Thus, die-pressing commonly leaves some remnant (possibly laminar) pore character that is not totally removed by subsequent isopressing and thus possibly limiting properties. Finally, note: 1) the similar trends for B and G for carbon [also shown by Green et al.'s (91) data for Al_2O_3], and that data for metals agree with the expected MSA models (briefly noted earlier and discussed in more detail in Section 8.3.1).

The MSA and some mechanistic models are not sufficiently precise to directly give the porosity dependence of v since they give the dependence of $E/E_0 \sim B/B_0 \sim G/G_0$; however, as shown below, further analysis provides guidance in refining the porosity dependence of E, B, and G, and thus also v. The P-dependence of v, as compiled by Boccaccini (52) (Fig. 3.12) generally shows v having little substantial decrease with increasing P, i.e., with b values varying from 0 to about 4 [consistent with Rice's earlier survey (33) giving b values

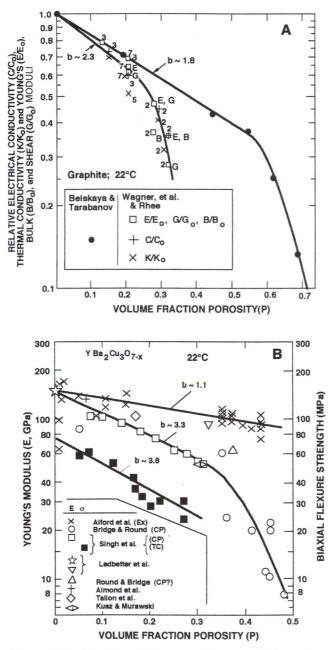

Figure 3.11 Elastic and other properties versus P for carbon and an oxide superconductor for various fabrications. A) Relative Young's, shear, and bulk moduli along with relative electrical and thermal conductivities of various graphite and carbon bodies at

generally between 0 and 2.5, Fig. 3.13B]. The only probable deviation from this was an indicated increase of v with increasing P, primarily at P > 0.7 from two overlapping studies of sol-gel derived SiO_2 (81–83), with v approximately independent of P to P ~ 0.5, and clearly increasing with P from P~ 0.65–0.9. This is discussed further in Section 3.4.

Modulus and velocity data from ultrasonic (mostly pulse echo) tests versus P are also reasonably consistent with expectations for b values (as discussed in Section 3.2.2). Thus, while there is no porosity characterization and often very limited data in several of the data sets surveyed by Roth et al. (105,106), evaluation of the four data sets having more data indicate b values for longitudinal waves of ~ 1, 1, 2, and 3, which would be consistent with b values for E of 3, 5, and 7. Earlier data for Al_2O_3 of Nagarajan (107) gave b values of ~ 0.8 and 0.7, respectively, for shear and longitudinal velocities indicating b values of 2.6 and 2.4 for the respective moduli. More recent b values respectively for shear and longitudinal velocities were 0.9 and 1.4, respectively, and ~ (4.0) 2.9, 2.8, and 1.2 for (B) E, G, and v, for die pressed ZnO (108) [and for die-pressed steel (109) 3.8, 3.5, and 0.85 and HIPed/ hot-pressed alpha titanium aluminide (110) ~2 and ~3, and ~ 3-5 and ~ 8; see also Section 8.3.1]. While the inherently lower nonlinearity of the velocity versus P data results in a greater extent of nearly linear behavior, this is less so for the moduli, e.g., as shown by the Al_2O_3 (107). Recent high-frequency evaluation of SiC (111) has shown the feasibility of differentiating areas of some differences in porosity or pore size. Again, as noted in Section 1.4, lower frequency measurements (e.g., < 10 KHz) in graphite give increasingly lower moduli as porosity increases above a threshold level (112). Whether such effects are unique to such highly anisotropic materials with weak (Van der Walls) bonding in one (the C) direction, or whether this simply effects the frequency and porosity levels where this occurs is not known.

3.3.3 Comparison with Other Models

Developers of various models (Tables 3.1–3.3) have generally evaluated them against data, but usually with a limited number of cases (e.g., 2–12), with no

---

22°C. Note the higher slope and lower indicated P_C value for typical pressed bodies (93–95) versus the lower slope (b ~ 1.8) and greater P_C value (> 0.7) for the extruded material of Belskaga and Tarabanov (96), the overall agreement of the various properties, and scatter probably reflecting porosity heterogeneity. B)Young's modulus and flexure strength data for $YBa_2Cu_3O_{7-x}$ superconductors (97–104) made by extrusion and die pressing. Note the distinctly lower Young's modulus decrease as P increases for extruded bodies (measured parallel with the extrusion axis) versus die-pressed material. Also note the similar trend for E and flexure strength for similarly processed material (97). Published with permission of J. Mat. Sci. (7).

Table 3.8 Fit of Various Property-Porosity Equations[a]

Property[b]	Comparison criteria[c]	1: $1 - AP$	2: e^{-bP}	3: $\dfrac{1-P}{1+AP}$	4: $1 - AP^n$	5: $(1-AP)^n$	6: $1 - A_1P + A_2P^2$	7: $e^{(-b_1P - b_2P^2)}$	Material	Investigator
E	Coefficient of variation	2.9	6	5	2.7				16 data sets for 11 oxides	Dean & Lopez (113)
G	Coefficient of variation	2	4.2	4.1	2.7				11 data sets for 8 oxides	
E	% Error	1.9	1.4	2.2					Dy_2O_3	Manning, et al. (114)
		2.2	2.3	2.4					Er_2O_3	
		1.9	1.9	2.0					Ho_2O_3	
		1.2	2.0	2.2					Y_2O_3	
G	% Error	1.8	1.5	1.6					Dy_2O_3	Manning, et al. (114)
		1.9	2.1	2.1					Er_2O_3	
		1.8	2.0	2.4					Ho_2O_3	
		1.6	2.0	2.1					Y_2O_3	
E	Rank	2	3	4			1		Al_2O_3	De Portu & Vincenzini (115)
E	Standard error	17.9	26.6	7100		12.2		13.7	Al_2O_3	Phani & Niyogi (116)
		21.3	15.2	14.8		11.8		10.9		
		10.6	6.8	12.7		1.8		—		
		11.1	4.2	65.5		4.6		4.8		
E	Standard error	0.75	0.61	0.65					β-Al_2O_3	Petrak et al. (117)
		0.39	0.30	0.31					CoO	
		0.14	0.15	0.17					CoO-MgO	
		0.28	0.32	0.38					$CoAl_2O_3$	
									$MgAl_2O_3$	

	Phani (118) ThO_2	Boocock et al. (119) UO_2	Phani & Niyogi (61), (116) UO_2	Dutta et al. (120) Si_3N_4	Salak et al. (121) Fe
E Standard error			5.18	5.16	
G Standard error			1.46	1.44	
E RMS	7	8.5	8.5		
G RMS		3.3			
K RMS		13			
E Correlation coefficient	0.977	0.9	0.998		
E X^2	563	521	3949	515	516
σ Rank	5	1	2	3	4

Source: ref. (7); published with permission of J. Mat. Sci.

[a] P = volume fraction porosity, b, b_1, b_2, and m = parameters.

[b] E = Young's modulus, G = shear modulus, K = bulk modulus, σ = tensile strength.

[c] RMS = root mean square; rank: 1 = best, 2 = next, etc.

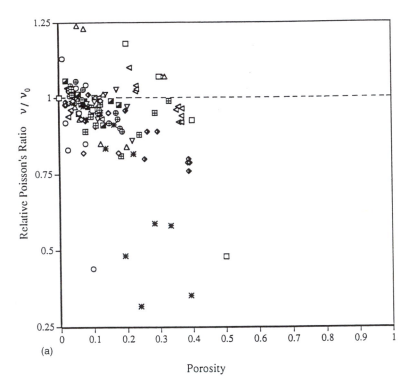

(a)

Porosity

Figure 3.12 Compilation of the dependence of Poisson's ratio (ν) versus P at 22°C after Boccaccini (51). (a) Data for materials ($\square$, Al_2O_3; $\diamond$, HfO_2; $\square$, Lu_2O_3; O, ZrO_2; $\triangledown$, Y_2O_3; $\triangle$, $MgAl_2O_4$; $\square$, Sm_2O_3; $\triangleleft$, Gd_2O_3; $\oplus$, Dy_2O_3; $\square$, HoO_3; $\ast$, ThO_2) with $\nu_0 > 0.25$. (b) Data for materials ($\square$, Al_2O_3; $\triangledown$, SiO_2; $\diamond$, MgO; $\square$, AlN; $\triangle$, SiO_2 from sol-gel processing; O, graphite; $\blacklozenge$, MgO; $\square$, Al_2O_3; $\oplus$, MgO) with $\nu_0 < 0.25$. With permission of Boccaccini and the J. Am. Cer. Soc.

basis given for the selection and no attempt to characterize the porosities involved. L. F. Nielsen tested his model against two sets of data for cement materials and one data compilation each for Al_2O_3 and MgO (21). While he showed good agreement in this limited evaluation, no discussion or evaluation was given of the pore-shape factors, ξ, used. Similarly L. E. Nielsen (24) tested his models against two limited data sets for polymer-based composites. Waugh et al. (42) fitted their equation to 13 data sets (with one not fitting well) using their exponent simply as a fitting parameter. In Dean's (47) application of composite models for spheroidal inclusions to porosity, he noted that resultant solutions of the coupled equations indicated Young's modulus was very close to being linearly dependant on porosity, while the shear modulus was slightly and the bulk modulus substantially curved downward when plotted versus

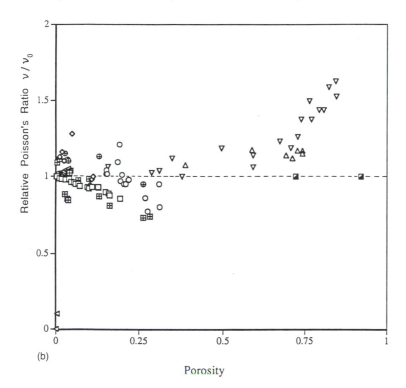

(b)

Porosity

P. Evaluating seven data sets using the aspect ratio of the spheroidal axes as an adjustable parameter gave good fits in six cases and one a poor fit for Young's and shear moduli, but two of the six cases gave poor fits for ν that were similar to but not as extreme as in the case of the poor fit for Young's and shear moduli. No attempts were made to justify the aspect ratios used in terms of known or expected pore character.

Besides the initial evaluations by the developers of the models discussed above, there have been other, sometimes more extensive, evaluations (113–121) (Table 3.8). Though individually often of limited or uncertain accuracy due to fitting and data quality issues (discussed later) these show: 1) MSA models (columns 1, 2, 4, especially column 2) and some mechanistic models (columns 1, 4, 5) are reasonable to good fits to data, 2) Hasselman's adaptation of Hashin's model (column 3) was sometimes a very poor fit, and 3) mostly empirical relations with more fitting parameters (columns 6, 7) offered no significant advantage. [Note that Wang's (31) modification of the exponential relation, column 7, while giving good fits in its limited tests, does not give outstanding results.] It should be noted that even more extensive analyses such as that of Dean and Lopez (113) must still be used with caution since issues of systematic errors, e.g., due to inhomogeneous porosity, are neglected. Thus, data for ThO_2 (122) and MgO (123) that had substantial impact on the agreement for fitting of some models was used without noting their deviations. These were inhomogeneous porosity with increasing P and varying, opposite trends of v with P (indicating possible porosity inhomogeneity, varying anisotropy, or both, Figs. 3.15, 3.17; Table 3.8). Also note that the largest deviation for fitting of e^{-bP} in Phani and Niyogis' evaluation (116) was for Wang's alumina data (31), which clearly presents fitting problems, not recognized in their evaluation (Fig. 3.18).

These evaluations have typically directly or indirectly used porosity coefficients or exponents, e.g., A and n, as adjustable parameters for data fitting and neglected the porosity level and shape that they were derived for (and do not address porosity heterogeneity). Similarly, other parameters that are in principle available but not obtained from data, e.g., P_C and ξ values in Nielsen's equation, have also been used as fitting parameters, as has v by Ramakrishnan and Arunachalam (48,49). Recent evaluation of the fits of some models showed that:

1) Limited data for aligned cylindrical pores stressed parallel to their axes versus $1 - P$ on log-log plots gave the slope (n) = 1, and data for bodies with at least nominally spherical pores gave n ~ 2.

2) Evaluations of the data sets used in evaluating their model by Ramakrishnan and Arunachalam (48,49) showed the average b values were 3.6 ± 0.8 for E and 3.3 ± 1.1 for G.

Since b for spherical pores is about 3, the latter shows that the model evaluated was reasonably consistent for spherical porosity for which it was derived, as well as the corresponding MSA model.

Rice (33) simultaneously compared mechanistic and MSA models with average b value data for elastic properties. Most of the models fall within the range of the average $b \pm 1$ standard deviation (i.e., between b = 2–4 for E/E_0; Fig. 3.13A). The deviations toward greater porosity decreases as P increased were: 1) Hashin's equation (Eq. 8 of Table 3.1) for high concentrations of spherical pores below P ~ 0.4, where it is not valid (above this it begins to approach data the MSA model for spherical pores), and 2) Dewey's, Hashin's (for low-concentration spherical pores and for aligned cylindrical pores stressed normal to their axes), and Budiansky's equations for pores between particles (using $v = v_0 (1 - P)$; equation 7 of Table 3.1), as well as Mackenzie's equation for P > 0.4 to 0.5. The latter deviations below the average b value may often be physically correct since properties commonly decrease toward their P_C value by or before this P range. This is clearly the case for Hashin's equation for aligned cylindrical pores stressed normal to their axes, which agrees fairly well with the corresponding MSA model (Fig. 3.2A). While this greater decrease at P > 0.4 to 0.5 for Budiansky's equation is also consistent with data and MSA models, it is

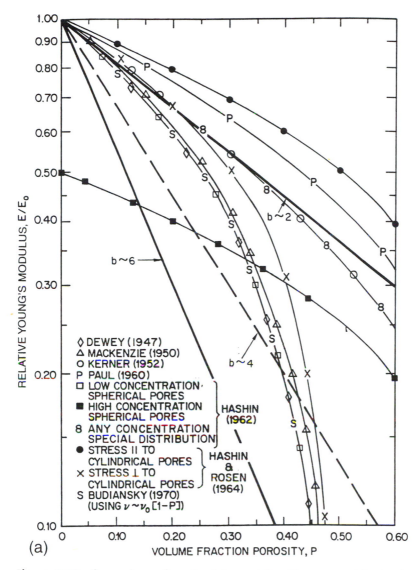

(a)

Figure 3.13 Comparison of mechanistic models with average data trends for (a) Young's modulus and (b) Poisson's ratio in terms of their b values, and hence to MSA models at 22°C (33). Dashed lines show the average b value and solid lines without symbols on either side show ±1 standard deviation. Published with permission of Academic Press (33).

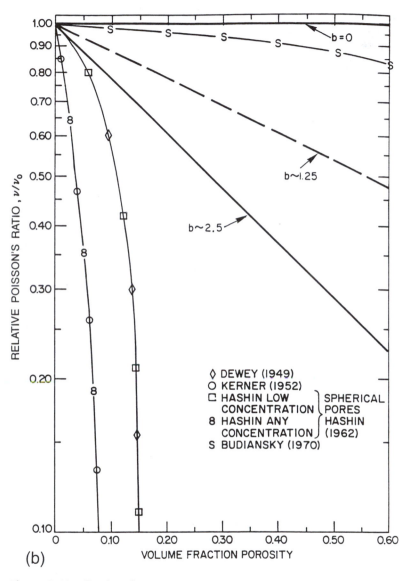

(b)

Figure 3.13 Continued

inconsistent for the spherical pore models of Dewey, Hashin, and Mackenzie, i.e., compare their curves, data in Fig. 3.13A, Table 3.7, and MSA models (Fig. 3.1); however, they are consistent with data and MSA models for P < 0.4. Models deviating toward less property decrease with increasing P are (in order of the least to greater decrease with increasing P): Hashin's model for aligned cylindrical pores stressed parallel to their axes (which is identical to the corresponding MSA model, and agrees with data), and the equations of Paul, Kerner, and Hashin (for any concentration of spherical pores). While these equations agree with data trends (and MSA models) for spherical pores at higher P, they are not particularly good fits at lower P levels. This is consistent with Hasselman's modification of Hashin's equation, which sometimes gives poor fits to data.

Comparisons of the lesser amounts of data for shear and bulk moduli overall show similar trends, but with some shifts (33). In particular Kerner's and Hashin's models, respectively, for (nominally) spherical and any concentration of spherical pores fall along the models for aligned cylindrical pores stressed parallel to their axes for shear modules, i.e., showing substantially less decrease of properties with P than found for the type of porosity for which they were derived; this seriously questions their utility. Comparison of the still smaller data base for the porosity dependence of Poission's ratio predicted from the mechanics-based models (with reasonable closed-form solutions) is summarized in Fig. 3.13B. Kerner's and Hashin's models, respectively, for (nominally) spherical and any concentration of spherical pores show precipitously high decreases in v. Dewey's and Hashin's models for limited concentrations of spherical pores show similarly rapid decreases with P. Only Budiansky's model was anywhere in the range of data. (Note that Budiansky's model gives a quadratic equation for v, having one real root.) Hasselman's adaptation of Hashin's equation (Eq. 9 of Table 3.1), while often being a competitive fit to data, also resulted in by far the worst fits in some cases, thus again indicating caution in its use.

There are thus cases where mechanics-based and MSA models are in approximate agreement with each other (and data), but there are also cases where they seriously disagree. In such cases comparison with data often favors MSA models, e.g., consider Rossi's model (43) based on Hashin's model and stress concentrations for aligned spheroids compared to the corresponding MSA model (126), as shown in (Fig. 3.14). Comparing the two models (having different P dependences) over P = 0.1 to 0.4 shows that the models agree exactly for all P values at the extreme of prolate spheroids, i.e., cylindrical pores (which, as noted earlier, have no stress concentration for stressing parallel to their length, as is the case here); however, as the spheroid shape changes to spheres, then to oblate spheroids of increasing eccentricity, Rossi's stress concentration-porosity coefficient goes to infinity, and hence zero properties at the extreme of oblate spheroids, i.e., cracks. Since elastic properties do not go to zero when some fine cracks are present (Table 3.4), this result appears incorrect.

Thus in summary, several mechanics-based models can be used with limited error at low P, i.e., they are generally similar to MSA models there; however,

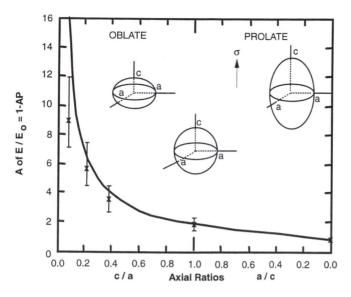

Figure 3.14 Comparison of Rossi's mechanistic model (based on stress concentrations, giving A ~ 5a/4c + 3/4, solid line, for v_0) (43) for Young's modulus of porous materials with MSA models for ordered spheroidal pores (126). Note that since the MSA model equation varies as $P^{2/3}$ (Table 2.4) and Rossi's as P, they must be compared for specific P values (here for P = 0.1 to 0.4, as shown by the range bars about the average of these values marked by the **X**'s). Published with permission of J. Mat. Sci. (126).

this is generally due more to properties not changing very rapidly with increasing P there, rather than to merits of the models. (A clear exception is Hashin's model for high concentration of spherical pores, but it is obviously not valid there for reasons based on its derivation.) At moderate and higher porosity the utility of many mechanics based models is more limited. There, the primary mechanics-based models that show good agreement with data are the models of Hashin for aligned cylindrical pores (stressed either parallel or perpendicular to the pore axis), and Budiansky's model, nominally for pores between particles; however note that the fit of the latter depends on the porosity dependence used for v, and appears most applicable for random packing of particles, which is an important case but not the complete picture for such bodies. Thus, the mechanics-based models in some cases agree with data and MSA models, but often fall well away from either.

Next consider the limited evaluations of finite element and other computer modeling of elastic properties as a function of porosity. Agarwal et al.'s (64) three-dimensional analysis of the effects of isolated spherical pores gives E and v decreasing with P to P ~ 0.45, i.e., to nearly the point of open porosity (P ~ 0.52). E decreases with increasing P similar to the MSA model for cubic stacking of spherical pores to P ~ 0.3 (i.e., an average b value of ~ 3), but then progressively deviates above this MSA model (i.e., to lower average b values) as P further increases (where the treatment as isolated, i.e., noninteracting, pores is less and less valid). Their E-P dependence also thus deviated to less reduction with increasing P in comparison with the limited data they compared their results to, consistent with their b values generally being greater than or equal to 3. They showed v decreasing from a value of 0.55 at P = 0 with increasing P (e.g., with an average b ~ 0.8). Their results thus corroborate MSA results over the lower P range over which their modeling is valid. Two-dimensional analysis of bodies with aligned tubular pores in various stackings for their elastic response in planes transverse to the pore axes also corroborates the E-P dependence of MSA models for such pores, as noted earlier. This modeling, based on a merging of atomistic and continuum methods, shows E/E_0 to be independent of the initial v, but complex porosity dependence of v. Thus, v decreased to a fixed value at P = 0.5 to 0.6, or initially decreased some with increasing P but then increased to a value of 1 at P = 0.9, with the differences being due to the P = 0 value of v, and especially the stacking of the cylindrical pores. This complex behavior must be a result of the substantial anisotropy of the mechanical behavior of such bodies with aligned, highly anisotropically shaped pores, as discussed below. The very limited number of pores in the analysis may also be a factor. Ramakrishnan and Arunachalam (49) analyzed elastic behavior of sheets with circular or irregular holes via two-dimensional finite element techniques to justify variations in v for better fitting of their modified elastic property-porosity model for spherical, i.e., three-dimensional pores. They showed substantial variation in the v-P dependence, i.e., being constant at ~ 0.25[*], or converging on this value as P → 0.5 or 1.0). They neglected: 1) Agarwal et al.'s results for spherical pores (64) (from three-dimensional analysis noted above) being contrary to their (two-dimensional), and other cited results, and 2) the probable significant effects of anisotropic behavior (due to the aligned tubular pores their two-dimensional analysis represents) has on the porosity dependence of v. For example, linear plots of E versus 1 − P for stress normal to the aligned axes of cylindrical (i.e., two-dimensional) pores in random stacking show distinct changes in slopes at P ~ 0.25. which must complicate v-P relations. Thus, at the minimum substantial caution must be exercised in application of two-dimensional results to other pore systems, especially isotropic ones. They also cited v variations for various MgO bodies (Fig.3.15) to support their proposed P-dependence of v varying with the level of v_0, neglecting the cited v variations possibly being due to: 1) porosity differences in different bodies, 2) inconsistencies, e.g., variations in elastic property data used to calculate v, and 3) possible varying

---

[*] A survey of polymer foam data (15,16) shows v approximately constant at about 0.33 to at least P, ~ 0.05 (but possibly decreasing at higher P, Fig. 8.4).

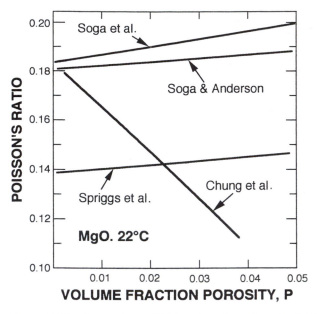

Figure 3.15 Comparison of Poisson's ratio values for MgO at 22°C after Soga and Schriber (123). Note the significant variations in the results for theses samples, most, or all, of which were hot pressed. Published with permission of J. Mat. Sci. (7).

processing induced anisotropy in the different MgO bodies. Any of these individual or combined factors could be significant since most or all of the MgO in their evaluations was hot pressed, probably under different conditions, and hence quite possibly of varying degrees of anisotropy. Thus, in summary, since elastic anisotropy, e.g., from anisotropicly shaped pores, significantly alters the P-dependence of ν, caution must be exercised in evaluating data to discern generic ν-P behavior.

Finally, consider proposals that elastic moduli (and fracture energy and toughness and to some extent strength) are linear versus $1 - P/P_C$, as noted in Section 3.1. Background to this proposal is given in Section 3.4, so only results are given here. Lam et al. (1) reported mechanical properties of their pressure cast alumina bodies were approximately linear versus $1 - P/P_C$ ($P_C \sim 0.5$–0.62; Table 4.2), and Nanjangud et al. (124) reported the Young's modulus of a commercial extruded alumina also being approximately linear versus $1 - P/P_C$ ($P_C \sim 0.42$–0.46), especially for $P \sim 0$–0.25, i.e., away from P_C. Evaluation of Young's modulus data for sintered glass beads of Coronel et al. (79) (Fig. 3.7A) and of Nanjangud and Green (125) (P: 0.36–0.17, $P_C \sim 0.36$) shows both also being approximately linear versus $1 - P/P_C$ (also indicated for biaxial flexure strength in the latter case). They also cited the Si_3N_4 data of Arato et al. (36) for

Young's modulus and other mechanical properties which were all ~ linear versus $1 - P$. While this does not guarantee linearity versus $1 - P/P_C$, their data is approximately linear versus $1 - P/P_C$; however, besides the deviations noted above for the extruded alumina, limited evaluation shows deviations are common and often large. Thus, for example, the bulk modulus data of Fig 3.7B deviates considerably, and the strength and fracture toughness data of Fig. 3.7A deviates, respectively, some and greatly from a linear dependence on $1 - P/P_C$. Further, as shown in Fig. 2.12, most MSA model curves, which have been shown to be consistent with a large amount of data, are not normalized to a linear trend versus $1 - P/P_C$. All except the model for tubular pores stressed parallel with their axes, which is intrinsically linear versus $1 - P$, fall below the $1 - P/P_C$ trend; however, some of the more extreme deviations noted above are well above the linear $1 - P/P_C$ trend. There are thus a variety of issues in the scope of the application of this P-dependence (discussed in Section 3.4).

3.3.4 Models for Effects of Microcracks on Elastic Properties

Data on the microcrack dependence of elastic (and other) properties are limited by the challenges of determining the presence of such cracks and their parameters, i.e., their numbers, size, shape, and stacking. Some microcracks can frequently be seen by higher magnification microscopy (e.g., ref. 127), but accurate quatification is difficult at best; however, there are four methods of demonstrating their presence and effects. The first two are based on the fact that much microcracking occurs as a result of stresses built up between grains from cooling as a result of strain mismatches from thermal expansion anisotropy (TEA) between grains or phase transformation of grains. Whether these cause cracking is critically dependent on grain size (and physical properties such as E and TEA; see eqn. 10.1). Thus the first method is comparing bodies below and above the threshold grain sizes for microcracking to provide direct comparison of the effect of such cracks. A key example of this is Manning's (128,129) study of hot-pressed and sintered Nb_2O_5 showing respectively the normal, substantial decreases of E and G with increasing P, and almost no dependence on P, despite overlapping substantially in P levels (Fig. 3.16A). This difference is attributed mainly or exclusively to finer grain sizes being below the cracking threshold in the hot-pressed material and above it in the sintered material (e.g., ~ 8 versus ~ 20 μm). (Note the consistency of the E- and G-P data for hot-pressed bodies is consistent with MSA expectations for hot pressed materials, i.e., b values of ~ 4.7 for E and slightly less for G.) The role of grain size in microcracking is corroborated by substantial studies of other materials, e.g., HfO_2 (130) (discussed further below) and pseudobrookite oxides (127). The latter corroborate the differences between sintered and hot-pressed Nb_2O_5, i.e., showing E rapidly decreasing 4–5 fold to an approximately asymptotic limit as grain size progressively increases above the threshold for microcracking.

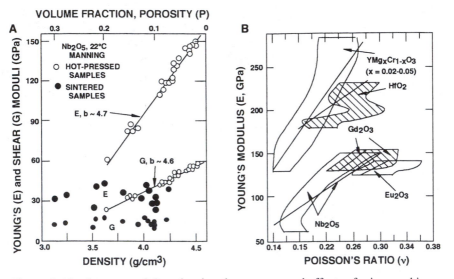

Figure 3.16 Summary of data showing the presence and effects of microcracking on elastic properties. A) Summary of Manning's (128,129) E-, G-P data for hot-pressed and sintered Nb_2O_5 showing: 1) the high moduli of the former decreasing rapidly with increasing P (as per MSA models) and no microcracks, and 2) low moduli, with limited P-dependence for sintered bodies with microcracks (as shown by other tests and observations). Published with permission of W. R. Manning and J. Am. Cer. Soc. B) Summary of the experimental data for E and v of ceramics with known microcracking (due to properties and grain sizes) showing E and v changes correlating from Case's (66) analysis. Bounds of the v variations are shown along with the trend lines for Budiansky and O'Connels' model (67).

The second method of demonstrating the presence and effects of microcracking due to TEA is to follow properties as a function of temperature since such microcracks close with increasing temperature and open or reform with decreasing temperature (exhibiting hysteresis). As discussed further in Chapters 9 and 10, such temperature cycling clearly shows properties such as elastic moduli increasing with increasing temperature until the cracks close up, where the normal decrease with increasing temperature takes over, and vice versa with decreasing temperature. Such opposite changes with temperature (and associated hysterisis) were shown for the sintered but not hot-pressed Nb_2O_5 by Manning (128,129) (Fig. 9.3). Such effects of thermal cycling above the critical grain size for microcracking are again corroborated by considerable study of other material such as HfO_2 (130), which also illustrate two complications applicable to many materials. First, microcracking was shown to occur (via reduction of Young's modulus) under the influence of moisture, i.e., showing that the microcracking, although driven by TEA stresses, was also environmentally dependent. Thus, microcracking did not occur in vacuum, and was time dependent in air, e.g., taking ~ 60 hours to saturate in specimens exposed to air after vacuum annealing to increase the grain size above the threshold (> ~ 2 μm). Second, as the HfO_2 grain size was further increased above the threshold size, a condition termed "unstable microcracking" was observed to emerge. Such microcracking showed much less closure or disappearance of the microcracks and their reductions in moduli in tests at higher temperatures. This reduction in the recovery of moduli with increased measurement temperatures implies that the nature of the microcracks changed with the grain growth that occurred reducing closure, healing, or both. This implies reduction in the registry of the originally mating microcrack faces, e.g., due to their size (especially if they are no longer confined to a single grain), microcrack branching, or other interactions.

The third method of demonstrating the presence and effects of microcracks also deals with their opening under tensile stresses and closure and opening due to the respective application or release of a compressive stress especially a hydrostatic one. The latter has been most commonly applied to rocks, e.g., as noted by Case (66). Rocks with low elastic moduli due to microcracks increase their elastic properties with increasing compressive stress, and decrease again as it is reduced, whereas no changes occur in rocks with no microcracks. (Note that some similar effects are indicated with some pores, e.g., between particles or balloons, Fig. 3.9.) Such increases in the elastic moduli of rocks and the accompanying increase in ν is consistent with Cases's analysis (see Section 3.2.3).

The effects of stress application to microcracking materials can, however, be complicated, at least (or especially) in highly anisotropic materials such as graphite. Thus, Brocklehurst in his review of graphite (131) noted decreases in elastic moduli due to prior application of either a uniaxial tensile or compressive stress that varied with the microstructure and stress, and that could be fully or partially recovered by subsequent treatments, especially by annealing. The decreases in moduli increase nonlinearly and sometimes irregularly as the previous stress level increases, e.g., reaching reductions of > 30%. Greater decreases may occur from compressive stressing (possibly in part because of the greater stress levels achievable in compression) than with tensile stressing, and other differences occur, e.g., density and thermal expansion reduced from tensile loading and increased from compressive loading. Other properties such as electrical conductivity are also effected, but strengths are often reduced little or none. Major sources of these complications in graphite are effects of grain and pore parameters, e.g., microcracks originating and branching from, as well as termination on, pores have been reported.

The fourth method of demonstrating the presence and effects of microcracks is by correlating related properties with similar, or preferably the same, microcrack-modeled dependence per the analysis of Case (66), e.g., as noted above for rocks. A summary of Cases's evaluation of ceramics with microcracks (Fig. 3.16 B) clearly shows E changes correlating with ν changes. These changes are also reasonably consistent with a microcrack model of Budiansky and O'Connel (67), which was also reasonably consistent with other models reviewed (66).

Zimmerman (132) corroborated and extended (as well as made a minor correction to) Case's analysis to eliminate the difficult to determine the crack density parameter for one elastic property by using that of another related elastic property. As noted above and earlier, Case eliminated the crack density parameter in relations for Young's modulus via similar relations for Poisson's ratio, thus obtaining an E-ν relation. Zimmerman instead used relations for the dependences of E and G on microcrack density to eliminate the latter. He argued correctly that the resultant E-G relations are a more accurate test since experimental values of G are more accurate than those of ν. He showed that the E-G correlations for microcracked materials was even better between the various models and experimental data (including additional data not used by Case).

3.4 SELF-CONSISTENCY TESTS AND REFINEMENT OF MODELS

Having evaluated models against one another and especially against data, it is important to test them for self-consistency. Although almost never done, this is important to indicate defects or improvements in the models that may not be clear from their derivations and evaluations against data (e.g., because of scatter and uncertainties in it as discussed below). Evaluating the self-consistency of models for elastic properties as a function of porosity involves seeing that the proposed dependences are consistent with the inherent interrelation of the elastic properties, e.g., per Eqs. 3.1–3.3, and basic physics. A related step is to evaluate their consistency in terms of the property derivatives as a function of P. The derivative of ν, i.e., $d\nu/dP$, is important since it determines whether values of ν are consistent as the properties approach zero as P approaches P_C (per Eq. 3.1). The need for and the importance of such evaluations is indicated by significant variations of ν with increasing P, especially in view of some proposals that ν either remains constant at an intermediate value ($\sim 0.2–0.25$), or increases or decreases from respectively lower or higher values to this intermediate value at P ~ 0.5 or 1 (48,49). Thus ν is an important but not the only component of discussion.

Consider first basic trends for isotropic materials, where Eqs. 3.1–3.3 are fully applicable. The P-dependence of the moduli being equal, as assumed in some models, is consistent with Eqs. 3.2 and 3.3; however, this would make ν

independent of P (per Eq. 3.1) and hence inconsistent with models (and most data) showing ν dependent on (decreasing with) P (3). Also, if E, B, and G all had the same value as they approached zero at the same P_C value, they would indicate inconsistent values of 0.5 and -0.125 (per Eq. 3.1). More specifically, identical P-dependence for the moduli, and hence no P dependence for ν means that $d\nu/dP$ must be zero. However,

$$\frac{d\nu}{dP} = \left[G\left(\frac{dE}{dP}\right) - E\left(\frac{dG}{dP}\right) \right] (2G^2)^{-1} =$$

$$18\left[G\left(\frac{dB}{dP}\right) - B\left(\frac{dG}{dP}\right) \right] (6B + 2G)^{-2}$$

(3.11)

Having $d\nu/dP = 0$ requires that $GdE/dP = EdG/dP$ and $GdB/dP = BdG/dP$, which is inconsistent with the P-dependence of E, G, and B being the same (since $E \neq B \neq G$) except for the mathematical but physically unreal case of no P-dependence of E, G, and B. Thus, the P dependence of E, B, and G cannot be identical for isotropic materials, and hence the similarity of the P-dependence of these properties assumed in MSA and some mechanics-based models (i.e., Eqs. 1, 2, 6, 9 of Table 3.1), can only be approximate.

In order to evaluate the direction and level of differences in the P-dependence of the moduli and ν, consider $d\nu/dP$ further and the physical fact that while E, B, and G are all positive values, all of their derivatives with respect to P are negative. Having positive values of $d\nu/dP$ thus requires (in terms of absolute values) $EdG/dP > GdE/dP$ and $BdG/dP > GdB/dP$, and for $d\nu/dP$ to be negative requires that the inequalities be reversed. Both are feasible mathematically. For reference G is commonly $\sim 40\%$ of E and B is commonly $\sim 75\%$ of E, i.e., requiring that dE/dP be at least 2.5 times dG/dP and that dB/dP be about 2 or more times dG/dP for $d\nu/dP$ to be positive and the inverse of these ratios be 0.4 and 0.5 for $d\nu/dP$ to be negative. However, a positive $d\nu/dP$ also requires that G have a greater P-dependence than E or B, which is generally not observed, while having $d\nu/dP$ be negative requires that G have somewhat less, and B somewhat greater P-dependence than E, which is consistent with data (e.g., Table 3.6 and ref. 33). Experimental data is consistent with the above analysis, showing that b values for B are $\sim 10\%$ greater than, and for G $\sim 10\%$ less than those for E, and b for ν is approximately the difference in the b values for B and G. Thus, the only case clearly consistent with the basic mathematical relations and data is for ν to decrease with P and for there to be a limited decrease in the P-dependence of B, E, and G, in the order listed, at least for isotropic materials. Therefore, models such as those of Banno (133) and Dunn et al. (134) that assume the P-dependence of ν is that of the elastic moduli are in error.

Consider ν further since it can be a sensitive indicator of trends and some unusual behavior has been suggested for it. Poisson's ratio normally reflects lateral strains of opposite sign to those along the stress axis, which is referred to as a positive ratio. The physics of this typical response can be seen based on dense materials attempting to conserve volume in a uniaxial elongation or compression due to the three-dimensional interconnection of bonds directly coupling axial and normal strains, especially in isotropic materials. Thus, for an axial strain of 0.1, two normal lateral strains of opposite sign and ~ 45% of this axial strain closely preserve volume, i.e., the volume of an initially 1 cm cube under tension is 1.1 $(0.955)^2$ ~ 1. Since typically $0.2 < ν < 0.4$ some increase in volume in tension (and decrease in compression) occurs, which is expected from and must be consistent with the bulk modulus (the measure of the change in volume per unit of hydrostatic tensile or compressive stress and hence further illustrates the interrelation between the elastic properties). Values of $ν > ~0.6$ mean a volume change of $\geq 3\%$ and values $< ~0.3$ mean an opposite volume change of $\geq 3\%$. Thus, ν values outside the normal range for dense materials require further understanding.

The claim of greater decreases of ν from initial values > 0.25, or increases from initial values < 0.25 (49), with increasing porosity raises three basic questions. The first is the source of initially high or low ν values at $P = 0$ as noted above. The second question is, while it is logical to consider ν decreasing from higher values with increasing P similar to the moduli, why would it increase from lower values to (if not already at) an intermediate value (hence closer to volume conservation) at higher P, especially at P ~1. The third question is why should there be measurable values of ν as P $\rightarrow$ 0.5, and especially 1. Approaching the latter limit was addressed by the above analysis, since the claimed, essentially asymptotic approach of ν to an intermediate fixed value at $P = 0.5$ or 1, implies $dν/dP = 0$, which was shown to not be consistent with known (isotropic) behavior. Similarly ν (often steeply) increasing to a fixed value at $P = 0.5$ or 1 raises serious questions of the implied large positive and negative values of $dν/dP$.

Such trends also differ from predictions of other analysis. Thus, Zimmerman (135), setting properties of one phase in two composite models to 0 obtained $ν = 0.2$ or 0.2–0.3 at $P = 1$ depending on the model and the solid ν. Further, simple considerations suggest other results. Thus, as P $\rightarrow$ 1 axial tensile strains can be large, but lateral strains cannot be, indicating ν decreases, quite possibly going to zero. In compression, axial strains can be high, but not ∞, while lateral strains will likely be less since, for example, buckling can occur, limiting lateral expansion. This again indicates ν decreases, possibly going to zero (but possibly at different rates for tensile versus compressive stressing). Thus, claims of limiting ν values other than 0 require scrutiny in view of both uncertainties in the approaches and lack of experimental characterization, particularly in view of possible microstructural effects.

As an example of other microstructural effects, consider atypical materials with opposite lateral strain behavior (i.e., "negative" values of ν so bodies expand latterly in tension). Thus, progressive addition of such negative ν behavior with increasing P could more rapidly decrease initially higher net lateral strains, thus possibly explaining ν decreasing faster, but not the initially high ν. Conversely, decreasing the extent of negative ν contributions in bodies as P increases could explain increasing ν values in bodies of initially low ν (and be the reason for their initially low values); however, a negative ν is related to unusual, typically open structures (136–139), as in; 1) mixtures of connected fibers and plates such that the plates rotate from parallel to the stress to normal to it as stress increases (136,137), or 2) reentrant foam structures (137–139), (or the counterpart of this in crystal structures, e.g., α-cristobalite (140,141) and possibly α-quartz (142) that require lateral expansion to allow axial extension; however, such atypical ν dependence typically occurs over limited, higher P ranges, e.g., negative ν behavior occurs over the P range where material plates can rotate, or re-entrant foam struts can extend. It also appears to be primarily or exclusively associated with anisotropic structures, which commonly exhibit such behavior primarily in one direction, and typically in tension and not necessairly in compression. Thus, such mechanisms in tension do not necessarily have the opposite effect in compression, i.e., a contraction, and if the counterpart exists for compression, i.e., a contraction on compression, it may not result in the opposite behavior, i.e., an expansion, operative in tension. Such behavior usually results in changed microstructure in the material and is more common in more flexible materials such as polymer or metal foam materials with lower modulus, some ductility, or both. Similarly, mechanisms for lateral contraction in compression may not give lateral expansion in tension.

More broadly, increasing ν as P increases most likely reflects: 1) anisotropic properties, e.g., from anisotropic or unique pore structures, 2) porosity heterogeneities, or 3) special microstructural mechanisms outlined above that change axial versus lateral strains, with one and two being most probable. Thus, aligned cylindrical pores [which Ramakrishnan and Arunachalam (49) were actually analyzing the transverse behavior of cylindrical pores in their attempt to evaluate more general behavior] give respectively higher average axial and about average transverse stiffnesses, hence respectively lower and higher resultant strains for a given stress in comparison to each other and most other bodies of the same porosity. Thus, axial stressing of bodies with such pores would give higher values of ν, and transverse stressing lower values, but only due to the pores, and thus not in their absence. Heterogeneities of porosity and resultant elastic moduli versus P data scatter or variations can result in apparent increases in v. Citing MgO variable v data (123) (Fig. 3.15) as evidence supporting their proposed v trends (142) thus neglects possible heterogeneous porosity or anisitropy as causes of the variable v.

The only two cases of ceramic data reasonably indicating no decrease in v or a modest (81,82) or a clear increase (85) in v over P (respectively ~ 0.2–0.7, and 0.35–0.9) are for a sol-gel derived SiO_2 (Figs. 3.8, 3.12), thus is of doubtful relation to v for quartz or cristobalite. Such sol-gel bodies at higher porosities consist of tangled, interconnected chains of small SiO_2 particles, which suggests either of two special microstructural, rather than generic pore, origins of such behavior. The first is some possible anisotropy due to gravitational effects on the forming and bonding of the ball-and-chain structure. The second is that the ball-and-chain structure may result in some re-entrant structure or some differing degrees or effectiveness of "axial" versus "lateral" bonding of these chains could occur, e.g., due to directional drying in the axial direction. Thus, much more understanding, evaluation, and caution is needed in identifying behavior of Poission's ratio trends as P → 1, with much closer attention to data quality and variations as well as microstructural issues and characterization.

3.5 DATA QUALITY

Data quality is unfortunately generally either neglected or assumed to be good, resulting in far fewer and less satisfactory measurements. While less scatter and more data over a greater porosity range are generally advantageous for evaluating models, this is not always sufficient, e.g., due to heterogeneities of porosity. While this will often increase scatter, it need not since this can lead to systematic errors due to porosity inhomogeneity increasing with P. As discussed in Chapter 1, such errors can be determined by: 1) measuring the same properties (and porosities) on different size pieces from the same body, 2) measuring the same properties in the same bodies by use of different vibration modes and frequencies, and 3) measuring different but related properties to check for self-consistency (discussed above). Similarly, isotropy of properties is almost universally (explicitly or implicitly) assumed despite common sources of anisotropically shaped and oriented pores, or oriented arrays of equiaxed pores (as well as of some possible crystallographic orientation). While resultant elastic anisotropy can significantly effect properties and can be readily checked experimentally, this is also seldom done.

Another perspective on the variations, hence quality, of the data is given by comparing the statistical variations of porosity and property measurements. As discussed in Section 2.5, since the statistical variations of many mechanical properties is given via the Weibull modulus, it is also suggested for comparing the variation of the porosity and modulus values for the same data, e.g., that of ThO_2 (Table 3.9, see also Table 10.6 for similar data for other materials). Note

Table 3.9   Summary of the Weibull Modulus for Porosity and Young's and Shear Moduli of ThO_2[a]

Group/No.[b]	Porosity (%)	m(P)	m(E_1)	m(E_2)	m(E_3)	m(G)
I/29	3.73 ± 0.09	49.7	396	309	396	470
II/46	7.15 ± 0.53	15.8	427	320	367	502
III/39	17.27 ± 0.80	25.6	236	272	273	289
III/39	23.66 ± 0.33	86.3	241	293	72.9	292
III/40	26.19 ± 0.47	70.0	138	179	179	169
III/40	26.08 ± 0.23	137	149	200	200	180

[a] From data of Spinner et al. (122), Weibull modulus (m) for measurement of porosity (P), Young's modulus (three separate measurements, E_1, E_2, E_3) and Shear modulus (G).
[b] Group number/ number of specimens for each average.

three characteristics of the data. First, there is substantial variation in the Weibull moduli for both the porosity and elastic moduli measurements, reinforcing the issue of heterogeneity. Second, despite these variations there is a trend for a lower Weibull modulus, i.e., more variation, as porosity (P) increases (i.e., similar to Fig. 3.17). Third, the Weibull modulus for elastic moduli can be lower than for the porosity measurement, but are often larger, indicating that other factors than just the heterogeneity of the amount of porosity effect the elastic moduli, hence reinforcing effects of pore character. Note also that the same evaluation of data from the same study, but with fewer data points (e.g., 6) for each of a number of other samples show the same results, except somewhat wider variation, especially to lower m values at higher P.

That such data variations are important is shown by three examples. First, of two sets of data significantly impacting the ranking of fits of various equations in Dean and Lopezs' study (113), the ThO_2 data (122) explicitly showed increasing effects of inhomogeneous porosity by both methods 1 and 2 above (Fig. 3.17, Table 3.9), which is consistent with other demonstrations of inhomogeneity (143) and theoretical expectations (144). Further, the MgO data used showed considerable inconsistency in the results for Poission's ratio as a function of P (Fig. 3.15) as derived from evaluations of its other elastic properties. Much, possibly all, of this variation probably reflects effects of inhomogeneous porosity (hence illustrating method 3 above). Oriented, anisotropically shaped pores or crystallographic orientation, or both could also be sources of this variability of Poission's ratio (and hence also of the properties that determine it, which could also be the case for ThO_2). Porosity inhomogeneity effects are not restricted to older, sintered, or oxide samples. Wu et al. (145), observed that Young's modules decreases in hot-pressed B_4C (P-0–0.15) varied by 13% depending on whether the porosity was measured on the pieces of material from which individual specimens were obtained or on the specimens themselves. Clearly it is difficult to fully determine effects of inhomogeneously distributed

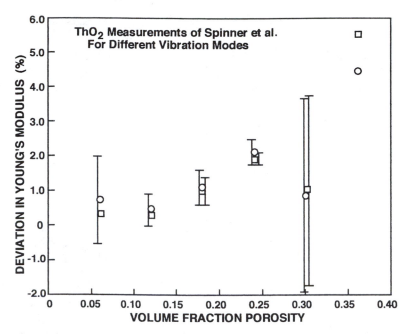

Figure 3.17 Deviations in Young's modules vs. P from Spinner et al.'s measurements of sintered ThO_2 samples tested in various vibration modes (at 22°C) (122). Their comparison was of vibrating bars edgewise vs. faltwise (circles) and endwise (i.e., longitudinal) vs. flatwise (squares) to determine Young's modulus. Since different modes stress different areas of the sample, this shows heterogeneity, which increases with P. Published with permission of J. Mat. Sci. (7).

and anisotropically shaped pores in evaluating models for the porosity dependence of properties, but two steps would appear to be a minimum. First, the impact of data known or suspected of reflecting such effects of inhomogeneous porosity should be noted and considered in the evaluations being made. Second, since tests for such effects of inhomogeneity have generally been totally neglected, it is important to use a large, rather than a small to moderate, number of studies for evaluating models for porosity effects on properties.

Two factors should be noted about the above evaluations. Though some, e.g., Dean and Lopez (113) used zero porosity values from single crystal data, this is often not done. Failure to use these can result in considerable fitting error, as illustrated in Fig. 3.18, for determining b values and fit of the exponential expression. Clearly, the impacts of not using such single-crystal–derived data (where available) is increased when data does not extend to nearly zero porosity, and problems also occur when data extends, especially only partly into the

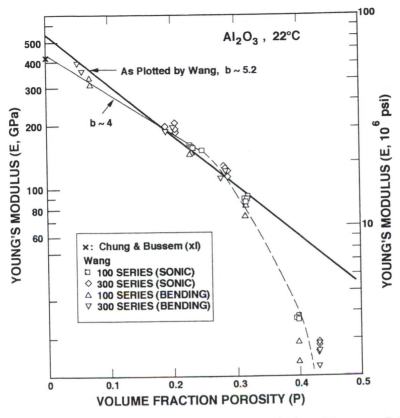

Figure 3.18   Comparison of Wang's (31) Al_2O_3 Young's modulus versus P data without (as plotted by Wang) and with single crystal (xl, sapphire) data. Note that the latter shows Wang clearly overestimated the slope (b value) of his data; which, in turn, even more clearly shows the property deviation below the linear trend at higher P. Such errors illustrate an important pitfall in comparing the fitting of such data with various, e.g., MSA, models. Published with permission of J. Mat. Sci. (7).

rollover region, (e.g., in Dean and Lopezs' study, the lower P levels ranged from 0 to 0.11, averaging ~ 0.04, and high P limits varied from P = 0.36 to 0.5, averaging 0.34). The upper level of P is important since the higher this is, the greater is the probability that the property values are more rapidly decreasing toward their P_C limits and thus no longer fitting the exponential relation, which is only valid before this transition begins. (Similar limits may apply to other models.) Note that the lack of P =0 data and limited data into the transition to P_C values can seriously exacerbate such fitting errors.

3.6   THE NEED FOR MORE COMPREHENSIVE STUDIES

Both a significant motivation and challenge in writing this book is the large number of studies that are so narrow or limited in one, often several, key aspects such as the breadth of: 1) theoretical perspective and evaluation, 2) porosity and other related body characterizations, and 3) property measurements, e.g., measuring more than one property and checking for gradients, anisotropies, and heterogeneities. The above discussions of self-consistency and data quality deal with important components of the problem. This section summarizes and illustrates other and broader aspects of the need to take a more comprehensive approach. Much of this is done by considering a few representative studies, all of which have made some useful contributions but do not address issues raised by the study, which if successfully addressed could have added much more to the field. Such considerations are particularly appropriate for elastic properties since they are basic and the most extensive area of porosity study, hence presenting the greatest challenge and illustration of the problem.

Thus, for example, Nagarajan (107), as noted earlier, usefully measured both longitudinal and shear velocities normal to the plane of thin, partially sintered alumina discs made by die- then isopressing (respectively at ~ 100 and 200 MPa) showing that dilute porosity models did not fit his data consistent with it extending to P ~0.4. (While Nagarajan's specimens were subsequently isopressed, it was shown earlier that such pressing does not necessarily eliminate probable laminar pores from previous die pressing.) He further noted that of the few models considered, one for stress normal to the axes of aligned cylindrical pores fit best. A number of investigators have used this data. None of these investigators, nor those using the common die pressing of thin discs for making samples have seriously considered that (depending on powder character, specimen dimensions, and pressing parameters) varying degrees of laminar pores can occur in such bodies. Such pores result in very similar effects to cylindrical pores, including substantial anisotropy of properties measured parallel versus perpendicular to the tubular or laminar pore axes. Thus, even limited quantities of such pores can have a substantial effect on properties, e.g., from the standpoint of anisotropy, so neglecting them can cause serious errors in interpreting data and models.

Phani (146) cited Nagarajan's study as a basis for deriving a modified model for aligned cylindrical pores stressed normal to their axes. He compared data for the sonic velocities and for E (with some of the former not fitting his model very well), along with E-P data from an earlier survey of alumina. From this he concluded that such cylindrical pore models provided a "better mechanical model for predicting the effects of porosity on elastic moduli and ultrasonic velocity at least in the case of the data on the alumina studied." On the other hand, Phani (147) also simultaneously applied a composite-derived model based on a spheroidal second phase to ultrasonic velocities of ceramics and metals as a function of the amount of porosity. He used the spheroid aspect ratio, α, i.e., of the minor to the major axis ($\alpha = 1$ for spheres and $\alpha < 1$ for both oblate spheroids which become disc cracks and prolate spheroids which become needle cracks at $\alpha = 0$), as an adjustable parameter to fit 10 sets of data for five ceramics and two metals, with two not fitting well (one being Nagarajan's alumina and the other data for a porcelain). Based on this fitting with $\alpha = 0.26$ to 0.72, he (correctly) concluded that pore shape is important although there was only limited porosity characterization for just one body (SiC). No distinction was made between possible prolate or oblate pores. He also concluded that such a model is suitable for a variety of materials although there was no justification of why such models of composites adapted to spheroidal pores would be applicable to bodies where the majority of the pores are likely to be better modeled as those between spheroidal particles, which entails a substantially different pore geometry. Finally, noted that no conclusion could be drawn whether any relation exists between the effective aspect ratio and the fabrication, except that he hints that it may depend on particle size (based on data for one material at two different particle sizes) and on hot-pressing versus reaction sintering of Si_3N_4.

Next consider further the issue of some data being linear versus $1 - P/P_C$, results of which were outlined at the end of Section 3.3.3. This functional dependence arose from modifying the $(1 - P)^n$ porosity dependence by substituting $AP = P/P_C$ for P (148), which, although logical, is essentially empirical. Model observations, e.g., Knudsen (4), were used to obtain $1 \leq 1/P_C \leq 3.85$ (e.g., for spherical pores to orthorhombic packing of uniform spherical particles, Figs. 2.10, 3.1). Then using $A = 1/P_C$ within the above bounds and n as fitting parameters to seven sets of rare earth oxide data $A = 1/P_C$ and n respectively were 2.32 ± 0.38 and 0.91 ± 0.27 (149). This was then cited as suggesting a $1 - P/P_C$ dependence (1) despite a nearly 30% variation in n. Next, instead of $E/E_0 = 1 - P/P_C$, it was proposed that

$$(E - E')(E_0 - E') = \frac{1 - P}{P_C} \tag{3.12}$$

based on the recognition that there is some nonzero modules in the unsintered body (150). How much of this might be due to binder and how much is due to particle interlocking and why the correction to E and E_0 should be the same was not addressed. Using E' as a fitting parameter not surprisingly, gave a better fit to the limited data, including that for a commercial alumina, primarily at > 0.5 P_C; however in the alumina case, the fact that it was extruded, and hence quite likely contained some (possibly substantial) approximately tubular porosity parallel with the extrusion axis, which inherently follows a $1 - P$, hence also likely a $1 - P/P_C$, dependence was not considered (124).

The above studies all make useful contributions; however, they are also representative of the extensive ad hoc nature of many expressions and the parameters in them, the absence of pore and body characterization available, and the limited extent of the evaluations. Thus, even limited microstructural examination, especially by stereological techniques, and use of general process-pore expectations, e.g., of possible laminations in die pressing, possible laminar gradients in pressure casting, and approximately tubular pores in extruded bodies, were not utilized. Similarly, no measurements of specimens of different size, orientation, or both were used to give a larger data base and possible variations in possible gradients and orientations of pores, as well as the opportunity to directly address possible resultant specimen anisitropies.

The key need besides more comprehensive measurements and suitable microstructural characterization is to first recognize that there is no universal porosity and no single model for porosity effects, even for a single property. Instead it is essential to recognize that an array of models is necessary to cover the diversity of pore (and microcrack) character, with the pertinent range of applicability of the models identified not only qualitatively, but also quantitatively in terms of equation parameters. Further both the microstructural characterization and the methodology to combine models to address (often changing) mixtures of pores not adequately encompassed by a single model.

3.7 SUMMARY

Modeling based on empirical, more on pore geometry (MSA), or more on mechanics, all commonly lead to one of three equation forms: e^{-bP}, $1 - AP^n$, and $(1 - AP)^n$. MSA models emphasize the first two forms, and mechanistic models the latter form. The first and third forms reflect respectively semilog and log-log plots, with the former being more advantageous. The second and third forms are identical when $n = 1$, and are the upper bound when $A = n = 1$. This is also approximately true for e^{-bP} with $b \sim 1$ at low P, e.g., the first and third expressions reduce to the same form at low P. The use of bounds to aid model development or selection is of limited value for porosity (as opposed to their use for composites) since the bounds for porosity are very broad because the lower bound is zero due to the properties of pores being zero. The upper bound is $1 - P$, which holds for aligned cylindrical and some other tubular pores stressed parallel to their axes.

The e^{-bP} dependence is highly advantageous for modeling based on minimum solid area (MSA) since it is a close approximation over the first 0.33 to 0.5 of the P range before properties start dropping rapidly to go to zero at P_C (Figs. 2.11, 3.1, 3.2A). This dependence logically leads to the use of semilog plots with MSA models and the attendant advantages of such plots, e.g., the direct applicability of substantial data from the past extensive empirical use of e^{-bP}; however, such use with MSA models provides insight into the previously unrecognized relation of b values, the semilog slopes, over the low–modest range of P, to pore character. Further, since both semilog plots and all MSA models cover the complete applicable range of P, they readily reveal percolation limits, i.e., P_C values, which have their own individual, logically varying dependences on pore character, and an inverse trend with b values inherent in these models. Such plots also allow the ready, logical, and simple, at least approximate, evaluation of combined and changing porosity, as well as potential porosity inhomogeneity, by interpolation between, or combination of, the various semilog curves. Similarly, anisotropy of properties due to varying degrees of alignment of anisotropically shaped pores are a very direct and visible result of these models (Tables 3.2, 3.3; Figs. 3.1, 3.2).

Another key advantage of MSA models, besides their simplicity and covering the complete applicable P range, is that they cover in a very visible fashion, the range of pore character encountered in most, if not all, processing. Thus, while being simpler in their mechanics, they are very practical and are the only models currently available with some truly predictive capability, i.e., giving explicit values of elastic moduli for a substantial range and amount of specific pore character. This assessment of the MSA family of models is based on their fitting surprisingly well data for bodies with a practical range of ideally shaped pores, as well as pore combinations. It is also based on their being consistent with the porosity trends of the range of pore character and mixes thereof for various ceramic forming operations. The latter data comparison is the most extensive relative to other evaluations (which have often seriously neglected data variability issues) by about an order of magnitude. MSA, as well as many other, models, do not directly give the (typically limited) variations of the P dependences of B, E, and G, nor the P dependence of ν; however, MSA models are consistent with data showing b values for B ~ 10% > for E and for G ~ 10% < for E, with b for ν being approximately the difference in b values of B and G. Comparison of models and data also supports the combination of pore models by interpolation between their semilog curves or MSA parameters, but needs more evaluation. Further, these models are similarly applicable to a variety of other properties, e.g., as discussed in Chapters 2, 4–8, which provides valuable intercomparison.

Mechanistic models, often being rigorous for at least bulk and sometimes also shear moduli with some uncertainty for Young's modulus and especially Poisson's ratio, are potentially the most accurate models; however, such modeling of porosity effects was almost exclusively from composite modeling with limited (undefined) concentrations of a single population of idealized spheroidal, especially spherical, second-phase particles of unspecified stacking. Uncertainties of adapting composite models to pores and the narrow pore character that can be handled by mechanical analysis (in contrast to the diversity of pore character addressed by MSA models) are serious limitations, e.g., as shown by frequent disagreements between mechanistic models for similar or the same pore character. Further, though derived for a few very specific pore types, such models were widely applied to any, usually unspecified, pore character, frequently by using equation parameters for fitting (as was also often the case for MSA models). It was claimed or implied that differing values of A and n were constant for a given pore character and systematically changed with such character changes despite at best limited demonstration of this and evidence to the contrary over larger P ranges (Fig. 2.14). Ad hoc introduction of P_C values < 1 and use of $1 - P/P_C$ dependence, as well as of porosity dependence of ν, and especially of ν_0 to allow better data fitting, although all but the latter are logical, reflect limitations of these models and their application. However, derivation and use of mechanistic models is evolving to significantly reduce these previous limitations. Substantial derivation of pore-shape effects and combinations of pore types, though mainly for two-dimensional (i.e., tubular) pores, is providing the basis for a useful family of models needed to address porosity effects. Such developments also generally agree with MSA results (e.g., Fig. 3.2A) and there are indications of correlations of n and b values.

Most evaluations of the porosity dependence of elastic (and other) properties have neglected three important factors: 1) self-consistency, 2) heterogeneity, and 3) anisotropy. Evaluations of self-consistency show that the porosity dependences of B, E, G for identical porosities in isotropic bodies cannot be identical as commonly assumed (at least as an approximation). Despite clear expectations and demonstrations that porosity often is heterogeneous, e.g., in its spatial distribution, experimentally checking for this, despite being quite practical, is almost never done. Effects of heterogeneity have been neglected in almost all evaluations of the accuracy of various models fitting data. Such effects of heterogeneity often increase in severity as the nonlinearity of porosity dependence of properties increases, i.e., with P itself. Similarly, the issue or anisotropy of properties due to aligned, anisotropically shaped pores commonly resulting from typical forming operations such as die pressing and especially extrusion, are almost universally neglected (as is possible anisotropy due to crystallographic orientation of grains). Anisotropies of properties are probably a major factor in many of the variations in the porosity dependence of Poisson's ratio, in addition to its inherent variations due to its being a small quantity. (Other evidence of anisotropic properties from aligned, anisotropically shaped pores will be shown in subsequent chapters).

In contrast to porosity, mechanistic models have been the only successful ones for micricracks, MSA modeling has not been successful. Difficulties of adequately characterizing microcrack populations places much more emphasis on model results which closely parallel results for porosity. Thus models for dilute populations of microcracks show properties decreasing lineraly as micro-crack content increases, but with nonlinear property decreases with increasing microcrack content and interaction. In either case the property dependence varies with the shape, stacking, and orientation of the microcracks, i.e., as for pores, and again shows that there is no universal microcrack model as there is no universal pore model. Thus, the focus is on identifying and applying appro-priate models covering the range and character of microcracks involved. Heterogeneities and anisotropy of microcrack populations, though generally neglected, need to be addressed.

The first of three critical needs for further development is much better characterization of microcracks and porosity, both in terms of character and amount, especially spatial distribution and heterogeneity of both of these to address problems noted above. This requires much more application of existing, especially stereological, measurements as well as measurements of MSA, parti-cle and pore or microcrack stacking (e.g., C_n), orientations, gradients and heterogeneity, and mixes of porosities or microcracks, or both. Since such char-acterization is typically lacking for almost all available data, a broad data base is needed in order to make reasonable evaluations of porosity trends. The second need, in part related to the first, is for more and better property measurements, e.g., anisotropy and heterogeneity, and of more properties, e.g., at least two or three, if not all four, of B, E, G, and v, and other mechanical, as well as nonmechanical, properties for comparison. One goal is to improve the accuracy of determining porosity dependences, e.g., b values beyond ~ ± 0.5. The third need is for improved models, e.g., mechanistic models for more three-dimensional pores, especially pores between particles and MSA models for more realistic sintering situations, e.g., accounting for varying particle and pore size, shape, orientation and spatial distributions, as well as for tapering foam cell walls and struts. Such computations, commonly via computers, should also include other forms of bonding particles, e.g., chemical bonding via reactions in gypsum, cement, etc., bodies, and via chemical vapor infiltration, which do not preserve body mass (as sintering, the basis of the current models, does). Broader use, comparison, and possibly marriage of MSA and mechanics- based models are key elements of this.

REFERENCES

1. D. C. Lam, F. F. Lange, and A. G. Evans, "Mechanical Properties of Partially Dense Alumina Produced from Powder Compacts," J. Am. Cer. Soc., **77** (8), pp. 2113–17, 1994.
2. R. W. Rice, "Evaluating Porosity Parameters for Property-Porosity Relations," J. Am. Cer. Soc., **76** (7), pp. 1801–08, 1993.

2. R. W. Rice, "Evaluating Porosity Parameters for Property-Porosity Relations," J. Am. Cer. Soc., **76** (7), pp. 1801–08, 1993.

3. R. W. Rice, "The Porosity Dependance of Physical Properties of Materials—A Summary Review," Porous Ceramic Materials, Fabrication, Characterization, Applications (D. M. Liu, Ed.), Trans Tech Publications, Zurich, Switzerland, pp. 1–19, 1996.

4. F. P. Knudsen, "Dependence of Mechanical Strength of Brittle Polycrystalline Specimens on Porosity and Grain Size," J. Am. Cer. Soc., **42** (8), pp. 376–388, 1959.

5. R. W. Rice and S. W. Freiman, "The Porosity Dependance of Fracture Energies" Ceramic Microstructures '76: With Emphasis on Energy Related Applications (R. M. Fulrath and J. A. Pask, Eds.), Westview Press, Bolder, CO, pp. 800–23, 1977.

6. R. W. Rice, "Evaluation and Extension of Physical Property-Porosity Models Based on Minimum Solid Area," J. Mat. Sci., **31**, pp. 102–8, 1996.

7. R. W. Rice, "Comparison of Physical Property-Porosity Behavior with Minimum Solid Area Models," J. Mat. Sci., **31**, pp. 1509–28, 1996.

8. C. A. Anderson, "Derivation of the Exponential Relation for the Effect of Ellipsoidal Porosity on Elastic Modulus," J. Am. Cer. Soc., **79** (8), pp. 2181–84, 1996.

9. R. W. Rice, "Extension of the Exponential Porosity of Strength and Elastic Moduli," J. Am. Cer. Soc., **59** (11–12), pp. 536–7, 1976.

10. M. Eudier, "The Mechanical Properties of Sintered Low-Alloy Steels," Pwd. Met., **9**, pp. 278–290, 1962.

11. S. D. Brown, R. B. Biddulph, and P. D. Wilcox, "A Strength-Porosity Relation Involving Different Pore Geometry and Orientation," J. Am. Cer. Soc., **47** (7), pp. 520–22, 1964.

12. O. Ishai and L. J. Cohen, "Elastic Properties of Filled and Porous Epoxy Composites," Intl. J. Mech. Sci., **9**, 539–546, 1967.

13. B. Paul, "Prediction of Elastic Constants of Multiphase Materials," Trans. Metal. Soc. AIME, **218**, pp. 36–41, 1960.

14. A. R. Boccaccini, G. Ondracek, P. Mazilu, and D. Windelberg, "On the Effective Young's Modulus of Elasticity for Porous Materials: Microstructure Modeling and Comparison Between Calculated and Experimental Values," J. Mech. Behav. Mat., **4** (2), pp. 119–28, 1993.

15. L. J. Gibson and M. F. Ashby, "Cellular Solids, Structure & Properties," Pergamon Press, New York, 1988.

16. L. J. Gibson and M. F. Ashby, "The Mechanics of Three-Dimensional Cellular Materials," Proc. Roy. Soc. Lond. A **382**, pp. 43–59, 1982.

17. J. M. Dewey, "The Elastic Constants of Materials Loaded with Non-Rigid Fillers," J. Appl. Phy., **18**, pp. 578–81, 1947.

18. Z. Hashin, "Elasticity of Ceramic Systems," Ceramic Microstructures '76: With Emphasis on Energy Related Applications (R. M. Fulrath and J. A. Pask, Eds.), Westview Press, Bolder, CO, pp. 313–41, 1977.

19. B. Budiansky, "On the Elastic Moduli of Some Heterogeneous Materials," Mech. Phys. Solids, **13**, pp. 223–27, 1965.

20. D. P. H. Hasselman, "On the Porosity Dependence of the Elastic Moduli of Polycrystalline Refractory Materials," J. Am. Cer. Soc., **54** (9), pp. 442–53, 1962.

21. L. F. Nielsen, "Elasticity and Damping of Porous Materials and Impregnated Materials, J. Am. Cer. Soc., **67** (2), 93–98, 1983.

22. L. F. Nielsen, "Strength and Stiffness of Porous Materials," J. Am. Cer. Soc., **73** (9), pp. 2684–89, 1990.

23. L. F. Nielsen, "Elastic Properties of Two-Phase Materials," Mat. Sci. & Eng., **52**, pp. 39–62, 1982.

24. L. E. Nielsen, "Generalized Equation for the Elastic Properties of Composite Materials," J. Appl. Phy., **41** (11), pp. 4626–27, 1970.

25. R. M. Christensen, "A Critical Evaluation for a Class of Micro-Mechanistic models," J. Mech. Phys. Solids, **38** (3), pp. 379–404, 1990.

26. D. P. H. Hasselman and R. M. Fulrath, "Effect of Cylindrical Porosity on Young's Moduls of Polycrystalline Brittle Materials," J. Am. Cer. Soc., **48** (10). p. 545, 1965.

27. Z. Hashin and B. W. Rosen, "The Elastic Moduli of Fiber-Reinforced Materials," J. Appl Mech., pp. 223–32, 6/1964

28. L. Gibson, "Modeling Mechanical Behavior of Cellular Materials," Mat. Sci. & Eng., **A110**, pp. 1–36, 1989.

29. S. T. Gulati, "Effects of Cell Geometry on Thermal Shock Resistance of Catalytic Monoliths," Soc. Auto. Eng. Congress and Exposition Paper 750171, Detroit, MI, 2/1975.

30. R. A. Helmuth and D. H. Turk, "Elastic Moduli of Hardened Cement Pastes," Symp. On Structure of Portland Cement Pastes and Concrete. Spec. Rept. No. 90, Highway Research Board, Washington, D.C., 1966.

31. J. C. Wang, "Young's Modulus of Porous Materials," J. Mat. Sci., **19**, pp. 801–8, 1984.

32. R. D. Sudduth, "A Generalized Model to Predict the Effect of Voids on Modulus in Ceramics," J. Mat. Sci., **30**, pp. 4451–62, 1995.

33. R. W. Rice, "Microstructural Dependence of Mechanical Behavior of Ceramics," Treatise on Materials Science and Technology, **11**, (R. McCrone, Ed.), Academic Press, pp. 199–381, 1977.

34. A. R. Day, K. A. Snyder, E. J. Garboczi, and M. F. Thorpe, "The Elastic Moduli of a Sheet Containing Circular Holes," J. Mech. Phys. Solids, **40** (5), pp. 1031–51, 1992.

35. K. A. Snyder, E. J. Garboczi, and A. R. Day, "The Elastic Moduli of Simple Two-Dimensional Isotropic Composites: Computer Simulation and Effective Medium Theory," J. Appl. Phys., **72** (12), pp. 5948–55, 1992.

36. P. Arato, E. Besenyei, A. Kele, and F. Weber, "Mechanical Properties in the Initial Stage of Sintering," J. Mat. Sci., **30**, pp. 1863–71, 1995.

37. A. T. Huber and L. J. Gibson, "Anisotropy of Foams," J. Mat. Sci., **23**, pp. 3031–40, 1988.

38. D. J. Green and F. F. Lange, "Micromechanical Model for Fiberous Ceramic Bodies," J. Am. Cer. Soc., **65** (3), pp. 138–41, 1982.

39. A. N. Gent and A. G. Thomas, "The Deformation of Foamed Elastic," Rubber Chem. Tech., **36**, pp. 597–610, 1963.

40. E. H. Kerner, "The Elastic and Thermo-Elastic Properties of Composite Media," Proc. Roy. Soc. (London) Ser. B, **69**, pp. 808–13, 1952.

41. J. K. Mackenzie, "The Elastic Constants of a Solid Containing Spherical Holes," Proc. Roy. Soc. Lond., B **63** (1), pp. 2–11, 1950.

42. A. S. Waugh, R. B. Poeppel, and J. P. Singh, "Open Pore Description of Mechanical Properties of Ceramics," J. Mat. Sci., **26**, pp. 3862–68, 1991.

43. R. C. Rossi, "Prediction of the Elastic Moduli of Composites," J. Am. Cer. Soc., **51** (8), pp. 433–9, 1968.
44. C. W. Bert, "Prediction of Elastic Moduli of Solids with Oriented Porosity," J. Mat. Sci., **20**, pp. 2220–24, 1985.
45. T. Mori and K. Tanaka, "Average Stress in Martix and Average Elastic Energy of Materials with Misfitting Inclusions," Acta Met., **21**, pp. 57–74, 1973.
46. G. T. Kuster and M. N. Toksoz, "Velocity and Attenuation of Seismic Waves in Two-Phase Media: I. Theoretical Formulation," Geophysics, **39**, pp. 587–606, 1974.
47. E. A. Dean, "Elastic Moduli of Porous Sintered Materials as Modeled by a Variable-Aspect-Ratio Self-Consistent Oblate-Spheroid-Inclusion Theory," J. Am. Cer. Soc., **66** (12), pp. 847–54, 1983.
48. N. Ramakrishnan and V. S. Arunachalam, "Effective Elastic Moduli of Porous Solids," J. Mat. Sci. **25**, pp. 3930–37, 1990.
49. N. Ramakrishnan and V. S. Arunachalam, "Effective Elastic Moduli of Porous Ceramic Materials," J. Am. Cer. Soc., **76** (11), pp. 2745–52, 1993.
50. J. G. Berryman and P. A. Berge, "Critique of Two Explicit Schemes for Estimating Elastic Properties of Miltiphase Composites," Mechanics of Materials, **22**, pp. 149–64, 1996.
51. M. Kachanov, I. Tsukrov, and B. Shafiro, "Effective Moduli of Solids with Cavities of Various Shapes," Micromechanics of Random Media (M. Ostoja-Starzewski and I. Jasiuk, Eds.), ASME Book. No. AMR 139S170, pp. S151–74, 1994. Appl. Mech. Rev., **47** (1, part 2).
52. A. R. Boccaccini, "Comment on Effective Elastic Moduli of Porous Ceramic Materials," J. Am. Cer. Soc., **77** (10), pp. 2779–81, 1994.
53. R. W. Rice, "Comment on Effective Elastic Moduli of Porous Ceramic Materials," J. Am. Cer. Soc., **78** (6), p. 1711, 1995.
54. A. R. Boccaccini and G. Ondracek, "On the Porosity Dependence of the Fracture Strength of Ceramics," Presented at the 3rd European Ceramic Soc. Conf., Madrid, 1993.
55. A. R. Boccaccini and Z. Fan, "A New Approach for the Young's Modulus-Porosity Correlation of Ceramics," Ceramics Intl. **23**, pp. 239–45, 1997.
56. Z. Fan, "A New Approach to the Electrical Resistivity of Two-Phase Composites," Acta. Metal. Mater., **43** (1), pp. 43–9, 1995.
57. B. I. Halperin, S. Feng, and P. N. Sen, "Differences Between Lattice and Continuum Percolation Transport Exponents," Phy. Rev. Lets., **54** (22), pp. 2391–94, 6/3/1995.
58. K. Sieradzki and R. Li, "Fracture Behavior of a Solid with Random Porosity," Phy. Rev. Lets., **56** (23), pp. 2509–12, 6/1996.
59. C. J. Lobb and M. G. Forrester, "Measurement of Nonuniversal Critical Behavior in a Two-Dimensional Continuum Percolating System," Phy. Rev. B, **35** (4), pp. 1899–1901, 2/1/1987.
60. J. Sofo, J. Lorenzana, and E. N. Martinez, "Critical Behavior of Young's Modulus for Two-Dimensional Randomly Holed Metallized Mylar," Phy. Rev. B, **36** (7), pp. 3960–62, 9/1/1987.
61. K. K. Phani and S. K. Niyogi, "Porosity Dependence of Ultrasonic Velocity and Elastic Modulus in Sintered Uranium Dioxide—A Discussion," J. Mat. Sci. Ltrs., **5** pp. 427–430 (1986).

62. A. K. Maitra and K. K. Phani, "Ultrasonic Evaluation of Elastic Parameters of Sintered Powder Compacts," J. Mat. Sci., **29**, pp. 4415–19, 1994.
63. M. Kupova, "Porosity Dependence of Material Elastic Moduli," J. Mat. Sci., **28**, pp. 5265–68, 1993.
64. B. D. Agarwal, G. A. Panizza, and L. J. Broutman, "Micromechanics Analysis of Porous and Filled Ceramic Composites," J. Am. Cer. Soc., **54** (12), pp. 620–24, 1971.
65. A. P. Roberts and M. A. Knackstedt, "Mechanical and Transport Properties of Model Foamed Solids," J. Mat. Sci. Let., **14**, pp. 1357–59, 1995.
66. E. D. Case, "The Effect of Microcracking Upon the Poission's Ratio for Brittle Materials," J. Mat. Sci., **19**, pp. 3702–12, 1984.
67. B. Budiansky and R. J. O'Connel, "Elastic Moduli of a Cracked Solid," J. Solids Struct., **12**, pp. 81–97, 1976.
68. N. Laws and J. R. Brockenbough, "The Effect of Micro-Crack Systems on the Loss of Stiffness of Brittle Solids," Intl. J. Solids Struct., **23**, pp. 1247–68, 1987.
69. D. P. H. Hassselman and J. Singh, "Analysis of Thermal Stress Resistance of Microcracked Ceramics," Am. Cer. Soc. Bul., **53** (8), pp. 856–60, 1979.
70. R. W. Rice, "Comparison of Stress Concentration Versus Minimum Solid Area Based Mechanical Property-Porosity Relations," J. Mat. Sci., **28**, pp. 2187–90, 1993.
71. D. P. H. Hasselman and R. M. Fulrath, "Effect of Small Fraction of Spherical Porosity on Elastic Moduli of Glass," J. Am. Cer. Soc., **47** (1), pp. 52–53, 1964.
72. D. R. Biswas, "Influence of Porosity on the Mechanical Properties of Lead Zirconate-Titanate Ceramics," Materials and Molecular Research Division Annual Report, Lawrence Berkeley Laboratory, University of California, Berkeley, pp. 110–113, 1976.
73. D. R. Biswas and R. M. Fulrath, "Strength of Porous Polycrystalline Ceramics," Trans. J. Brit. Cer. Soc., **79**, pp. 1–5, 1980.
74. J. B. Walsh, W. F. Brace, and A. W. England, "Effect of Porosity on Compressibility of Glass," J. Am. Cer. Soc., **48** (12), pp. 605–608, 1965.
75. N. Warren, "Elastic Constants Versus Porosity for a Highly Porous Ceramic, Perlite," J. Geophysical Research, **74** (2), pp. 713–719, 1969.
76. J. G. Zwissler and M. A. Adams, "Fracture Mechanics of Cellular Glass," Fracture Mechanics of Ceramics (R. C. Bradt, A. G. Evans, D. P. H. Hasselman, and F. F. Lange, Ed.) pp. 211–242, Plenum Press, New York, 1983.
77. H. Hagiwara and D. J. Green, "Elastic Behavior of Open-Cell Alumina," J. Am. Cer. Soc., **70** (11), pp. 811–15, 1987.
78. D. J. Green, "Mechanical Behavior of Lightweight Ceramics," Fracture Mechanics of Ceramics, vol. 8 (R. C. Bradt, A. G. Evans, D. P. H. Hasselman, and F. F. Lange, Eds.), Plenum Press, New York, pp. 39–59, 1986.
79. L. Coronel, J. P. Jernot, F. Osterstock, "Microstructure and Mechanical Properties of Sintered Glass," J. Mat. Sci., **25**, pp. 4866–72, 1990.
80. P. A. Berge, B. P. Bonner, and J. A. Berryman, "Ultrasonic Velocity-Porosity Relationships for Sandstone Analogs Made from Fused Glass Beads," Geophy., **60** (1), pp. 108–19, 1–2/1995.
81. D. Ashkin, R. A. Haber, and J. B. Wachtman, "Elastic Properties of Porous Silica Derived from Colloidal Gels," J. Am. Cer. Soc., **73** (11), pp. 3376–81, 1990.

82. D. Ashkin, "Properties of Bulk Microporous Ceramics" Ph.D. Thesis, Rutgers University, New Brusnwick, NJ, 1990.

83. S. C. Park and L. L. Hench, "Physical Properties of Partially Densified SiO_2 Gels," Sci. Cer. Chem. Proc., pp. 168–72, 1986.

84. T. Fujiu, G. L. Messing, and W. Huebner, "Processing and Properties of Cellular Silica Synthesized by Foaming Sol-Gels," J. Am. Cer. Soc., **73** (1), pp. 85–90, 1990.

85. T. Adachi and S. Saka, "Dependence of the Elastic Moduli of Porous Silica Gel Prepared by the Sol-Gel Method on Heat Treatment," J. Mat. Sci., **25**, pp. 4732–37, 1990.

86. Y. L. Krasulin, V. N. Timofeev, S. M. Barinov, and A. B. Ivanov, "Strength and Fracture of Porous Ceramic Sintered from Spherical Particles," J. Mat. Sci., **15**, pp. 1402–1406, 1980.

87. D. J. Green, "Fabrication and Mechanical Properties of Lightweight Ceramics Produced by Sintering of Hollow Spheres," J. Am. Cer. Soc., **68** (7), pp. 403–409, 1985.

88. D. J. Green and R. G. Hoagland, "Mechanical Behavior of Lightweight Ceramics Based on Sintered Hollow Spheres," J. Am. Cer. Soc., **68** (7), pp. 395–398, 1985.

89. J. N. Harris and E. A. Welsh, "Fused Silica Design Manual, **I**," Georgia Institute of Technology (Atlanta, GA) report for Naval Ordinance Systems Command Contract N 00017-72-C-4434, 5/1973.

90. G. M. Tomilov, "Changes in the Elastic Properties of Ceramic Quartz During Sintering," Tr. from Izvestiya Akademii Nauk SSSR, Neorganicheskie Materialy, **13** (1), pp. 117–119, 1/1977.

91. D. J. Green, C. Nader, and R. Brezny, "The Elastic Behavior of Partially-Sintered Alumina," Sintering of Advanced Ceramics, (C. A. Handwerker, J. E. Blendell, and W. Kaysser, Eds.), Am. Cer. Soc., pp. 347–56, Westerville, OH, 1990.

92. P. D. Wilcox and I. B. Cutler, "Strength of Partly Sintered Alumina Compacts," J. Am. Cer. Soc., **49** (5), pp. 249–252, 1966.

93. P. Wagner, J. A. O'Rourke, and P. E. Armstrong, "Porosity Effects in Polycrystalline Graphite," J. Am. Cer. Soc., **55** (4), pp. 214–219, 1972.

94. S. K. Rhee, "Porosity-Thermal Conductivity Correlations for Ceramic Materials," J. Mat. Sci. and Eng., **20**, pp. 89–93, 1975.

95. S. K. Rhee, "Discussion of Porosity Effects in Polycrystalline Graphite," J. Am. Cer. Soc., **55** (11), pp. 580–81, 1972.

96. E. A. Belskaya and A. S. Tarabanov, "Experimental Studies Concerning the Electrical Conductivity of High-Porosity Carbon-Graphitic Materials," Institute of High Temperatures, Academy of Sciences of the USSR, Tr. from Inzhenerno-Fizicheskii Zhurnal, **20** (4), pp. 654–659, 4/1971.

97. N. McAlford, J. D. Birchall, W. J. Clegg, M.A. Harmer, K. Kendall and D. H. Jones, "Physical and Mechanical Properties of YBa2 $Cu_3 O_{7-8}$ Superconductors," J. Mat. Sci., **23**, pp. 761–768, 1988.

98. B. Bridge and R. Round, "Density Dependence of the Ultrasonic Properties of High TC Sintered $YBa_2Cu_3O_{7-x}$ Superconductors," J. Mat. Sci., **8**, pp. 691–694, 1989.

99. G. Lewis, "Dependence of Modulus of Elasticity on Porosity for Polycrystalline $YBa_2Cu_3O_{7-x}$," J. Can. Cer. Soc., **62** (4), pp. 258–61, 1993.

100. J. P. Singh, H. J. Leu, R. P. Poeppel, E. Van Voorhees, G. T. Goudey, K. Winsley, and D. Shi, "Effect of Silver and Silver Oxide Additions on the Mechanical and

Superconducting Properties of $YBa_2Cu_3O_{7-x}$ Superconductors," J. Appl. Phy., **66** (7), pp. 3154–59, 1989.

101. H. M. Ledbetter, M. W. Austin, S. A. Kim, and M. Lei, "Elastic Constants and Debye Temperature of Polycrystalline $YBa_2Cu_3O_{7-x}$," J. Mat. Res., **2** (6), pp. 786–88, 1987.

102. H. M. Ledbetter, "Elastic Constants of Polycrystalline $YBa_2Cu_3O_{7-x}$," J. Mat. Res., **7** (11), pp. 2905–08, 1992.

103. H. M. Ledbetter and M. Lei, "Monocrystal Elastic Constants of Orthorhombic $YBa_2Cu_3O_{7-x}$," J. Mat. Res., **6** (11), pp. 2253–55, 1991.

104. R. Round and B. Bridge, "Elastic Constants of the High-Temperature Ceramic Superconductor $YBa_2Cu_3O_{7-x}$," J. Mat. Sci. Let., **6**, pp. 1471–72, 1987.

105. D. J. Roth, D. B. Stang, S. M. Swickard, and M. R. DeGuire, "Review and Statistical Analysis of the Ultrasonic Velocity Method for Estimating the Porosity Fraction in Polycrystalline Materials," NASA Tech. Memorandum 102501, Revised Copy, 7/1990.

106. D. J. Roth, D. B. Stang, S. M. Swickard, M. R. DeGuire, and L. E. Dolhert, "Review, Modeling, and Statistical Analysis of the Ultrasonic Velocity-Pore Fraction Relations in Polycrystalline Materials," Materials Eval., **49**, pp. 883–88, 7/1991.

107. A. Nagarajan, "Ultrasonic Study of Elastic-Porosity Relations in Polycrystalline Alumina," J. Appl. Phy., **42** (10), pp. 3693–96.

108. L. P. Martin, D. Dadon, and M. Rosen, "Evaluation of Ultrasonically Determined Elasticity-Porosity Relations in Zinc Oxide," J. Am. Cer. Soc., **79** (5), pp. 1281–89, 1996.

109. J. P. Panakkal, H. Willems, and W. Arnold, "Nondestructive Evaluation of Elastic Parameters of Sintered Iron Powder Compacts," J. Mat. Sci., **25**, pp. 1397–1402, 1990.

110. T. E. Matikas, P. Karpur, and S. Shamasundar, "Measurement of the Dynamic Elastic Moduli of Porous Compacts," J. Mat. Sci., **32**, pp. 1099–1103, 1997.

111. G. Y. Baaklini, E. R. Generazio, and J. D. Kiser, "High-Frequency Ultrasonic Characterization of Sintetred Silicon Carbide," J. Am. Cer. Soc., **72** (3), pp. 383–87, 1989.

112. R. Pampuch and J. Piekarczyl, "The Texture and Elastic Properties of Porous Graphites," Sci. Cer., (P. Popper, Ed.), Br. Cer. Soc. **8**, pp. 257–68, 1976.

113. E. A. Dean and J. A. Lopez, "Empirical Dependance of Elastic Moduli on Porosity for Ceramic Materials," J. Am. Cer. Soc., **66** (5), pp. 366–70, 1983.

114. W. R. Manning, O. Hunter, Jr., and B. R. Powell, Jr., "Elastic Properties of Polycrystalline Yttrium Oxide, Dysprosium Oxide, Holmium Oxide, and Erbium Oxide: Room Temperature Measurements," J. Am. Cer. Soc., **52** (8), pp. 436–442, 1969.

115. G. E. Portu and P. Vincenzini, "Young's Modulus-Porosity Relationship for Alumina Substrates," Ceramurgia Intl., **5** (4), pp. 165–167, 1979.

116. K. K. Phani and S. K. Niyogi, "Young's Modulus of Porous Brittle Solids," J. Mat. Sci., **22**, pp. 257–263, 1987.

117. D. R. Petrak, D. T. Rankin, R. Ruh, and R. D. Sisson, "Effect of Porosity on the Elastic Moduli of CoO, CoO-MgO Solid Solutions, and $CoAl_2O_4$," J. Am. Cer. Soc., **58** (1–2), pp. 78–79, 1975.

118. K. K. Phani, "Elastic-Constant-Porosity Relations for Polycrystalline Thoria," J. Mat. Sci. Ltrs., **51**, pp. 747–750, 1986.

119. J. Boocock, A. S. Furzer, J. R. Matthews, "The Effect of Porosity on the Elastic Moduli of UO_2 as Measured by an Ultrasonic Technique," Report: AERE-M2565, Process Technology Division, Atomic Energy Res. Establishment, Harwell, Berkshire, England, 1972.

120. S. K. Datta, A. K. Mukhopadhyay, and D. Chakraborty, "Young's Modulus-Porosity Relationships for Si3N4 Ceramics—A Critical Evaluation," Am. Cer. Soc. Bull., **48** (12), pp. 2098–2102, 1989.

121. A. Salak, V. Miskovic, E. Dudrova and E. Rudnayova, "The Dependence of Mechanical Properties of Sintered Iron Compacts Upon Porosity," Pwd. Met. Mat., **6** (3), pp. 128–132, 1974.

122. S. Spinner, F. P. Knudsen, and L. Stone, "Elastic Constant-Porosity Relations for Polycrystalline Thoria," J. Research, National Bureau of Standards, **67C** (1), pp. 39–46, 1–3/1963.

123. N. Soga and E. Schreiber, "Porosity Dependence of Sound Velocity and Poisson's Ratio for Polycrystalline MgO Determined by Resonant Sphere Method," J. Am. Cer. Soc., **51** (8), pp. 465–66, 1968.

124. S. C. Nanjangud, R. Brezny, and D. J. Green, "Strength and Young's Modulus Behavior of a Partially Sintered Porous Alumina," J. Am. Cer. Soc., **78** (1), pp. 266–68, 1995.

125. S. C. Nanjangud and D. J. Green, "Mechanical Behavior of Porous Glasses Produced by Sintering of Spherical Particles," J. Eu. Cer. Soc., **15**, pp. 655–60, 1995.

126. R. W. Rice, "Comparison of Stress Concentration Versus Minimum Solid Area Based Mechanical Property-Porosity Relations," J. Mat. Sci., **28**, pp. 2187–90, 1993.

127. J. J. Cleveland and R. C. Bradt, "Grain Size/Microcracking Relations for Pseudobrookite Oxides," J. Am. Cer. Soc., **61** (11–12), pp. 478–81, 1978.

128. W. R. Manning, "Anomalous Elastic Behavior of Polycrystalline Nb_2O_5," Ph.D. Thesis, Iowa State Univ., Ames, IA, 5/1971.

129. W. R. Manning, "Anomalous Elastic Behavior of Polycrystalline Nb_2O_5," J. Am. Cer. Soc., **56** (11), pp. 602–3, 1973.

130. S. L. Dole, O. Hunter, Jr., and D. J. Bray, "Microcracking of Monoclinic HfO_2," J. Am. Cer. Soc., **61** (11–12), pp. 486–90, 1978.

131. J. E. Brocklehurst, "Fracture in Polycrystalline Graphite," Chemistry and Physics of Carbon, vol. 13 (P. L. Walker and P. A. Thrower, Eds.), Marcel Dekker, Inc., New York, pp. 145–279, 1977.

132. R. W. Zimmerman, "The Effect of Microcracks on the Elastic Moduli of Brittle Materials," J. Mat. Sci. Let., **4**, pp. 1457–60, 1985.

133. H. Banno, "Effects of Shape and Volume Fraction of Closed Pores on Dielectric, Elastic, and Electromechanical Properties of Dielectric and Piezoelectric Ceramics-A Theoretical Approach," Am. Cer. Soc. Bull., **66** (9), pp. 1332–37, 1987.

134. M. L. Dunn and M. Taya, "Electromechanical Properties of Porous Piezoelectric Ceramics," J. Am. Cer. Soc., **76** (7), pp. 1697–706, 1993.

135. R. W. Zimmerman, "Behavior of the Poisson's Ratio of a Two-Phase Composite Material in the High-Concentration Limit," Micromechanics of Random Media

(M. Ostoja-Starzewski and I. Jasiuk, Eds.), ASME Book. No. AMR 139S170, pp. S38–44, 1994. Appl. Mech. Rev., **47** (1, part 2).

136. K. E. Evans and B. D. Caddock, "Microporous Materials with Negative Poission's Ratios: II. Mechanisms and Interpretation," J. Phy. D Appl. Phy., **22**, pp. 1877–82, 1989.

137. R. Lakes, "Deformation Mechanisms in Negative Poissions Ratio Materials: Structural Aspects," J. Mat. Sci., **26**, pp. 2287–92, 1991.

138. E. A. Friis, R. S. Lake, and J. B. Park, "Negative Poission's Ratio Polymeric and Metallic Foams," J. Mat. Sci., **23**, pp. 4406–14, 1988.

139. J. B. Choi and R. S. Lakes, "Non-Linear Properties of Polymer Cellular Materials with a Negative Poission's Ratio," J. Mat. Sci., **27**, pp. 4678–84, 1992.

140. A. Yeganeh-Haeri, D. J. Weidner, and J. B. Parks, "Elasticity of α-Cristobalite: A Silicon Dioxide with a Negative Poission's Ratio," Science, **257**, pp. 650–52, 7/1992.

141. N. R. Keskar and J. R. Chelikowsky, "Negative Poission's Ratio SiO_2 from First-Principles Calculations," Let. to Nature, **358**, pp. 222–24, 7/1992.

142. N. Ramakrishnan and V. S. Arunachalam, "Reply to 'Comment on Effective Elastic Moduli of Porous Ceramic Materials'," J. Am. Cer. Soc., **78** (6), p. 1712, 1995.

143. S. M. Lang, "Properties of High-Temperature Ceramics and Cermets, Elasticity and Density at Room Temperature," Nat. Bureau of Standards Monograph 6, U.S. Govt. Printing Office, Washington D.C., 3/1960.

144. R. W. Rice, "Effects of Inhomogeneous Porosity on Elastic Properties of Ceramics," J. Am. Cer. Soc., **58** (9), pp. 458–9, 1973.

145. C. Cm. Wu and R. W. Rice, "Porosity Dependance of Wear and Other Mechanical Properties on Fine-Grain Al_2O_3 and B_4C," Cer. Eng. Sci. Proc., **6** (7–8), pp. 977–94, 1985.

146. K. Phani, "Porosity-Dependence of Elastic Properties and Ultrasonic Velocity in Polycrystalline Alumina—A Model Based on Cylindrical pores," J. Mat. Sci., **31**, pp. 262–66, 1996.

147. K. Phani, "Porosity-Dependence of Ultrasonic Velocity in Sintered Materials—A Model Based on the Self-Consistent Spheroidal Inclusion Theory," J. Mat. Sci., **31**, pp. 272–79, 1996.

148. K. K. Phani and S. K. Niyogi, "Young's Modulus of Porous Brittle Solids," J. Mat. Sci., **22**, pp. 257–63, 1987.

149. K. K. Phani and S. K. Niyogi, "Young's Modulus-Porosity Relations in Polycrystalline Rare-Earth Oxides," J. Am. Cer. Soc., **70** (12), pp. C-362–66, 1987.

150. D. Hardy and D. J. Green, "Mechanical Properties of a Partially Sintered Alumina," J. Eu. Cer. Soc., **15**, pp. 769–75, 1995.

4

POROSITY AND MICROCRACK DEPENDENCE OF FRACTURE ENERGY AND TOUGHNESS AND RELATED CRACK PROPAGATION AT 22°C

KEY CHAPTER GOALS

1. Show that overall fracture energy and toughness trends, and frequently individual data sets, are consistent with MSA models and E-P trends, but there are significant variations which must be considered when evaluating the porosity dependence of these properties.
2. Demonstrate that variations are mainly due to crack interactions (e.g., branching and bridging) with various combinations of larger grains and pore distributions on a scale substantially larger than that of the microstructure.
3. Show that porosity and microcracks often have similar effects, but issues of pore-microcrack interactions, stress/crack-generated versus preexisting microcracks, and behavior of some (larger or multigrain?) microcracks remain.

4.1 INTRODUCTION

The focus of this chapter is primarily the porosity, and secondarily the microcrack, dependence of fracture energy (γ) and fracture toughness (K_{IC}) nominally at ~22°C, where the bulk of the data is. These properties are an important transition from the basic elastic properties, determined by nondestructive means, to fracture properties such as tensile and compressive strengths determined by destructive means. Fracture energy and toughness also introduce grain-size effects and their interaction with porosity effects as well as effects of crack-size-to-microstructure ratio effects that can be extreme. The latter typically reflect effects of larger scale crack propagation, especially wake effects, such as bridging. These, while having received considerable study in dense materials, have previously received little or no attention to effects of porosity and microcracking on them. Data showing probable crack branching from porosity, a previously neglected phenomenon, and limited observations showing grain boundary pores aiding intergranular fracture are presented. Little or no study exists of the effects of porosity or microcracking on slow crack growth due to environmental effects. Thus, the possibility that open porosity might affect crack propagation by providing an alternate route for active species, e.g., H_2O, to the crack tip other than by their diffusion only along the crack itself, hence not limiting crack growth rate to that of the diffusion, has not been studied; however, the one study of environmentally driven slow crack growth in rigidized glass fiber felts is briefly discussed.

4.2 CONCEPTS AND MODELS

4.2.1 Porosity Dependence

While fracture energy, and especially fracture toughness, are central to the fracture mechanics approach to the strength and fracture of ceramics, there has been limited attention to developing theoretical relations for the porosity-dependence of these properties. This is due in part to the development of fracture mechanics at a later time than when there was more interest in porosity effects on properties. Thus, there has been much less consideration of these properties from either an empirical, analytical, or experimental standpoint; however there is limited but useful quantitative and qualitative guidance on the porosity-dependence of fracture energy and fracture toughness, one component of which is correlation with Young's modulus.

Fracture energy (γ) and toughness (K_{IC}) are intimately related to each other and to Young's modulus (E) for simple linear elastic fracture by the equation:

$$K_{IC} \sim (2E\gamma)^{\frac{1}{2}} \qquad (4.1)$$

(the $\sim$ symbol is used, since depending on whether the applied stress condition is one of plain strain or stress the equation is respectively exact, or the terms under the square root are divided by $(1 - v^2)$). The relation to E is increased by the fact that γ is related to E, i.e., a basic theory (1) gives:

$$\gamma \sim \left(\frac{E}{l}\right)\left(\frac{r}{\pi}\right)^2 \qquad (4.2)$$

where l is the equilibrium lattice constant normal to the plane of fracture and r is the range of the atomic bonding forces. Neither l nor r vary widely for many ceramics, are constant for a given material, and do not depend on porosity. Thus, γ is closely related to E, especially for a given material with varying porosity, i.e., E is the only factor in γ that is dependent on porosity.

The minimum solid area (MSA) and other reasonable models for the porosity dependence of E are thus relevant to that of γ and K_{IC} whenever Eqs. 4.1 and 4.2 are at least approximately applicable, i.e.:

$$K_{IC} \sim [(2E\, f(P)\, \gamma\, f(P)]^{\frac{1}{2}} \sim K_{IC0}\, f(P) \qquad (4.3)$$

where K_{IC0} = the value at P = 0, and f(P) is the functional dependence on P, e.g., e^{-bP} or $(1 - AP)^n$ for E.

Thus, whenever crack propagation is controlled by simple linear elastic fracture, i.e., when Eq. 4.1 holds along with Eq. 4.2 (or a similar relation), the porosity dependence of γ and K_{IC} will be the same as that of E (and commonly also σ, Chapter 5). Such correlation will be shown to frequently be the case, but there are variations that affect a number of tests and bodies, with deviations from E-P trends generally greatest for γ, next for K_{IC} (and less for σ, Chapter 5).

The MSA approach also has direct application to the porosity dependence of γ and K_{IC} since it is based on the minimum solid area supporting stress. Thus, as previously argued (2–4), fracture energy and toughness are measures of the resistance to a propagating crack and the resultant fracture surface generation. Therefore, since fracture is typically a weak-link process, it will seek out the minimum area to fracture, i.e., indicating direct correlation with MSA, not just via its correlation with E (Fig. 4.1).

Direct application of MSA models, and application of models for the porosity-dependence of Young's modulus implies the same or similar bounds for γ and K_{IC} as for E; however, these are of limited value as for E (Eqs. 2.3, 2.4, 3.4). Further difficulty of the use of bounds and more general evaluation is compounded by the P = 0 values of γ and K_{IC} often being dependent on the test method, material, microstructure, or various combinations of these. At the other extreme of P, P_C values for E-P trends are generally applicable, except they will commonly be somewhat higher for low values, e.g., for $P_C \lesssim 0.5$ and somewhat lower at high P_C values. This arises since as porosity heterogeneity increases it respectively decreases the regularity of particle packing at lower P_C and of pore stacking at high P_C values. Because fracture is a weak-link process, more porous areas have greater impact on P_C values than for properties determined nondestructively, such as elastic moduli. The limited guidance on bounds combined with the limited data and microstructural characterization often limit identification of probable trends, which are fortunately generally corroborated by contrast and similarities with E-P and σ-P trends.

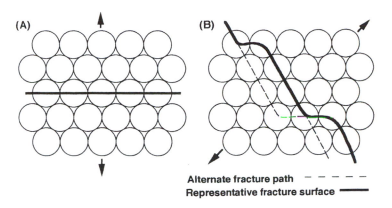

Alternate fracture path   – – – – –
Representative fracture surface ━━━

Figure 4.1 Examples of fracture paths in idealized simple cubic stacking of uniform spherical bubbles. A) Representative MSA fracture path for stressing in a <100> direction, i.e., fracture through bubble walls, along their equators for this stacking. Note that for such simple, perfect stacking the crack cannot jog onto another layer of bubbles without an increase in fracture area and hence energy. B) A representative MSA fracture path and an alternate path for a <110> stressing direction. Such alternate paths are a natural consequence of the misorientation of ideal fracture paths and stacking planes (as stacking defects would also be). Such alternate paths are probable basic sources of crack branching and bridging, besides frequent, larger scale porosity heterogeneities (see Fig. 4.2).

Correlation with Young's modulus and its porosity dependence and direct applicability of MSA models thus provides an important basis for evaluating the porosity dependence of fracture energy and toughness; however, it is also of value to consider the few models that have been developed for these properties, often for limited types or ranges of porosity. At least one model has been derived for porous metals that may be applicable to some ceramics as outlined in Section 8.2.1. Those derived specifically for ceramics, usually by investigators considering porosity dependences of other properties, are considered here. Wagh et al. (5) adapted their approach of replacing aligned, rectangular grains in dense bodies by cylinders of the same length as the grains, but of varying smaller radius to introduce porosity as was used for their derivation of the P dependence of E. This gave, mainly for low-moderate P,:

$$\frac{\gamma}{\gamma_0} = (1 - P)^{n+1} \qquad (4.4)$$

$$\frac{K_{IC}}{K_{ICO}} = (1 - P)^{n+0.5} \qquad (4.5)$$

where n = the exponent in the Eq. $E/E_0 = (1 - P)^n$. Their equations thus do not show that the P-dependence of E, γ, and K_{IC} is the same, but rather show more porosity dependence of γ, not less, as is often observed when the porosity dependence of γ, K_{IC}, and E are not the same. Their tests of Eq. 4.5 on five data sets based on the logical assumption that tensile strength also has the same porosity dependence as K_{IC} (see Chapter 5), showed n values from 2.1 to 4.1, and exponents for strength being 0.4 to 1.5 higher (averaging ~ 0.5 higher as predicted); however, Arato (6) recently pointed out an error in the derivation of Eqs. 4.4 and 4.5 that removes the discrepancies of the exponents from that of E, i.e., the correct exponents for Eqs. 4.4 and 4.5 are both n. Other sources of varying porosity dependence of γ and K_{IC} are noted later.

A model for γ and K_{IC} was derived by Yasuda et al. (7) based on analysis of changes in the Gibb's free energy of a body with a main crack and spherical pores (which is small in comparison to the crack half-length, c) as a function of P giving:

$$\frac{K_{IC}}{K_{IC0}} = (1 + AP)^{-\frac{1}{2}} \tag{4.6}$$

from which they derived:

$$\frac{\gamma}{\gamma_0} = (1 + AP)^{-1} \tag{4.7}$$

where A is given in terms of ν, c, and half of the specimen dimension normal to the crack plane (L). For $\nu = 1/3$ and $\frac{1}{2}$ A only changes respectively from 0.53L/c to 0.58L/c. This makes the dependence on specimen and crack dimensions explicit, which raises a concern for this model. Two other concerns are: 1) the equations may often not go to zero at P = 1 or less, and 2) they are not consistent with the average γ - P and K_{IC} - P trends being similar to those of E-P.

Arato et al. (8) derived a model for the toughness of a body of partially sintered particles with average particle and sintered neck radii of R and r, respectively, (the latter clearly relating to MSA; Tables 2.6, 2.8), giving:

$$K_{IC} = A\sigma_n R^{1/2} \left(\frac{r}{R}\right)^2 \tag{4.8}$$

where σ_n = the strength of the sintered neck. In their derivation A =1, while in a similar derivation Green and Hardy (9) obtained the same expression, but with A ~ 1.4 $[1/4(\pi)^{3/2}]$. The explicit dependence on r can be eliminated by using similar relations for Young's modulus (Table 2.8), giving:

$$K_{IC} = A\sigma_n R^{1/2} \left(\frac{E}{E_0}\right)^n \qquad (4.9)$$

where A and n are both 1 for Arato et al. (8) and, respectively, $A(C_n)^{-2}$ and 2 for Green and Hardy (9). The differences arise primarily from different dependences of E on the exponent n for the ratio r/R (Table 2.8). Thus Arato et al.'s (8) expression is consistent with fracture toughness and Young's modulus (as well as hardness and strength) having the same porosity dependence while that of Green and Hardy (9) is not.

Green and Hoagland (10) presented a model for the fracture toughness of bodies of sintered balloons (i.e., for moderate to high P), fitting data for their glass balloon bodies (Fig. 3.9B). While this model is for a rather specialized structure, it is an example of marrying aspects of load-bearing and detailed mechanical analysis since it focuses on mechanical analysis of the fracture of bonds between balloons, i.e., of their MSA.

There have been a few models derived for fracture energy or toughness, or both of foams (11–13) and of rigidized glass fiber felts (14) (Table 4.1), i.e., for particular high porosity bodies. These are basically MSA models, e.g., assuming (uniform) struts of open-cell foams determine the behavior, neglecting tapering cross-sections toward their centers. Modeling of the effects of foam anisotropy on K_{IC} has been done by Huber and Gibson (15) in conjunction with that for E (Eq. 3.5) giving the ratios of the toughnesses on fracture planes normal versus parallel to the direction of foam rising (and assumed elongation); $K_{IC} = R$ (=H/L) when the crack is also propagating normal to the rise direction, and $K_{IC} = R^{3/2}$ when the crack propagation is parallel to the rise direction. For cracks whose plane is parallel to the rise direction the ratio of toughness for their propagation being normal versus parallel to the rise direction, is $R^{1/2}$.

The P-dependences of K_{IC} and E for the above foam models are not necessarily the same, implying that the porosity dependence of γ is $1 - P$ for the

Table 4.1 Fracture Toughness-Porosity Relations for Foam and Felt Ceramic Bodies

Material[a]	Normalized toughness[b]	~ C_i Value[c]		$E/E_0 \propto$	Source
		Theo.	Exp.		
Closed cell foam	$C_1(1-P)^2$	1	0.65	$(1-P)^3$	Maiti et al. (11)
Open cell foam	$C_2(1-P)^{3/2}$	1	0.65	$(1-P)^2$	Maiti et al. (11)
Glass felt	$C_3(1-P)^{3/4}$	—	—	—	Green and Lange (14)

[a] Glass felt has fibers bonded at their junctions.

[b] Some earlier normalization was based on dividing K_{IC} by $[\sigma_s(\pi l)^{1/2}]$, where σ_s = foam strut strength and l = cell dimension for open cell foam, but the model of Maiti et al. (11) instead used the K_{IC} of the cell struts for normalization.

[c] Approximate theoretical or experimental values of C_1 or C_2.

foams, thus giving E, γ, and K_{IC} with different dependences on P; differing some from expectations discussed above for K_{IC} and substantially for γ; however, recall that these models are neither necessarily exact nor universal [e.g., some models for differing cell arrangements gives $E/E_0 = 1 - P$; see Fig. 3.3; Gent and Thomas (16)]. Some data for foams and related materials and data for much less porous bodies later will be shown to often be in agreement with the same or similar P dependence of K_{IC} and E; however, considerable variation and some differences of $\gamma - P$ and $K_{IC} - P$ versus $E - P$ trends will be shown, but for differing reasons and over a limited P range. An important question for all fracture properties is that of the effect of crack to cell size. Studies indicate that the fracture toughness of brittle foams is independent of crack size provided that this is ≥ 10 times the foam cell size (17).

4.2.2 Microcrack Dependence

As outlined in Chapters 1 and 2, microcracks commonly occur from mismatch strains arising from differences in elastic moduli, phase transformations, and especially thermal expansion mismatches between grains or particles in a body. While the resulting stresses are independent of the grain or particle scale, the forming of microcracks occurs only above a critical grain or particle size, D_S, dependent on the local properties controlling the specific strain mismatches (Eq. 10.1). This grain- or particle-size dependence of the occurrence (and character) of microcracks thus is a substantial and complex subject, part of which is treated in subsequent considerations of these important microstructural parameters. The broader trends related to the microcracks affecting properties in fashions pertinent to the porosity dependence of properties are treated here.

Microcracks are idealized as simple arrays of simple planar shapes such as penny cracks of zero thickness, but giving a discontinuity of properties normal to their surface. Models for the effects of such cracks on elastic properties (Section 3.2.3) are in principle directly applicable to fracture energy and toughness (Eqs. 4.1 and 4.2) in the same way that models for the porosity dependence of elastic properties are, as discussed in the previous section. There are, however, frequent complications in a simple direct correlation of γ- and K_{IC}- microcrack dependence with those of Young's modulus for the same microcracks, similar to those for the corresponding porosity dependences. Some of the former are discussed below and in the next section (along with those for porosity dependences). Fundamental complications of treating microcracking effects on fracture properties arise from the fact that the microcrack population is often not fixed as is typically the case for porosity (except under creep conditions). A key result of this is that the microcrack population frequently changes substantially with changes of stress or temperature. Thus, while preexisting microcracks, such as those that are of primary effect on most elastic property measurements, are important in fracture processes involved in crack propagation, they are frequently only part of the net microcrack population affecting crack propagation. This arises since stresses near the tip of a macrocrack subjected to external applied stress can generate new microcracks (and may modify existing microcracks). Thus, Fu and Evans (18) reported modeling showing that high crack-tip stress fields extend the occurrence of microcracking down to $\sim 40\%$ of D_S, thus adding to the microcrack population during stressing.

A second major complication in modeling effects of microcracking on macrocrack propagation is the effect of the former on the latter. Some aspects of this are discussed in the next section, but the key issue of propagation of the failure-causing crack by its linking with microcracks ahead of it is discussed here. The net result is that models of fracture toughness effects of microcracking associated with increasing grain size (D) (18,19) leads to toughness independent of D until $D \sim 0.4D_S$, then increasing with increasing D due to increasing microcracking, reaching a maximum at D_S. Toughness then decreases with further increase of grain size, ultimately falling below the fine grain levels due to enhanced crack propagation via microcrack linking. The first of two key aspects of such dependence is that the trends outlined are in reasonable agreement with experimental data (see Fig. 4.7) for measurements with large, but not necessarily small, cracks. Second, such toughness effects reduce or eliminate direct correlation of toughness and (the bulk or global) Young's modulus since the microcrack population interacting with the macrocrack is changing significantly in the toughness measurements above $D \sim 0.4 D_S$, decreasing the local E.

A third complication is the primary source of the increased toughness. Earlier views were that the effects of microcracking arose primarily from their formation at and ahead of the crack tip (20–22), e.g., due to general energy considerations or deflection or branching of the main crack. Subsequently, the focus shifted to effects of the microcracks in the wake of the crack, i.e., behind the crack tip reducing the driving force for extension of the main crack (18,23). While such wake effects raise questions of crack size and shape (24), they give better predictions of large crack results (and are also more consistent with treatment of transformation toughening).

A fourth complication is the combination and interaction of pore and microcrack effects, since both are often present. While it is commonly assumed that pores and microcracks are independent elements with independent effects, this is not necessarily so. Clarke (25) proposed a model where (spherical) grain boundary pores act as nuclei of microcracks forming from grain mismatch strains. More generally there can be pore-microcrack interactions (e.g., Fig. 7.10A).

In addition to the above brief outline of a complex and developing subject, other models and related specialized topics and representative references include: 1) other models, e.g., based on the same analysis for other properties (26), statistical aspects of microcrack linking (27), and computer modeling (28–

30). The first model (26) shows microcracking actually increasing K_{IC} above that of the uncracked matrix at small microcrack sizes and spacings, e.g., in the nanometer range. Another reference shows that some of the toughening from microcracking also results from the local reduction of Young's modulus around the crack tip (29). Another example of adapting other models is that of Yasuda et al. (7). They used the same approach for pores to obtain nearly identical equations for bodies with fine, nominally randomly oriented penny-shaped microcracks. The only difference was that the A factor, while similar to that for pores in Eqs 4.6 and 4.7, was more complex due to it also depending on the microcrack diameter and thickness, i.e., opening.

4.2.3 Deviations of Fracture Energy and Toughness Trends from those of Young's Modulus

While there are both theoretical and practical grounds for expecting the porosity and microcrack dependence of E, γ, and K_{IC} (as well as σ) to be the same or similar, as discussed above, there are also grounds for expecting deviations. There are two intrinsic and one extrinsic reasons. The probable broadest intrinsic reason arises from the possibility of crack branching, bridging, and related phenomena from crack interactions with pores or microcracks, as outlined in Figs. 4.1, 4.2. Thus, for example, a body of partially sintered uniform spherical particles or pores in simple cubic packing stressed in a <100> direction would normally fracture along a {100} plane (i.e., using Miller indices in analogy with crystal structure designations, Fig. 4.1 A); however, significant perturbations from this idealized fracture path can be seen in model bodies, for other stressing directions (2,3,31) (Fig. 4.1 B). Similarly, fracture of bodies with denser ideal (or defective) stackings for any stressing direction present a more complex picture. These complexities arises from the fact that while cracks would like to follow paths of minimum bond (solid) area, they face dilemmas in that, except for very ideal cases such as those of Fig. 4.1 A, the paths for minimum solid fracture areas on the scale of a single pore or particle and on a multiple particle or pore scale are not necessarily the same. This means that fracture will commonly generate more net fracture area than in idealized cases (i.e., contrast Figs. 4.1A and B). Similar effects should occur for arrays of microcracks. Further, alternate fracture paths are more likely to occur and be more extensive, extreme, or both in real bodies due to heterogeneous location and character of pores, and their changes with grain growth (Fig. 4.2). All of these variations are likely to result in compromises in the fracture path, i.e., not necessarily always fracturing just the minimum bond areas. However, note that there is likely to be a pore size dependence to some of these mechanisms. Such size dependence should result when there is a significant dependence of the mechanism on the level of stress concentration ahead of the main crack. In such cases smaller pores will experience less stress decrease, hence higher net stress, across the pores, thus favoring cracking of smaller pores.

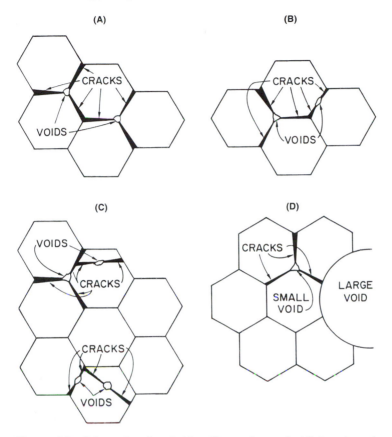

Figure 4.2 Schematic of probable effects of pore (void) location, shape, size, and spatial distribution relative to grain structure on crack branching (hence also possibly bridging). Schematics of crack linking and branching from A) pores at grain triple points, B) similar effects due to a pore on a grain boundary and one at a triple point, C) intra- and intergranular pores, D) a triple-point pore and a larger pore (e.g., from particle bridging), and E) a crack front (solid line) shown approaching some spherical pores from the left. Dashed lines reflect subsequent cracks above or below the plane of the original crack, hence crack branching, while the connected solid line reflects continuation of the original crack in essentially its original plane.

These effects of varying fracture paths from crack interactions with pores or microcracks are likely to lead to crack branching, bridging, or both (e.g., Fig. 4.2E). Some direct evidence of this has been previously shown (Fig. 4.3) (32). Evaluation of the porosity dependence of γ and K_{IC} (4) strongly indicate that such crack phenomena is a fairly common occurrence which can have variable, often significant effects on some fracture behavior, as discussed further in this

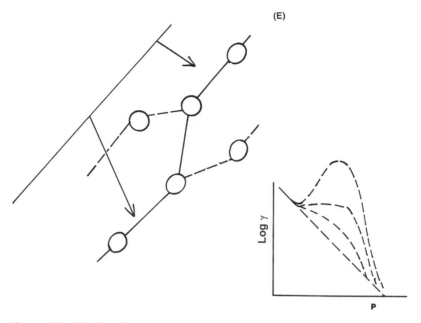

Figure 4.2   Continued

chapter. Heterogeneous porosity is likely to play an important role in some of these effects. The net effect of these qualitative evaluations is that while a general correlation of the porosity dependences of E, γ, and K_{IC} is expected, considerable variations may occur that are not necessarily related to those reflected in Eqs: 4.1–4.3 (or 4.4, 4.5 as corrected). Crack branching and bridging will increase γ, often significantly, while the effects on K_{IC} will be mitigated by the effect of E, since it is less affected by the porosity distribution; however, these effects can be significantly affected by the nature of the γ- or K_{IC}-test methods since this impacts the opportunity and extent of branching and bridging, especially the scale of the initial crack and its propagation. Evidence for such effects, which implies deviations for all fracture energy- and toughness-P, including MSA, models will be shown later.

There are two further effects on fracture energy and toughness that can intrinsically or extrinsically effect their porosity dependences. Both effects are associated with larger grains. The first is the location of pores, primarily in polycrystalline bodies (which also affects their shape, size, and homogeneity of distribution). In bodies of partially sintered glass or crystalline particles, the grain-particle size and the amount and size of the pores are intimately related in the same fashion since the pores are the primary demarcations of the boundaries between the particles or grains; however, as densification proceeds there begins

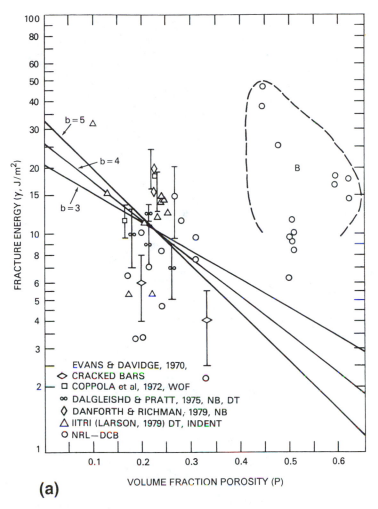

(a)

Figure 4.3 Anomalously high fracture energy of reaction sintered Si_3N_4 at 22°C. a) Fracture energy versus P. Note the data set partly outlined by a dashed line for bodies with a dual pore population of natural and artificially introduced, larger spherical (20–200 μm dia.) pores versus material with only natural pores to the left and below (with slopes of b = 3–5 through the centroid of that data). b) Microradiograhic photos of cracks in bodies with the dual-pore population and high fracture energy. The three larger photos (A, B, C) are higher magnifications of crack positions shown in the thinner photo. Note crack branching and possibly bridging associated with the larger, artificially introduced pores. Published with permission of J. Mat. Sci. (32).

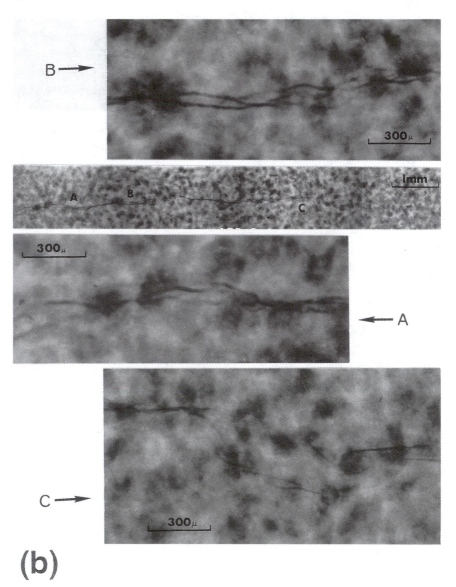

(b)

Figure 4.3 Continued

to be increasing differentiation between the relation of the pores and the particles or grains. In both types of bodies, increased sintering, i.e., reduction of porosity, results in particle or grain growth enveloping some pores within the particles or grains with differing results (Figs. 1.1, 1.2). In sintered glass bodies there are no grain boundaries or crystalline anisitropies, so pores enveloped within particles can have shapes and sizes more like those that are between particles, and there is less, or no, inherent differentiation in trans- versus inter-particle fracture; however, in polycrystalline bodies grain boundaries and related anisotropy result in differing grain-boundary and intragranular pore shapes so envelopment of pores within grains results in significant differentiations (Figs. 1.3, 1.4). Smaller pores are more easily enveloped by grain growth than are larger pores, and pore growth along grain boundaries is easier than within grains, so grain-boundary, i.e., intergranular, pores commonly are, or become, larger than intragranular pores. Further, pores within grains typically vary from polyhedral to approximately spherical in shape, while intergranular (i.e., boundary) pores often become lenticular or cusp shaped, especially as grain size becomes larger than the pores and P decreases (e.g., Figs. 1.3, 1.4). Finally, there is typically some differentiation between intergranular and transgranular fracture, so fracture is often predominantly intergranular or transgranular, with intergranular pores frequently shifting the balance toward intergranular fracture, especially at higher temperatures (33). The frequent larger pore size (and hence also often greater amount of porosity) along grain boundaries combined with its frequent greater boundary coverage due to its common lenticular shape increases the effects of pores in enhancing intergranular fracture.

The overall impact of the differentiation of intragranular and grain-boundary pores is that the fracture cannot fully interact with the full cross-section of the porosity. Thus, transgranular fracture will reflect direct interaction with both a fair amount of intragranular pores and also some grain-boundary pores. On the other hand, intergranular fracture reflects direct interaction with only grain-boundary pores. Further, since there are many fewer intergranular than transgranular paths, intergranular fracture probably reflects a biased sampling (i.e., toward more grain-boundary coverage of pores) of the total pore population. Clearly, these effects can be further complicated if there is mixed mode fracture, i.e., some intergranluar and some transgranular fracture, with this being even more complicated if the fracture mode changes as a function of crack size, location, velocity, etc. (e.g., as discussed further in Chapter 5), as well as heterogeneities in the distribution of either, or both, types of porosity. Since γ extensively reflects this interaction, it will again be affected most, while the effects on K_{IC} will be mitigated by the more general effects of the pores on E, which will be affected by both types of pores regardless of the resultant fracture mode. Interaction with less than the full extent of porosity (e.g., following intergranular, but missing most or all intragranular, pores) will intrinsically give higher γ and K_{IC} values for a given porosity (hence lower b values). On the other hand, following intergranular paths due to limited porosity only or mainly along the fracture path will result in lower γ and K_{IC} values, hence higher b values. These effects may be respectively enhanced or partly countered by crack branching due to pores (or microcracks). An extrinsic increase in γ and K_{IC} at low P will often occur if grain size is substantially larger there, giving an apparent increase in the porosity dependence (i.e., higher b values).

Finally, the variations and uncertainties of the techniques of measuring γ and K_{IC} must be considered in evaluating effects of porosity on these properties, since techniques for measuring them are subject to more variations than in measuring elastic moduli and often also than for measuring strength. While measuring strength depends on test methods and parameters, the presence and causes of these variations are generally better recognized and addressed. A complete evaluation of the problems of measuring γ and K_{IC} cannot be given here both because of its length and scope as well as incomplete understanding, but a few key points will be made. First, results are often dependant on both the starting crack size, as well as the extent of its propagation, especially where bridging occurs (e.g., Figs. 4.1, 4.2). This depends on both the specific technique used (e.g., crack sizes typically increase in the following approximate order: values from fractography, indentation or indentation fracture, notch beam and work of fracture and other chevron notch methods, and double-cantilever and double-torsion tests) (34), as well as parameters of the specific test. Second, there are a number of partially identified or understood factors for various individual techniques. Thus there are (variable) residual stress effects on indentation techniques, and of the presence and character of cracks in the notch-beam method (35).

4.3   COMPARISON OF DATA AND THEORETICAL EXPECTATIONS

4.3.1   Porosity

There is limited data on the crack propagation and fracture energy and toughness for ceramics in general and less for special materials, e.g., graphites, as a function of porosity in part because of the limited range of porosity commonly available. However, Brocklehurst (36) noted that cracks frequently originate at pores in graphite and propagate through and sometimes terminate on them. These crack-pore interactions were more frequent for larger pores or pore clusters, especially for cracks propagation through or terminating on them. One of the few studies of the porosity dependence of fracture toughness in graphite is that of Kimura et al. (37). They studied two graphites (P ~ 0.13–0.21) that

showed considerable scatter (with as much as or more for E as for other properties), but had approximately linear decreases in E, K_{IC}, and σ versus P. The b
values for the respective properties for the two graphites were: 0.6, 2; 1.9, 3.1;
and 3.2, 3, i.e., some similar and different values for E and σ.

Consider now γ- or K_{IC} - P data for other ceramic materials tested against
specific models. Green and Hardy (9) experimentally tested their Eq. 4.8 for K_{IC}
for a limited range of porosity (P: 0.36–0.42) in sintered alumina. They showed
that over most of this limited porosity range that the exponent, n (of E/E_0 in Eq.
4.9) was ~ 1.2, but increased to ~ 2 at the higher P level; however, they did not
consider that the change to a higher exponent may well have simply reflected
the rollover toward P_C. Thus, overall results were closer to trends predicted by
Arato et al. (8) than by Green and Hardy; however, such K_{IC} - E evaluation of γ-
and K_{IC}-P trends is aided by comparison with E-P trends.

Consider now broader γ- or K_{IC}-P trends from general data that was not
generated to test models. While fracture energy and toughness data can be interrelated with that of E (Eqs. 4.1, 4.2), the specific porosity dependence of E is not
available for some of the few specific bodies for which γ- or K_{IC}-P data is available and suitable porosity characterization is lacking. Also, there is very little
data for pores of near model character for γ, and little for K_{IC}. Thus, there is
greater uncertainty in discerning γ- and K_{IC}-P trends; however, careful evaluation of the data combined with other pertinent observations (including data in
Chapter 5) indicates important, logical trends. To do this the limited actual γ-P
data is presented and discussed, then key examples of this are converted to K_{IC}-
P data (using actual or typical E-P dependences) for comparison with the larger
but still limited K_{IC}-P data base. Previous surveys (2,3) showed that γ-P (and
hence also K_{IC}-P) data generally is linear on a semilog plot over the typically
limited to moderate P range covered. Thus, although there are variations (discussed below), the semilog slope (b value) is used as a characterization tool, e.g.,
to compare the P dependence of E, γ, and K_{IC} (as well as σ or other properties).

Much of the very limited γ-P data [in and beyond earlier surveys (2–4)] is
summarized in Table 4.2 along with the K_{IC}-P data amenable to characterization
by b values. Though limited and variable, three trends are indicated by this data.
The first and fairly consistent trend is for the overall P-dependence of γ, K_{IC} and
E (and often also σ, Chapter 5) to be similar as expected if simple elastic fracture occurs, as discussed earlier. Six of the 16 data sets are most consistent with
this, two for oxides (38,46), and four for nonoxides (8,32,39,50), and four data
sets suggest this based on similarity of b values for σ and E or σ and K_{IC}
(8,32,38,43,44). These were typically in finer grain bodies. Lam et al.'s (38)
measurements of the P dependence of σ, E, and γ (with a chevron-notched
beam, then calculating K_{IC} from measured γ and E values from Eq. 4.1) for their
partially sintered Al_2O_3 gave the most similar b values for all of these properties. The colloidal processing may imply greater homogeneity that would be

Table 4.2 Comparison of the Porosity Dependence of Fracture Energy (γ) and Toughness (K) with that of Young's Modulus (E) and Tensile Strength (σ)[a]

Material	Fabrication[b]	GS[c] (μm)	P[d]	Slope[e] (b value) for: γ	K	E	σ	Investigator
Al_2O_3	PC	0.5	0.04–0.42	3.3 (NB)	3.4	3.4	3.4	Lam et al. (38)
Al_2O_3	HP	~1	0.02–0.40	1.5 (DCB)	2.2	2.6	—	Wu and Rice (39)
Al_2O_3	HP	~1	~0–0.9	0 (WOF)	2	—	—	Cappola and Bradt (40)
Al_2O_3	S	5–50	~0–0.44	~3 (NB)	~3.4[f]	2	—	Pabst (41)
Al_2O_3	HP/S	2–20	~2–0.46	- (NB)	<4.2[f]	2.1	—	Claussen et al. (42)
Al_2O_3	S	3	0.05–0.50	- (SEPB)	2.5	—	2.9	Evans and Tappin (43)
Al_2O_3	S	~1	0.02–0.47	- (SEPB)	3	3.8	3.8[g]	Knechtel et al. (44)
MgO	RHP	<1	0.07–0.18	11 (NB)		~6		Baddi et al. (45)
PZT	S	~5	0.02–0.15	- (DT)	2.4	2.6	3.4	Biswas and Fulrath (46)
B_4C	HP	5	~0–0.15	- (DCB)	2.3	2.7	3.9	Wu and Rice (39)
B_4C	HP	5	~0–0.15	0 (DT)	3.4	—	—	Hollenberg and Walther (47)
αSiC	S	5–10	0.01–0.07	(V)	3–5	2.2	4.5	Seshardi et al. (48)
HfN	HIP	—	0.05–0.14	(NB)	1.7	5.3	3.3	Desmaison-Burt et al. (49)
Si_3N_4	HP/S	~2–4	~0–0.10	- (I)	5.2	5.4	—	Mukhoyadhyay et al. (50)
Si_3N_4	HP/S	~2–4	~0–0.50	3–5[h] (V)	2.4[h]	3.7[h]	4[h]	Rice et al. (32)
Si_3N_4	S	≥11	0.05–0.40	- (I)	~5	5	~5[i]	Arato et al. (8)

[a] Much of this data is from ref. (4).
[b] Fabrication: PC = pressure cast, S = sintered, HP = hot-pressed, RHP = reaction hot-pressed.
[c] Grain size.
[d] Volume fraction porosity.
[e] Slope of approximately linear region of property on a semilog plot versus P at low to intermediate P level; γ and K_{IC} test technique shown in ( ); V = various.
[f] Finer grain size at higher P and larger grain size at lower P is a major reason for higher b values for γ, K_{IC}, or both, than E.
[g] Average value shown agrees well with that for E, but strength data was more scattered.
[h] For natural porosity, see Fig. 4.3A for data with additional, artificially introduced porosity.
[i] For both 3- and 4-point flexure despite differences in the strength levels.

consistent with this. This trend is also supported by Rice's (51,52) K_{IC}-P (hence also γ-P) data calculated from σ-P data and fracture-initiating flaw fractography observations of hot-pressed boron and Al_2O_3 normalized to the P = 0 value (~ 3 and 4 MPam$^{1/2}$ respectively), showing similar b values of ~ 4 for E, K_{IC}, and flexure strength.

The second probable and more variable trend is for higher b values for γ and K_{IC} versus E. The four cases of this in Table 4.2, two for oxides (44,46), and two for nonoxides (39,49) might just be scatter and variations that are unfortunately common to much ceramic data. However, two cases involve NB tests which can give anomalously higher values as discussed below. Furthermore, some of these are in bodies with either larger grains, in materials prone to exaggerated grain growth, or both, at lower P indicate this may be a real property trend. The first and commonly the most important aspect of this increase arises from increasing grain size as P decreases since larger grains commonly give higher γ and K_{IC} values, as noted earlier. This commonly gives an apparent, variable reduction of the porosity dependence due to frequent increases of fracture energy or toughness values as due to increasing grain size as P decreases, thus giving higher apparent b values. There may also be a second aspect of the changes with increasing grain size as P decreases, a real effect of porosity, due to some pores becoming intragranular with grain growth. While this effect can have variable effects on fracture energy and toughness, it may often increase them, again increasing b values.

The third trend is for lower b values for K_{IC}, and especially γ relative to those for E, i.e., as indicated by two oxides (39,40) and two nonoxides (32,47). While there are fewer cases for this in Table 4.2, two are extreme, i.e., no P dependence of γ for hot-pressed Al_2O_3 (40) and B_4C (47) to the P limits investigated of P = 0.09. and 0.15, respectively. These are consistent with other at least, extreme results shown below. Note that three of these four cases encompass three of the five tests with large cracks (i.e., WOF, DCB, and DT). Also, some of the above cases for similar b values for the different properties may be part of this second group, especially for PZT (46) and B_4C (39). This is logical since: 1) there should be a continuum between the two trends, and 2) these would add the only other two cases with large crack (DCB and DT) tests to this group. That significant variations are real but dependent on the amount and type of porosity is shown first by the NB data for Gd_2O_3 of Case and Smyth (53). This shows no P dependence, or possibly a limited increase in γ for sintered Gd_2O_3 to P ~ 0.15 before decreasing with further P increases (Fig. 4.4).

An even more extreme case of fracture energy and toughness deviations also indicates the mechanism (Fig. 4.3). Thus, while RSSN with natural pores showed normal E, γ and K_{IC} dependence on P (Table 4.1), bodies with added, larger (spherical) pores had anomalously high γ values over a moderate to higher P range (Fig. 4.3A). This was shown to be associated with crack branching (and probably bridging) that occurred in these bodies (Fig. 4.3B), but not normal RSSN. Clearly either the anomalous bodies by themselves, or their combination with the rest of the data, would significantly lower b values. Heterogeneity and other aspects of porosity (e.g., a dual pore population as in the RSSN case) may cause such branching and related phenomena, as discussed in the previous section. Increased crack branching and bridging with larger crack sizes is consistent with its more frequent association with tests involving larger cracks such as in DCB and DT tests. Expected variations between trends

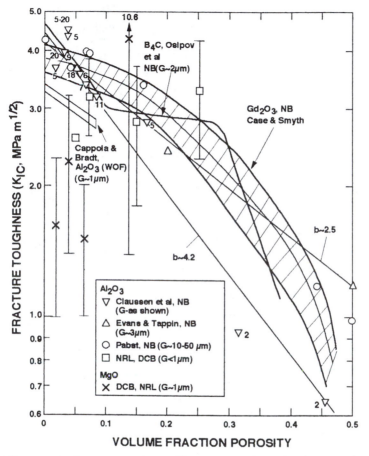

Figure 4.4 Fracture toughness, K_{IC}, versus volume fraction porosity, P, for various ceramics. Al_2O_3 data (40) and Gd_2O_3 data (53) were converted from their original fracture energy values using typical P dependence of E. Higher slopes of Pabst's (41) and Claussen et al.'s (42) Al_2O_3 data are attributed to increasing grain sizes (D, values next to the points in μm) as P decreases, e.g., compare data of Evans and Tappin (43); other extreme cases discussed in the text. Published with permission of J. Mat. Sci. (4).

are indicated by the following consideration of some of the above, and other, cases in terms of K_{IC}-P trends.

Haggerty et al. (54) reported data for an unusually fine microstructure RSSN (average grain size ~ 0.03 μm with some grains or agglomerates to 0.7μm and average pores ~ 0.2 μm with some occasionally to 5–70 μm and many in the nm range) for P = 0.22 and 0.33 with limited scatter. They reported K_{IC} values ~

50% higher than those of Rice et al.s' survey (32), which extrapolated to a value of nearly $K_{IC} = 4$ at $P = 0$, consistent with other data for dense Si_3N_4 without additives [which increase fracture toughness (32)]. While their limited data was consistent with a linear trend, it gives a b value of ~ 5 (and ~ 6 for flexure strength, where their individual values were again higher than normal data, e.g., by ~ 100%). This limited RSSN data with fine grain and pore sizes appears to be similar to two other sets of limited data. The first is data of Kodama and Myoshi (55) for SiC hot-pressed or HIPed from polycarbosilane-derived powder that gave low b values of ~ 1.8 and ~ 2 (P = 0–0.2) for their fine grain bodies of ~ 0.25 and ~ 0.9 μm, respectively. While there may be some uncertainty in whether these are distinct slopes, they are clearly low, raising questions of what mechanisms are operative for this at such fine grain and hence presumably at fine pore size. Strengths also tended to follow the microstructural dependence of fracture toughness. Second is limited data of Hurley and Youngman (56) for sapphire with nanometer pores from neutron irradiation at elevated temperatures. Thus, as irradiation temperatures increased from 925 to 1015 to 1100°K, K_{IC} and flexure strength (σ) were ~ 1.2, 1.3, and 2.1 (K_{IC}) or 1.7 (σ) relative to unirradiated samples. Corresponding pore sizes were 3.6, 5.2, and 9.0 nm. Though limited, these data raise the question of whether such fine microstructures result in other behavior, or a shift in normal behavior, e.g., crack branching, but with increased strength due to the fine microstructure, as discussed later.

Next, consider K_{IC}-P data for natural porosity beyond that of Table 4.1 as well as some data from there for reference in Fig. 4.4. Thus use of typical E-P dependence for Al_2O_3 (4,40) and Gd_2O_3 (4,53) noted above for their atypical, approximately zero γ-P dependence to P = 0.1–0.15 results in lower P dependence of K_{IC} over the range of the atypical P dependence of γ (i.e., relative to that of other properties, especially E). Also shown are three other sets of Al_2O_3 data from Table 4.1 for reference along with K_{IC}-P data for Al_2O_3 and MgO (4,57) and B_4C (4,58) that are not simply characterizable by b values. Thus, note: 1) the significant interruption of the decrease of K_{IC} for B_4C (4,58), i.e., similar to the trend for the RSSN showing an anomalous γ-P maximum, and 2) data for hot-pressed Al_2O_3 and MgO showing no decrease, or an increase in K_{IC} with P (4,57) While specific trends are uncertain due to scatter, it is significant that K_{IC} values for some of these porous bodies were clearly at least as high as values for bodies with less or no porosity.

The P-dependence of K_{IC} for the few known studies involving mainly model type porosity show some similarity to, and considerable difference from, E-P trends. Those of: 1) Zwisler and Adams (Fig. 3.5) for foamed glass (13,59), 2) Green (Fig. 3.9B) for sintered glass balloons (60), and 3) Cornel et al. (Fig. 3.7A) for sintered glass beads (4,61) all showed a trend for less K_{IC} dependence on P (except possibly at high P). This deviation is particularly extreme for the

Table 4.3 Summary of Fracture Energy and Related Tests of Magnesia Refractories

Density (gm/c^2)	Approximate[a]		Young's modulus (10^6 psi)	Flexure strength (10^3 psi)	Fracture energy (J/m^2)	
	Porosity percent	Average grain size (µm)			WOF	NB
2.98	17	3000 (5000)	6.2	2.0	144	4.9
2.98	17	1500 (4000)	9.2	2.6	135	9.5
2.92	18.5	200 (2000)	11.2	2.8	102	6.0
2.86	20	100 (1500)	10.5	3.3	132	11.7
2.76	23	75 (300)	10.9	3.0	68	13.2
2.71	24.5	50 (150)	12.8	3.1	44	10.2

Source. From data of ref. (65). Copyright ASTM, reprinted with permission.
[a] Porosity is based on density of refractory and theoretical MgO density (3.58 g/cm) and hence is approximate because effects of a few percent of other phases are neglected. Grain size is based on the mixed mesh sizes of grain used to make the refractory since little grain growth should occur. The values shown are the approximate values at 50 and 20 weight percent on a cumulative weight-grain size distribution curve, and give some idea of the differing but generally broad grain size distribution used.

glass bead bodies where there is no overlap of K_{IC} data with that for E and σ over most of the P range. On the other hand, while the σ-P trend for the glass beads is somewhat higher than that for E-P, there is extensive overlap of these, thus being another case of more deviation for K_{IC} than for σ (Chapter 5). The study of Krauslin et al. (62) of bodies sintered from ZrO_2 beads (mostly 40–100 µm dia.) is not as definitive because of lack of detailed characterization, but it shows similar slopes of E and K_{IC} (and flexure strength) versus bead diameter, and hence with porosity (Fig. 3.9B).

Many refractories were developed with substantial porosity for both insulating and thermal shock functions, and hence should be important sources of data on effects of porosity of fracture energy and toughness. While, such development was mostly done before extensive development and use of fracture mechanics, there have been a few subsequent studies, especially by Bradt and colleagues on magnesia based (63) and alumina-based (64) refractories and additional evaluation of these (65) that provide useful data. Thus, NB tests of magnesia refractories showed the resultant γ's paralleling the trends with E. i.e, both about double as porosity probably increases some (e.g., from ~ 0.17–0.24) and grain sizes decrease over an order of magnitude (Table 4.3); however, NB values varied from ~ 50–100% of values for fully dense MgO, thus indicating both similar γ-P and E-P trends, but anomalously high γ values. (Strengths show a similar but reduced trend, i.e., increasing by about 50% with the microstructural changes. The effect of grain size on strength, despite some probable increase in P, is expected.) The apparent correlation of increasing E with decreasing grain size probably reflects the use of a broader range of particle sizes with larger grains to get better particle packing, hence probable lower porosity, but not necessarily more bonding or MSA. In contrast to these NB values, WOF gave much higher γ values, which had an inverse correlation with E and changed by over threefold, i.e., 4–14-fold the value for fully dense MgO. Both tests probably reflect crack deflection, branching, bridging, and related processes, but much more in the WOF tests than the NB tests.

Similar trends are indicated in more extensive tests of a variety of alumina-silica based refractories, varying from 45–99% alumina. Not surprisingly all mechanical properties generally increased as the alumina content increased, i.e., E ~ 14 to 120 MPa (and strength ~ 2 to 30 MPa), with E having the best but still scattered correlation. There is also some of the expected correlation with porosity, i.e., reasonable, but scattered decrease of E as P increased, and some but less definitive and more scattered decrease in strength with increased P. NB fracture toughness, not surprisingly, showed better correlation with E than the resulting γ, which varied from ~ 1 to 100% of values for dense aluminas. Again WOF γ values were higher (by ~ 3–14-fold), but probably correlated better with E, i.e., most were at least as great as values for dense polycrystals. While the trends are more mixed, probably due to the greater complexity of these bodies from a broader composition range, there were again variable results as a function of porosity for the same alumina content—i.e., some γ values decreased with increased porosity, some did not, and some increased with increasing P, with more variation again seen in WOF results.

Four other significant deviations of γ and K_{IC} as a function of porosity should be noted. First, hot-pressed MgO with a few percent or less porosity showed γ similar to that for fully dense bodies at modest grain size, but increased significantly above these values as P decreased (but not to zero) and grain size increased (4,66–68). This is associated with intragranular porosity and transgranular fracture with resultant cleavage steps (e.g., Figs. 1.4, 1.6) and correlates with reduced crack branching and some possible increase in flexure strength (67) indicating that such intragranular pores increases γ. This is also shown by Forwood's estimated large (e.g., two orders of magnitude) increase in γ from decreases in cleavage crack velocity in NaCl crystals with fine porosity (69). Although this is now seen as a gross overestimate, e.g., actually being a factor of 2 or less (4), it is still of significance. Second, Cottom and Mayo (70) showed fracture toughness of fine grain (0.06–0.4 μm) ZrO_2 –3 mole% Y_2O_3 varying from 2.3 to 4.1 MPam$^{1/2}$ over the P = 0 to 0.1 range. While there is considerable scatter, there is a probable maximum in the P = 0.03–0.08 range, which is in contrast to a continuous increase in hardness with decreasing P for the same bodies. While these toughness results by themselves would be dismissed as scatter, e.g., due to heterogeneity in the bodies, the toughness deviations noted above for low-moderate P indicate this is another case of such deviations. Third, note the closer agreement between the P-dependence of E and strength than with γ or K_{IC}, discussed further in Chapter 5. A fourth deviation can arise from microcracking, whether by itself or in conjunction with porosity effects.

Two other cases showing significant deviations at moderate porosity are from related materials. First, Hoagland et al. (71) found fracture energies of a fine grain ($\sim$ 1 μm) limestone increased to a modest plateau of $\sim$ 200 J/m^2 with crack extension, which is higher than for dense ceramics, indicating some enhancement, however, that for a (somewhat more porous, P $\sim$ 0.15–0.2, larger grain, $\sim$ 20 μm, calcite bonded) sandstone was much more extreme, rising to nearly 1600 J/m^2 in DCB tests. The much greater rise in the latter rock material was attributed to microcracking occurring around and along the crack in the DCB specimens (based on both acoustic emission and direct examination; Fig. 4.5). Such microcracking is likely to also result in some crack branching, bridging, or both, in any event reducing the porosity dependence of γ and K_{IC} over the P (and other microstructural) range where it occurs. [Note: 1) Crack propagation through the calcite bonding phase makes the gross crack propagation essentially intergranular, which should favor bridging, 2) Fracture energies and toughnesses of both the limestone and sandstone tested showed significant orientation dependence, e.g., by 2–4-fold in peak values, probably due mainly to porosity anisotropy, and 3) coupling of porosity and microcracking effects is also shown in Nb_2O_5, Fig. 7.10A, and in ceramic composites, Section 8.3.6.] The second case is data for metals, which shows both similarity and differences from E-P trends Section 8.3.1, including a maximum of K_{IC}-P for Be (Fig. 8.6), and direct evidence of crack branching in porous Ni (72).

Finally, there are some experimental results for ceramic foams and rigidized glass fiber felts. Thus, Brezny and Green (73) studied the toughness of 3 commercial alumina-based open cell foams (P $\sim$ 0.76–0.92, with cell sizes of $\sim$ 1.4.–3.4 mm). They corroborated the $(1 - P)^n$ dependence, but finding n varied from 1.73 to 2.3 (averaging 2), i.e., higher than the predicted n = 1.5; however, they reported that correction for the hollow nature of the struts reduces n to $\sim$ 1.5 and approximately the same as for E (Table 4.1). (They also found similar as-measured and corrected n values for flexure strength, again consistent with the model.) On the other hand, they noted that the proportionality constant in the pertinent Eq. of Table 4.1 (for open cell foam) was 0.13–0.23, i.e., about 1/3 the value of 0.65 estimated from polymeric foams. Note that the proportionality

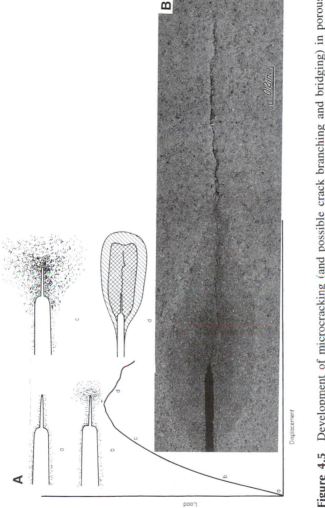

Figure 4.5 Development of microcracking (and possible crack branching and bridging) in porous sandstone after Hoagland et al. (71). A) Schematic of development of an extensive microcrack zone around a crack from a notch as a function of marked positions (a, b, and c) on the load-displacement curve. B) Micrograph of the notch-crack region (greater polymer impregnation in the microcrack zone highlights the microcracking). Published with permission of Springer-Verlag.

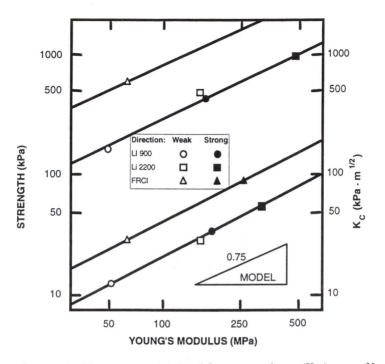

Figure 4.6 Flexure strength (σ) and fracture toughness (K_{IC}) versus Young's modulus (E) for rigidized glass felts after Green (75). Note that data for both the weak and strong directions (respectively open and solid symbols) for each of the two materials falling on the line for that material. Since data at different P levels was not available, plotting σ and K_{IC} vs. E indicates that n of $(1 - P)^n$ is ~ 0.75 for σ and K_{IC}, since that for E is predicted to be ~1. Published with permission of Plenum Press.

constant for elastic moduli was also only ~ 1/3 that expected from the model. They also noted that cell strut strengths were important, as theoretically predicted, and found Weibull moduli (m) for the individual struts (by directly testing individual struts) varying from 1.5–1.9, 1.9–2.9, and 2.1–3.5 for the three foams. These m values for the struts were about half the values for the complete foam (Section 5.3.5). Brezny and Green (74) also showed the importance of strut strength in glassy carbon foams. Huang and Gibson (17) found general consistency of data for glassy carbon open-cell foams (P ~ 0.97, cell dimensions ~ 0.25–2.5 mm) with the open-cell model (Table 4.1). They found m for the struts to be 4.5–6.2 for the different foams. At the lower m levels they predict a modest decrease in K_{IC} as cell size increases, but none at m = 6. [Note that the sintered ZrO_2 bead bodies of Krasulin et al. (62), which have some similarity to closed-cell foams, showed a significant decrease in K_{IC} (and flexure strength) as bead size increased.]

Green (75) showed that both K_{IC} and flexure strengh of rigidized glass fiber felts (Space Shuttle tiles) followed an $\sim (1 - P)^{3/4}$ dependence (Fig. 4.6) as predicted by the model of Green and Lange (14) (Table 4.1). This was true for both the strong and weak directions resulting from some anisotropy due to some fiber orientation during felt forming. Finally, note that dynamic fatigue studies of these glass felts showed that slow crack growth due to stress corrosion occurred due to effects of moisture on individual fibers (76).

4.3.2 Microcracking

Addressing the microcrack dependence of fracture energy and toughness requires further addressing their character, distribution in the body, and the grain-size dependence of their occurrence and size. As discussed in Chapters 1–3, although more challenging to directly observe, at least some preexisting microcracks can be observed by higher magnification microscopy than needed for most porosity (77). Such microscopy shows that much microcracking in graphite is transgranular (via basal plane cleavage) rather than intergranular (36,78) as commonly assumed and seen (78). While the extent of such transgranular microcrack character in graphite probably reflects the high anisotropy and resultant basal-plane cleavage, at least some transgranular microcracking is observed in some ceramics, e.g., TiO_2 (79). Microcracks generated by larger propagating crack are reasonably assumed to be the same or similar to those observed as preexisting based both on the mechanism of their generation and their effects on impacted properties as a function of key parameters of grain size and test temperature. The grain-size dependence of microcracking is now well established (77,79-81), but important issues remain, especially of preexisting microcracks versus those generated by crack propagation (Fig. 4.7).

The occurrence and effects of microcracking as a result of crack propagation are complicated by specimen and test factors as a function of grain size (4) and related phenomena such as crack bridging and R-curve effects (82), and unexplored probable crack shape effects (24). Tests allowing substantial scale of crack propagation relative to the scale of microcracking in noncubic materials (or composites) with sufficient differences in thermal expansion show fracture energies and toughnesses passing through maxima as a function of grain size (Fig. 4.7). This occurs for both of two sets of materials: 1) very anisotropic materials such as the dititanates of MgO, Fe_2O_3, and Al_2O_3 microcracking at finer grain sizes (e.g., a few microns) and 2) less anisotropic materials such as the single oxides: Al_2O_3, BeO, and TiO_2. While both types of materials show similar fracture energy and toughness maxima as a function of grain size (D), they have important differences in corresponding Young's modulus and strength behavior as a function of D.

The less anisotropic single oxides with sufficient fine grain sizes in relatively dense bodies generally show fracture energies and toughnesses independent of

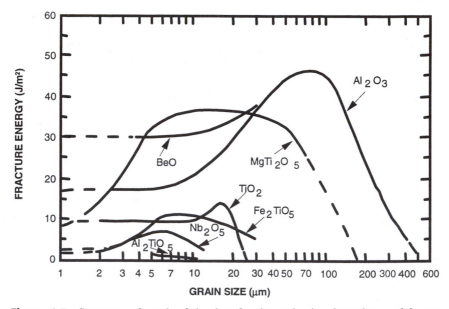

Figure 4.7 Summary of much of the data for the grain-size dependence of fracture energy for noncubic ceramics with intermediate to high thermal expansion anisotropy. Fracture toughness shows the same trend, but less extreme since it is simply $\sim (2E\gamma)^{1/2}$. Data summary after Rice (65). Reprinted with permission of ASTM.

D at finer grain sizes—i.e. less than $\sim 0.4\ D_s$, too small to generate micro-cracks—but begin to rise when $D \sim 0.4\ D_s$. Such rises in fracture energies and toughnesses are related to R-curve effects which give increasing toughness with increasing crack propagation to a saturation limit. The maxima in energies and toughnesses occurs at $D \sim D_s$, then values decrease, approaching zero at sufficiently large grains. These trends are very consistent qualitatively and reasonably consistent quantitatively with available models (18,19). The rise in fracture energies and toughnesses with increasing grain size is attributed to increasing numbers and sizes of microcracks. The subsequent decrease, and hence intervening maxima, are attributed to subsequent increasing ease of prop-agation of the failure-causing macrocrack by increasing linking of the micro-cracks. The observations of Hoagland et al. (71) (Fig. 4.5), though complicated some by the presence and probable effects of pores, are a vivid demonstration of microcracking associated with a macrocrack.

The more anisotropic dititanates have been obtained in relatively dense bodies only with grain sizes near their D_s values, so some only show decreasing fracture energies and toughnesses as D increases. Bodies with correspondingly finer grains relative to their D_s values as those for the single oxides have not been obtained. There is also limited strength and Young's modulus data for the single oxides for D greater than D_s. It is tempting to assume that the behaviors of these two sets of materials would be the same at comparable grain sizes relative to their D_s values; however, there is both some clear, as well as less complete, evidence that there are some basic differences in the behavior of these two sets of materials. Thus, the dititanates show similar rapid drops in both flexure strengths and Young's moduli (77) (along with increased damping) (83) as grain sizes increases; also, γ and K_{IC} values first increase then decrease (Fig. 4.7), indicating that their microcracks are predominantly preexisting. The subsequent, approximately asymptotic behavior of these properties indicates a saturation of microcracking at $\gg D_S$, but there has been little or no study of this. On the other hand, there is a clear lack of a decrease in E and only limited decreases in flexure strengths of Al_2O_3, BeO, and TiO_2 with increasing grain sizes at $< D_S$ indicating that microcracking is associated primarily with propagation of large cracks used for fracture energy and toughness versus strength measurements. Whether the significant decreases in γ and K_{IC} values for these materials above D_S values is accompanied by comparable decreases in Young's moduli and flexural strengths, and hence presumably by mostly pre-existing microcracks is uncertain. However, these two sets of materials clearly show fundamentally opposite trends of γ and K_{IC} from those of Young's moduli and flexural strengths as grain size increases toward γ and K_{IC} maxima. This marked contrast in fracture energy and toughness versus Young's moduli and tensile strengths parallels that of porous bodies showing temporarily reduced or reversed decreases in γ and K_{IC} values (Fig. 4.4).

Insight to some of the uncertainties and gaps in the above results comparing Young's modulus and fracture energy and toughness trends are indicated by more comprehensive data from Tomaszewski's (84) study of Al_2O_3, in which he measured E by both static and dynamic methods. Static measurements of E (in conjunction with strength tests) showed E decreasing with increasing grain size, indicating progressive microcrack formation, which was confirmed by direct microscopic observations, while the dynamic measurements showed E remaining constant over his broad range of D ($\sim$ 1 to nearly 500 μm). This clearly indicates that considerable microcracking was generated by the stresses during strength testing as opposed to no changes in dynamic testing where applied stresses are typically trivial. Despite such stress-generated microcracking, there still appears to be significant differences between materials of moderate anisotropy, such as alumina, versus very anisotropic materials, such as the dititanates.

Tomaszewski's (84) Al_2O_3 study also provides additional useful information showing the significant payoff from a more comprehensive approach, especially the benefits of additional comparative measurements as stressed in Section 1.4.1. Thus, he showed a similar relative maxima in K_{IC} (via an NB test), also at

D ~ 100 µm, consistent with the fracture energy data in Fig. 4.7. He also corroborated microcrack formation by direct observation (and correlated this with fracture mode), as well as reporting that the ratio of the microcrack size to the grain size changed from ~5 for fine grains to ~ 1.2 at large grains. This again indicates the complexity of the microcracking process, for example, suggesting larger relative crack sizes to those of the grains at finer grains, i.e., near the onset of microcracking, and lower relative sizes at larger grain sizes, which would be consistent with interactive effects reducing the relative microcrack extent at larger grain size where microcracking should be more extensive.

Besides studies as a function of grain size demonstrating effects of microcracking on crack propagation, there are beginning to be more quantitative studies of microcracking and its effects, e.g., those of Ohaya and colleagues (85,86) on Al_2TiO_5. They showed that analysis of thermal expansion data of polycrystals versus single crystal (i.e., lattice) expansion can yield the net microcrack volume, which was shown to increase with $D^{1/2}$. Such data can be used to estimate the volume of individual microcracks for comparison with direct observations. More directly to crack propagation, these authors report that microcracking caused deflection, blunting, and branching of macrocracks which more than doubled fracture energy (but decreased both Young's modulus and flexure strength) and reduced crack velocities to 1/2000 of those in the absence of microcracks.

While microcracking due to thermal expansion differences between grains (whether due to TEA of the same phase or to different chemical phases), it can also occur due to crystallographic phase changes of a given chemical composition. Key examples of this are the tetragonal-monoclinic transformation of ZrO_2 and HfO_2. Again, the occurrence of microcracking from such transformations is a function of the material-phase properties and grain size. Studies of Hunter et al. (87) of the fracture energy and toughness of HfO_2– 1 mole % Er_2O_3 shows these properties respectively increasing from ~ 20 to 40–70 J/m^2 and from under 2 to 4–6 $MPam^{1/2}$ as the fraction of tetragonal phase in the body before test increased from under 0.1 to 0.25–0.35. A key question is the extent to which such increases are due to transformation toughnening due to the transformation itself, or due to microcracking from the transformation. Microcracking clearly occurred in these samples as shown by decreases in Young's modulus and was thus judged to be an important factor by the authors, but more investigation with better characterization of microstructure (e.g., grain size) and more comprehensive property measurement is needed to clarify the extent of the relative roles.

Two further observations should be made regarding such microcracking effects. The first is on the nature of the microcrack zone relative to the macrocrack around which they are generated by stressing. Evidence clearly shows that earlier concepts of microcrack zones forming ahead of the macrocrack tip, with more above and below the plane of the macrocrack (20,21) is not an accurate view. Results are closer to the microcracks being more in the wake zone, but not entirely so. Thus, Hoagland et al.'s (71) results (Fig. 4.5) showed much wake microcracking, but also considerable microcracking ahead of and around the macrocrack, but much less in spatial extent than first proposed; however, the extent to which the porosity in his sandstone was a factor in this broader spatial distribution of microcracks is unknown. Ultrasonic probing of the microcrack character around a stressed macrocrack by Swanson (88) in both granite and alumina samples shows results similar to those of Hoagland et al., but with microcracking being confined closer to the macrocrack and more in the wake region. Thus, predominate formation of a microcrack zone ahead of and around the macrocrack is not supported, and a considerable, but not exclusive, wake effect is indicated. Some contribution of other mechanisms such as crack deflection, branching, and tortuosity may also be involved, depending on material and microstructural parameters, especially grain size and porosity.

The second observation to note is the substantial similarities and limited differences between the fracture energy and toughness trends as a function of porosity (Figs. 4.3, 4.4) and as a function of grain size in noncubic materials (Fig. 4.7). The close parallel between the two sets of data is seen when it is recalled that the size and density of microcracks, hence their volume fraction associated with a propagating macrocrack is directly related to the grain size. Thus, both plots reflect very similar functional dependences on the volume fractions of pores or microcracks (a special type of pore); however, pores typically have a volume distribution, while microcracks may be primarily associated with a macrocrack. The similarities are further seen when it is recognized that in both cases a variation, which can be substantial, is superimposed on a basic trend for fracture energy or toughness. For porosity, the basic trend is decreasing fracture energy and toughness with increasing P, similar to that for elastic moduli. For microcracking, the basic trend is little or no dependence on grain size. In both cases these normal trends are interrupted by first increases of fracture energy and toughness above the normal levels as porosity or grain size (hence often microcrack content) further increase beyond some level to a maximum, then decreasing. The increases, maxima, and decreases are all functions of the material and microstructure, e.g., fracture energy and toughness independent of grain size when $D \ll D_s$, rising to a maximum at D_s, then decreasing. The primary differences are; 1) the changes with increasing porosity are a perturbation from a continuous decrease in properties versus an increase then a decrease from otherwise constant properties with microcracking, 2) porosity being globally distributed and static in its population versus microcracking often initiating at some grain size, then continuously increasing, in association with the propagating macrocrack, and 3) the latter increase becoming so extreme that fracture energy and toughness can approach zero. The basic alterations of the trends are very similar, however, reinforcing the observation the these deviation in γ-P and K_{IC}-P from E-P trends are related to complexities of crack propagation such as crack branching and tortuosity, that are probably also a factor in microcracking effects.

4.4 GENERAL DISCUSSION

The primary aspect of the summarized γ-P and K_{IC}-P dependences needing discussion is that there is: 1) often overall similar trends with those for E-P (and σ-P) trends as expected for simple, linear, elastic crack propagation, but also 2) deviations from this similarity of trends that can be substantial in extent and frequency of occurrence. The similarity of trends with P is consistent with theoretical expectations discussed in Sections 4.1 and 4.2. The deviations, which can be for more or less P dependence from that of E, may be systematic or variable, e.g., increasing scatter. In either case they can arise from a number of sources, reflecting important microstructural phenomena, inadequate experimental technique, or both. These deviations and their known or probable sources and factors are summarized in Table 4.4 and the following paragraphs.

Consider first the experimental issues, which entail mainly inadequate microstructural characterization and recognition or control of crack character, or both. Key examples of inadequate microstructural characterization are failure to recognize anisotropic, heterogeneous, or variability of the porosity, as well as of significant changes in grain structure that often accompany significant reductions of porosity, with or without significant associated pore size, shape, and location changes (Figs. 1.2–1.4). Variations in any of these are major reasons for much of the scatter and variations, especially in view of the typical absence of any reasonable pore characterization. Anisotropic shaped pores (and their degree of orientation) and grain-size effects are selected as particular examples of these in Table 4.4. The first two entries there reflect 1) known grain-size trends with some documentation for γ-P and K_{IC}-P trends (Table 4.1, Fig. 4.4), and 2) expectations, but an almost total lack of characterization, especially of pore anisotropy. Failure to identify such microstructural variations is compounded by porosity effects generally being nonlinear, which is true for elastic properties and fracture energy and toughness, but with greater sensitivity of fracture processes to such nonlinearity, hence of γ and K_{IC} (and frequently even more for tensile strength, Chapter 5); however, deviations of the porosity dependence of fracture energy and toughness from that of elastic properties can be an important indicator of heterogeneity, anisotropy, or variability of the porosity, and may influence, positively of negatively, crack branching and bridging discussed below.

Consider next specific aspects of γ and K_{IC} tests, which entail primarily false assumptions of the crack character, and especially failure to account for crack-

Table 4.4   Probable Sources of γ-P and K_{IC}-P Variations from E-P Trends[a]

Porosity dependence[b]	Known or probable source(s)	Known or probable factors
Variable	Larger grain sizes at lower P (giving apparent increased P-dependence; see Fig. 4.3, 4.4). Heterogeneous or oriented aniso-tropy of porosity, or both.	γ, K_{IC} dependence on grain size (often dependent on test method and parameters). Greater effects of such porosity on fracture; differing stress axes relative to anisotropy.
Reduced	Branching/bridging (usually over lower P range); e.g., Figs. 4.1, 4.2. Increased pore character/location changes due to increased intra- instead of all intergranular porosity from grain growth with sintering to lower P.	Pore character; test-crack size and extent of propagation. Crack size and extent of propagation, γ and K_{IC} grain size (and test factor; Ch.10) dependence, and fracture mode: a) reduced crack-pore inter-actions with such pore changes, or b) mainly crack branching/bridging, secondarily fracture tails (for TGF[c]).

[a] Primarily for bodies with low to intermediate P levels, e.g., as from sintering.
[b] Porosity dependence of γ and K_{IC}.
[c] TGF = transgranular fracture.

size effects relative to the scale of microstructural character and crack interac-tions Thus, it has been shown that the assumed slit crack at the base of the notch in the notch-beam test is not always accurate, e.g., in both glass (35) and in ZrO_2 crystals (89) samples some failures occurred from cracks much closer to a half penny-crack, about doubling K_{IC} values calculated assuming a slit crack. Such deviations toward failure from cracks closer to half-penny instead of slit cracks seem much more likely in the bodies of partially sintered glass beads such as those of Cornel et al. (31,61), and thus is a likely significant (but probably not the total) cause of their high K_{IC} values, i.e., reduced P-dependence over much of the P range. More generally the starting and final crack sizes of the various tests are likely to be important factors. This is suggested by other tests on refractories as a function of crack size (35) (e.g., Table 4.4), and the branching-bridging possibility and the similar dependence of bridging associated with grain size and dispersed phases. The limited γ- and K_{IC}-P data is consistent with this, i.e., the b values most consistent with those for E and flexure strength were those from fractography (34), and hence the smallest crack sizes, while greater deviations are found for tests with the larger cracks (4), e.g., the most extreme difference was with a double-torsion test of B_4C (Table 4.2). All of these indications are useful guidelines for the substantial study of the effects of porosity on crack propagation that should be conducted.

While variability of porosity trends is an important issue, there are also frequent trends for lower b values; and more extreme, a temporary lack of porosity dependence or an increase of γ and possibly K_{IC} versus P, usually over low to intermediate porosity ranges for some bodies. These more extreme variations of porosity dependence of fracture toughness and especially energy combined with the frequent deviations toward less P dependence of γ or K_{IC}, show that deviations from the relations expected from Eq. 4.1 are frequent, possibly even a majority. While the data by itself are not sufficient to clearly define the source(s) of the variations, thus requiring more study, two important, often related, factors are indicated. One is the shift of some pores from inter- to intragranular pores, and associated changes in pore character with grain growth and possibly frequent changes in the extent of crack-pore interactions. No attention has been given to such (and other pore) factors in the experimental and little in theoretical considerations, which is not surprising in view of the very limited attention to porosity effects on fracture energy or toughness. Effects of pore location and associated shapes were noted for MgO (and NaCl), where possible greater effect resulted from slip associated with stress-pore interactions and/or formation of cleavage steps, was noted (Fig. 6.16). While such slip effects are limited, the issue of effects of pore character-location is much broader.

The other, more significant factor in reduced porosity dependencies is that of crack branching and bridging. This may occur primarily due to the pore character (e.g., mixed or heterogeneous; Figs. 4.1 and 4.2), but may also be aided by the above inter- to intragranular pore shift. Crack deflection and bridging mechanisms due to pores are qualitatively consistent with results and supported by one set of direct observations in RSSN (Fig. 4.3). Similar observations have been made in porous metals [Ni(72)] and rocks (Fig. 4.5), the latter also involving microcracking. Limited porosity has also been shown to be critical to microcracking and resultant crack complexity in duplex composites (90). Such crack effects are clearly a logical [and possibly the only] mechanism that could increase γ and K_{IC} values to or above those at P = 0, and thus also be the most significant source of differences with comparable comparison of tensile strength- P versus E-P trends (see Table 5.2). Contributions of pores to microcracking or bridging might thus be modeled as pores having a nonzero contribution to fracture energy and toughness, and thus improve bounding of results, since then the lower limit would no longer be zero.

Besides the interaction of porosity and microcrack effects noted above and the effects on thermal shock discussed below, possible variations of these effects with crack-microstructure scales should be noted. Thus limited data for fine, e.g., nanometer, pores (and grains) all indicate similar trends of fracture toughness and strength with porosity in RSSN (54), SiC (55), and sapphire (56) with differing sources of their fine porosity. The latter also indicated increased toughness and strength above those for dense sapphire. These results suggest that crack branching and bridging may be determined by crack sizes relative to the microstructural scale rather than absolute crack sizes, so strengths in these finer than normal microstructures follow toughnesses. Thus, further investigation of such fine porosity is warranted not only to expand our limited data and understanding but also because of possible benefits if crack branching and bridging or other mechanisms can benefit strengths and along with fracture toughnesses. Such exploration of fine microstructures is also suggested by simple microcrack models (26) indicating increases of fracture toughness above uncracked matrix levels for fine, closely spaced microcracks.

Crack branching and bridging are probably important factors in the effect of porosity on thermal shock resistance, Ts_r, so such resistance supports the occurrence of these phenomena. The introduction of substantial porosity often results in large increases in Ts_r e.g., as used in refractories (63,64), and shown in glasses by Nakashima (91); however, the specific mechanism(s) of this marked improvement has not been well explained, especially on a quantitative basis. The dilemma is readily illustrated by the following basic relation:

$$Ts_r \propto \sigma(\alpha E)^{-1} \qquad (4.10)$$

where σ = tensile strength, α = coefficient of thermal expansion, and E = Young's modulus; however, note that (as discussed earlier) the P dependence of σ and E are often similar or identical, and α typically has no porosity dependence, so the introduction of porosity would have little or no effect per the above equation. In the cases where thermal conductivity plays a significant role, it also goes in the numerator. This increases the dilemma since thermal conductivity also decreases with porosity (Chapter 6), thus indicating less Ts_r with increased P rather than the increase commonly observed; however, recalling that $\sigma = Y(2E\gamma)^{1/2}(c)^{-1/2}$ and that the P-dependence of γ can be considerably reduced, or temporarily reversed, over some range of P readily suggests a quantitative explanation for improve Ts_r with increased porosity. Since this effect on γ is attributed to crack branching and bridging, the above observations of Ts_r give added impetus to the study of such phenomena in porous materials. Effects of porosity on thermal shock resistance are discussed further in Chapter 9.

Another important issue is what γ and K_{IC} values to use for fracture from isolated pore clusters and especially isolated pores (Chapter 5). In many such cases, especially from isolated pores, γ and K_{IC} values measured with large cracks are often not appropriate since they typically involve effects of several to many such pores or clusters. On the other hand, failure from such pore origins typically involves no crack propagation through other such pores since catastrophic crack propagation commonly occurs before another large pore or cluster is reached. Thus, the γ or K_{IC} values appropriate to such failure are much closer to that of the (often nearly, or fully, dense) material surrounding the pore or cluster, as discussed further in Chapter 5.

Finally, note key observations and issues regarding microcracking. A key difference between microcracking and porosity is that the pore population is normally constant in a given body, while the microcrack population can increase in association with a propagating macrocrack; however, there are significant similarities in the effects of some porosities and of microcracks increases on fracture energies and toughnesses, namely increases rising to maxima relative to their normal levels, then decreases back to, or below, the normal levels for the associated respective levels of porosity or grain size. Whether the relative increases for porosity involves microcracking between pores or other common effects such as crack branching, bridging, etc., or both is not known. In any case, however, interaction of pores and microcracks is an important issue that has received almost no attention. Both similarities and differences also occur in the effects of microcracks on thermal shock. An important difference is that microcracking can significantly improve thermal shock resistance due to the significant reductions in bulk thermal expansion due to microcracking (Chapter 9); however, the extent to which microcracking also improves thermal shock resistance by improving resistance to crack propagation as indicated for porosity is not known.

Important, partially related microcracking issues are: 1) the extent of preexisting microcracks (distributed throughout the body) and microcracks generated in association with a propagating crack, and 2) the closure and apparent healing of preexisting microcracks on heating. Highly anisotropic materials such as some dititanates (Fig. 4.7, which microcrack at fine grain sizes) show precipitous drops in both Young's modulus and tensile strength as γ or K_{IC} versus D increase, implying that the microcracks are preexisting—i.e. were in the body prior to testing; they are commonly formed on cooling from fabrication. On the other hand, less anisotropic materials—i.e. tested single oxides, while showing similar γ or K_{IC} versus D trends at larger D values—do not show corresponding decreases in Young's moduli or tensile strengths at $D < D_S$, implying that the microcracks are mainly generated in association with macrocrack propagation. However, Tomaszewski's (84) Al_2O_3 clearly shows the value of comparing static and dynamic moduli measurements, since the former, but not the latter, will reveal stress generated microcracks. (Limited data on these properties at $D > D_S$, also raise questions of their comparable behavior to more anisotropic materials microcracking at finer grain sizes.) Two other possibly related and complicating issues are the observations on HfO_2 that much microcracking is associated with stress corrosion effects (requiring a few days to reach saturation) and that there is a change in character-behavior of microcracks at larger grain size, i.e., less propensity to fully close/heal at high temperatures. How general these two observations are and their relevance to the above differences is uncertain; however, microcracking at larger grain sizes would release more strain per microcrack, which may make registry for closure more difficult (and probably, at the minimum, hindering healing), especially so if microcracks are not confined to a single grain or grain facet. Greater strain release per microcracks with larger grains may also effect the spatial distribution of microcracks, e.g., their continuity, which in turn may effects the impact of stress corrosion. Finally, while effects of pores and microcracks are typically treated as independent, as noted earlier and later (Fig. 7.10A), this is not necessarily true. Thus, much remains to be understood about microcracking and its effects.

4.5   SUMMARY

Evaluation of the porosity dependence of crack propagation, γ, and K_{IC} as a function of P suffer more extensively from the same problems as the study of elastic properties versus P; however, the data base is much smaller (almost zero for crack propagation), the effects of inhomogeneous porosity have not been addressed, there is almost no cross-correlation of data and there are no models for other related properties that check for self-consistency, except as presented here (and Chapter 5). Only a few models have been directly derived for the porosity dependence of γ or K_{IC}, most present problems or uncertainties, and none address indications of crack branching and bridging, which have been almost totally neglected experimentally. Despite such limitations, evaluation of models, data, and self-consistency provide important conclusions and implications. The few models specifically developed for γ, and K_{IC}, while offering some ideas and results for further development raise basic questions of self-consistency with the porosity dependence of E (and σ).

The one model set that is at least approximately consistent with a fair amount of data and the frequent self-consistency with the P-dependence of E, thus being a reference for the deviations discussed, are the MSA models. The MSA approach thus presently provides the broadest porosity application and the only one having demonstrated some predictive ability, and is thus recommended for further use and development; however, substantially more development and evaluation are needed, with attention to modeling of other pore and particle stackings, and heterogeneities, gradients, and anisotropy are needed along with much better porosity characterization.

There are major deviations from any single model, which appear to depend substantially on test method and especially crack size relative to pore sizes and spacings. The least, commonly no significant, deviations occur with γ and K_{IC} values from fractography, i.e., reflecting the crack sizes typically controlling strengths and often moderate to fine grain size also means less propensity for crack bridging. Greater deviations are indicated as the crack size increase with other tests, e.g., in the approximate order-indent, indentation fracture, notch beam, double-cantilever beam, and double torsion. Deviations are to lower P dependence of K_{IC} and still lower dependence of γ, mainly at low or moderate P, with extremes being temporary plateaus or maxima of K_{IC}, and especially γ, with the maxima often being at levels similar to or above the values at P = 0. Such deviations are attributed to factors such as microcracking, and especially crack branching and bridging, and are based on: 1) limited experimental observations, 2) reasonable qualitative models of crack-pore interactions showing extensive opportunities for branching, and hence also bridging, and 3) the effects that such branching and bridging should have being consistent with observed deviations. The latter includes the important case of such phenomena being consistent with and providing a more concrete explanation for the high thermal shock resistance of many porous bodies such as refractories. Heterogeneous and multimodal pore character are also suggested as factors possibly enhancing crack branching and bridging.

The limited data on foams and rigidized felts provide some useful guidance, at least as a starting point, but more attention to cell stacking and wall or strut variations is needed, along with much better characterization.

Many obvious improvements are needed, including a better data base, i.e., more and better data and characterization, including effects of porosity heterogeneity and anisotropy, as well as cross-correlation with other measurements, especially of E and σ; however, addressing branching and bridging, both from an analytical and an experimental standpoint, is also a key need, as is the study of the effects of crack size and test method on these, and identifying crack-size–test–pore-character effects. An important question needing further study is the issue of the fracture energy or toughness pertinent to failure from a pore or pore cluster since this often involves fracture of dense material around the pore, not that reflected by normal tests that propagate a crack through the solid material and the pores.

Microcracking results in maxima of fracture energy and toughness of noncubic materials as a function of grain size (and hence of microcrack content) very similar to effects with some porosities. Such microcrack effects are generally consistent with models; however, there are several basic questions raised by differing effects of microcracking on strength and Young's modulus, implying differences between effects of preexisting microcracks and those generated in association with macrocrack propagation and other factors impacting microcrack effects. The latter include effects indicated at larger grain size and effects of stress corrosion. Again, more comprehensive studies such as those of Tomaszewski (84) are needed, for example, to discern differences between preexisting and stress-generated microcracks and to characterize the microcracks and correlate their effects on various properties.

Both limited data as well as some models and concepts indicate the need and opportunity for investigation of fine (e.g., nanometer) pores and microcracks (hence also generally fine grains). Besides extending understanding to a finer scale, such studies may result in further increases in fracture toughness and strength for such bodies.

REFERENCES

1. J. J. Gilman, "Cleavage, Ductility, and Tenacity in Crystals," Fracture (B. L. Averbach, D. K. Felbeck, G. T. Hahn, and D. A. Thomas, Eds.), Tech. Press of MIT, Boston, MA, pp. 193–222, 1959.
2. R. W. Rice and S. W. Freiman, "The Porosity Dependance of Fracture Energies" Ceramic Microstructures '76: With Emphasis on Energy Related Applications (R. M. Fulrath and J. A. Pask, Eds.), Westview Press, Bolder, CO, pp. 800–23, 1977.
3. R. W. Rice, "Microstructural Dependance of Mechanical Behavior of Ceramics," Treatise on Materials Science and Technology, 11. Properties and Microstructure (R. K. MacCrone, Ed.), Academic Press, New York, pp. 191–381, 1977.
4. R. W. Rice, "Grain Size and Porosity Dependance of Fracture Energy and Toughness of Ceramics at 22°C," J. Mat. Sci., 31, pp. 1969–83, 1996.
5. A. S. Waugh, J. P. Singh, and R. B. Poeppel, "Dependence of Ceramic Fracture Properties on Porosity," J. Mat. Sci., 28, pp. 3589–93, 1993.
6. P. Arato, "Comment on 'Dependence of Ceramics Fracture Properties on Porosity'," J. Mat. Sci. Let., 15, pp. 32–33, 1996.
7. K. Yasuda, Y. Matsuo, S. Kimura, and T. Mori, "Analysis of Dependence of Volume Fraction and Micropore Shape on Fracture Toughness by the Equivalent Inclusion Method," J. Cer. Soc. Jap., Intl. Ed., 99, pp. 720–27, 1991.
8. P. Arato, E. Besenyei, A. Kele, and F. Weber, "Mechanical Properties in the Initial Stage of Sintering," J. Mat. Sci., 30, pp. 1863–71, 1995.
9. D. J. Green and D. Hardy, "Fracture Toughness of Partially-Sintered Brittle Materials," J. Mat. Sci. Let., 15, pp. 1167–68, 1996.
10. D. J. Green and R. G. Hoagland, "Mechanical Behavior of Lightweight Ceramics Based on Sintered Hollow Spheres," J. Am. Cer. Soc., 68(7), pp. 395–398, 1985.
11. S. K. Maiti, M. F. Ashby, and L. J. Gibson, "Fracture Toughness of Brittle Cellular Solids," Scripta Met., 18, pp. 213–17, 1984.
12. A. T. Huber and L. J. Gibson, "Anisotropy of Foams," J. Mat. Sci., 23, pp. 3031–40, 1988.
13. L. J. Gibson and M. F. Ashby, "Cellular Solids, Structure & Properties," Pergamon Press, New York (1988).
14. D. J. Green and F. F. Lange, "Micromechanical Model for Fibrous Ceramic Bodies," J. Am. Cer. Soc., 65(3), pp. 138–41, 1982.
15. L. J. Gibson and M. F. Ashby, "The Mechanics of Three-Dimensional Cellular Materials," Proc. Roy. Soc. Lond. A 382, pp. 43–59, 1982.

16. A. N. Gent and A. G. Thomas, "The Deformation of Foamed Elastic," Rubber Chem. Tech., **36**, pp. 597–610, 1963.

17. J. S. Huang and L. J. Gibson, "Fracture Toughness of Brittle Foams," Acta Metall. Mater., **9**(7), pp. 1627–36, 1991.

18. Y. Fu and A. G. Evans, "Microcrack Zone Formation in Single Phase Polycrystals," Acta Metall., **30**, pp. 1619–25, 1982.

19. R. W. Rice and S. W. Freiman, "Grain-Size Dependence of Fracture Energy in Ceramics: II, A Model for Noncubic Materials," J. Am. Cer. Soc., **64**(6), pp. 350–54, 1981.

20. A. G. Evans and K. T. Faber, "Toughening of Ceramics by Circumferential Microcracking," J. Am. Cer. Soc., **64**(7), pp. 394–98, 1981.

21. W. Kreher and W. Pompe, "Increased Fracture Toughness of Ceramics by Energy-Dissipative Mechanisms," J. Mat. Sci., **16**, pp. 694–706, 1981.

22. L. R. F. Rose, "Effective Fracture Toughness of Microcracked Materials," J. Am. Cer. Soc., **69**(3), pp. 212–14, 1986.

23. D. R. Clarke, "A Simple Calculation of Process-Zone Toughening by Microcracking," J. Am. Cer. Soc., **67**(1), pp. C-15–16, 1984.

24. R. W. Rice, "Crack-Shape-Wake-Area Effects on Ceramic Fracture Toughness and Strength," J. Am. Cer. Soc., **77**(9), pp. 2479–80, 1994.

25. F. J. P. Clarke, "Residual Strain and the Fracture Stress-Grain Size Relation in Brittle Solids," Acta Metall., **12**, pp. 139–43, 1964.

26. D. P. H. Hassselman and J. Singh, "Analysis of Thermal Stress Resistance of Microcracked Ceramics," Am. Cer. Soc. Bul., **53**(8), pp. 856–60, 1979.

27. R. A. Hunt, "A Theory of the Statistical Linking of Microcracks Consistent with Classical Reliability Theory," Acta Metall., **26**, pp. 1443–52, 1978.

28. R. G. Hoagland and J. D. Embury, "A Treatment of Inelastic Deformation Around a Crack Tip Due to Microcracking," J. Am. Cer. Soc., **63**(7–8), pp. 404–10, 1980.

29. H. Cai, B. Moran, and K. T. Faber, "Analysis of a Microcrack Prototype and Its Implications for Microcrack Toughening," J. Am. Cer. Soc., **70**(11), pp. 849–54, 1987.

30. G. D. Bowling, K. T. Faber, and R. G. Hoagland, "Computer Simulations of R-Curve Behavior in Microcracking Materials," J. Am. Cer. Soc., **74**(7), pp. 1695–98, 1991.

31. R. W. Rice, "Evaluation and Extension of Physical Property-Porosity Models Based on Minimum Solid Area," J. Mat. Sci., **31**, pp. 102–18, 1996.

32. R. W. Rice, K. R. McKinney, C. C. Wu, S. W. Freiman, and W. J. McDonough, "Fracture Energy of Si_3N_4," J. Mat. Sci., **20**, pp. 1392–1406, 1985.

33. R. W. Rice, "Ceramic Fracture Mode-Intergranular vs. Transgranular Fracture," Fractography of Glasses and Ceramics III (J. R. Varner, V. D. Frechette and G. D. Quinn, Eds.), Am. Cer. Soc., Westerville, OH, pp. 1–53, 1996. Ceramic Transactions, vol. 64.

34. R. W. Rice, "Fractographic Determination of KIC and Effects of Microstructural Stresses in Ceramics," Fractography of Glasses and Ceramics II (V. D. Frechette and J. R. Varner, Eds.) Am. Cer. Soc., Westerville, OH, pp. 509–45, 1991. Ceramic Transactions, vol. 17.

35. K. R. McKinney and R. W. Rice, "Specimen Size Effects in Fracture Toughness Testing of Heterogeneous Ceramics by the Notch Beam Method," Fracture Mechanics Methods for Ceramics, Rocks, and Concrete, ASTM STP 745

(S. W. Freiman and E. R. Fuller Jr., Eds.), ASTM, Philadelphia, PA, pp. 118–26, 1981.

36. J. E. Brocklehurst, "Fracture in Polycrystalline Graphite," Chemisty & Physics of Carbon, vol. 13 (P. L. Walker and P. A. Thrower, Eds.), Marcel Dekker, Inc., New York, pp. 145–279, 1977.

37. S. Kimura, M. Ishizaki, K. Yasuda, and Y. Matsuo, "The Dependence of Porosity on the Fracture Toughness of Polycrystalline Graphites," TANSO, **148**, pp. 134–41, 1991.

38. D. C. C. Lam, F. F. Lange, and A. G. Evans, "Mechanical Properties of Partially Dense Alumina Produced from Powder Compacts," J. Am. Cer. Soc., **77**(8), pp. 2113–7, 1994.

39. C. Cm. Wu and R. W. Rice, "Porosity Dependance of Wear and Other Mechanical Properties on Fine-Grain Al_2O_3 and B_4C," Cer. Eng. Sci. Proc., **6**(7–8), pp. 977–94, 1985.

40. J. A. Cappola and R. C. Bradt, "Effects of Porosity on Fracture of Al_2O_3," J. Am. Cer. Soc., **56**(7), pp. 392–3, 1973.

41. R. F. Pabst, "Determination of KIC-Factors with Diamond-Saw-Cuts in Ceramic Materials," Fracture Mechanics of Ceramics, **2** (R. C. Bradt, D. P. H. Hasselman, and F. F. Lange, Eds.), Plenum Press, New York, pp. 555–65, 1974.

42. N. Claussen, R. Pabst, and C. P. Lahmann, "Influence of Microstructure of Al_2O_3 and ZrO_2 on K_{IC}," Proc. Brit. Cer. Soc., **25**, pp. 139–49, 5/1975.

43. A. G. Evans and G. Tappin, "Effects of Microstructure on the Stress to Propagate Inherent Flaws," Proc. Brit. Cer. Soc., **20**, pp. 275–97, 6/1972.

44. M. Knechtel, C. Cloutier, R. Bordia, and J. Rodel, "Mechanical Properties of Partially Sintered Alumina," J. Am. Cer. Soc., in press.

45. R. Baddi, J. Crampon, and R. Duclos, "Temperature and Porosity Effects on the Fracture of Fine Grain Size MgO," J. Mat. Sci., **21**, pp. 1145–50, 1986.

46. D. R. Biswas and R. M. Fulrath, "Strength of Porous Polycrystalline Ceramics," Trans. & J. Brit Cer. Soc., **79**, pp. 1–5, 1980.

47. G. W. Hollenburg and G. Walther, "The Elastic Modulus Fracture of Boron Carbide," J. Am. Cer. Soc., **63**(11–12), pp. 610–3, 1980.

48. S. G. Seshadri, M. Srinivasan and K. Y. Chia, "Microstructure and Mechanical Properties of Pressureless Sintered Alpha-SiC," Silicon Carbide (J. D. Cawley & C. E. Semler, Eds.), Am. Cer. Soc., Westerville, OH, pp. 215–225 (1989). Ceramic Transactions, vol. 2.

49. M. Desmaison-Burt, J. Montinin, F. Valin, and M. Boncoeur, "Mechanical Properties and Oxidation Behavior of HIPed Hafnium Nitride Ceramics," J. Eur. Cer. Soc., **13**, pp. 379–86, 1994.

50. A. K. Mukhopdhay, D. Chakrborty, and J. Mukerji, "Fractographic Study of Sintered Si_3N_4 and RBSN," J. Mat. Sci. Let., **6**, pp. 1198–1200, 1987.

51. R. W. Rice, "Porosity Effects on Machining Direction-Strength Anisotropy and Failure Mechanisms," J. Am. Cer. Soc. **77**(8), pp. 2232–36, 1994.

52. R. W. Rice, "Effects of Ceramic Microstructural Character on the Machining Direction-Strength Anisotropy," Machining of Advanced Materials, Proc. Intl. Conf. Machining of Advanced Materials 7/20–22/93 (S. Johanmir, Ed.), NIST Spec. Pub. 847, U.S. Govt. Printing Office, Washington, D.C., pp. 185–204, 6/1993.

53. E. D. Case and J. R. Smyth, "Room Temperature Fracture Energy of Monoclinic Gd_2O_3," J. Mat. Sci., **16**, pp. 3215–7, 1981.
54. J. S. Haggerty, A. Lightfoot, J. E. Ritter, and S. V. Nair, "High Strength, Porous, Brittle Materials," Mechanical Properties of Porous and Cellular Materials, (K. Sieradzki, D. J. Green, and L. J. Gibson, Eds.), Mat. Res. Soc. Symp. Proc., **207**, pp. 71–76, 1991.
55. H. Kodama and T. Miyoshi, "Study of Fracture Behavior of Very Fine-Grained Silicon Carbide," J. Am. Cer. Soc., **73**(10), pp. 3081–86, 1990.
56. G. F. Hurley and R. A. Youngman, "Toughening of Single Crystal Al_2O_3 by Neutron Irradiation," Third Annual Prog. Report on Special Purpose Materials for Magnetically Confined Fusion Reactors, U.S. DOE Report, DOE/ER-0113, pp. 51–54, 11/1981.
57. R. W. Rice, K. R. McKinney, and C. C. Wu, Unpublished Data on Al_2O_3 and MgO at the U.S. Naval Res. Lab.
58. A. D. Osipov, I. T. Ostapenko, V. P. Podtykan, N. S. Poltavsev, and I. A. Snezhko, "Effect of Porosity on Cracking Resistance of Hot-Compacted Boron Carbide," Tr. from Poroshkovaya Metallurgiya, **10**(322), pp. 48–52, 1989.
59. J. G. Zwissler and M. A. Adams, "Fracture Mechanics of Cellular Glass" (R. C. Bradt, A. G. Evans, D. P. H. Hasselman, and F. F. Lange, Eds.), Plenum Press, New York, **6**, pp. 211–242, 1983, Fracture Mechanics of Ceramics, vol. 6.
60. D. J. Green, "Fabrication and Mechanical Properties of Lightweight Ceramics Produced by Sintering of Hollow Spheres," J. Am. Cer. Soc., **68** [7], pp. 403–409 (1985).
61. L. Coronel, J. P. Jernot, and F. Osterstock, "Microstructure and Mechanical Properties of Sintered Glass," J. Mat. Sci., **25**, pp. 4866–72 (1990).
62. Y. L. Krasulin, V. N. Timofeev, S. M. Barinov, and A. B. Ivanov, "Strength and Fracture of Porous Ceramic Sintered from Spherical Particles," J. Mat. Sci., **15**, pp. 1402–1406 (1980).
63. J. J. Uchno, R. C. Bradt, and D. P. H. Hasselman, "Fracture Surface Energies of Magnesite Refractories," Am. Cer. Soc. Bul., **55**(7), pp. 665–68, 1976.
64. D. R. Larson, J. A. Cappola, D. P. H. Hasselman, and R. C. Bradt, "Fracture Toughness and Spalling Behavior of High-Al_2O_3 Refractories," J. Am. Cer. Soc., **57**(10), pp. 417–21, 1974.
65. R. W. Rice, "Test-Microstructural Dependance of Fracture Energy Measurements in Ceramics," Fracture Mechanics Methods for Ceramics, Rocks, and Concrete (S. W. Freiman and E. R. Fuller, Jr., Eds.) ASTM STP 745, pp. 96–117, 1982
66. F. J. P. Clarke, H. G. Tattersall, and G. Tappin, "Toughness of Ceramics and Their Work of Fracture," Proc. Brit. Cer. Soc., **6**, pp. 163–72, 1966.
67. J. B. Kessler, J. E. Ritter Jr., and R. W. Rice, "The Effects of Microstructure on the Fracture Energy of Hot Pressed MgO," Surfaces and Interfaces of Glass and Ceramics (V. D. Frechette, W. C. LaCourse, and V. L. Burdick, Eds.), Plenum Press., New York, pp. 529–44, 1974. Materials Science Research, vol. 7.
68. R. W. Rice, "Strength and Fracture of Hot-Pressed MgO," Proc. Brit. Cer. Soc., **20**, pp. 329–63, 6/1972.
69. C. T. Forwood, "The Work of Fracture in Crystals of Sodium Chloride Containing Cavities," Phil. Mag., **17**(148), pp. 657–67, 1968.

70. B. A. Cottom and M. J. Mayo, "Fracture Toughness of Nanocrystalline ZrO_2–3mol% Y_2O_3 Determined by Vickers Indentation," Scripta Mats., **34**(5), pp. 809–14, 1996.

71. R. G. Hoagland, G. T. Hahn, and A. R. Rosenfeld, "Influence of Microstructure on Fracture Propagation in Rock," Rock Mech., **5**, pp. 77–106, 1973.

72. P. Bompard, D. Wei, T. Guennouni, and D. Francois, "Mechanical and Fracture Behavior of Porous Materials," Eng. Fract. Mech., **28**(5/6), pp. 627–42, 1987.

73. R. Brezny and D. J. Green, "Fracture Behavior of Open-Cell Ceramics," J. Am. Cer. Soc., **72**(7), pp. 1145–52, 1989.

74. R. Brezny and D. J. Green, "Factors Controlling the Fracture Resistance of Brittle Cellular Materials," J. Am. Cer. Soc., 74(5), pp. 1061–65, 1991.

75. D. J. Green, "Mechanical Behavior of Lightweight Ceramics," Fracture Mechanics of Ceramics, vol. 8 (R. C. Bradt, A. G. Evans, D.P. H. Hasselman, and F. F. Lange, Eds.), Plenum Press, New York, pp. 39–59, 1986.

76. D. J. Green, J. E. Ritter, Jr., and F. F. Lange, "Fracture Behavior of Low-Density Fibrous Ceramics," J. Am. Cer. Soc., **65**(3), pp. 141–46, 1982.

77. J. J. Cleveland and R. C. Bradt, "Grain Size/Microcrack Relations for Pseudo-brookite Oxides," J. Am. Cer. Soc. **61**(11–12), pp. 478–81, 1978.

78. R. Stevens, "Fracture Behavior and Electron Microscopy of a Fine Grained Graphite," Carbon, **9**, pp. 573–78, 1971.

79. R. W. Rice and R. C. Pohanka, "Grain-Size Dependence of Spontaneous Cracking in Ceramics," J. Am. Cer. Soc. **62**(11–12), pp. 559–63, 1979.

80. E. D. Case, J. R. Smyth, and O. Hunter, "Grain-Size Dependence on Microcrack Initiation in Brittle Materials," J. Mat. Sci., **15**, pp. 149–53, 1980.

81. R. W. Rice, S. W. Freiman, and P. F. Becher, "Grain-Size Dependence of Fracture Energy in Ceramics: I, Experiment," J. Am. Cer. Soc., **64**(6), pp. 345–50, 1981.

82. R. W. Rice, "Comment on 'Role of Grain Size in the Strength and R-Curve Properties of Alumina'," J. Am. Cer. Soc., **76**(7), pp. 1898–99, 1993.

83. J. A. Kuszyk and R. C. Bradt, "Influence of Grain Size on Effects of Thermal Expansion Anisotropy in $MgTi_2O_5$," J. Am. Cer. Soc., **56**(8), pp. 420–23, 1973.

84. H. Tomaszewski, "Influence of Microstructure on the Thermomechanical Properties of Alumina Ceramics," Cer. Intl., **18**, pp. 51–55, 1992.

85. Y. Ohya and Z. Nakagawa, "Measurement of Crack Volume Due to Thermal Expansion Anisotropy in Aluminum Titanate Ceramics," J. Mat. Sci., **31**, pp. 1555–59, 1996.

86. K. Hamano, Y. Ohya, and Z. Nakagawa, "Crack Propagation Resistance of Aluminum Titanate Ceramics," Int. J. High Tech. Cer., **1**, pp. 129–37, 1985.

87. O. Hunter, Jr., R. W. Scheidecker, and S. Tojo, "Characterization of Metastable Tetragonal Hafnia," Ceramurgica, Intl., **5**(4), pp. 137–, 1979.

88. P. L. Swanson, "Tensile Fracture Resistance Mechanisms in Brittle Polycrystals: An Ultrasonic and In Situ Microscopy Investigation," J. Geophy. Res., **92**(B8), pp. 8015–36, 1987.

89. R. P. Ingel, U.S. Naval Res. Lab., Washington, D.C., private communication.

90. E. H. Lutz, M. V. Swain, and N. Claussen, "Thermal Shock Behavior of Duplex Ceramics," J. Am. Cer. Soc., **74**(1), pp. 19–24, 1991.

91. T. Nakashima, M. Shimizu, and M. Kukizaki, "Mechanical Strength and Thermal Resistance of Porous Glass," J. Cer. Soc. Jap., Intl. Ed., **100**, pp. 1389–93, 1992.

5

POROSITY EFFECTS ON TENSILE STRENGTH (AND RELIABILITY) AT LOW TO MODERATE TEMPERATURES

KEY CHAPTER GOALS

1. Show that a family of models is necessary and that MSA models are better for general porosity.
2. Demonstrate that major variations occur due to: 1) grain size increases with decreasing P, 2) porosity heterogeneity constraining lengths of normally elongated machining flaws, and especially 3) individual and clustered pores acting as fracture origins approximately consistent with mechanistic models, but with important material-related and other deviations, especially pores as blunter flaws in glasses.
3. Show that overall P-dependence of strength is closer to that of E than of K_{IC}, mainly due to crack branching and bridging having much less impact on strength since flaw sizes are often too small relative to the microstructure.
4. Show strengths decrease rapidly initially, then more slowly, with increasing contents of preexisting microcracks.

5.1 INTRODUCTION

This chapter addresses the porosity dependence of tensile (mostly uniaxial flexure) strength, with attention also given to the very limited data on porosity effects on reliability (via Weibull moduli), at 22°C. Unfortunately, there is not sufficient known data to define the effect of porosity on failure from miltiaxial stresses; however, the porosity dependence of thermal shock resistance, which is typically related to that of tensile strength at modest temperatures but involves at least some biaxial stress, is discussed in Chapter 9. There is also no data on porosity effects on tensile fatigue, e.g., to evaluate possible effects of crack branching and bridging due to interaction of large cracks with pores (as discussed in Chapter 4).

Tensile strength is proportional to K_{IC}, and hence to E (see Eqs. 4.1–4.3), and inversely dependent on flaw size (per Eq. 5.3 below), so there are three cases of porosity dependence. The first case is where both pores and grains and their heterogenieties are all small with respect to the failure causing (e.g., machining) flaws. The second case is where grains are still small relative to the flaw size, but individual pores, pore clusters, or serious porosity heterogeneities significantly impact flaw character, e.g., are large enough to be either a measurable component of are or the complete flaw causing failure. Such cases are particularly pertinent to polycrystalline bodies, but can also occur in glasses and single crystals. The third case is where the sizes of individual pores and grains, or clusters of these are both significant components of failure causing flaws. (Large grains or clusters of larger grains acting as fracture origins have either associated pores or cracks for them to be fracture origins). The first two cases are considered in detail in this chapter, and some aspects of the third case are discussed. In all three cases the flaw size (and secondarily, the flaw character) determine the level of strength, while effects of porosity on K_{IC} (and hence on E) are a major factor in and commonly dominate the porosity dependence of strength. Cases where flaw size (and character) change substantially with P, hence substantially changing the porosity dependence, while limited, are important and are illustrated and discussed.

The similarities and interrelations, as well as important differences between the effects of porosity on elastic moduli and fracture energy or toughness with those of strength are discussed. The weak-link nature of fracture is the primary factor in these differences, i.e., failure from pores or pore clusters from the main stream of the porosity distribution in the body, or more commonly a manifestation of extremes of the heterogeneity of porosity being all, or a significant portion, of flaws causing failure. Such pore-initiated failure accentuates the anisotropy in strength resulting from anisotropic pore shapes or clusters that can result from forming operations such as extrusion and die pressing (Chapter 2). Thus the sensitivity of brittle fracture to the weakest link commonly makes pore heterogeneity and anisotropy effects on strength more pronounced, as opposed to fracture toughness, and especially elastic behavior.

5.2 MODELS FOR POROSITY (AND MICROCRACK) DEPENDENCE

Consider first the case where pores, grains, or serious clustering of them, are all small relative to the flaw size. Historically, many of the same basic approaches to modeling the porosity dependence of elastic moduli have been applied to modeling such dependence of tensile strength, as previously reviewed (1). Thus, there have been the same three basic classes of approaches, namely: 1) empirical, 2) more geometrically based analysis of load bearing area (primarily MSA), or 3) detailed mechanics-based analysis, e.g., focused on stress concentrations. The empirical approaches were commonly based on log-log or semilog plots giving straight lines and thus respective equations for the relative tensile strength (σ/σ_0, where again the subscript 0 refers to the strength at $P = 0$) of:

$$\frac{\sigma}{\sigma_0} = (1 - AP)^n \tag{5.1}$$

$$\frac{\sigma}{\sigma_0} = e^{-bP} \tag{5.2}$$

Eq. 5.1 has usually been used with $A = 1$, but has sometimes been generalized empirically by letting $A = (P_C)^{-1}$ (which is required to fit much data since P/P_C is often substantially less than 1; Figs. 2.11, 3.1).

Tensile strength follows the traditional Griffith-Irwin equation:

$$\sigma = YK_{IC} (c)^{-\frac{1}{2}} \tag{5.3}$$

where Y is a factor dependent on the shape and location of the flaw causing failure and c is the flaw radius. Thus, if fracture only entails simple linear elastic processes, and the porosity does not significantly effect c, then per Eqs. 4.1–4.3, the P dependence of σ will be that of E, which will be shown here to commonly be the case for porous bodies. Variations of the porosity dependence of γ and K_{IC} (Table 4.2), though sometimes magnified (due to weak-link failure emphasizing extremes), are often reduced, or are not a factor in σ-P relations due to large crack effects on γ and K_{IC} having less or no applicability to the smaller cracks controlling much strength behavior, as discussed in Chapter 4 and later. These variations set further limitation on the use of bounds for σ-P behavior implied by the E-P bounds, which are of limited use themselves because of their breadth, i.e., the upper bound = 1-P and the lower bound = 0 (Eqs. 2.3, 2.4, 3.4). The P = 0 limit of σ is even more variable and uncertain than that of K_{IC} since σ depends as much and often more than K_{IC} on grain size (and other grain parameters), but not always in the same fashion. Further, there are often effects of porosity on c that can also effect the extrapolation of σ to P = 0. Again, the upper bound of P, P_C, for σ will often be approximately that for E, but can vary, e.g., be less at high P_C and somewhat higher at low P_C, due to heterogeneity effects on the weak-link nature of fracture.

The load-bearing–MSA models discussed for elastic moduli were often originally derived for strength, e.g., those of Eudier (2) and Knudsen (3) (the latter showing that Eq. 5.2 was a good approximation over the first $1/3$ to $1/2$ of the pertinent P range, i.e., of P/P_C). Thus, all of the MSA models of Tables 3.1–3.3 and Figures 2.11 and 3.1 have been applied to strength and will be considered here. Again, the array of models and their combinations allow this approach to be applied to probably any type and combination of porosity. Brown et al. (4) also proposed using the array of MSA models and their combination for strength as follows:

$$\frac{\sigma}{\sigma_0} = 1 - \Sigma\, \upsilon_i P_i \qquad (5.3)$$

where υ_i is the ratio of the MSA to that of the unit cell for a particular pore type of volume fraction P_i (which represents a somewhat different method of combining different porosities than discussed in section 2.4.3 and Fig. 2.11). Other load-bearing models were derived by Schiller (5) and Harma and Satava (6) (both for gypsum-based materials), and Bache (7) (for cement-based materials), the latter two involving parameters, e.g., for pore shape, that are often not calculable. The former two models both recognized and included P_C, i.e., Schiller's equation was of the form:

$$\frac{\sigma}{\sigma_0} = 1 - \left(\frac{P}{P_C}\right)^n \qquad (5.4)$$

where $n = 3$ for spherical, and 2 for cylindrical pores.

For the case where the flaw size $\gg$ pore spacing Carniglia (8) presented, based on load bearing arguments, the equation:

$$\frac{\sigma}{\sigma_0} = \left[(1-P)\frac{E}{E_0}\right]^{1/2} \qquad (5.5)$$

For the case where the flaw size $\ll$ pore spacing, he presented a more complex equation, with a pore-shape factor based on both load-bearing and stress-concentration arguments. In discussing Schiller's paper, Millard suggested the equation (5):

$$\frac{\sigma}{\sigma_0} = \left[1 - \left(\frac{P}{P_C}\right)^q\right]\left[1 + k\left(\frac{P}{P_C}\right)^n\right] \qquad (5.6)$$

where k, q, and n are empirically determined parameters, noting the first set of terms is based on load-bearing area and the second set on stress concentrations.

Another model not readily fitting a particular classification, that of Wagh et al. (9), is based on introducing cylindrical grains for rectangular grains to generate pores of random size, and then evaluating the bonding between the resultant grains. This gives Eq. 5.1 with A = 1 and n = q + 1/2 for both γ and K_{IC}, where q is the exponent for the porosity dependence of E; however, as discussed in Chapter 4, Arato (10) has shown an error in the derivation, so the exponents are all the same for E, K_{IC}, and σ, i.e., n = q. Lam et al. (11) proposed an equation based on addressing the porosity dependence of K_{IC} and of the flaw size, c. They assumed that the P-dependence of K_{IC} was simply $1 - P/P_c$, based on their limited data for E, γ, and K_{IC} fitting this expression; however, this expression is valid only for some porosity (Fig. 2.13, Section 3.4.3), so their data fitting is fortuitous since neither the character of porosity for which the expression is valid nor that of their specimens was identified. They also assumed that characteristic flaws are introduced in the powder consolidation process, and then shrink in size in proportion to the overall shrinkage of the body. This concept warrants attention, but it is open to question in their case since, though their samples were machined, they neglected probable effects of resultant machining flaws and their interaction with pores. Their equation fit their data to P ~ 0.2, below which there were serious deviations (but it also fit MSA models; Table 4.2). The latter deviations probably reflect in part approximations made to simplify the equation, as well as possible changing pore-machining flaw interactions.

Some models have been derived for the porosity dependence of strength based on mechanics analysis, e.g., based on the assumption that pore stress concentrations increased the stress on nearby (unspecified) flaws that were the actual cause of failure (1). A more recent modeling approach that combines some load-bearing and stress-concentration aspects has been presented by Boccaccini and Ondracek (12) similar to their modeling of elastic properties (13). They also obtained Eq. 5.1 with A = 1 and n = the stress concentration of the pore radius of curvature at the point of tangency of the stress direction and the pore surface, assuming, as have others, that other pore structures are equivalent to one of spheroidal (i.e., closed) pores of the same mean shape and orientation. For a body of aligned cylindrical (or prismatic) pores stressed parallel to their axes n = 1, consistent with that derived from load-bearing (and other) concepts (Table 2.5); however, it is also consistent with the exponent being the stress concentration factor, which for this case is 1, i.e., no stress concentration. For spherical pores n = 2, consistent with the stress concentration for spheres under uniaxial stress. They also noted that the model is only applicable to noninteracting pores, i.e., dilute levels of porosity and that most data lacks information to calculate the stress concentration exponent; however, their claim that dilute concentrations of pores can be up to P = 0.4 appears to be an over estimate by a factor of two or more (Section 2.4.2; Eq. 2.5).

All of the cautions of mechanics approaches to modeling the porosity dependence of elastic properties (Section 3.4) apply here. Thus, there is the issue of how real the assumption of representing pores as spheroids is, especially for the important case of pores between powder particles where there are basic differences of shape and typically of interaction, e.g., of open versus closed pores. Further, emphasis on stress concentration raises the issue of their dependence on global and local stress state (Table 2.4). These issues apply to all porosity, e.g., where pores and grains are small with respect to flaws causing failure, as well as where several or many clustered pores are a significant part of the failure causing flaw; however, even if the stress state dependence issue could be handled, it is again difficult to see how this has much effect for pores that are small in comparison with the crack interacting with them. Stress concentrations reflecting pore-shape effects would at first seem more applicable to the cases where pores, especially individual ones, are a significant factor in the failure causing flaws; however, the basic question is the effect of cracks postulated to be associated with such pores or clusters of them acting as the flaws (discussed below) on stress concentrations.

Consider now models for pores or clusters of them acting as the failure causing flaws. Evans and Tappin (14) made a major contribution by presenting and using specific models (which significantly mitigate the role of stress concentrations). This concept of pores as the flaws was a basic departure from that of pores as a source of added stress on other, unspecified, flaws causing failure. That larger pores or pore clusters are indeed typically the origin of failure when present, is extensively shown (15–30) and illustrated later. Such failure also questions stress concentrations playing a central role in the effect of pores on strength since such concentrations are a function of pore shape, not size.

Evans and Tappins' model was developed by treating pores with cracks using Bowie's model for a cylindrical hole with a crack emanating from one or both sides of the hole in a radial direction in a plane bisecting the circular cross-section (Fig. 5.1A) (31), as also suggested by Hasselman (32,33). A cylindrical pore-crack model, since improved (34), was used at that time since no solutions were available for spherical holes with cracks, which would be more appropriate in many, but not all, cases. Spherical pore-crack models were subsequently developed by Barrata (35) and refined by him (36) and others (37,38) (Fig. 5.1B). In either case, the concept was that a peripheral crack forms along part or all of one or both sides of a cylindrical pore or partly or fully around a spherical pore (at the equator, Fig. 5.2A,B). The focus was on polycrystalline bodies where it was postulated, originally by Evans and Tappin, and then others, that the peripheral crack around the pore formed, e.g., due to the applied stress, along the first layer of grain boundaries intersecting the pore surface, i.e., about half a grain depth from the pore surface (Fig. 5.2 B). This pore-crack combination, which was focused on pores substantially larger than the surrounding (fine) grain size, was then seen as the failure-causing flaw. Rice (39) postulated that

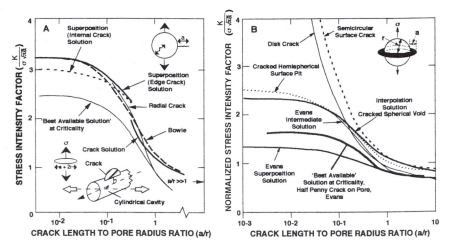

Figure 5.1 Normalized stress intensity versus the ratio of the crack radius from the pore surface (as sketched; see also Fig. 5.2) to radius of the pore for A) cylindrical and B) spherical pores. Published with permission of the J. Am. Cer. Soc. (34,37)

this model could be conceptually applied to pores smaller than the grain size (Fig. 5.2C,D) including intragranular pores (Figs. 1.4, 4.2C). (Some of these cases also provide opportunities for crack branching, and hence bridging, but this typically requires substantial extent of crack propagation.)

Evans and Tappin also proposed that a crack could grow subcritically between clustered pores to become a failure-causing flaw. Their model of subcritical crack connection of nearby pores was based on treating a simple cubic array of pore-crack combinations, i.e., as has been used for modeling thermal shock, giving the ratio of the stress intensity and the square root of the flaw size in the Griffith equation multiplied by two factors: 1) the ratio of the stress to propagate the pore-crack combination to that for a sharp flaw of the same size as the pore plus crack, and 2) that for activating propagation of cracks in the crack array; however, as shown by Rice (15,39) and later here, the first factor is typically 1 for polycrystals, and the second factor is also commonly ~1, e.g., being > 0.95 to P = 0.25 and > 0.5 to P > 0.8. Note that stress concentrations of such pore-crack fracture origins rapidly become that of the crack, greatly reducing stress concentrations of the pores alone.

Next, briefly consider models for effects of crack arrays on strength, i.e., similar to models for their effects on elastic properties (Chapter 3). Nielsen (40) presented a model for spherical or cylindrical pores having two sets of cracks emanating a distance a from the pore surfaces along two orthogonal diameters, giving net pore-crack diameters of 2 l. He calculates the "pore shape factor" ξ_0 (Table 3.4) as :

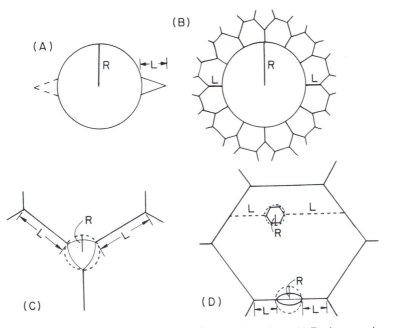

Figure 5.2   Schematic of pore-crack flaw combinations. A) Basic, generic model, e.g., for glasses. B) Schematics of possible effects of grain size in polycrystalline bodies constraining crack depths from the pore walls to ~ ½ the size of the surrounding grain(s), as originally suggested by Evans and Tappin (14) for cylindrical pores, but conceptually applicable to other, e.g., spherical, pores. C, D) Schematic of possible extensions of this concept (39). Note that while this is a very useful concept, there is substantial uncertainty in specifics of its application as discussed in the text. See also Fig. 4.2.

$$\xi_0 = \left(\frac{\pi}{4}\right)\left(1 - \frac{a}{l}\right)^3 \tag{5.7}$$

He then derived two approximate equations for strength of bodies with differing arrays of such pore-crack combinations, which he in turn reduced to

$$\frac{\sigma}{\sigma_0} \sim \left(1 - \frac{P}{P_C}\right)^{n+P/2} \tag{5.8}$$

where n is the same exponent in his equation for the pore shape factor, ξ_0, (Eq. 5.7) for any $P < 0.4$ with limited variation in the lengths of the cracks. For low P:

$$\frac{\sigma}{\sigma_0} \sim e^{-bP} \tag{5.9}$$

was given where $b = (n + P/2)(1/P_C)$.

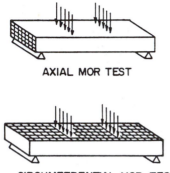

AXIAL MOR TEST

CIRCUMFERENTIAL MOR TEST

Figure 5.3 Schematic of flexure (tensile) testing of two basic orientations of square cell honeycomb, along with theoretical predictions (41). Fracture stress, $\sigma_f = \sigma_w (2t/L) = \sigma_w (1 - P)$ $\sigma_f = \sigma_w (t/L) = 1/2\sigma_w (1 - P)$ (σ_w = wall material strength, t = cell wall thickness, and L = cell spacing).

Consider now models specifically for the porosity dependence of highly porous materials, starting with honeycomb materials. Gulati and Schneider (41) showed that the flexural (tensile) strength of square cell honeycomb for both axial and normal stressing (relative to the extrusion axis, Fig. 5.3) both had a 1-P dependence, as does E, and were directly proportional to the strength of the wall material (σ_w). Their derivation gave the strength for normal versus parallel stressing to the extrusion axis (i.e., axial versus circumferential per Fig. 5.3) as ½; however, they noted that actual strengths for parallel stressing were only 50–70% of predicted values, and those for normal stressing were even lower, i.e., they had less than the predicted half the strength for parallel stressing due to variations in cell geometry and wall quality. Note that the identical P dependence for strength and E for square cell geometry implies this also holds for the other cell geometries in Table 3.3.

Turning to foams and rigidized fiber felts, the porosity dependence of tensile strengths should be that of K_{IC} (Table 4.1). While this often differs some from that of E (or G) for models of these highly porous materials, there are uncertainties in these (as in many) models, e.g., in this case due to cell stacking arrangements. A key example of this is the difference in values of n for the P-dependence of E, i.e., $(1 - P)^n$, being 1 versus 2 for the two cell geometries shown respectively in Figs. 3.3A,B. Thus, in the absence of a model, the P-dependence of E should be either a good guide, or at worst, an upper bound to the P-dependence of strength.

While tensile fatigue, whether enhanced by stress corrosion effects, as has been demonstrated in studies of dense ceramics, is also a concern, no theoretical or experimental studies of this are known (though there have been a few tests of repeated thermal shock of ceramic foams).

No models have been derive directly for general effects of only microcracks on strength, except for a simple cubic array of aligned slit cracks (42) (as used for modeling thermal shock), because of the complexities involved. This cubic-array model shows that microcracks progressively reduce strengths as the ratio of microcrack spacing to size decreases, as expected, except possibly at very small ratios for very small cracks. This model also reduces to the normal Griffith equation for a single flaw. If microcracks do not significantly effect the size of failure-causing flaws nor change K_{IC}, then models for microcrack effects on elastic moduli or fracture toughness may be applicable to strength.

5.3 DATA AND MODEL COMPARISON (LOW TO INTERMEDIATE P AT 22°C)

5.3.1 General Strength-Porosity and Strength-Microcrack Comparisons

Consider first bodies covering the most common range of porosity, e.g., P ~ 0–0.5, beginning with a comparison of data for bodies with porosity approximating that assumed in pertinent MSA models. Ali et al.'s (43) flexure strengths of glass bodies sintered from powder and glass balloons (~30 – 300 μm dia.) agree quite well with the MSA model for random stacking of spherical pores, i.e., the

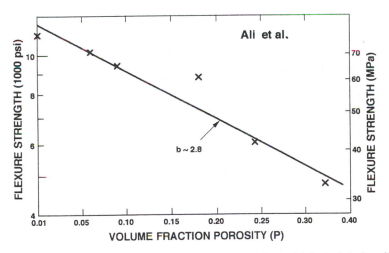

Figure 5.4 Flexure strength of a silicate-based glass with included glass balloons (42) showing good agreement with the b value for MSA models for spherical pores. Note that later analysis (see Fig. 5.12) indicates that the pores in the larger, ~ 300 μm diameter, balloons act as fracture origins.

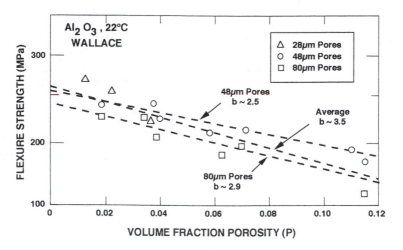

Figure 5.5 Flexure strength of Al$_2$O$_3$ with various volumes and sizes of artificially introduced spherical pores (via burnout of spherical plastic beads) (44). Note the overall good agreement of the b values with the value expected from the MSA model for random spherical pores of ~ 3, as well as the trend for lower strength at any P as pore size increases (see also Fig. 5.16).

b value of 2.8 (Fig. 5.4) agrees almost exactly with the MSA model for random (i.e., approximately simple cubic) stacking of pores (Fig. 2.11, 3.1). This agreement is attributed to pores in the larger balloons acting as an approximately constant flaw for all of these bodies independent of P (Fig. 5.12). Similarly, flexure strengths of Al$_2$O$_3$ (44) and PZT (45,46) with mainly larger spherical pores (from burnout of organic spheres from 25 to 80 or 150 μm dia.) also agree with the same MSA model, i.e., b ~3 (Figs. 3.4 and 5.5). Again, this consistancy is attributed to failure flaws being larger pores and hence not dependent on P (e.g., note the decrease of strengths with increasing pore size (Figs. 5.5, 5.16). Variations probably reflect the presence of other pores and especially the heterogeneous pore distribution (Sections 5.3 and 5.4).

The PZT data has a somewhat higher b value for σ than for E or K$_{IC}$ (Fig. 3.4, which may reflect in part that they have respectively ~ ½ and 1/4 of the P range for σ), but all are in reasonable agreement with one another, and collectively probably reflect common scatter about the MSA model. This general agreement and the absence of a clear dependence of strength of pore size is attributed to clusters of pores being frequent fracture origins, with larger clusters of smaller pores (Fig. 5.24) maintaining an approximately constant flaw size (Fig. 5.16). The same PZT study also included limited data for normal pores (2–3 μm) between the particles and artificially introduced acicular pores giving lower respective b values of 1.6–2.4. The lower value for the acicular pores is expected from some probable alignment of them with the tensile axis. The unusually low b value for the natural pores may also reflect some lamination, or the few data points and limited P range.

The porosity dependence of strengths for bodies sintered from only glass balloons (47,48) (dia. 36 ± 11 μm) showed similar σ-P with K_{IC}-P and especially E-P trends (Fig. 3.9B). Fracture was through the sintered necks of the balloons, which is consistent with MSA concepts.

Turning to bodies of sintered spherical particles, Krasulin et al. (49) studies of sintered (6 wt.%) CaO stabilized ZrO_2 beads (mostly 40–100 μm dia., giving P = 0.3–0.4) showed tensile strength decreasing somewhat faster with increased bead diameter (hence also P) than E or K_{IC}, which had similar dependences to each other (Fig. 3.9A). The strengths of Coronel et al. for sintered glass beads deviate considerably above the trend for E and the MSA model for random packing of uniform spheres, but were less than half as much as K_{IC} (50) (Fig.3.7A). The higher scatter for σ almost totally overlaps with the E data, but almost not at all with K_{IC}. Again, fracture was through the sintered necks of these bodies, consistent with MSA concepts.

Consider next some strength data for sintered SiO_2 gels (51–54) [typically consisting of chains of very fine (nanoscale) spherical particles (51)]. Tensile strengths of these bodies from different investigators and measuring techniques are reasonably consistent with one another (Fig. 3.8). The one probable deviation is Fijiu et al.'s (53) data, which is not surprising since it involved large spherical pores from foaming in addition to the pores from entangling the chains of small spherical particles. Lack of strength data for fully dense bodies in all but Ashkin's study, combined with effects of differing surface finishes and test methods, make specific quantitative comparisons uncertain; however, the overall trends and positions of the strength data relative to their P levels are reasonably consistent with the E-P data. Although the absolute strength levels, especially for Ashkin where there is the most accurate comparison, fall above the E-P trend, especially at lower P, overall they tend to be between cubic packing of spherical and cylindrical pores at high P as expected.

Turning now to tensile (flexure) strength data for ceramics with natural porosity from incomplete sintering of powders, much of this is summarized in Table 3.7 (the pertinent portion of which is reproduced as Table 5.1) and Table 5.2. While such strength data is more limited and often more scattered than for elastic moduli, the overall trends for both are similar, but possibly with somewhat higher b values for tensile strength versus those for elastic moduli. Thus, except for the three higher b values for extruded and one higher value for isopressed bodies, the data is consistent with this and the expected trends with processing-pore character (Table 5.1). Also, the data for cold-pressed, colloidally processed, and hot-pressed bodies follow the expected processing-pore character trends discussed for elastic properties. Deviations likely reflect heterogeneous porosity effects, larger pores, or clusters as fracture origins, or

Table 5.1 Summary of Ceramic Property Porosity Dependence for Different
Processing[a]

Property	Extruded	b Values Cold (die) pressed	Iso pressed	Colloidal	Hot pressed
Young's modulus (E)	2.3 (1)	3.6 ± 0.8 (25)	4±0 ± 1.2 (14)	3.9 ± 1.2 (4)	4.5 ± 0.7 (31)
Shear modulus (G)	2.2 (1)	3.2 ± 0.9 (8)	3.9 ± 1.3 (12)		4.3 ± 0.6 (17)
Bulk modulus		4.0 (1)			5.1 ± 0.8 (8)
Poisson's ratio		0.5 ± 0.5 (3)	0.5 ± 0.3 (3)	0 (1)	0.4 ± 0.6 (2)
Tensile strength (σ)	5.1 ± 2.1 (3)	4.0 ± 1.1 (3)	7 (1)	4 (1)	4.5 ± 1.2 (15)

[a] Values shown are averages and ± 1 standard deviation where more than one value was available.
Values in ( ) = number of studies averaged.

both, as discussed later. Data in Tables 4.2 and 5.2, some of which is included in Tables 3.7 and 5.1, also supports these trends. An earlier, substantial survey (1) also showed b values averaging 4 ± 2 for both flexure strength and Young's modulus. Other individual sets of data also showed this overall similarity, e.g., for Young's modulus of die-pressed and flexure strength of tape-cast Ba-Y-Cu oxide superconductor (Fig. 3.11B). Similarly, Al_2O_3 data (Figs. 3.10B, 5.5) is consistent with trends expected from MSA models, i.e., for die pressing b ~ 5 for E (71) and ~ 7 for strength [from a different investigation and with higher pressing pressure (72)], but ~ 3 for strength of extruded material (73). Finally, two other cases show lower b values for testing samples with some tubular or platelet pores that were aligned approximately parallel to the tensile axis as shown for E (Fig. 3.2A; Tables 3.2, 3.3). Hot-pressed Bi-Pb-Sr-Cu oxide superconductor with a considerable fraction of its porosity of a laminar character (from considerable orientation of the platy grains) gave a lower b, ~ 2.6 (P ~ 0.1–0.3) for flexure strengths for stressing parallel to the plane of lamination (74). More extensive is data gathered by Johnson (75) on extruded samples from a ceramics-forming class showing substantially lower b values for flexure strength parallel to the extrusion axis. Since there was no evidence of crystallographic orientation in this clay-based body, the lower b value is attributed to elongated pore structure from the extrusion.

Data for other materials, including metals, is also generally consistent with that of ceramics (see Chapter 8). As a particular example of the latter, Salak et al. (76) found the exponential relation was by a substantial margin the best fitting relation for the 834 data points for the tensile strength of sintered iron. Their b value of 4.3 for this data is a very reasonable value for sintered bodies based on MSA models (Figs. 2.11, 3.1).

Table 5.2 Comparison of b Values for Young's Modulus and Tensile Strength, and some Other Properties for a Given Material by the Same Investigator

Material	Processing[a]	Porosity	Grain size (μm)	b-E[c]	b-σ_t[b]	b-H[b]	b-σ_c[b]	Investigator
Al_2O_3	SC	0.15–0.5	23[c]	2.8	4			Coble and Kingery (55)
B_6O	HP	0–0.8	1	4.0	5		5.4	Petrak et al. (56)
BeO	EX	0.2–0.17	~20	2.3	2.8			Chandler et al. (57)
BeO	CP	0–0.7	~5	3.7	4			O'Neal and Livey (58)
MgO	CP	0.5–0.30	<5[d]	3.7	4.6			Biddulph (59)
ThO_2	CP	0.5–0.41	60	3.4	2.7			Curtis and Johnson (60)
βAl_2O_3	IP	0.01–0.4	<5[e]	5	7			Evans et al. (61)
SiB_4	HIP	0–0.19	40[f]	5.0	6.3[e]			Trembly and Angers (62)
TiB_2	HP	0.03–0.3	20–100	3.7	3.2		4.5	Mandorf and Hartwig (63)
B_4C	HP	0–0.15	≤5	4	3.9			Wu and Rice (64)
B_4C	S,HP	0–0.25	5–10	3.4	2.6			Schwetz et al. (65)
B_4C	HP	0–0.25	10–30	5	5			Champaign and Angers (66)
SiC	S[g]	~0–0.7 ~0–0.14	2–10	2.2	4.5	5–7		Seshardi et al. (27)
ZrC	Ex	0.01–0.3	2–20[h]		5.7		5.1	Bulychev et al. (67)
AlN	HP	0–0.2	20–50	5	3.5	6–8		Boch et al. (68)
HfN	HIP	0.05–0.14		5.3	3.3	5.4		Desmaison-Burt et al. (69)
TiN	HP	0.02–0.21	>2	5.0	1.6	2.3		Moriyama et al. (70)

[a] SC = slip cast, HP = hot pressed, EX = extruded, CP = cold (die) pressed, IP = isopressed, HIP = hot isostatically pressed.

[b] b values for the properties indicated; E = Young's modulus, σ_t = tensile strength, H = Vickers hardness, and σ_c = compressive strength.

[c] Made from fused grain.

[d] Estimated.

[e] Isolated grains to ~ 100 μm at lower P, all had transgranular fracture.

[f] For HIPing (of coarse grain material, particle size 2–100, av. ~ 10 μm) at 1300°C, b drops to 0.9; for HIPing at 1350°C as G increases to 60–120 μm and fracture becomes predominantly transgranular.

[g] Included die pressing, isopressing, and injection molding.

[h] Considerable intragranular porosity in most bodies, especially as G increased, as expected.

Much of the scatter and variation of data (e.g., Tables 3.7, 5.1, and 5.2) is due to deviations, some of which are systematic, e.g., with increased grain size, as well as random, systematic effects or both, due to porosity heterogeneity, orientation, or gradients. More specific aspects of some of these, especially heterogeneity, are discussed below in terms of reliability. Here we consider the effects of grain size. There is a general trend for bodies with finer grains to have b values for strength that are approximately the same as, or somewhat higher than for, E, and b values for bodies with larger grains to be the same as, or frequently lower for strength than for E. The latter is typically in bodies having substantial grain growth, and hence a larger average grain size or a substantial presence of larger grains, e.g., B_4C, HfN, TiN, AlN, SiB_4, and SiC of Table 5.2. It at first might be thought that the latter lower b trend reflects effects of crack bridging, i.e., as indicated for fracture energy and toughness (see Sections 4.2 and 4.3), but several factors argue against this.

The above effects with larger grains are attributed mainly to the effects of increased grain size on strength, and secondarily to changes in pore character accompanying grain growth, or both, and not to crack branching-bridging effects for three reasons. First, there is a definite decrease in strength with increased grain size, frequently even if the grain size increase is heterogeneous (1). Second, while shifting some pores to intragranular locations reduces the overall extent of crack-pore interactions, especially with frequently increased transgranular fracture (77), grain boundary coverage by pores often increases, at least locally (Fig. 1.3), frequently lowering strengths, especially as grain sizes increase beyond ~ 10 μm. The third, and most important reason is the weak-link nature of tensile failure. This greatly enhances the impact of even scattered, larger grains, with or without more grain-boundary pore coverage determining strength, precluding much, or all of the opportunity for crack branching or bridging. A particularly pronounced example of the effects of increased grain size on porosity dependence is that HIPing SiB_4 at 1350°C instead of 1300°C decreased the b value for strength from ~ 6 to ~ 1, consistent with grain size increased from ~ 40 to 60–120 μm. There has been almost no direct study of this grain size effect on porosity dependence; however, Knudsen (3) study showed b values for strengths of ThO_2 decreased from ~ 5.5 to ~ 3.7 as grain size increased by ~ 50% (e.g., to ~ 50 μm). Analysis of Broussaud et al.'s (78) data shows the b value for strength of fine grain (average ~ 3 and maximum ~ 6 μm) SiC at 22°C was ~ 4.3, but ~ 3 for coarser grain (average > 4 and maximum ~ 20 μm) bodies. The latter bodies of course had lower strengths and less total porosity consistent with the grain growth and more sintering accompanying obtaining larger grains. Also there is no apparent differentiation of cubic (e.g., ThO_2, ZrC, HfN, and TiN) and noncubic materials (e.g., TiB_2 and B_4C)., but the latter favor effects of crack branching-bridging.

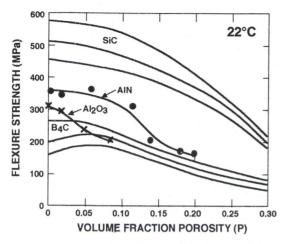

Figure 5.6 Examples of decreases in the porosity dependence of flexure strength (σ) at low P. The three curves each for B₄C and SiC are the average and the upper and lower bounds of the data, respectively, of Champaign and Angers (66) and Prochazka et al. (79). Actual data points and mean curves are shown for data of Boch et al. (68) for AlN (●) and Biswas (80) for Al₂O₃ (**X**). These all show less increase of σ at low P (best seen in such linear plots rather than semilog or log-log plots), attributed to grain size increases and pore flaw character effects.

Strengths at P = 0 are generally consistent with those at higher P, i.e via extrapolation, as for elastic properties (e.g., Figs. 3.4 3.11B, 5.4, 5.5); however, a number of exceptions to this occur, typically to lower than extrapolated strength at P = 0, often modestly so, but can in some cases be more pronounced (best seen on linear plots). Strengths at P = 0 different from those extrapolated to P = 0 can arise from changes in failure-causing flaw character as P→0. Thus, samples failing from large pores or pore clusters larger than other flaws in the body will exhibit higher strengths at P = 0 due to the loss of such pores as fracture origins. More commonly, however, strengths extrapolated to P = 0 are higher than actually measured there, which is typically due to effects of grain growth as P→0. Thus, less σ increase of SiC data of Prochazka et al. (79) with decreased P at low P (Fig. 5.6) is probably, at least partly, due to frequent opposite grain size effects on γ and K_{IC} versus on σ. Similar trends seen for AlN (68) and B₄C (66) probably also reflect effects of grain growth; however, there can also be other sources of such reduced P-dependence at low P from changes in flaws controlling σ. Thus, elimination of larger spherical (~ 40 μm diameter), artificial pores in alumina with flaws from 2.5 kg Knoop indents gave less σ increase as P decreased (80) indicating the increased σ was due to the elimination of these large pores was partly counteracted by the indent flaws no longer being as constrained in their elongation.

There is much less evaluation of the limited mechanics and other non MSA models in the literature; however, as noted earlier, Nielsen (40) showed good agreement between his model and data for Al_2O_3 and two cementatious materials. (He also showed similarly good fits of the exponential, i.e., MSA, model to this data.) Also, as noted earlier, Waugh et al. (9) showed good agreement of their model with five sets of data (and also showed agreement with the exponential MSA model). The latter modeling approach has recently been criticized by Boccaccini (81). Such agreement is reasonable since the form of most of these equations [i.e., $(1 - P)^n$] clearly fits some data for E, and hence will fit much of the same or similar strength data (when heterogeneities of porosity and failure from larger pores or pore clusters, discussed below, is not significant); however, no attention has been paid to pore character in these limited evaluations, which is essential to really evaluate the scientific validity of these models as well as to make them useful from an engineering stand point.

Boccaccini and Ondracek (12) tested their model by considering pores as spheroids (with aspect ratios based on pore character) and determining the resultant stress concentrations, which is the exponent n in the relative strength $(1 - P)^n$. Four cases fit with n = 2 for approximately spherical porosity (e.g., that of Figs. 3.4, and 5.4), but only one other pore case was fit with a calculated n value of 5.5 due to the sparsity of data with suitably characterized, nonspherical porosity for such evaluations. This and related approaches of approximating pores as spheroids is amenable to modeling (e.g., Chapter 3) and deserves further evaluation; however, serious questions remain, such as:

1. How realistic it is to apply such models to general porosity, especially open pores, as typically result from sintering until low porosity is reached?
2. Can such spheroidal pore characterization be consistently related to real porosity and reflect adequate property differentiation?
3. Since the derivation is only valid at lower P, can the different behaviors there for different pore types be related to the continuation of their behavior at higher P?

Strength data as a function of microcrack content is at best very limited; however, the presence of significant preexisting microcrack contents is usually, if not always, associated with lower strength, consistent with expectations outlined earlier. This is clearly shown in studies of Cleveland and Bradt (82) of dititanates of MgO, Fe_2O_3, and Al_2O_3 which have high TEA (increasing in the order listed). These show precipitous initial decreases in tensile strengths closely following those for Young's moduli as grain size increases above the critical values for the onset of microcracking, i.e., with increasing microcracking. As grain sizes are further increased, the decreases in strengths and moduli saturate, approaching asymptotic values, e.g., of ≤ 20 MPa. The strength decrease before saturation is far greater than the normal strength decrease with increasing grain size without preexisting microcracks, e.g., respectively for Al_2O_3 and $MgAl_2O_4$ (1,83). However, note that Al_2O_3 data of Tomaszewski (84) show some microcracking occurring during stressing that increased as grain size increased and thus was probably a factor in the rate of strength decrease as grain size increased, but still at a rate less than for more anisotropic dititanates.

Similar strength decreases with increasing toughness from microcracks are seen in composites (see Sections 8.3.6, 9.3.2). The marked differences in fracture energy-toughness versus strength (and Young's modulus) behavior is attributed to crack-size effects similar to those observed with some porosity, i.e., crack branching, bridging, or both, with larger cracks commonly used in fracture energy-toughness measurements versus much smaller cracks typically controlling strength. Note that the approximate saturation of strength and Young's moduli losses implies approximate saturation of microcracking at larger grain sizes relative to the critical values for microcracking, but this region of behavior has received no significant study.

5.3.2 Reliability and Porosity Heterogeneity

First, consider the effects of porosity on mechanical reliability, which has received limited attention. Such data exists mainly for porous bodies from incomplete sintering, mostly from three studies of the effect of porosity on the Weibull modulus (m). Recent studies by Knechtel et al. (85) (Table 4.1) and Nanjangud et al. (72) on partially sintered, fine Al_2O_3 powders showed m increasing from 6 to 8 at P = 0.25–0.4 to ~ 9–13 at P = 0.05 (Fig. 5.7), i.e., m increasing as strength increased and P decreased. Similarly, data of Pissenberger and Gritzner (86) for sintered ZrO_2 + 10–18 m/o Nb_2O_5 or Ta_2O_5, though more scattered (probably due to the use of two different powder preparations and some variation in the level of additive), showed a similar trend, but with a more rapid initial decrease in m as P increased, and then probably less decrease with further P increase. These and other results (e.g., in conjunction with Fig. 5.10) suggest a trend for decreased Weibull modulus as P increases, however, recent studies of Kishimoto et al. (87) on partially sintered TiO_2 are not fully consistent with this. They found m values for flexural strength for the following P levels of: 0.4–8.4, 0.3–5.3, 0.2–5.7, 0.1–4.3, and 0.05–6.2, i.e., m = 6 ± 1.5, which were nearly identical for dielectric breakdown (Section 7.3.2). The scatter of these m values for flexure appears similar to the scatter in the average strength-P trend giving an average b of ~ 5.

Two other studies of the Weibull moduli of porous bodies are noted. A few studies of ceramic foams (Section 5.3.5) and highly porous CVI- fiber composites (see Fig. 8.15) with (P ~ 0.7–0.95 gave variable m values of 2–13, i.e., mainly in the same range of values as in Fig. 5.7. Many of these measurements are based on only 5–7 specimens so individual values, especially extreme ones, should be viewed with skepticism, but average trends appear to be of some use, indicating limited further decrease in m as P decreases at higher P levels.

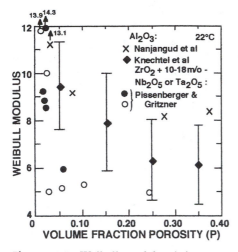

Figure 5.7 Weibull modulus (m) versus volume fraction porosity (P) at 22°C for sintered ceramics. Sintered materials of Al_2O_3 (die-pressed) of Knechtel et al. (85) and (extruded rods) Nanjangud et al. (72), and of isopressed ZrO_2 + 10–18 m/o Nb_2O_5 or Ta_2O_5 of Pissenberger and Gritzner (86) made using two different powder preparation routes are in the left portion of the plot.

The other study, that of Biswas and Fulrath (45,46), was on otherwise dense doped PZT, with either limited natural porosity (P = 0.024 to 0.15) or large pores from burnout of organic spheres (P = 0.09, diameter ~ 100 μm). Both increased m from ~ 10 to ~ 14, though with reduced strength levels as P or the pore size increased; however, note that in bodies with artificially introduced spherical pores failure initiated from clusters of such pores (Fig. 5.24). Thus, if there had been more uniform dispersion of the plastic beads used to generate the spherical pores, the increase in Weibull modulus would probably have been greater and the strength decrease less. Results of this study, though of opposite trend of m versus the direction of P change (i.e., Fig. 5.7) are not necessarily contradictory, e.g., due to differing processing and both showed higher m at modest P. They again illustrate the variable effects that can occur as the amount, character (e.g., scale of the pores relative to the grain or flaw sizes) and homogeneity of the porosity changes. The above trend for reduced m as P increases over a substantial range indicated in three studies of sintered powders, as well as some modest m increases at low P with larger pores should ultimately be understood by considering effects of porosity levels and character on pertinent properties E, γ, and K_{IC} and on flaws. There is limited but more information on effects of porosity level, P (Fig. 5.7), and less on character (e.g., size, geometry, and heterogeneity) which can be related to P, as noted below.Thus, evidence of increased heterogeneity of porosity and associated increased variation of E were shown earlier (Fig. 3.17, Table 3.9). Such E variations contribute to increased strength variations and thus reduced Weibull modulus. Similarly, the significant variability of γ and K_{IC} that can occur, especially at low P, can also contribute to strength variations and hence lower m values; however, as discussed in Chapter 4 and below, while some of these variations may impact strength, lowering m, much of the variations of γ and K_{IC} do not necessarily affect strengths due to their arising from larger crack sizes than commonly control strength.

Effects of porosity on flaws can be particularly significant for strength due to it being a weak-link phenomena, and hence especially sensitive to extremes of porosity heterogeneity from either of two ways. The first is pores or pore clusters being all, or an important portion of, the failure-causing flaws (i.e., emphasizing pore or cluster size) is discussed in the next section. Most other effcts of pore character on reliability have not been studied, but there are some indications of effects of porosity heterogeneity, which is the second way to effect strength and its variability via impact on the character of other flaws causing failure, especially machining flaws, which are again often associated with larger pores and porous areas. Such pore-crack combinations are often a factor in the pores being fracture origins (as discussed in the next section); however, pores or highly porous areas are also natural termini for machining flaws, i.e., a flaw intersecting a pore or porous area of similar size as the flaw from one side will generally not propagate around the pore or across porous area (Fig. 5.8).

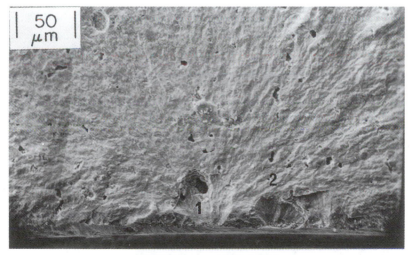

PZT σ_f~15 KSI

Figure 5.8 SEM of fracture origin (bottom center, $\sigma \sim 105$ MPa) of a flexure test at 22°C of a commercial PZT from a machining flaw, 1, and an isolated pore. Note the termination of the machining flaw, 1, by the pore (and two other similar machining flaws to the right, 2).

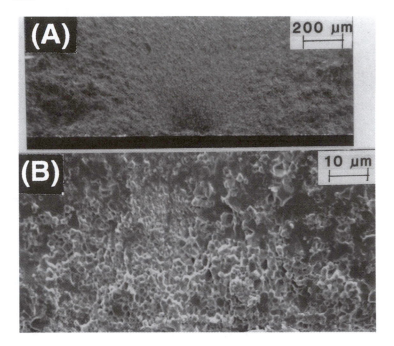

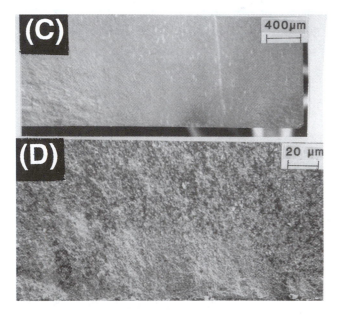

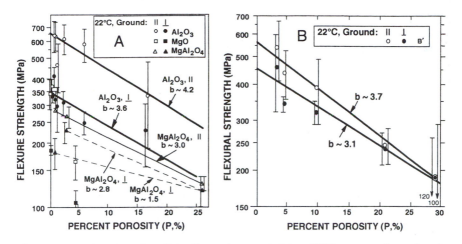

Figure 5.10 Semilog plot of the flexural strength (σ) at 22°C versus volume fraction porosity (P) for: A) hot-pressed Al_2O_3 and MgO, and sintered $MgAl_2O_4$, and B) hot-pressed boron samples machined parallel (‖) or perpendicular (⊥) to the bar and tensile axis after Rice (88,89). Note that the latter strengths show less P-dependence than the former as the heterogeneity of the porosity increases (see Figs. 5.9, 5.11). This decreased P-dependence is attributed to the heterogeneous porosity increasing with P progressively reducing the normal elongation of the machining flaws that form parallel with the abrasive motion and hence control strengths for stressing normal to the machining direction (Fig. 5.9) versus approximately half-penny flaws controlling strengths for machining and stressing parallel to the tensile axis (88–92). Published with permission of the Am. Cer. Soc.

The above machining flaw-pore interaction, while often variable, can be systematic due to increasing heterogeneity of porosity as P increases, giving systematic flaw-shape effects from machining. Machining bodies parallel with the subsequent tensile axis typically gives substantially higher strengths than when they are machined perpendicular to the tensile axis. This strength difference arises from strengths being controlled by the two sets of machining flaws typically introduced by machining (88–92). Flaws formed parallel and perpendicular to the machining direction both have similar depths, but are, respectively, typically near-half-penny and more elongated cracks. The elongated flaws formed parallel to the direction of machining control give lower strengths when activated by stresses essentially normal to them. Such flaw elongation

Figure 5.9 Examples of fracture origins from elongated flaws from machining porous hot-pressed alumina bars perpendicular to their tensile axes from strength-porosity studies in Fig. 5.10A. A) Lower and B) higher magnifications of the fracture origin (bottom center) in a body with P = 0.008. C) Lower and D) higher magnifications of the fracture origin (bottom center) in a body with P = 0.165. Note the similar, elongated flaw shapes and the relatively homogeneous character of the bodies.

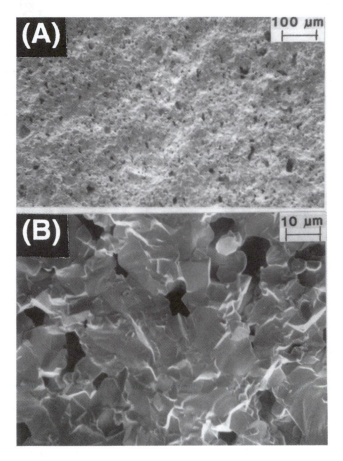

Figure 5.11 Examples of heterogeneous porosity in the bodies of Fig. 5.10. A) Lower and B) higher magnifications of $MgAl_2O_4$ with P ~ 0.26. C) Established and D) probable fracture origins from clustered pores in hot-pressed boron, with P = 0.098 and 0.21, at 353 and 215 MPa, respectively.

effects occur even in porous bodies, as shown by both fractography (Fig. 5.9) and strengths as a function of stress versus machining direction (Fig. 5.10). It has been recently shown that heterogeneous porosity affects this anisotropy (88,89), but homogeneous porosity does not (Fig. 5.10). Thus, the strength of Al_2O_3 ground parallel and perpendicular to the tensile axis shows very similar P dependence, i.e., nearly parallel σ-P trends, but a slight trend for convergence; however, boron and $MgAl_2O_4$ both showed convergence of the σ-P trends at P = 0.2 to 0.3. This convergence correlated with greater porosity heterogeneity (Fig. 5.11). The convergence of σ-P trends resulted from a reduced P-dependence of the specimens ground perpendicular to the tensile axis, i.e., the b values for

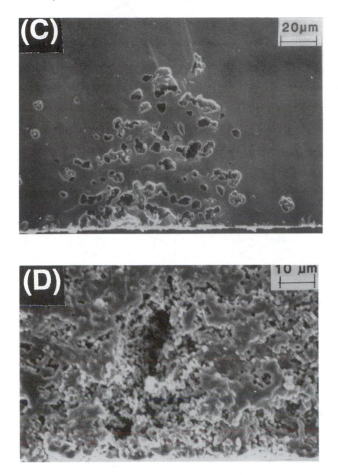

specimens ground parallel with the tensile axis are reasonable for the expected porosity, but those for specimens ground perpendicular to the tensile axis are lower. The logical conclusion is that increasing heterogeneity of porosity as P increases reduced the elongation of flaws forming parallel with the machining direction, and hence the strength anisotropy as a function of machining direction.

The above machining direction results and preceding Weibull analysis of partially sintered bodies are consistent with increased heterogeneity of porosity as P increases in two ways. First, lower Weibull moduli as P increases is likely to reflect increased heterogeneity of the porosity as noted earlier. Second, use of Eq. 2.2 to calculate Weibull moduli for the data points in Fig. 5.10 also shows a similar trend. Thus, while the database for each point is limited, the resultant m values at the higher P levels are typically $1/2$ to $1/3$ those at low P, again indicating increased heterogeneity of the porosity via reduced m values as well as via the reduction of effects of transverse machining noted earlier.

As noted earlier, Lam et al. (11) polished their alumina samples, but neglected the possible presence and effects of the resultant machining flaws on failure. Their data, which generally fit MSA models (Table 4.2), showed significant deviation from their model below P ~ 0.2. While this may reflect deviations from their assumptions, it may also reflect changing effects and interactions of the machining flaws and the porosity, especially its heterogeneity.

5.3.3 Failure Originating from Individual Pores

FRACTURE ORIGINATING FROM SINGLE PORES IN GLASSES

That large pores, pore clusters, or more porous areas are frequent sources of failure, either by themselves or in conjunction with other flaws is well established (1,15–30) (e.g., Figs. 1.10, 1.12, 5.8, 5.14, 5.17–19, 5.21–24). The challenge is to sort out the specifics of these failures for comparison with models. This is often complicated by lack of information on specific failures, e.g., pore shapes and sizes, effects of stress gradients (in flexure with larger pores) and whether other, e.g., machining, flaws are involved. A basic question

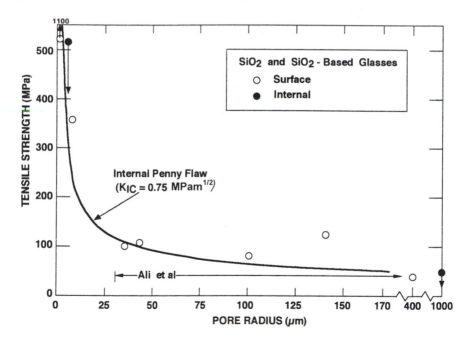

Figure 5.12 Tensile strength versus pore radii at fracture origins in silica-based glasses. Note the close agreement of most data with the curve treating the pores as sharp flaws (i.e., Eq. 5.3, with $K_{IC} = 0.75$ MPam$^{1/2}$) and the general agreement of the data of Ali et al. (43) at their larger balloon size (300 μm). Some points may be shifted some to correct the tensile stress from the surface to the pore surface or center, particularly for the second smallest pore (smallest internal pore origin, arrow; see Fig. 5.14 A).

is whether extensive to complete peripheral-equatorial cracking occurs around the pore before or during the failure process.

The results of the few studies of failure from isolated pores are summarized here. For guidance and some contrast, consider first tensile or flexure strengths of SiO_2 and SiO_2-based glasses (15) versus the radius of the pore at the fracture origin (Fig. 5.12). Note the limited overall data scatter despite reflecting different tests, specimen sizes, surface finishes (ground or flame polished), and compositions. This includes data of Ali et al. (43) for their sintered bodies with glass balloons (30–300 µm) at low P levels (Fig. 5.4), where failure from individual pores is likely to occur, especially from larger balloons, for which agreement with the other data is best. Note that most of the data generally agrees quite well with failure from a sharp crack having the same radius as the pore at the origin using a representative K_{IC} (0.75 $MPam^{1/2}$), i.e., the internal penny-flaw curve of Fig. 5.12. Two of the four more significant exceptions are the two internal pore origins, but correction of the surface stresses for the linear stress gradient into the sample depth to estimate the actual stress at the internal location of the pores, results in good agreement for these two points. The spherical pore-equatorial crack model also fits most of the data well using the same K_{IC} (0.75 $MPam^{1/2}$) and a crack-to-pore radius, a/r, ratio of 0.1 almost identically follows the internal penny-flaw curve (a few and 10–20 % above that curve respectively at smaller and larger pore size). Using a/r = 0.5 gives a curve that is only slightly more below the penny-flaw curve at larger r and somewhat more as r decreases. Note that the corresponding curve for a surface half-penny or equatorially cracked surface hemispherical pore (pit) flaw would only be about 10% lower, so the agreement is also good for failure from such surface flaws as well.

While most of the glass strength data for failure from isolated pores is close to that for failure from sharp flaw or pore-crack models, there are important and significant variations. Two data points clearly lie above the internal penny-flaw curve (for the smallest and the third largest pores, Fig. 5.12), indicating that these pores, while the fracture origins, acted as less than sharp flaws, and hence may follow pore-crack models with substantially smaller a/r ratios. There are other important variations, especially for data not plotted in Fig. 5.12. These include clear cases of failure not occurring from pores, where failure instead occurred from machining flaws smaller than the pores that were near the pores, but not necessarily close enough to the pores to be significantly effected by the pores (Fig. 5.13). In such cases there was no evidence of any cracks directly associated with the pores. Thus, pores in glasses without preexisting associated cracks from machining or other processes are less severe sources of failure than sharp flaws of the same or smaller size than the pore, e.g., indicating smaller, probably variable a/r ratios.

Fractographic evaluation (for which such glasses are typically quite good) for fractures originating from isolated pores (Fig. 5.12) shows important varia-

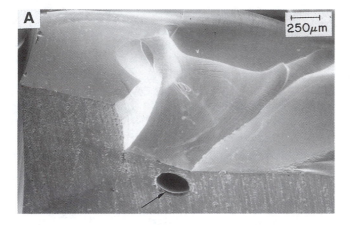

Figure 5.13 Examples of failure not occurring from pores at the surface of diamond ground glass specimens. A) Failure clearly not occurring from a large (~ 300 μm diameter) pore (arrow) exposed on the surface at 53 MPa, $\sigma_p/\sigma_a > 1.1$. Note that the higher SEM magnification shows chipping around the pore-machined tensile surface intersection, but no evidence of cracks penetrating into the glass from this intersection. B) Failure from a surface machining flaw (F) at 82 MPa despite a substantially larger pore (bubble) just below the flaw and inside of the tensile surface, i.e., note the deflection of the crack to avoid the bubble (B) and the resultant emanating fracture marking). C) Both the pore and flaw are seen in the matching fracture half (via additional side lighting in the optical microscope). D) Higher magnification of failure-causing machining flaw (F). Note the symmetrical flaw and mirror, normal mirror to flaw size (e.g., vertical tic marks in C, $R_m/c \sim 10$), and $\sigma_p/\sigma_a > 1.7$. Published with permission of J. Mat. Sci. (15)

tions, especially from the models based on the formation of an extensive peripheral equatorial crack around the pore as a key step in the failure (Figs. 5.1, 5.2). Thus, while fracture mirrors from the few pore failures in the interior of samples were fairly symmetric, as is typical for symmetric flaws in a reasonably uniform stress field (93,94) (e.g., Fig. 5.14A), all most all fracture mirrors from pores exposed on machined surfaces were asymmetric, usually substantially so (Fig. 5.14B–E). This asymmetry was associated with cracks of differing sizes at the intersection of the exposed pore and the machined tensile surface, with cases of a crack and subsequent failure initiation from only one side of the pore exposed on the machined tensile surface (15). Thus, when there were machining cracks associated with pores at fracture origins, fracture markings clearly showed that failure proceeded from machining cracks at the intersection of the pore and the machined tensile surface, and not by first generating a peripheral equatorial crack as in the pore-crack models (Figs. 5.1, 5.2). Furthermore, pores not acting as failure sources showed no evidence of associated machining flaws. Refining the fractographic observations to clarify many of these important details is

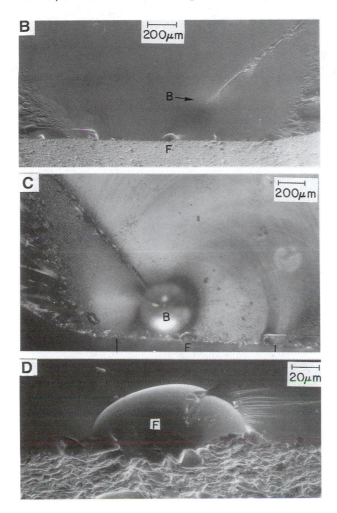

probably, at least partly, feasible in glasses (but is more challenging and often less feasible in polycrystalline ceramics).

Another key fractographic observation on glass fractures originating from pores, besides the fracture mirror shape, is the sizes of the fracture mirrors around the origins versus the sizes of the fracture origin. Thus, evaluating the ratio of the radius of the mirror (R_m) to the flaw size, c, causing failure, Rice (15) showed that the R_m/c ratio will change for blunter versus sharp flaws (based on the well-known relations for the failure stress, σ_f):

$$\sigma_f = A\,(c)^{-\frac{1}{2}} = B\,(R_m)^{-\frac{1}{2}} \qquad (5.10)$$

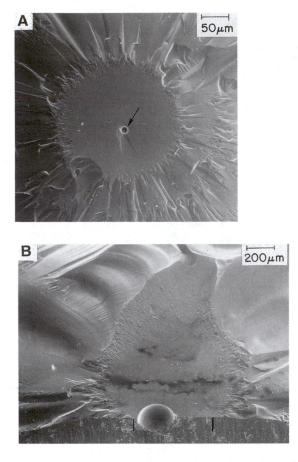

Figure 5.14 Examples of glass failures from: A) an internal pore and B–E) surface pores. A) Flame-polished fused SiO_2 (530 MPa, ~ 410 MPa at pore center), R_m/R_p ~ 16, σ_p/σ_a ~ 8.8. B) Ground glass bar failing from a pore exposed on the tensile surface (128 MPa). Note the substantial asymmetry of the fracture mirror (vertical tick marks) indicating that the fracture initiated at the tensile surface on the right side of the pore, R_m/R_p ~ 4.4, σ_p/σ_a ~ 2.2. C, D) Failure from another pore, P, on the surface of a ground bar (81 MPa), again showing an asymmetrical mirror (M) and a fracture tail (T) due to the meeting of two cracks starting from opposite sides of the bubble, as schematically illustrated in E). The tail being on the right side shows the crack at the intersection of the tensile surface and the left edge of the pore was the larger (D) and propagated further around the pore than the smaller crack on the right side of the pore where it intersects the tensile surface. R_m/R_p ~ 8.7, σ_p/σ_a ~ 2.1. Published with permission of J. Mat. Sci. (15).

where A and B are constants for a given material. The first equation is simply the Griffith Equation, i.e., with A being the product of the K_{IC} and the flaw-geometry factor [which is appropriately taken as that for a circular flaw with c

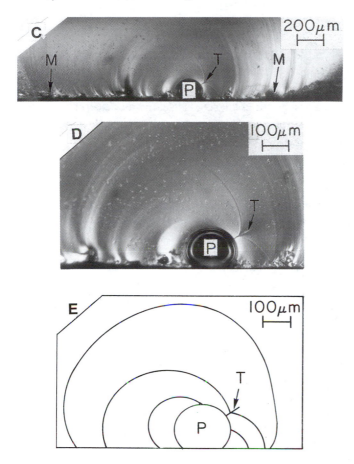

then taken as the radius of the penny or half-penny flaw of the same area as the flaw at the fracture origin (95)]. For normal failure from sharp flaws, the mirror to flaw size ratio is thus:

$$\frac{R_m}{c} = \left(\frac{B}{A}\right)^2 \tag{5.11}$$

which is typically ~ 12 for glasses and some times similar or less for single- and polycrystals (21,96,97). For failure from a blunter flaw, the second half of Eq. 5.10 will be the same since the mirror is still formed by the propagation of a sharp flaw, but the first half must be modified to reflect failure initiation from a blunter flaw such as some pores. Thus, the first equality in Eq 5.10 must be multiplied by the ratio of the stress for failure from a pore to that for failure from a sharp flaw, σ_p/σ_f, which thus gives a different mirror-to-flaw size ratio:

$$R_m/c = (B/A)^2 (\sigma_p/\sigma_f)^{-2} \qquad\qquad (5.12)$$

The mirror sizes of the glass samples clearly reflected the indicated range of R_m/c ratios found by calculation of σ_p/σ_f from the fracture stresses and the pore sizes at the fracture origins. (The exact flaw size for failure from pores is generally not known since the associated crack size is usually not discernable, but, as indicated above, the size of the associated crack is generally small relative to the pore size, so $R_p \sim c$). Thus, mirror-to-flaw size ratios for failures from pores ranged from those from machining flaws to smaller ratios (per Eq. 5.12) as the extent of the associated machining flaw diminished, then was not existent (15). These variations are corroborated by the shapes of the mirrors (Figs. 5.13, 5.14). Even in the absence of any associated machining flaws, i.e for internal pore origins, the mirrors are not necessarily fully symmetrical with the pore shape; this indicates that uniform equatorial cracks do not necessarily form and that failure is probably sensitive to local pore character.

The only known occurrence of equatorial cracks occurring around pores is in a glassy material, namely amber (98). A microphotograph (scale not given) showed complete equatorial cracks (with $a/r \sim 0.5$) associated with only the two largest pores (and none of the other pores of somewhat, to substantially, smaller size, regardless of close proximity of associated pores). Conditions for the formation of the approximately equatorial cracks around the two pores are unknown, e.g., whether gas (i.e., hydrostatic) pressure in the pore, or other pressures in or on the sample during or after its transformation to amber contributed to the forming the two large equatorial cracks.

Finally, before proceeding to failures from pore clusters, it is important to note that the K_{IC} pertinent to failure from isolated pores or pore clusters is often not that measured on the porous body. The latter reflects effects of all pores encountered, generally reducing K_{IC} (see Chapter 4); however, cracks for failure from pores or pore clusters generally propagate through denser material surrounding the pore or cluster origin. When such pores or pore clusters are substantially isolated, as is often the case as for glass fractures studied, K_{IC} measured on the actual versus a fully dense body will not differ significantly. In failure from single or clustered pores, however, where there are a number of these in the body, K_{IC} measured on the body will be too low due to crack interaction with several isolated pores and will generally give lower toughness values. While the appropriate value can be variable, a value about halfway between the latter and the fully dense value should often be a reasonable estimate. On the other hand, the K_{IC} for failure from part of a pore cluster may be below even that for the body itself due to the initial crack propagation through the remaining pores of the cluster. These issues are pertinent to many of the failures in the next sections.

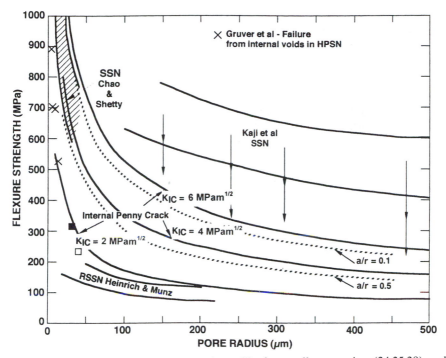

Figure 5.15 Flexure strength versus the radii of naturally occurring (24,25,29) and artificially introduced (from burnout of plastic beads) (30) pores at fracture origins in various silicon nitride bodies at 22°C. Note: 1) curves for sharp, internal penny flaws for various K_{IC} values and for pore-crack models with a/r ratios of 0.1 and 0.5, 2) the progressive downward adjustment of Kaj et al.'s (30) data (arrows) to account respectively for the distances of the pore surface (closest to the tensile surface) and center from the tensile surface, and 3) the approximate agreement of data with the curve treating the pores as equivalent with sharp flaws of the same size, c, and less agreement with pore-crack models.

FRACTURE ORIGINS FROM SINGLE PORES IN POLYCRYSTALLINE SAMPLES

Consider next data for failure from isolated voids in reaction sintered, sintered, and hot pressed Si_3N_4 (Fig. 5.15). Rice's two flexure data points for RSSN with natural pores (19) agree reasonably well with flexure data of Heinrich and Munz (25) for RSSN (grain size ~ 3μm) with artificially introduced pores by the burnout of wax spheres varying in diameter to about 200 μm. The limited data identifying surface and internal pore origins showed no clear differentiation between them (Heinrich and Munz did not designate the location of the pores, but imply that most or all of the pores at origins were at the surface, i.e., approximately hemispherical pits). All of these data agreed reasonably well with the pores acting as sharp flaws of the same size using $K_{IC} = 2$MPam$^{1/2}$ (the authors measured 1.5 MPam$^{1/2}$). Use of K_{IC} somewhat higher than measured is consistent with correction as noted above; however, analysis also shows that while the pore-crack models do not fit the data for cracks that are approximately 1 grain depth out from the pore, they do fit for cracks a few to several grains deep from the pore walls (99). This is not surprising since increasing the size of the crack around the pore reduces the (typically limited) difference from a sharp flaw.

Flexure data of Chao and Shetty (29) was for dense (3.29 g/cc) sintered Si$_3$N$_4$ (grain size ~ 2μm) failing from natural pores (i.e., somewhat irregular in shape, but roughly spherical pores, apparently at or near the tensile surfaces, and often containing some material such as intergrown whiskers). This data is in reasonable agreement with treating the pores as sharp flaws with a K_{IC} of 4–6 MPam$^{1/2}$ (K_{IC} 4.5 MPam$^{1/2}$ measured), as well as for pore-crack models with a/r ratios of 0.1 to 0.5, as also shown by Chao and Shetty. The limited data of Gruver et al. (23) for hot-pressed Si$_3$N$_4$ (corrected for the location of the mainly internal pore origins) agrees reasonably with data of Chao and Shetty. As plotted, flexure data of Kaji et al. (30) for their material (density 3.2 g/cc, made by placing individual organic spheres varying in diameters from ~ 100 to 500 μm near the tensile surface in die pressing the bodies) does not agree with either the other data for dense Si$_3$N$_4$ or penny or pore-crack models. While Kaji et al.'s data was for failure well inside of the tensile surface (they rejected surface connected failures), however, their stress values are for failure from the tensile surface, i.e., neglecting the linear stress gradient to zero stress at the neutral axis. They gave the average depths of the origins for each of the four different average-sized spheres they used, thus allowing the failure-stress at the pore to be estimated from their surface failure stresses. Thus, correction of the average failure stress from that for failure at the tensile surface to the stress at the pore surface closest to the tensile surface, and to the stress at the equator of the spherical pores results in progressively better agreement with the other data for dense Si$_3$N$_4$, as well as the penny-flaw and pore-crack models.

Next consider the two pertinent studies of sintered SiC. Seshardi and Srinivasan (27) evaluated flexure failures of their sintered α-SiC (grain size ~ 8μm) at 22°C (and 1260°C), and compared the resultant failures from pores and other flaws with both a circular sharp flaw of the same area to that of the spherical pore-equatorial crack model. Scatter was substantial in comparison with each type of flaw, with the equivalent circular flaw giving less scatter at 22°C, but more at 1260°C. At 22°C most of the data for circular flaw and equatorial crack-pore models were bounded by K_{IC} values of 2 to 4 or 5 MPam$^{1/2}$ respectively (in comparison with measured indentation values of 4.5–5 and NB values of 3.5 MPam$^{1/2}$). At 1260°C the total scatter for either model was less, and the corresponding K_{IC} bounds were 2 to 4 and 3 MPam$^{1/2}$ respectively, consistent with some limited decrease in K_{IC} at the higher temperature. Ohji et al. (28) also evaluated failures of their sintered β SiC (3.12 g/cc, grain size ~ 4 μm) tested (in true tension) at 22 and 1300°C against both penny-flaw and pore-crack models. Their strengths agreed quite well with those of Seshardi and Srinivasan over their limited pore size range (20 to 40 μm). They showed that the peripherally cracked spherical pore model fit their data better than the penny-flaw model at both temperatures (with no obvious differences in strengths at the higher temperature); however, this model fitting was based on their use of K_{IC} = 2.3 MPam$^{1/2}$ (based on PCNB measurements of 2.2 to 2.5 MPam$^{1/2}$ and an equatorial crack depth of half the grain size out from the pore wall (i.e., ~ 2 μm). Use of a higher value, e.g., 2.5 to 4 MPam$^{1/2}$, would result in the equivalent penny-flaw model being the better fit (optimum agreement being at K_{IC} ~ 3 MPam$^{1/2}$); however, use of K_{IC} ~ 2.1MPam$^{1/2}$ for the equatorially cracked spherical pore model would give similar agreement with the data for the same shallow,~ 2 μm, crack depth, as would use of a higher K_{IC} with a higher crack depth, i.e., a few or more times the grain size.

Next consider flexure data for failure from isolated (natural) pores reflecting part of the typically few percent porosity in bodies of partially stabilized (TZP) ZrO_2, PZT, and Al_2O_3 at 22°C, and one set of Al_2O_3 data at −196°C (Fig. 5.16). While much of the data is over a limited range of pore sizes, three observations can be made despite or within the bounds of the scatter: 1) there is no obvious difference in values between failure identified from pores at the tensile surface or inside of it (provided that stresses are corrected for pore location, unless very near the surface), 2) there is general agreement between data from different investigators on similar but not identical bodies [including extrapolations of the Al_2O_3 of Wallace, Fig. 5.4, and PZT (Fig. 3.3) of Biswas and Fulrath to P ~ 0], and 3) the trends of the data over larger pore size ranges (i.e., Al_2O_3 and PZT data (18,20,100) are reasonably consistent with the trends for treating pores as either penny-flaw or pore-crack models with reasonable K_{IC} values and a/r ratios of 0.1 to 0.5. The numerical values of most of the rest of the data are consistent with the latter trends. There are, however, some questions, e.g., the differences between the Al_2O_3 data at 22 and −196°C, discussed later.

The above observations are for nominally equiaxial pores (though with considerable shape variation); however, as noted earlier, some common forming processes can result in substantial anisotropy of pore shape. Figure 5.17 is an important example of this and summarizes more extreme results from the earlier commercial production of PZT sonar rings by die pressing (22) (see also Fig. 1.10; subsequently, isopressing cylinders and slicing into rings became common). This shows significant anisotropy of strength (and hence also indirectly of E, γ, and K_{IC}) due to whether the laminar lenticular pores were stressed to cause failure either in the plane of the lens, or perpendicular to it. The former strengths averaged 20 to 40% lower than the latter over a range of

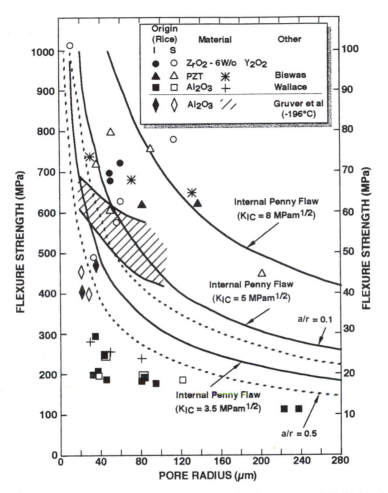

Figure 5.16 Flexure strength versus the radii of naturally (18,20,23,24,100) and artificially (from burnout of organic beads) (44–46) occurring pores at fracture origins in various oxide bodies of various investigators. Note: 1) left scale is for all but the PZT data, for which the the right scale applies exclusively, 2) curves for sharp, internal penny flaws for various K_{IC} values and for pore-crack models with a/r ratios of 0.1 and 0.5, and 3) agreement of extrapolating PZT data of Biswas and Fulrath (45,46) (Fig. 3.4), and especially Al_2O_3 data of Wallace (44) (Fig. 5.5) to P = 0.

pore-shape anisotropies and sizes, but were often half or less for more extreme pore anisotropies shown in Fig. 5.17. Clearly, the strength anisotropy is much less than the anisotropy in the stress concentrations associated with the pores in the two stressed orientations, e.g., as indicated by the differences in the radii of curvature respectively of the dome portion and the edge of the lenticular pores. Note that scaling the strengths as inversely proportional to the square root of the

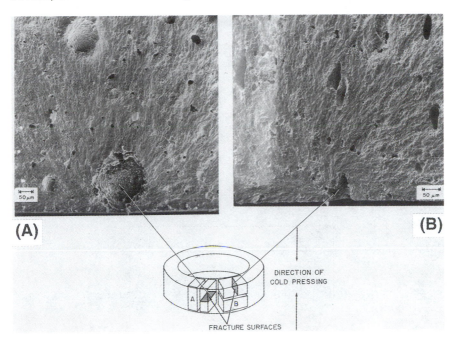

Figure 5.17 SEM fracture origin photographs of test bars machined from earlier commercial PZT sonar transducer rings made by die pressing for testing strengths parallel or perpendicular to the pressing direction. A) Fracture surface normal to the pressing direction. Note: 1) the large, lenticular-laminar pores, especially the one bottom center from which fracture initiated, and 2) the fracture markings around it indicating fracture propagated from the bottom half of the pore rather than first forming a peripheral crack as proposed by the pore-crack models. B) Fracture surface parallel to the pressing direction. Note the distinctive, aligned, lenticular character of the pores showing they were laminar, especially the large pore at the fracture origin (bottom center), and again fracture markings indicating that fracture initiation was from the left, more irregular side of the pore, rather that first forming a peripheral crack.

cross-sections of the lenticular pores in the plane of fracture results in reasonable approximation of the strength anisotropy.

Now consider observations on fracture mode around pores at fracture origins. In the substantial portion of polycrystalline bodies with reasonable to substantial transgranular fracture (77), fracture of grains immediately surrounding pores at fracture origins is also predominantly transgranular (15) (e.g., Figs. 5.17–5.19), not intergranular as originally proposed (14). There are at least two possible reasons for this. First, there are typically other stresses around a pore surface, some unique to pores and some modified by the pore. These include stresses from thermal expansion anisotropy (only in noncubic materials) and

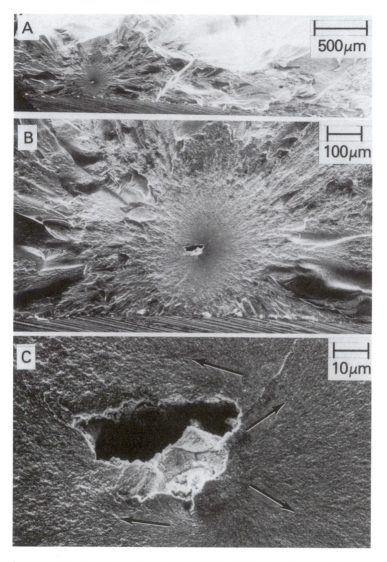

Figure 5.18 Fracture initiation in a (CVD) W-C body from an isolated internal pore. A–C) Increasing SEM magnifications of the origin. Stresses at failure were 1430 and 930 MPa, respectively, at the tensile surface and the pore. Note the probable transgranular fracture and that the resultant fracture markings (arrows in C) show the fracture proceeding from the lower right portion of the pore, and not from an initially formed peripheral crack around the pore. Copyright ASTM, published with permission (21).

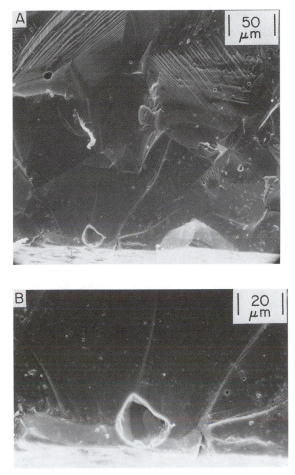

MgAl$_2$O$_4$ $\sigma_f \sim$ 30 KSI

Figure 5.19 Fracture initiation from an isolated pore at the tensile surface of a dense MgAl$_2$O$_4$ (at 210 MPa). A) Lower and B) higher magnification SEM photographs of the origin from a pore (~bottom center) smaller than the associated large grain. Note the nearly exclusive transgranular fracture and the single crystal mist/hackle features in grains above the origin (near the top of A) indicating that any slow crack growth across the grain in which the pore occurred was limited. Treating either the pore alone or the pore plus the section of the grain boundary exposed along the bottom edge of the grain the pore is in as the flaw gives K$_{IC}$ of 1.2–1.3 MPam$^{1/2}$, respectively, which is consistent with the expected single-crystal values for failure from a flaw smaller than the grain.

stress concentrations due to elastic anisotropy (in essentially all crystalline materials) and smaller nearby pores (Fig. 5.20). These, as well as the intragranular location of some smaller pores, may provide mechanisms for transgranular fracture around pores at fracture origins. Second, the concept of easier intergranular fracture due to weaker grain boundaries is often incorrect due to greater fracture area and local mixed mode failure for intergranular fracture compensating, at least partially, for the lower fracture energy for single-grain boundary facets versus transgranular fracture of grains (15,21,77); however, the role of fracture mode around pores is another important area of uncertainty requiring further study of pores as fracture origins.

The above observations of pores as fracture origins are all based on failure of bulk samples, hence entailing pores of dimensions of tens of micrometers, i.e., on the same scale as other failure-causing defects with which they must compete as fracture origins. This raises the issue of failure from finer pores and greater grain-to-pore size ratios. The issue of the extent of failure from finer pores and whether such failures scale in stress with decreasing pore size consistent with extrapolations of bulk failure (Figs. 5.12, 5.15, 5.16) is partially answered by fractography observations on ceramic fibers. Although not a particularly common source of fiber failure, pores have been observed as fracture origins in a variety of fibers. These include for example polymer-derived SiC-based fiber (102) (Fig. 5.21A), sol-derived yttrium aluminum garnet (103), as well as melt-derived eutectic $Y_3Al_5O_{12}$-Al_2O_3 (104), and other alumina (Fig. 5.21B,C) fibers. Though the extent of data is probably not sufficient to quantitatively demonstrate that these pore origins in fibers are an extrapolation of pore failures in bulk bodies, this is clearly qualitatively indicated. [While the finer scale of pores at fracture origins in fibers means that most such origins are from larger, isolated pores, some indication of fiber failure from a few associated, finer pores is also indicated (105), Fig. 5.21B.]

Pores can be fracture origins in bulk single crystals, since even with the best processing there will often be some small isolated pores. Some crystal growth processes are more prone to produce voids than others, e.g., Verneuil growth of sapphire results in more and larger pores (as well as other microstructural defects) than processes such as Czochralski growth. Fine pores in crystals can be revealed as fracture origins by chemical polishing the surfaces of crystal specimens then mechanical testing so that surface flaws no longer limit strengths to levels below those at which most pores (or other internal defects) would cause failure. Such testing readily reveals the differences in ultimate strengths of such processes, e.g., 1500 ± 200 GPa and 5600 ± 200 GPa respectively for Verniul and Czochralski sapphire (106). Edge-defined, film-fed growth (EFG) of sapphire filaments results in some pores (107,108). While these pores (e.g., 0.1 to 2 μm diameter) are more readily seen as frequent sources of failure at higher temperatures (due to lower strengths, Chapter 8), there is also evidence of this at 22°C, with pores being at least approximately

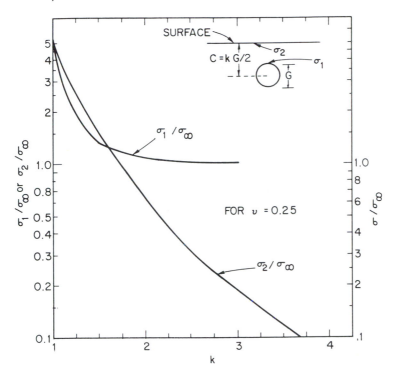

Figure 5.20 Plot of local stress enhancements at the surfaces of a large pore and a nearby (spherical) grain having mismatch stress with the surrounding material, e.g., due to thermal expansion or elastic anisotropy, based on analysis of Wachtman and Dunduras (101). σ_1 is the stress at the surface of a spherical grain (diameter G), σ_2 is the stress at the surface (in this case the surface of a pore) directly above the grain, and σ infinity is the value of σ_1 when $C \to \infty$, i.e., when the grain is far below the surface. Published with permission of J. Mat. Sci. (15).

equivalent to sharp flaws (109–111). This raises the question of why pores in single crystals are apparently more serious flaws than is often the case for glasses. Such failures are also sometimes indicated in some polycrystalline specimens with larger average or local grains, so a single pore is associated with a single grain, e.g., MgAl$_2$O$_4$ (Fig. 5.19) and Al$_2$O$_3$ [Fig. 46 and ref. 21 in (112)]; however, there is often some uncertainty due to other factors, e.g possible effects of machining flaws or grain boundaries (Fig. 5.19).

Three key observations and resultant implications and conclusions should be noted from the single-crystal data and discussion presented above. First, while there are clearly similarities in pores acting as fracture origins in poly- and singlecrystals, as well as in glasses, there are also important differences. The frequent occurrence of pores as fracture origins in polycrystalline bodies occurs with much less or no occurrence of pores acting as blunt flaws in them in

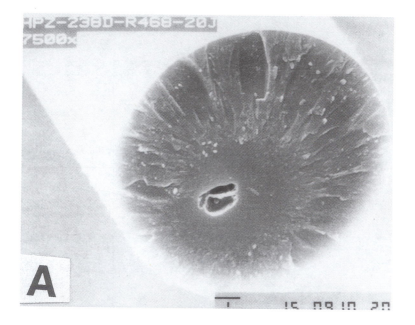

Figure 5.21 Failure initiation from pores in ceramic fibers. A) A polymer-derived, SiC-based fiber (HPZ, 11.7 μm diameter. Photo courtesy of Dr. J. Lipowitz, Dow Corning Corp.) B, C) Coarser grain alumina-based fibers (104) failing at ~ 388 and 382 MPa, respectively, from a central pore and about three smaller, intergranular pores (v, arrows in B). Photo published with permission of Academic Press (1).

contrast to glasses; this supports the concept that there are intrinsic mechanisms associated with the grain structure for forming associated cracks. Pores are much less frequently observed as fracture origins in single crystals or glasses, as normally produced and surface finished, because of the infrequent occurrence of pores in them, especially of sizes to compete with other normal surface flaws; however, limited crystal data shows that where pores are sufficiently large relative to other flaws present, they are sources of failure as sharp flaws, or approximately so. Pores can clearly be fracture origins in glasses where they act as blunter flaws unless associated cracks are separately introduced, e.g., by machining, giving more variable behavior and illustrating that glasses do not have the same mechanism of forming associated cracks. Thus, pores are not always fracture origins in glasses, despite being larger than similarly stressed machining flaws, and show variable mirror-to-flaw size ratios, M/c, consistent with variable degrees of pore-crack sharpness or bluntness. Basic differences impacting pores as fracture origins in polycrystals, single crystals, and glasses are seen as: 1) grain boundary grooving in polycrystals and surface-step formation on both polycrystalline grain and single-crystal pore surfaces versus

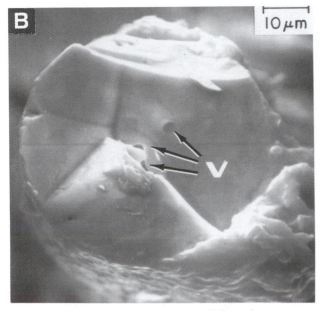

Al_2O_3 $\sigma_f \sim 55\,ksi$

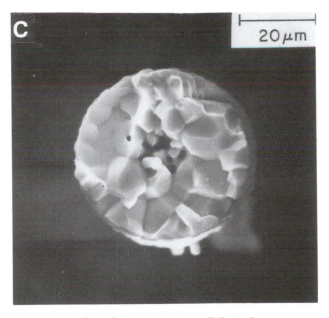

Al_2O_3 $\sigma_f \sim 56\,ksi$

atomically smooth glass pore surfaces, and 2) crystalline anisotropies in single- and polycrystals versus isotropy in glasses. However, the depth, character, and peripheral extent of such cracks and their mechanism of formation are neither well defined nor consistent with the models. For example, the originally proposed intergranular, peripheral fracture around the pores is generally not found (unless that is the normal fracture mode for that body). Similarly, crack formation to a depth of the grain boundary facets abutting the pore, as originally proposed is not supported. Some have proposed variable depths, e.g., increasing with pore size, but fractography provides no evidence of this and questions the formation of such a complete peripheral crack.

Second, strength-pore size data for similar materials from different investigators agrees fairly well (when corrected to the stress at the pore). Such data shows that the stress for failure from pores in any of the materials studied clearly decreases as pore size increases, but is not significantly sensitive to variation in pore shape, i.e., data for artificially introduced and natural pores are similar. This is contrary to the concept of stress concentration from pores causing other nearby flaws to propagate as the mechanism of pores influencing failure. If the latter were the case the pores would not necessarily be part of the fracture origin, and there should be no dependence of failure stress on pore size, but rather a substantial dependence on pore shape. (However, fracture stress is sensitive to the pore depth in from the tensile surface in flexure tests.) Both penny-flaw and pore-crack models fit the strength-pore size data reasonably well, implying substantial peripheral crack-to-pore size (a/r) ratios (if nearly, or fully complete peripheral cracks form, see below). Similar fitting of penny- and pore-crack flaw models supports the use of the former, as commonly done, for the practical reason of its simplicity over the latter. The similarity of these two models in fitting the data is consistent with the crack sizes and resultant crack-to-pore size ratios (a/r) since they are very similar for a/r greater than or equal to 0.1 (Fig. 5.1); however, implied greater crack depths from the pore wall is much more than half the grain size first proposed by Evans and Tappin (14). Furthermore, the pore-crack model may require the crack depth from the pore wall to increase as the pore size increases (99). Both of these modifications require further examination (discussed below).

Third, fractography is a critical factor in identifying pores as fracture origins and factors and mechanisms involved. It clearly shows that in glasses complete, or nearly complete, peripheral, equatorial cracks typically do not form as a critical step in the failure process. While the absence of such peripheral cracks is observed in stress gradients (in flexure tests), the pores were less than 10% of the specimen thicknesses, so the absence of extensive peripheral cracks appears to be intrinsic or due to modest stress gradients. Polycrystalline fractographic observations, although much less definitive, also suggest that substantial peripheral cracks do not form as a step to failure. Note, however, that Evans et al. (34,37) theoretically considered partial cracks giving similar trends (Fig. 5.1), so the apparent general absence of complete peripheral cracks does not mean the models are not valid. Fractography also clearly shows: 1) in glasses, failure occurs from pores plus machining flaws causing resultant variable mirror-to-flaw size ratios, and 2) in polycrystals, extensive transgranular, instead of the originally proposed intergranular, fracture around pores. Thus, better understanding of the indicated preference for small, partial cracks and their relation to the grain structure in polycrystalline bodies is needed.

5.3.4 Failure from Pore Clusters

Having considered failure from single, isolated pores in the previous section, and again starting with glasses, consider failure origins from clusters of two or more pores. Such origins are again commonly found in bodies with limited porosity. Consider first two key examples of such fracture origins from two pores, which show that: 1) there may be multiple mechanisms, and 2) the proposed pore-crack linkage is not at all universal, and may not occur. In Fig. 5.22 a larger and a somewhat smaller, adjacent pore at the fracture origin would suggest linkage of the two pores by cracks forming on one or each side of the web between the two pores. The fracture tail on the larger pore would be consistent with this; however, the origin, especially the fracture tail, would also be consistent with two other possibilities, each with its uncertainties. One is failure starting from a surface flaw between the two pores on the side near the tensile surface, but this is uncertain since the specimen had a flame-polished surface (which limits but does not preclude the possibilities of surface flaws). The other possibility is failure from the smaller pore, which would require some aspect of it to make it a more severe flaw despite its smaller size. The mirror-to-flaw size ratio (M/c) and the fracture tail would be consistent with the latter possibility and inconsistent with the larger pore being the origin as a blunt flaw as expected for a pore without an associated crack. Thus, a M/c ratio of ~12 would be consistent with that for a sharp flaw about the size or smaller than the smaller pore, e.g., a limited connection of the two pores before failure occurred. On the other hand, the mirror for the failure shown in Fig. 5.22B,C from a larger and a substantially smaller, adjacent pore in a ground specimen shows that failure did not initiate from the web between the two pores, but from the opposite side of the large pore from the small one, probably from an associated machining flaw. This implies that fracturing of the web between the two pores is not an easy process, and that it is readily preempted by the presence of a flaws, e.g., from machining.

Turning to glass failures with more than two pores in the origin area, Fig. 5.23A,B shows three similar pores (and at least one smaller one) in the origin area. The fracture tails on these three pores indicate that more than one crack may have been involved, i.e., that a single crack may not have formed to connect the pores. Similarly, the mirror size, which is clearly far too large for failure from a blunt pore flaw, is consistent with a sharp flaw about the size of

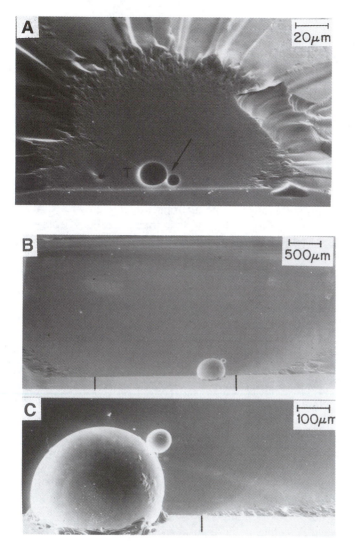

Figure 5.22 Examples of fracture origins from two touching pores in silicate-based glasses. A) Flame-polished SiO_2 (510 MPa). Note the fracture tail (T) and some irregularities in the mirror associated with the tail, $R_m/R_p \sim 6$. B) Lower and C) higher magnification SEM photos of a machined silicate glass bar (105 MPa). Note the significant asymmetry of the mirror (vertical tic marks) indicating that fracture started from the opposite side of the larger pore from the smaller one, $R_m/R_p \sim 4.7$. Published with permission of J. Mat. Sci. (15)

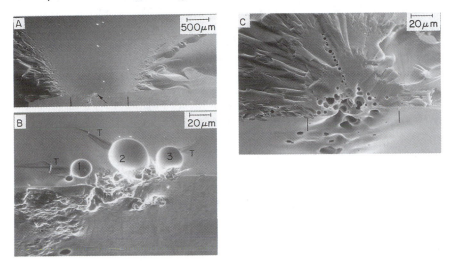

Figure 5.23   Fracture origins in silicate-based glasses with three or more clustered pores. A) Lower and B) higher magnification photos of the origin in machined bar (92 MPa). Note the fracture tails (T) on all three of the large pores in the origin area, the fairly symmetrical mirror, and that R_m/R_p for the center pore (2) is ~ 44, equivalent area of the three pores is ~ 23, and the envelope of the three pores is ~ 19. C) Flame-polished SiO_2 failure from many pores (654 MPa). Note the relatively symmetrical mirror (vertical tic marks), and that R_m/R_p for the center pore is ~ 7.7. Published with permission of J. Mat. Sci. (15).

one of the pores plus a partial crack around it. The glass results are similar to those for polycrystalline origins from two or more pores discussed below, with one important exception. Glass samples failing from pores, but with flame polished surfaces (Figs. 5.14A, 5.22A, 5.23C) have higher failure stresses than those that had machined surfaces (Figs. 5.14B–E, 5.23A,B). This clearly shows that cracks, especially from machining, play a substantial role in failure of glasses from two or more pores, as with single pores as origins.

In considering failures from two or more pores in glass samples, Rice (15) considered three generic flaw types :1) the envelope of the pores (reflecting linkage of the pores by a crack), 2) a sharp, circular flaw of the same area as that of the intersection of the pores and the fracture surface, and 3) the individual pores (as sharp or blunt flaws). The envelope approach always gave poor results, except where there was little difference between it and a larger pore (e.g., Fig. 5.23), thus arguing against a crack joining the pores. The net-pore-area approach was the next best, but generally not a good approximation of the flaw size, which was usually much more consistent with failure from one pore or parts of a few pores. Thus, the glass samples again seriously question many aspects of the models proposed, especially the concept of a crack connecting closely spaced pores, and strongly indicates substantial variability in the failure process from multiple pores. Where two or more pores are at or very near the fracture origin, there are two possible counter effects. The simple absence of material due to pores very near the origin could lower the local fracture toughness for failure beyond the limited effects noted earlier for single pores. On the other hand, crack branching and bridging might occur to increase the local toughness when cracks around pores and the fracture plane are not all coplanar, which is particularly likely when there are greater than three pores. The former lowering of the fracture toughness is likely to dominate and to agree with smaller fracture origins than the envelope of the pore cluster.

Consider now data for failure from areas having clustered pores in polycrystals. Data are most extensive for PZT samples with pores from burnout of organic spheres (45,46). (Note that while these samples were made with the intention of having single pore origins and their mechanical behavior was initially analyzed on this assumption (45,46), subsequent fractography (15), illustrated below, showed failure was from areas of clustered pores, reflecting effects of probable attraction of the organic spheres in mixing.) Analysis of the 12 PZT origins observed showed that only four of them were consistent with the envelope of the clustered pores being the size of a sharp flaw that caused failure, and in one of these cases, the sharp flaw size was also reasonably consistent with either the equivalent area of the two pores and one of the pores. In three other cases a sharp flaw whose size equaled the combined areas of the clustered pores was close to the expected flaw size, but in one case was competitive with one of the pores being a sharp flaw. In one case the sharp flaw size was between that of the equivalent pore area and one pore. In three cases the size of one of the larger pores in a logical position to be the origin was the size of the expected sharp flaw. Specific examples illustrate this variability.

Figure 5.24A,B shows failure from the tensile surface associated with one complete and two partial pores. Both fracture markings and analysis indicate that one whole pore dominated failure. The envelope of the clustered pores is clearly too large, and the equivalent area was somewhat large for the flaw size. Figure 5.24C shows failure from the corner of a sample with four pores. Both the fracture markings and analysis indicate that failure occurred from pore closest to the corner. In Fig. 5.24D the expected flaw size lay between the size of the envelope and the equivalent area of the four clustered pores. Thus, these PZT studies show the same key trends for failure from clustered pores as the glass results did, namely: 1) the results are variable, 2) connection of close, even intersecting, pores by a sharp crack prior to failure is, at best, an uncommon occurrence, i.e., subcritical cracking between pores is not an easy or common process, and 3) failure from individual pores in a cluster appears to occur, but some cooperative failure of pores is indicated by flaw sizes that are a combination of the areas of pores in the fracture origin area. Therefore, the same issues of possible counter-effects on the local toughness controlling failure due to: 1)

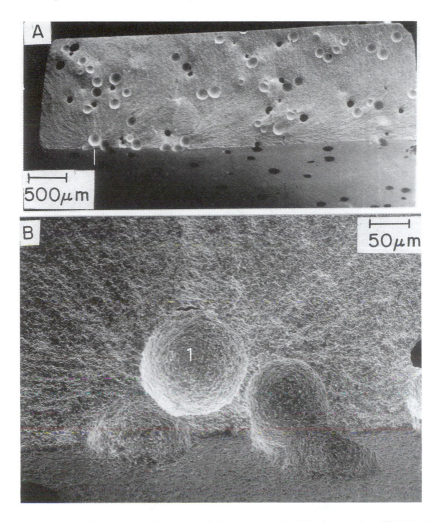

Figure 5.24 Examples of fracture origins associated with clustered, artificially introduced, spherical pores in PZT (15,45,46). A) Lower and B) higher magnification SEM photos of fracture originating from three clustered pores (vertical tic mark in A) at 62 MPa. Analysis indicates that the flaw area for initiation was between that of the main pore (1) and the area of all three, while fracture markings indicate failure from the main pore (1), primarily the bottom half. C) Lower and D) higher magnifications of fracture origin from an area of four pores at 58 MPa. Analysis and fracture markings indicates failure originated from pore 1, the most separated and closest to the surface. E) Origin from an area with four clustered pores at 75 MPa. Analysis indicates that the flaw area was between that of the envelope and the area of the four pores. Published with permission of J. Mat. Sci. (15)

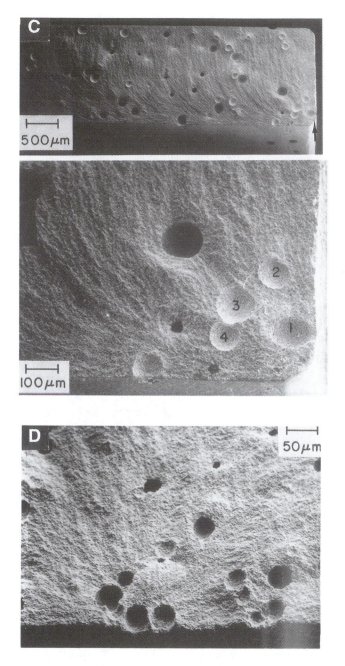

Figure 5.24 Continued

reduced toughness from the absence of material due to other pores and 2) possible increased toughness due to crack branching, etc. due to noncoplanar fracture of pores need to be considered. Again, the former is probably more important and favors flaws smaller than the pore cluster envelope, but the latter adds variation and uncertainty.

Anisotropy of strength and other properties readily results from preferred orientation of anisotropically shaped pores, especially larger ones, as illustrated above. However, property anisotropy also commonly occurs from clusters of smaller pores that individually may or may not have any significant anisotropy of shape, but have some to substantial preferred orientation of the cluster(s). Thus, while not extensively documented by systematic studies, significant laminar arrays of pores from lamination processes (e.g., Fig. 1.11) are seen as an important factor in the substantial property anisotropy common in such bodies in addition to varying preferred grain orientation that is also common in such processes. The substantial strength anisotropy that has been documented in several studies of hot pressed Si_3N_4 provides some insight. Lange (113) showed flexure bars oriented for fracture parallel to the hot pressing direction had strengths ~ 35% higher than bars oriented for fracture perpendicular to that direction. He attributed this to greater difficulty of fracture around the elongated grains that had a preferred orientation normal to the hot pressing direction, but his subsequent fracture energy tests only partially supported this, showing only 18% anisotropy in fracture energy (thus probably less for toughness) (114). Weston (115) corroborated the type of elongated grain texture found by Lange, as well as similar strength anisotropy (26–38% higher for fracture parallel versus perpendicular to the hot pressing direction) of another hot pressed S_3N_4, but an opposite and somewhat greater anisotropy of toughness. A probable resolution of differences between his and Lange's results was provided by his fractography showing flaws initiating fracture normal to the hot pressing direction averaged ~ 2.7 times the size of flaws initiating fracture parallel to the hot pressing direction and that the former fracture was predominately intergranular, while that for fracture parallel to the hot pressing direction was mixed inter- and transgranular. At least some of the greater intergranular fracture in the origin area for flaws in the plane of pressing probably reflects some laminar character of residual porosity, of which Fig. 1.12 represents an extreme case.

5.3.5 Porosity Dependence of Highly Porous Materials

Consider here data for highly porous i.e., honeycomb, foam, and felt materials. As noted earlier, Gulati and Schnieder (41) confirmed the P-dependence of extruded coordierite honeycomb with square cells was $1 - P$ (the same as for E) as predicted. They also noted that actual strengths were only 50–70% of predicted values for the stronger direction (stress parallel to the cell and extrusion direction) and even less for the weaker direction, so the latter strengths were < one-half the former (Fig. 5.3). The lower strengths were attributed to variations in the cell geometry and wall material, contrary to the assumption of all cells being the same.

Brezny and Green (116) studied flexure strengths of three alumina-based, open-cell foams, along with their K_{IC} (Section 4.3) and Weibull modulus, m. Their strengths for the three foams showed similar trends with one another as a function of P and in actual values, i.e., significant overlap, giving P dependencies of $(1 - P)^n$, with n = 1.45, 1.70, and 2.15 (average ~ 1.8, very similar to that for K_{IC}) in contrast to predicted values of n = 1.5 for K_{IC} and strength, and n = 2 for E and G. Again they reported that correction for the triangular tubular pore in the struts brings all the n values for strength to ~1.5, as for K_{IC}; however, in other work on a similar mullite foam, these authors (117) found that both E and compressive strength (Chapter 6) showed a modest increase with cell size (not indicated in the above flexure strength data), and that this could not be explained by the tubular strut pores. Thus, the issues of the n value, including whether or not it is the same or different from that for E, and of cell-size effects require more study.

Their m values for the foams ranged from 3.2 to 7.6, with some indication of a possible modest decrease with increasing P (116). This, and a similar indication in Fig. 5.7, suggests that the foam data is an approximate extension of denser, sintered bodies, i.e., indicating an overall trend for m to decrease with increasing P. While there may be such an overall trend, it is clear that there are at least significant variations of this. The presence of larger pores or pore clusters acting as fracture origins (in PZT) and the effects of porosity on machining direction-flaw elongation effects (Figs. 5.8, 5.10) would be examples of these. Weibull moduli foam values for the individual strut strengths were typically ~ 50% of the foam values. This would be logical where the flaws were large enough to involve several struts, since if the flaws entailed primarily one strut, the reliability would be closer to that of the struts. One of the two alumina bodies had calculated flaw sizes ~ 2.5 to ~7 times the cell size, hence was consistent with this; however, the other alumina and the alumina-zirconia foam had calculated flaws sizes that typically varied from only ~ 50–100% of the cell size, which appear to be inconsistent with higher Weibull moduli for the foam than for the individual struts. The flaw sizes were calculated, as usual, from the strengths and fracture toughnesses, thus raising questions as to the validity or meaning of the K_{IC} values, at least as they relate to strength, as is the case whenever K_{IC} measurements reflect values for a crack scale much greater than that determining strengths. Thus their Weibull modulus data for the foams also indicate the need for further study.

Brooks and Winter (118) also measured strengths and Weibull moduli of a variety of commercial ceramic foams of Al_2O_3-ZrO_2, fully and partially stabilized ZrO_2, Si_3N_4, and SiC made by sintering powders (as well as by CVI of SiC fibers). While their data by itself is not sufficient to determine the P-dependence of their properties, it is useful in conjunction with other data above. Thus, their

Al_2O_3-ZrO_2 data corroborates similar data of Brezny and Green (116). Further, their Weibull modulus data also suggests an overall trend to decrease with increasing P (Fig. 5.7, focusing on the average trend, since their use of only 5–7 specimens is too few to indicate anything other than overall trends).

Since Green and colleagues measured different properties on a few foam ceramics, it is useful to briefly compare their P-dependence to that of strength. The P-dependence of elastic moduli is generally followed $(1 - P)^n$, with n = 2, consistent with Gibson and Ashbys' model (119). The porosity dependence of fracture toughness also approximately followed this (n ~ 2.0 ± 0.3), as did flexure strength (n ~ 1.8 ± 0.3), inconsistent with theories predicting n = 1.5. Compressive strengths tended to have n values higher than tensile strength, but still are ~ 2 (117). Thus, data indicate a greater consistency of P-dependence for elastic moduli, fracture toughness, and strengths than the models do, but more study and evaluation is needed. Besides more tests, comparisons, and characterization, further evaluation of foam-cell stacking is needed, e.g., as indicated in Fig. 3.3.

Data for rigidized glass fiber felts showed the predicted $(1 - P)^{3/4}$ dependence for strength and K_{IC} (Fig. 4.6), i.e., indicating consistency of these properties with simple linear elastic fracture.

5.4 DISCUSSION

Comparison of γ-, K_{IC}-, and σ- versus E-P-dependences shows considerable similarities as expected for simple linear elastic fracture mechanics, but also substantial variations for γ- and K_{IC} versus E-P trends (Table 4.4), and some for σ-P trends, with the latter often less than and different from the former variations. Thus, eight data sets of both K_{IC} and σ on the same bodies (Table 4.2) gave average b values respectively of 3 ± 1 and 3.9 ± 0.6, while 24 data sets (8 and 16, respectively, from Tables 4.4 and 5.2) for measurement of the P-dependence of both E and σ on the same bodies gave b values of respectively 3.8 ± 1.0 and 4.0 ± 1.2. Although not statistically certain, these results indicate more variation of the P-dependence of K_{IC} than of σ versus that for E, which is consistent with substantial variations found for fracture toughness (showing both higher and lower values as P → 0, not just lower values as for strength). The broader data base for properties measured on different bodies (Table 5.1) also shows b values for elastic moduli and tensile strength about the same (neglecting some deviations toward lower strengths as P → 0 for some bodies; Fig. 5.6), but probably slightly less for E. Again, differences are indicated by individual, more extreme cases, e.g., the limited MgO data of Baddi et al. (120) is extreme, showing b for σ ~ 25–50% of that for γ (10–11, with increasing grain size, from ~ 0.3 to 0.7 µm, with decreasing P).

A broader perspective on porosity dependence of strength and understanding the above differences and variations requires addressing four partially interre-

Table 5.3   Variations of σ-P from E-P Trends

Variation	Cause (s)	Occurrences
Lower overall σ-P dependence	Reduction of flaw severity, especially elongation	Elongated flaws from machining normal to tensile axis, and (e.g., Knoop) indents
	Lower γ- and K_{IC}-P dependence from crack branching, etc.	With cracks large relative to pore and grain structures e.g., retention of lower starting strengths in thermal shock (Chapters 4 and 9)
Lower σ-P dependence at lower P	Decreased strength with increased grain size	With substantial homogeneous, heterogeneous, or exaggerated grain growth
	Increased porosity coverage of grain boundaries	Vacancy condensation, and especially decomposition of grain boundary impurities or additives
	Increased intragranular, hence reduced inter-granular porosity	Increased grain growth, but intergranular fracture as P decreases at lower P
Higher P dependence at lower P	Fracture initiation from pores or pore clusters	Where such pores occur at lower P, due to heterogeneities, pore coarsening, or both
Higher overall σ-P dependence	Greater effects of mixed, heterogeneous and changing porosity and increased non-linear P dependence of σ at high P	Bodies where pores individually or collectively increase flaw sizes or locally aid initial crack growth at high P
Mixed	Variations of above	Bodies with variable microstructure within or between samples, or both

lated aspects of porosity dependence: 1) heterogeneous porosity, 2) crack branching-bridging, 3) pores (or grains) and pore clusters as fracture origins, and 4) effects of grain size on apparent or real porosity dependences. Sources of deviations in the porosity dependences of γ and K_{IC} summarized in Tables 4.4 and 5.3 arise from the above microstructural factors and the scale and nature of crack propagation

Consider first differences in the effects of increased grain size as P decreases, with or without associated inter- to intragranular pore changes. Larger average or maximum grain size can either decrease or increase the apparent porosity dependence of γ and K_{IC}, the later commonly due to various combinations of larger scale crack propagation at limited velocities and substantial crack-surface interaction which may result in effects such as microcracking and crack branching and bridging; however, tensile strength failure is typically controlled by much smaller cracks, i.e., weak links, and hence predominantly or exclusively decreases with increasing average or maximum grain size, the later arising from large individual or clustered grains (plus associated pores or cracks) being failure sources. Such smaller strength-determining cracks generally entail much less extent of comparable crack propagation at moderate velocities and more propagation into the bulk than along the surface (121,122) versus larger cracks for determining most fracture energy and toughness. Another complication is that increasing intragranular pores with grain growth can reduce surface machining damage from slip, twinning, or microfracture associated with pores. This may effect tensile strength, e.g., as suggested by Rice (123) in studies of larger grain MgO with limited porosity (much intragranular), and is also likely to be a factor in wear (see Fig. 6.16).

Differences in crack scale effects between most γ and K_{IC} versus most strength tests noted above for effects of larger grains also result in the temporary reduction or reversal of the decreases of γ and K_{IC} with increasing P (Chapter 4) not being generally applicable to strength, and thus being another source of σ-P trends being different from those of γ- and K_{IC} and closer to those of E; however, as discussed in Chapters 4 and 9, the larger scale crack effects can be pertinent to severe thermal shock exposure where crack propagation commonly occurs on a scale to allow crack branching and bridging. Also, as noted in Chapter 4, there is limited data for a RSSN and a polymer-derived SiC body, each with substantially finer pore (and grain) structures as well as neutron irradiated sapphire with very fine (nanometer scale) pores where strength trends follow those of fracture toughness, including increases with introduced porosity. These results suggest that fine microstructures can allow crack branching and related effects (normally restricted to large cracks such as those of most γ and K_{IC} tests) to occur with typically smaller, strength-controlling cracks.

Next, consider heterogeneity of porosity which typically has greater effects on strength, either increasing or decreasing its apparent or real porosity dependence depending on whether these effects are at higher or lower P levels. Increased P-dependence is typically significant at higher P due to the increasingly nonlinear decrease of strength with P; however, decreased P-dependence as P increases occurs in some specimens machined normal to the tensile axis (Fig. 5.10). This is attributed to heterogeneities of porosity increasingly limiting the normal elongation of machining flaws formed parallel with the direction of abrasive motion (Figs. 5.8–5.10) that control strengths with the tensile axis normal to the machining direction. These machining-direction effects and their variability appear to be consistent respectively with some trends for decreasing Weibull modulus as P increases (Fig. 5.7) and other different or more mixed effects of porosity on Weibull modulus. All of these again show the need for more characterization and study of porosity heterogeneity (and anisotropy).

At lower porosity there can be apparent decreases in the P-dependence due to the weak link aspect of strength when isolated single or clustered pores remain to increasingly dominate failure (Sections 5.3.3, 5.3.4; Figs. 5.11, 5.13, 5.14, 5.17–5.19, 5.21–5.24). This is also consistent with introduction of larger, more uniform pores, while lowering strengths some and increasing the Weibull modulus some. Overall, while there is more sensitivity of strength to these and other heterogeneities, their differing trends as outlined above result in less differences between σ- and E-P dependences versus those of γ and K_{IC}.

Numerical calculations of flaw failures from isolated pores are generally consistent with models for a peripheral crack, especially for single pores; however, this is probably due to the fact that there is generally limited differences between such pore-crack models; over a range of peripheral cracks, or by simply treating the pore as a sharp flaw. Fractographic evidence is generally inconsistent with general formation of extensive peripheral cracks; i.e., crack propagation directions clearly are not in radial directions from the pores (Figs. 5.14B–D, 5.17, 5.18, 5.24). Figure 5.17 also vividly illustrates the impact of pore-shape anisotropy on strength anisotropy.

Failure from at least some internal pores in glasses may be consistent with formation of such equatorial cracks prior to or in the earlier stages of failure (Fig. 5.14A) since failure generally progresses in a radial fashion from the pore. Clear evidence of this has not been obtained, however, and is likely to be challenging to obtain since this entails discerning possibly subtle changes in the propagation velocity rather than the direction. The only cases where some equatorial cracks have been observed is around some bubbles in amber and in single crystals at higher temperatures, where slow crack growth is clearly their source (see Fig. 9.5A). The very limited amber observations, where conditions of their formation are unknown, but probably complex, showed such equatorial cracks only around two larger pores and not around several somewhat to substantially smaller pores even with other nearby pores. Furthermore, in glasses there are clear cases of pores not acting as failure origins despite being larger than nearby flaws (Fig. 5.13). The association of other, especially machining, flaws with pores in glasses appears to be an important factor the pores acting as failure origins. This general lack of an extensive or complete peripheral crack raises the issue of effects of limited stress gradients, e.g., Fig. 5.25. Limited data for single crystal failures suggest that pores also act as sharp flaws in single crystals.

There are even more issues of failure initiation from pore clusters since the concept of crack connection of close pores to form an "envelope" crack is seriously questioned by experimental data (e.g., the local fracture mode, noncoplanar character of fracture of some close pores, and possible local effects on key properties of E and K_{IC}). Thus, evaluation of fracture origins from two or more close or joined pores in glasses and especially polycrystalline bodies shows that equivalent sharp flaw sizes for failure varied. Occasionally they were

PORES IN TENSILE STRESS
GRADIENTS

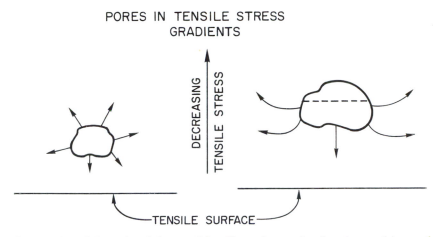

Figure 5.25   Schematic of the possible effect of pore size, location, and interaction with stress gradients on resultant initial crack propagation from pores. On the left is the smaller pore, closer to the tensile surface or with less (or no) stress gradient; hence the whole periphery pore uniformly leads to failure. On the right is the larger pore, further from the tensile surface or with a steeper decrease in tensile stress from the tensile surface; failure initiates from the portion of the pore closest to the tensile surface.

nearly those obtained from taking the envelope of the pores as the origin, mainly where one pore was large, so there was not a great difference between it and the envelope as the flaw. Taking the flaw size as that for the area encompassed by the pores involved more commonly agreed with the failure stress, but such agreement was most common with taking the flaw size as that corresponding to the area of the one or two (often joined) pores on the fracture surface. The latter, of course, had higher failure stresses and appears inconsistent with proposed linking of pores by cracks. Two possible counter-effects in such failures need further study: 1) reduction of local fracture toughness due to other pores very near the origin, and 2) possible increased local toughness due to noncoplanerity of cracks from such pores. Fractography of specific fracture origins from a larger pore and an attached or adjacent smaller pore in glasses indicates that fracture initiation from the web between the pores may occur, but that this is readily superseded by the presence of machining flaws on part of the larger pore. This is consistent with the significant effect of surface finish, i.e., flame polishing versus machining, on failure from pores in glasses. Furthermore, fracture origins from clustered or connected pores were dominant in polycrystalline bodies assumed to have failed from artificially introduced pores, showing the need for fractography to check assumptions based on expected, as opposed to verified, process results. More attention to possible effects of nearby pores reducing the stress intensity requirements for failure from another pore or pore group is needed along with attention to oriented pore arrays.

In polycrystalline samples there are also questions of the relation of cracks from pores to the grain structure. It was originally postulated that the cracks would form along grain boundaries intersecting the pores to ~ 0.5–1 grain depths. This agrees with considerable data, at least where the grain size is small relative to the pore size, but use of a greater crack depth, e.g., several grains, and scaling with pore size may improve agreement with some data, possibly because this also reduces any differences between crack-pore and normal flaw models. Fractography clearly shows that cracking around pores at fracture origins is generally not intergranular as originally proposed. Effects of other stresses, e.g., from elastic or thermal expansion anisotropy, or other (finer) pores may possibly explain the mainly transgranular fracture; however, fractography has not defined, and is not likely to define, such equatorial cracks in polycrystals, except in special cases, such as those suggested by Gruver et al. (23) based on fracture mode changes in finer grain alumina. Corroboration of this—especially broader understanding of the mechanism of such growth, e.g., via mirror to flaw size (M/c) ratios—is needed. In any event, further development of pore-crack models may be important for cases where there is much less overlap with sharp flaws, i.e., at lower a/r ratios (Fig. 5.1).

Finally, note three things. First, differences in pore-crack origin behavior between poly- and singlecrystals and glasses appear consistent with differences in material character. Thus, frequent pore origins in polycrystals is consistent with crystalline anisotropy compounded by grain misorientation and its ramifications such as grain boundary grooves. Similarly crystalline anisotropy and resultant polyhedral pores and surface steps are consistent with apparent pore-fracture origins in single crystals. On the other hand, complete isotropy and smooth, spherical pore surfaces in glasses are consistent with pores needing some other flaw, e.g., from machining, connected to pores for them to be fracture origins. Second, note; 1) the very similar fracture-energy toughness versus tensile strength dependences on microcrack content typically occur for the same crack-scale reason (Section 4.3.3) as found for pores, and 2) predictions of fracture toughness and strength increases at fine microcrack sizes (Section 4.2.2) similar to those indicated by limited data for very fine pore-microstructures. Third, note the critical absence of significant data for other stressing conditions as a function of porosity. These include bi- or multiaxial stresses and tensile or thermal fatigue. Some aspects of thermal stress failures are briefly discussed in Section 4.4, and somewhat more extensively in Chapter 9. Some guidance to tensile fatigue can be obtained from studies of metals and composites, e.g., as reviewed by Bombard et al. (124), some aspects of which are briefly noted in Chapter 8 on effects of porosity on high temperature properties, where such models and behavior may more readily be adapted for ceramics.

5.5 SUMMARY

Models for the general porosity dependence of tensile (and flexure) strength have usually been derived directly rather than via E-P and γ-P, or K_{IC}-P relations, which, in part, reflects little attention to the latter two as discussed in Chapter 4. Directly derived models, whether based on MSA or other concepts, are less rigorous mechanically than some models for elastic moduli; however, MSA models were originally derived directly for strength, i.e., assuming that strength directly correlated with the ratio of the MSA of a cell versus the cell cross-sectional area (or the integrated average MSA to the body area), which in turn leads to analogous assumptions for elastic properties, γ, and K_{IC}, so the same models also result from deriving them from basic σ-E-P relations. MSA models are again the most accurate and versatile, including handling mixed and changing porosity, and the full range of porosity, as well as being the only ones thus far providing true prediction of properties for practical ranges of the amount and character of porosities, but improvements are needed.

While much σ-P data follows MSA models, and hence also often E-P trends, there are significant variations and complications in the porosity dependence of strength, as summarized in Table 5.3. There are also similarities and greater variations of K_{IC}-P, and especially γ-P, trends. The primary sources of these variations are the typical weak-link control of strength and hence more sensitivity to microstructural (i.e., pore and grain) heterogeneities and the frequent larger crack-scale effects such as crack branching and bridging.

Effects of heterogeneous porosity on σ-P relations are more diverse, occurring over most or all of the P range, but often approximately balancing out to give closer similarity to E-P than K_{IC}-P trends. At intermediate and higher porosity there can be two counter-effects of heterogeneous porosity on strength: 1) increasing the porosity dependence due to both nonlinear decreases of E and γ as well as more porous areas being failure-causing flaws, and 2) decreasing the porosity dependence due to heterogeneities limiting elongation of flaws, e.g., some indent and especially machining flaws (formed parallel to the machining direction, hence dominating strength for stressing normal to the machining direction (Fig. 5.10). Since effects of reduced flaw severity outweigh those of increased P, the porosity dependence decreases, as does the strength anisotropy due to the differences in machining flaw geometry with machining direction. These effects are consistent with a trend for the Weibull modulus to sometimes increase with P (Fig. 5.7).

At low porosity, failure from isolated pores or pore clusters frequently occurs, reducing strength, thus giving an apparent reduction in the P-dependence (similar to that due to effects of large grains). The basic concept of isolated pores acting as sources of failure by the formation of an equatorial crack, so that the pore-crack combination becomes the failure-causing flaw, is generally validated, but variations occur. Such pore-crack combinations: 1) reduce effects of pore-stress concentrations (since these are at least partially superseded by even a partial crack around the pore), and 2) increase the impact of isolated more extreme pores as well as their anisotropy of shape on strength anosotropy (Fig. 5.17) due to effects of pore shape on net flaw shape. Pore clusters are also frequent sources of failure, but with broader variation in both their character and in their variations from original concepts of such flaws.

Both material and mechanism variations occur in such individual pore and pore-cluster fracture origins. They generally, if not universally, occur in poly-crystals, and also appear to occur in single crystals when they are present in sufficient size. They also occur in glasses, but apparently only in combination with other sharp, e.g., machining, flaws, especially for isolated pores. Cases of isolated pores in glasses not being the source of failure despite being larger than nearby failure-causing flaws have been observed, as has a range of effective flaw sharpnesses and associated fractographic variations for failures from pore-machining flaw combinations. Such material differences are consistent with various crystalline anisotropies giving both pore-surface steps (and grain-boundary grooves in polycrystals) and anisotropic properties effecting fracture versus complete isotropy in glasses giving smooth pores and isotropic proper-ties. In all materials, however, fractography does not support either the general formation of an extensive or complete equatorial crack around isolated pores or the general formation of an envelope crack around pore clusters. In polycrystals, the fracture mode immediately around pores origins is often transgranular instead of intergranular as originally postulated, and the concept of the crack depth being 0.5–1 grain deep from the pore is seriously questioned.

Despite the above deviations in the specifics of the crack character associ-ated with the pores, taking the pore size as the size of a sharp flaw generally agrees fairly well with most data. The primary exceptions are deviations to higher failure stresses (with smaller mirror-to-flaw size ratios for pores acting as blunter flaws in glasses). Though refinement is needed, general agreement is reasonably consistent with equatorial crack to pore size, a/r, ratios of ~ 1 that are often indicated, i.e., where models for sharp flaws alone and pore-cracks are approximately the same (Fig. 5.1). It is also likely that a partial equatorial crack has nearly the same effect of a full equatorial crack for a single pore and that outer pores in pore clusters reduce the stress intensity needed locally for failure, so complete envelope cracks around the cluster do not form and control failure.

The much more limited models and data for foams and rigidized felts are frequently consistent in overall porosity dependence, the major exceptions being that models predicting different P-dependences for elastic moduli and strength are in question; this is consistent with the expected consistency of such dependence. In terms of absolute agreement, experimental results indicate substantially lower coefficients of proportionality than predicted theoretically (see Chapter 10).

There are again major similarities between the porosity dependences of strength on porosity and microcracks, in particular showing both a much closer correlation with the corresponding dependences of elastic moduli and an often marked contrast to those of fracture energy and toughness. Two basic differences in porosity versus microcrack effects are: 1) more rapid initial decreases in properties with initial increases in microcrack versus pore contents and 2) a saturation of property decreases with increasing microcrack contents (e.g., at larger grain sizes). Other differences between microcrack and porosity effects on mechanical properties, as well as various uncertainties about microcracking and its effects were discussed in Chapter 4.

Thus, overall there is reasonable guidance theoretically and experimentally for both porosity and microcracks, but much remains to be done, and there is substantial uncertainty as to how accurate theory can be in covering the range of variations that occur. Experimentally observed variations and theoretical uncertainties raise serious concerns regarding the accuracy of predicting strengths via nondestructive evaluations.

REFERENCES

1. R. W. Rice, "Microstructural Dependance of Mechanical Behavior of Ceramics," Treatise on Materials Science and Technology, vol. 11: Properties and Microstructure (R. K. Mac Crone, Ed.), pp. 191–381, Academic Press, New York, 1977.
2. M. Eudier, "The Mechanical Properties of Sintered Low-Alloy Steels," Pwd. Met., **9**, pp. 278–290 (1962).
3. F. P. Knudsen, "Dependence of Mechanical Strength of Brittle Polycrystalline Specimens on Porosity and Grain Size," J. Am. Cer. Soc., **42** (8), pp. 376–388 (1959).
4. S. D. Brown, R. B. Biddulph, and P. D. Wilcox, "A Strength-Porosity Relation Involving Pore Geometry and Orientation," J. Am. Cer. Soc., **45** (9), pp. 435–38, 1964.
5. K. K. Schiller, "Porosity and Strength of Brittle Solids (with Particular Reference to Gypsum)," Mechanical Properties of Non-Metallic Brittle Materials (W. H. Warson, Ed.), Interscience Pub. Inc., New York, pp. 35–49, 1958.
6. P. Harma and V. Satava, "Model for Strength of Brittle Porous Materials," J. Am. Cer. Soc., **57** (2), pp. 71–73, 1974.
7. H. H. Bache, "Model for Strength of Brittle Materials Built Up of Particles Joined at Points of Contact," J. Am. Cer. Soc., 53 (12), pp. 654–58, 1970.
8. S. C. Carniglia, "Working Model for Porosity Effects on the Uniaxial Strength of Ceramics," J. Am. Cer. Soc., **55** (12), pp. 610–18, 1972.
9. A. S. Waugh, J. P. Singh, and R. B. Poeppel, "Dependence of Ceramic Fracture Properties on Porosity," J. Mat. Sci., **28**, pp. 3589–93, 1993.

10. P. Arato, "Comment on 'Dependence of Ceramics Fracture Properties on Porosity'," J. Mat. Sci. Let., **15**, pp. 32–33, 1996.
11. D. C. C. Lam, F. F. Lange, and A. G. Evans, "Mechanical Properties of Partially Dense Alumina Produced from Powder Compacts," J. Am. Cer. Soc., **77** (8), pp. 2113–7, 1994.
12. A. R. Boccaccini and G. Ondracek, "On the Porosity Dependence of the Fracture Strength of Ceramics," Third Euro-Ceramics, **3** Engineering Ceramics (P. Duran and J. F. Fernandez, eds.), Faenza Editrice Iberica, Castellon, Spain, pp. 895–900, 1993.
13. A. R. Boccaccini, G. Ondracek, P. Mazilu, and D. Windelberg, "On the Effective Young's Modulus of Elasticity for Porous Materials: Microstructure Modeling and Comparison Between Calculated and Experimental values," J. Mech. Behav. Mat., **4** (2), pp. 119–28, 1993.
14. A. G. Evans and G. Tappin, "Effects of Microstructure on the Stress to Propagate Inherent Flaws," Proc. Brit. Cer. Soc., **20**, pp. 275–97, 6/1972.
15. R. W. Rice, "Pores as Fracture Origins in Ceramics," J. Mat. Sci., **19**, pp. 895–914, 1984.
16. R. W. Rice, "Processing Induced Sources of Mechanical Failure in Ceramics," Processing of Crystalline Ceramics (H. Palmour III, R.F. Davis, and T.M. Hare, Eds.), Plenum Press, New York, pp. 303–19, 1978.
17. R. W. Rice, "Fractographic Identification of Strength-Controlling Flaws and Microstructure" (R. C. Bradt, D. P. H. Hasselman, and F. F. Lange, Eds.), Plenum Press, New York, pp. 323–43, 1974, Fracture Mechanics of Ceramics, vol. 1.
18. R. W. Rice, "Fractographic Determination of K_{IC} and Effects of Microstructural Stresses in Ceramics," Fractography of Glasses and Ceramics II (V. D. Frechette and J. R. Varner, Eds.), American Ceramics Society, Westerville, OH, pp. 509–45, 1991. Ceramic Transactions, vol. 17.
19. R. W. Rice, S. W. Freiman,, J. J. Mecholsky, Jr., R. Ruh, and Y. Harada, "Fractography of Si_3N_4 and SiC," Ceramics for High Performance Applications-II (J. J. Burke, E. N. Lenoe, and R. N. Katz, Eds.), Brookhill Pub. Co., Chestnut Hill, MA, pp. 669–87, 1978.
20. R. W. Rice, R. C. Pohanka, and W. J. McDonough, "Effect of Stresses from Thermal Expansion Anisotropy, Phase Transformations, and Second Phases on the Strength of Ceramics," J. Am. Cer. Soc. **63** (11–12), pp. 703–10, 1980.
21. R. W. Rice, "Ceramic Fracture Features, Observations, Mechanisms, and Uses," Fractography of Ceramic and Metal Failures ASTM STP 827 (J. J. Mecholsky Jr. and S. R. Powell Jr., Eds.) ASTM, pp. 5–103, 1984.
22. B. K. Molnar and R. W. Rice, "Strength Anisotropy in Lead Zirconate Titanate Transducer Rings," Am. Cer. Soc. Bul., **52** (6), pp. 505–9, 1973.
23. R. M. Gruver, W. A. Sotter, and H. P. Kirchner, "Variation of Fracture Stress with Flaw Character in 96% Al_2O_3," Am. Cer. Soc. Bul., **55** (2), pp. 198–204, 1976.
24. H. P. Kirchner, R. M. Gruver, and W. A. Sotter, "Characteristics of Flaws at Fracture Origins and Fracture Stress-Flaw Size Relations in Various Ceramics," Mat. Sci. & Eng., **22**, pp. 147–56, 1976.
25. J. Heinrich and D. Munz, "Strength of Reaction-Bonded Silicon Nitride with Artificial Pores," Am. Cer. Soc Bul., **59** (12), pp. 1221–2, 1980.
26. M. Watanabe and I. Fukuura, "The Strength of Al_2O_3 and Al_2O_3–TiC Ceramics in Relation to Their Fracture Sources," Ceramic Sci. & Tech. at the Present and in the Future, Japan, pp. 193–201, 1981.

27. S. G. Seshadri, M. Srinivasan and K. Y. Chia, "Microstructure and Mechanical Properties of Pressureless Sintered Alpha-SiC," Silicon Carbide (J. D. Cawley and C. E. Semler, Eds.), Am. Cer. Soc., Westerville, OH, pp. 215–225 (1989). Ceramic Transactions, vol. 2.

28. T. Ohji, Y. Tamauchi, W. Kanematsu, and S. Ito, "Tensile Rupture Strength and Fracture Defects of Sintered Silicon Carbide," J. Am. Cer. Soc., **72** (4), pp. 688–90, 1989.

29. L. Y. Chao and D. K. Shetty, "Extreme-Value Statistics Analysis of Fracture Strengths of a Sintered Silicon Nitride Failing from Pores," J. Am. Cer. Soc., **75** (8), pp. 2116–24, 1992.

30. M. Kaj, S. Yoshimura, and M. Nishimura, "Strength Estimation for Silicon Nitride Specimens with a Spherical Void," J. Mat. Sci., **29**, pp. 5947–52, 1994.

31. O. L. Bowie, "Analysis of an Infinite Plate Containing Radial Cracks Originating at the Boundary of an Internal Circular Hole," J. Math. Phy., **35** (1), pp. 60–71, 1956.

32. D. P. H. Hasselman, "Griffith Flaws and the Effect of Porosity on Tensile Strength of Brittle Ceramics," J. Am. Cer. Soc., **52** (8), p. 457, 1969.

33. D. P. H. Hasselman, "Correction—'Griffith Flaws and the Effect of Porosity on Tensile Strength of Brittle Ceramics'," J. Am. Cer. Soc., **56** (8), p. 491, 1973.

34. A. G. Evans, D. R. Biswas, and R. M. Fulrath, "Some Effects of Cavities on the Fracture of Ceramics: I, Cylindrical Cavities," J. Am. Cer. Soc., **62** (1–2), pp. 95–100, 1979.

35. F. I. Baratta, "Stress Intensity Factor Estimates for a Peripherally Cracked Spherical Void and a Hemispherical Surface Pit," J. Am. Cer. Soc., **61** (11–12), pp. 490–93, 1978.

36. F. I. Baratta, "Refinement of Stress Intensity Factor Estimates for a Peripherally Cracked Spherical Void and a Hemispherical Surface Pit," J. Am. Cer. Soc., **64** (1), pp. C3–4, 1981.

37. A. G. Evans, D. R. Biswas, and R. M. Fulrath, "Some Effects of Cavities on the Fracture of Ceramics: II, Spherical Cavities," J. Am. Cer. Soc., **62** (1–2), pp. 101–6, 1979.

38. D. J. Green, "Stress Intensity Factor Estimates for Annular Cracks at Spherical Voids," J. Am. Cer. Soc., **63** (5–6), pp. 342–44, 1980.

39. R. W. Rice, "Relation of Tensile Strength-Porosity Effects in Ceramics to Porosity Dependence of Young's Modulus and Fracture Energy, Porosity Character and Grain Size," Mat. Sci. & Eng., **A112**, pp. 215–24, 1989.

40. L. F. Nielsen, "Elasticity and Damping of Porous Materials and Impregnated Materials, J. Am. Cer. Soc., **67** (2), 93–98 (1983).

41. S. T. Gulati and G. Schneider, "Mechanical Strength of Cellular Ceramic Substrates," "Enviceram 88" Proc. Ceramics for Environmental Protection, Cologne Ger., 12/7–9/88, Deutsche Keramische Gesellschaft e. V., Köln, Germany.

42. D. P. H. Hassselman and J. Singh, "Analysis of Thermal Stress Resistance of Microcracked Ceramics," Am. Cer. Soc. Bul., **53** (8), pp. 856–60, 1979.

43. M. A. Ali, W. J. Knapp, and P. Kurtz, "Strength of Sintered Specimens Containing Hollow Glass Microspheres," Bull. Am. Cer. Soc., **46** (3), pp. 275–277 (1967).

44. J. S. Wallace, "Effect of Mechanical Discontinuities on the Strength of Polycrystalline Aluminum Oxide," Materials and Molecular Research Div., Lawrence Berkeley Laboratory, Annual Report LBL-6016, UC-13, TID-4500–R65, pp. 110–113, 1976.

45. D. R. Biswas and R. M. Fulrath, "Strength of Porous Polycrystalline Ceramics," Trans. & J. Brit Cer. Soc., **79**, pp. 1–5, 1980.
46. D. R. Biswas, "Influence of Porosity on the Mechanical Properties of Lead Zirconate-Titanate Ceramics," Materials and Molecular Research Division, Lawrence Berkeley Laboratory, Annual Report.
47. D. J. Green and R. G. Hoagland, "Mechanical Behavior of Lightweight Ceramics Based on Sintered Hollow Spheres," J. Am. Cer. Soc., **68** (7), pp. 395–398 (1985).
48. D. J. Green, "Fabrication and Mechanical Properties of Lightweight Ceramics Produced by Sintering of Hollow Spheres," J. Am. Cer. Soc., **68** (7), pp. 403–409 (1985).
49. Y. L. Krasulin, V. N. Timofeev, S. M. Barinov, and A. B. Ivanov, "Strength and Fracture of Porous Ceramic Sintered from Spherical Particles," J. Mat. Sci., **15**, pp. 1402–1406 (1980).
50. L. Coronel, J. P. Jernot, and F. Osterstock, "Microstructure and Mechanical Properties of Sintered Glass," J. Mat. Sci., **25**, pp. 4866–72 (1990).
51. D. Ashkin, "Properties of Bulk Microporous Ceramics," Ph.D. thesis, Rutgers University, New Brusnwick, NJ, 1990.
52. S. C. Park and L. L. Hench, "Physical Properties of Partially Densified SiO_2 Gels," Science of Ceramic Chemical Processing (L. L. Hench and D. R. Ulrich, Eds.), John Wiley & Sons, New York, pp. 168–72 (1986).
53. T. Fujiu, G. L. Messing, and W. Huebner, "Processing and Properties of Cellular Silica Synthesized by Foaming Sol-Gels," J. Am. Cer. Soc., **73**, pp. 85–90 (1990).
54. T. Woignier and J. Phalippou, "Mechanical Strength of Silica Aerogels," J. Non-Crystalline Solids, **100**, pp. 404–408 (1988).
55. R. L. Coble and W. D. Kingery, "Effect of Porosity on Physical Properties of Sintered Alumina, J. Am. Cer. Soc., **39** (11), pp. 377–385 (1956).
56. D. R. Petrak, R. Ruh, and G. R. Atkins, "Mechanical Properties of Hot-Pressed Boron Suboxide and Boron," Bull. Am. Cer. Soc., **53** (8), pp. 569–573 (1974).
57. B. A. Chandler, E. C. Duderstadt, and J. F. White, "Fabrication and Properties of Extruded and Sintered BeO," J. Nuc. Mat., **8** (3), pp. 329–347, (1963).
58. J. S. O'Neill and D. T. Livey, "Fabrication and Property Data for Two Samples of Beryllium Oxide," United Kingdom Atomic Energy Authority Research Report #AERER 4912 (1965).
59. R. B. Biddulph, "Rupture Strength and Young's Modulus as Functions of Porosity in Polycrystalline Magnesium Oxide," M.S. Thesis, Dept. of Ceramic Engineering, University of Utah, Salt Lake City, Utah, June, 1964.
60. C. E. Curtis and J. R. Johnson, "Properties of Thorium Oxide Ceramics, J. Am. Cer. Soc. **40** (2), pp. 63–68 (1957).
61. P. A. Evans, R. Stevens, and S. R. Tan, "A Study of the Sintering of β-Alumina: Microstructure and Mechanical Properties," Br. Cer. Trans. J., **83**, pp. 43–49 (1984).
62. R. Tremblay and R. Angers, "Mechanical Characterization of Dense Silicon Tetraboride (SiB_4)," Ceramics Intl., **18**, pp. 113–117, 1992.
63. V. Mandorf and J. Hartwig, "High Temperature Properties of Titanium Diboride," High Temperature Materials, vol. 11 (G. M. Ault, W. F. Barclay, and H. P. Munger, Eds.), pp. 455–467, Interscience Publishers, New York, 1963.
64. C. C. Wu and R. W. Rice, "Porosity Dependence of Wear and Other Mechanical Properties on Fine-Grain Al_2O_3 and B_4C," Cer. Eng. and Sci. Proc., **6** (7–8), pp. 977–994 (1985).

65. K. A. Schwetz, W. Grellner and A. Lipp, "Mechanical Properties of HIP Treated Sintered Boron Carbide," Inst. Phys. Con L. Ser., **75**, pp. 413–419 (1986).

66. B. Champagne and R. Angers, "Mechanical Properties of Hot-Pressed B-B4C Materials," J. Am. Cer. Soc., **62** (3–4), pp. 149–153 (1979).

67. V. P. Bulychev, R. A. Andrievskii, and L. B. Nezhevenko, "The Sintering of Zirconium Carbide," Poroshkovaya Metalurgia, **4** (172), pp. 273–275.

68. P. Boch, J. C. Glandus, J. Jarrige, J. P. Lecompte, and J. Mexmain, "Sintering, Oxidation and Mechanical Properties of Hot Pressed Aluminum Nitride," Ceramics Intl., **8** (1), pp. 34–38 (1982).

69. M. Desmaison-Burt, J. Montinin, F. Valin, and M. Boncoeur, "Mechanical Properties and Oxidation Behavior of HIPed Hafnium Nitride Ceramics," J. Eur. Cer. Soc., **13**, pp. 379–86, 1994.

70. M. Moriyama, K. Kamata, and Y. Kobayashi, "Mechanical and Electrical Properties of Hot-Pressed TiN Ceramics Without Additives," J. Cer. Soc. Jap., Intl. Ed., **99** 275–, 19.

71. D. J. Green, C. Nader, and R. Brezny, "The Elastic Behavior of Partially-Sintered Alumina," Sintering of Advanced Ceramics (C. A. Handwerker, J. E. Blendell, and W. Kaysser, Eds.), Am. Cer. Soc., Westerville, OH, pp. 347–56, 1990.

72. P. D. Wilcox and I. B. Cutler, "Strength of Partly Sintered Alumina Compacts," J. Am. Cer. Soc., **49** (5), pp. 249–252 (1966).

73. S. C. Nanjangud, R. Brezny, and D. J. Green, "Strength and Young's Modulus Behavior of a Partially Sintered Porous Alumina," J. Am. Cer. Soc., **78** (1), pp. 266–8 (1995).

74. N. Murayama, Y. Kodama, S. Sakaguchi, and F. Wakai, "Mechanical Strength of Hot Presssed Bi-Pb-Sr-Cu-Superconductor," J. Mat. Res., **7**, pp. 34–37, 1992.

75. P. Johnson, Alfred Univ. Private communication, 1994.

76. A. Salak, V. Miskovic, E. Dudrova, and E. Rudnayova, "The Dependence of Mechanical Properties of Sintered Iron Compacts Upon Porosity," Pwd. Met. Mat., **6** (3), pp. 128–132 (1974).

77. R. W. Rice, "Ceramic Fracture Mode-Intergranular vs, Transgranular Fracture," Fractography of Glasses and Ceramics III (J. R. Varner and G. Quinn, Eds,), Am. Cer. Soc., Westerville, OH, pp. 1–53, 1996.

78. D. Broussaud, J. P. Dumas and Y. Lazennec, "Interdependence of Composition, Processing Parameters, Final Properties and Microstructure of Hot Pressed SiC," Ceramic Microstructures '76, (R. M. Fulrath and J. A. Park, Eds.), pp. 679–688, Westview Press, Boulder, CO, 1977.

79. S. Prochazka, C. A. Johnson, and R. A. Giddings, "Investigation of Ceramics for High Temperature Turbine Components," GE report for NADC contract N62269–75–C-0122, 12/ 1975.

80. D. R. Biswas, "Crack-Void Interaction in Polycrystalline Alumina," J. Mat. Sci., **16**, pp. 2434–38, 1981.

81. A. R. Boccaccini, "Comment on 'Dependence of Ceramic Fracture Properties on Porosity'," J. Mat. Sci. Let., **13**, pp. 1035–37, 1994.

82. J. J. Cleveland and R. C. Bradt, "Grain Size/Microcrack Relations for Pseudo-brookite Oxides," J. Am. Cer. Soc. **61** (11–12), pp. 478–81, 1978.

83. R. W. Rice, "Review: Ceramic Tensile Strength-Grain Size Relations: Grain Sizes, Slopes, and Branch Intersections," J. Mat. Sci., **32**, pp. 1673–1692, 1997.

84. H. Tomaszewski, "Influence of Microstructure on the Thermomechanical Properties of Alumina Ceramics," Cer. Intl., **18**, pp 51–55, 1992.

85. M. Knechtel, C. Cloutier, R. Bordia, and J. Rodel, "Mechanical Properties of Partially Sintered Alumina," to be published.

86. A. Pissenberger and G. Gritzner, "Preparation and Properties of Niobia-and Tantala-Based Orthorhombic Zirconia," J. Mat. Sci. Let., **14**, pp. 1580–82, 1995.

87. A. Kishimoto, K. Koumoto, H. Yanagida, and M. Nameki, "Microstructure Dependence of Mechanical and Dielectric Strengths—I. Porosity," Eng. Fract. Mech., **40** (4/5), pp. 927–30, 1991.

88. R. W. Rice, "Porosity Effects on Machining Direction-Strength Anisotropy and Failure Mechanisms," J. Am. Cer. Soc. **77** (8), pp. 2232–36, 1994.

89. R. W. Rice, "Effects of Ceramic Microstructural Character on the Machining Direction-Strength Anisotropy," Machining of Advanced Materials, Proc. Intl. Conf. on Machining of Advanced Materials 7/20–22/93 (S. Johanmir, Ed.), NIST Spec. Pub. 847, pp. 185–204, U.S. Govt. Printing Office, Washington, D.C., 6/1993.

90. J. J. Mecholsky, Jr., S. W. Freiman, and R. W. Rice, "Effect of Grinding on Flaw Geometry and Fracture of Glass, J. Am. Cer. Soc., **60**, 3–4 (1977).

91. R. W. Rice and J. J. Mecholsky, Jr., "The Nature of Strength Controlling Machining Flaws in Ceramics," The Science of Ceramic Machining and Surface Finishing II, (B. J. Hockey and R. W. Rice, Eds.), pp. 351–78, U.S. Gov't Printing Office, Washington, D.C. (1979). NBS Special Publication 562.

92. R. W. Rice, J. J. Mecholsky, Jr., and P. F. Becher, "The Effect of Grinding Direction on Flaw Character and Strength of Single Crystal and Polycrystalline Ceramics," J. Mat. Sci., **16**, pp. 853–862 (1981).

93. J. J. Mecholsky, Jr., R. W. Rice, and S. W. Freiman, "Prediction of Fracture Energy and Flaw Size in Glasses from Measurements of Mirror Size," J. Am. Cer. Soc., **57** (10), pp. 440–43, 1974.

94. J. J. Mecholsky Jun., S. W. Freiman, and R. W. Rice, "Fracture Surface Analysis of Ceramics," J. Mat. Sci., **11**, pp. 1310–19, 1976.

95. G. K. Bansal, "Effect of Flaw Shape on Strengths of Ceramics," J. Am. Cer. Soc., **59** (1–2), pp. 87–88, 1976.

96. R. W. Rice, "The Difference in Mirror-to-Flaw Size Ratios Between Dense Glasses and Polycrystals," J. Am. Cer. Soc., **62** (9–10), pp. 533–35, 1979.

97. R. W. Rice, "Perspective on Fractography," Fractography of Glasses and Ceramics, (J. R. Varner and V. D. Frechette, Eds.), Am. Cer. Soc., Westerville, OH, pp. 3–56, 1988. Adv. in Ceramics, vol. 22.

98. P. A. Zahl, "Golden Window on the Past," National Geographic, **152** (3), pp. 423–35, 9/1977.

99. F. I. Baratta, "Comment on 'Strength of Reaction-Bonded Silicon Nitride with Artificial Pores'," J. Am. Cer. Soc., **65** (2), pp. C32–34, 1982.

100. R. W. Rice, "Specimen Size-Tensile Strength Relations for a Hot-Pressed Alumina and Lead Zirconate/Titanate," Am. Cer. Soc. Bul., **66** (5), pp. 794–98, 1987.

101. J. B. Wachtman, Jr. and J. Dunduras, "Large Localized Surface Stresses Caused by Thermal Expansion Anisotropy," J. Am. Cer. Soc., **54** (10), pp. 525–26, 1971.

102. L. C. Sawyer, M. Jamieson, D. Brikowski, M. I. Haider, and R. T. Chen, "Strength, Structure, and Fracture Properties of Ceramic Fibers Produced from Polymeric Precursors: I, Base-Line Studies," J. Am. Cer. Soc., **70** (11), pp. 798–810, 1987.

103. B. H. King and J. W. Halloran, "Polycrystalline Yttrium Aluminum Garnet Fibers from Colloidal Sols," J. Am. Cer. Soc., **78** (8), pp. 2141–48, 1995.

104. J. M. Yang, S. M. Jeng, and S. Chang, "Fracture Behavior of Directionally Solidified $Y_3Al_5O_{12}/Al_2O_3$," J. Am. Cer. Soc., **79** (5), pp. 1218–22, 1996.

105. F. H. Simpson, "Continuous Oxide Filament Synthesis (Devitrification)," Tech. Report AFML-Tr-fi-135, 1971.

106. R. W. Rice, P. F. Becher, and W. A. Schmidt, "The Strength of Gas Polished Sapphire and Rutile," The Science of Ceramic Machining and Surface Finishing, NBS Special pub. 348, U.S. Govt. Printing Office, Washington, D.C., pp. 267–69, May, 1972.

107. J. T. A. Pollock, "Filamentary Sapphire, Part 3. The Growth of Void-Free Sapphire Filaments at Rates Up to 3.0 cm/min," J. Mat. Sci., **7**, pp. 787–92, 1972.

108. J. T. A. Pollock and G. F. Hurley, "Dependence of Room Temperature Fracture Strength on Strain-Rate in Sapphire," J. Mat. Sci., **8**, pp. 1595–1602, 1973.

109. S. A. Newcomb, "Temperature and Time Dependence of the Strength of C-Axis Sapphire from 800–1500°C," MS Thesis, Pennsylvania State University, College Park, PA, May, 1992.

110. S. A. Newcomb and R. E. Tressler, "Slow Crack Growth of Sapphire at 800 to 1500°C," J. Am. Cer. Soc., **76** (10), pp. 2505–12, 1993.

111. R. W. Rice, "Corroboration and Extension of Analysis of C-Axis Sapphire Filament Fractures from Pores," J. Mat. Sci., **16**, pp. 202–205, 1997.

112. H. P. Kirchner and R. M. Gruver, "Fractographic Criteria for Subcritical Crack Growth Boundaries in 96% Al_2O_3," J. Am. Cer. Soc., **63** (3–4), pp. 169–74, 1980.

113. F. F. Lange, "Relation Between Strength, Fracture Energy, and Microstructure of Hot-Pressed Si_3N_4," J. Am. Cer. Soc., **56** (10), pp. 518–22, 1973.

114. F. F. Lange, "Fracture Toughness of Si_3N_4 as a Function of the Initial α-Phase Content," J. Am. Cer. Soc., **62** (7–8), pp. 428–30, 1979.

115. J. E. Weston, "Origin of Strength Anisotropy In Hot-Pressed Silicon Nitride," J. Mat. Sci., **15**, pp. 1568–76, 1980.

116. R. Brezny and D. J. Green, "Fracture Behavior of Open-Cell Ceramics," J. Am. Cer. Soc., **72** (7), pp. 1145–52, 1989.

117. R. Brezny and D. J. Green, "Uniaxial Strength Behavior of Brittle Cellular Materials," J. Am. Cer. Soc., **76** (9), pp. 2185–92, 1993.

118. D. L. Brooks and E. M. Winter, "Material Selection of Cellular Ceramics for a High Temperature Furnace," Columbia Gas System Service Corp., Columbus, OH, Report, cira 1989.

119. L. J. Gibson and M. F. Ashby, "Cellular Solids, Structure & Properties," Pergamon Press, New York (1988).

120. R. Baddi, J. Crampon, and R. Duclos, "Temperature and Porosity Effects on the Fracture of Fine Grain Size MgO," J. Mat. Sci., **21**, pp. 1145–50, 1986.

121. R. W. Rice, "Grain Size and Porosity Dependance of Fracture Energy and Toughness of Ceramics at 22°C," J. Mat. Sci., **31**, pp. 1969–83, 1996.

122. R. W. Rice, "Microstructural Dependance of Fracture Energy and Toughness of Ceramics and Ceramic Composites Versus That of Their Tensile Strengths at 22°C," J. Mat. Sci., **31**, pp. 4503–19, 1996.

123. R. W. Rice, "Machining, Surface Work Hardening, and Strength of MgO," J. Am. Cer. Soc., **56** (10), pp. 536–41, 1973.

124. P. Bompard, D. Wei, T. Guennouni, and D. Francois, "Mechanical and Fracture Behavior of Porous Materials," Eng. Fract. Mech., **28** (5/6), pp. 627–42, 1987.

6

POROSITY DEPENDENCE OF HARDNESS, COMPRESSIVE STRENGTH, WEAR, AND RELATED BEHAVIOR AT 22°C

KEY CHAPTER GOALS

1. Show the general theoretical and experimental correlation of the porosity dependence of hardness, compressive strength, wear, and related properties, but with variations.
2. Demonstrate that their porosity dependences are consistent overall with MSA models, the primary models available.
3. Show that deviations come from differing dependences on grain size and pore changes accompanying porosity reduction and local penetration, crushing, and collapse of pores.

6.1 INTRODUCTION

Hardness, compressive strength, and wear are important properties, and thus their porosity dependence is important. Of these properties, wear is of the broadest importance from a technological standpoint, both because of the importance of individual applications and their diversity and scope. The latter results from wear being a collection of behaviors, e.g., reflecting various combinations of mechanical, chemical, and thermal effects, thus making it a greater challenge to study and understand from both scientific and engineering standpoints. The relation of wear to other behavior such as machining and erosion adds to its importance. The very high levels of compressive strengths achievable in ceramics (e.g., 5–10 fold those in tension) make them of technological interest and their relation to basic characteristics such as bonding and deformation of scientific interest. Hardness is important since it correlates with wear and related behavior such as machining, erosion, and compressive strength. It is also important since it can be determined by a simple, nondestructive test of small samples with simple, not very expensive, equipment. Its dependencies on load and surface character, as well as microstructure make it more complicated, but also potentially add to its value as a probe of both surface and bulk behavior.

Hardness, wear, compressive strength and related behavior, which all involve significant compressive stressing and resultant local tensile stresses, are considered together since they are related qualitatively (and sometimes quantitatively). The resultant property correlations, especially with hardness are of significant practical value, e.g., hardness often being an approximate guide to the wear-resistance of materials. Both the importance and the complexities of these properties make it important to determine their substantial decreases with increasing porosity, and hence to understand and control one of their important variations. The microcracking dependence of these properties is also of some importance, though microcracked materials would not commonly be used for many applications involving significant hardness, compressive strength, or wear requirements. There is very little information on microcracking effects, but what little there is will be presented.

6.2 THEORETICAL ASPECTS

There are few or no specific models for the porosity dependence of hardness, compressive strength, or wear, erosion resistance, and related properties; however, both the basic concepts and especially the correlations of these properties with each other and with elastic moduli indicate the applicability of MSA (minimum solid area) and other models for Young's modulus. While there are important variations (1), MSA models are often quite applicable. Nielsen (2) proposed that his exponential (e^{-bP}) model for the dependence of tensile strength on pores and cracks at low P (Eq. 5.9) was also applicable to compressive strength. Sammis and Ashby (3) modeled dilute quantities of isolated, uniformly arrayed, spherical or cylindrical pores (aligned normal to the stress axis) and the resulting crack nucleation and growth parallel with the applied uniaxial stress (Fig. 6.1a, d). They also considered the important case for uniaxial compression under a supporting radial or hydrostatic pressure, showing that ductile type deformation can be obtained (Fig. 6.1b, c), as known for dense ceramics (4). Their model for local crack development and growth from the local tensile stresses that are approximately parallel with the compressive axis is basically the same as other models for compressive failure from flaws, e.g., of Sines and colleagues (5,6), but they explicitly treated pores as the source of cracks.

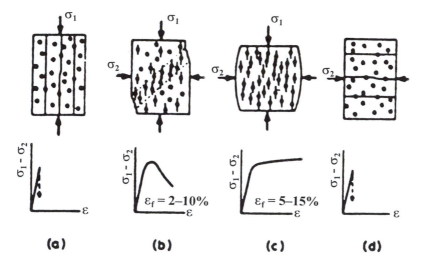

Figure 6.1 Schematic of the uniaxial compressive failure of brittle materials with dilute concentrations of uniformly distributed spherical or cylindrical pores (the latter aligned normal to the stress axis). a) and d) show failure from uniaxial compressive vertical and horizontal stresses, respectively, without a supporting stress (e.g., hydrostatic stress) providing radial support to the uniaxial stress, and b) and c) show failure with such supporting stress (respectively $\sigma_2 < \sigma_1$, $\sigma_2 \sim \sigma_1$). Figure used with permission of Sammis and Ashby (3) and Acta Metallurgica.

Gibson and Ashby (7,8) derived models for both hardness and compressive (crushing) strength of foams (Fig. 3.3B) following the same, essentially MSA, approach as for properties in earlier chapters. These models again give the porosity dependence as $(1 - P)^n$, with n = 3/2, i.e., the same as for fracture toughness and tensile strength, but different from their derivation of n = 2 for Young's modulus. (The model for hardness appears only applicable for tests with large punches, since the size of indentations for ceramics would typically be less than or equal to a single foam cell.) For closed-cell foams, they obtained n = 2 for compressive strength. These authors also defined a deformation-mode map with axes of the compressive stress normalized by the Young's modulus of the strut material versus the compressive strain [Maiti et al. (9)]. They defined three, generally sequential, types of behavior with progressive loading: 1) linear elastic, 2) cell collapse (i.e., fracture for brittle materials), and 3) crushing and compaction of debris at high strains (often with increased stress).

Gulati and Schneider (10) derived relations for the compressive strength of extruded square-cell honeycomb as a function of direction relative to the structure (Fig. 6.2). This shows that the P-dependence is $(1 - P)^n$ with n = 2 for the axial and diagonal loading directions, and $\sim$ 2 for the normal loading direction.

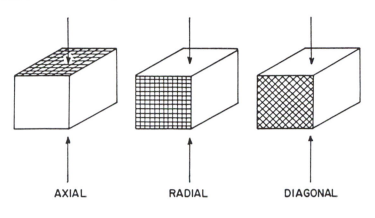

AXIAL   RADIAL   DIAGONAL

Normal loading

$$\sigma_n = t^2 (1+L)(6L\epsilon_n)^{-1}\sigma_w \propto (1-P)^2$$

Axial loading

$$\sigma_a = t^2(3L\epsilon_a)^{-1}\sigma_w \propto (1-P)^2$$

$$\frac{\sigma_n}{\sigma_a} \sim \frac{1}{2}\left(\frac{\epsilon_a}{\epsilon_n}\right)$$

Diagonal loading

$$\sigma_d = (2/3)\sigma_w(t^2)L^{-1} \sim \frac{\sigma_w}{6(1-P)^2}$$

$$\frac{\sigma_d}{\sigma_n} \sim 4\epsilon_n\,(L)^{-1}$$

Figure 6.2 Schematic of the compressive (crushing) strength of extruded square-cell honeycomb for the three loading directions, after Gulati and Schneider (10): t = cell wall thickness; L = cell spacing; σ = compressive strength; ∈ = compressive strain; and subscripts a, n, and d refer, respectively, to the three left-to-right loading directions.

This n = 2 dependence was also derived for hexagonal-cell honeycomb (7,8). Comparison of these porosity dependences with that for Young's modulus (Table 3.3) again shows different predicted P-dependences of Young's modulus and tensile strength for such porous honeycombs.

Consider now strengths of individual solid or hollow spheres since these, especially the former, are of importance individually, e.g., as bearings and milling balls, and both are of use collectively to make porous bodies (Section 10.5). Uniaxial compressive loading of a solid sphere is very similar to a diametral test, i.e., radial loading of a short cylinder along a diameter (Fig. 1.6C), and is amenable to an exact solution from Hiramatsu and Oka (11):

$$\sigma_c = \sigma_t = \frac{0.9F}{d^2} \tag{6.1}$$

where F = the force at failure and d = the sphere diameter (or for an irregular body, for which this equation is still often accurate, it is the separation of the two loading, contact, points). For thin-wall hollow spheres (i.e., balloons) Chung and colleagues (12,13) cite Bratt et al. (14) for the compressive failure stress, σ_c, as:

$$\sigma_c = F\,(2\pi t^2)^{-1} \tag{6.2}$$

where F = the compressive force applied at failure and t = the wall thickness, which was corroborated in tests of glass balloons. In terms of P, the denominator of Eq. 6.1 becomes:

$$2\pi r(1 - P^{1/3})^2 \tag{6.3}$$

where r = the outer radius of the balloon; however, by finite element analysis they found that:

$$\sigma_c = A\sigma_0(1 - P)^2 \tag{6.4}$$

where A = a constant that significantly depended on (e.g., could vary over an order of magnitude with) the extent of contact and deformation at the loading poles, and σ_0 = the strength of the balloon-wall material. Note that this gives the same P-dependence as found for closed-cell foams.

Chung and colleagues (12,13), following a model of Swanson and Cutler (15) derived for failure of arrays of ceramic balloons used as proppants with uniaxial and hydraulic loading, considered arrays of balloons loading each other over various extents of contact. Their finite element analysis showed that the stiffness of an array of spheres (bonded at the contact points) was proportional to $(1 - P)^2$. The compressive failure of such arrays was more complex, but was approximated by two similar functions of $(1 - P)^2$ at lower and higher strengths respectively:

$$\sigma_c = A_1\sigma_0(1 - P)^2 \tag{6.5}$$

$$\sigma_c = A_0 + A_2\sigma_0(1 - P)^2 \tag{6.6}$$

where the As are constants. The shift from the first to the second equation occurred at lower density and strength as the extent of contact between spheres decreased.

Arato et al. (16) proposed an MSA model for random packing of uniform particles as a function of their degree of sintering in conjunction with their modeling of Young's modules, fracture toughness, and strength. Their equation for hardness (H) was simply $H = 3\,\sigma_t\,(E/E_0)$, where σ_t was the strength of the sintered neck, and E/E_0 = the ratio of Young's modules at the P value of interest to that at $P = 0$. Note, the direct correlation with the P-dependence of E via the E/E_0 term which is replaceable with microstructural terms (Table 2.8). These models were all consistent with the same P-dependence for all properties as for E.

In view of there being few models, primarily for specialized cases, there is increased need for bounds and correlations with other properties; however, bounds are the same as for properties considered in previous chapters, i.e upper and lower bounds respectively of $(1 - P)$ and 0, which are thus again of limited value, especially the lower bound. Basic considerations, as well as interrelation of these and other properties, as discussed below, indicate that the upper limits to the P-dependence, i.e., P_C values where properties go to zero, will often be at $P \leq 1$, as for other properties. Property limits at $P = 0$ may in some cases be obtained by appropriate averaging of single-crystal values, e.g., as for elastic moduli, but property dependence on grain size is often a limitation. This is often reasonably feasible, at least approximately, for hardness for which there is a reasonable amount of single-crystal data (e.g., see refs. 17, 18, and especially Chapters 3, 11–18) since its dependence on grain size is limited; however, for compressive strength, wear, erosion, and related properties such use of single-crystal values is both theoretically quite uncertain and lacking in data. A problem in accurate determination of $P = 0$ limits is that all of these properties depend on grain parameters, e.g., shape, orientation, and especially size, which commonly increases as P decreases, especially at lower P levels. Thus, unless such changes are accounted for they can cause considerable variation in apparent $P = 0$ values, e.g., as discussed for fracture toughness and tensile strength in Chapters 4 and 5. This problem is generally modest for hardness, very substantial for compressive strength, and variable, but generally intermediate for wear and related properties. Another complication is the phase content present, either due to prior processing or in some cases to phase transformation during test, e.g., during hardness indentation. Thus, although earlier work on in situ (hence more difficult and uncertain) measurements of RSSN (19) indicated that β-Si_3N_4 was harder than α-Si_3N_4, more recent studies of single crystals (20) and dense polycrystalline bodies (20,21) show α-Si_3N_4 being substantially (e.g., 25%) harder than β-Si_3N_4.

While there is limited or no direct theoretical guidance for the general porosity dependence of these properties, there are relations between the properties of this chapter amongst themselves, as well as with elastic moduli. Thus, hardness (H) of ceramics and intermetallics correlates over a substantial range of materials with their bulk moduli, and for metals with their Young's moduli (22), with average log H-log moduli slopes of 3/2 and 1 respectively. More extensive plotting of data shows a broad correlation of H with both B and E, but with the latter slope being the higher one for ceramics (Fig. 6.3). There are of

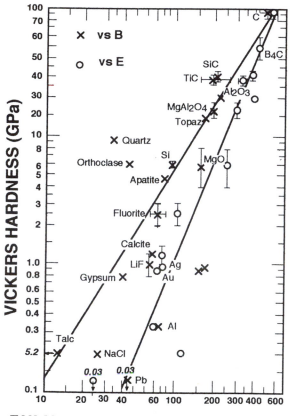

BULK or YOUNG'S MODULUS (GPa)

Figure 6.3 Log-log plot of hardness versus Young's (E) and bulk (B) modulus at 22°C. Note the higher H-E versus H-B slope in contrast to Gilman's correlation for metals (22).

course several factors that vary the H-E correlations, mainly due to some grain-size and surface-finish dependence of H and none of E (or B), and typically greater sensitivity of hardness to effects of other constituents, whether in solution or as second phases; however, studies of Si_3N_4 materials have also shown a H-E correlation over the narrower range of their property changes (23,24). Thus, the porosity dependence of hardness should at least approximately reflect the porosity dependence of Young's modulus, possibly closely if other factors are not significant. One possibly important difference is that pores or hetero-geneities of porosity on the scale of the indent can much more significantly effect H than bulk dependences such as that of Young's modulus.

Compressive strength (σ_c) is now known to be related to hardness and hence to also generally correlate with bulk and Young's moduli (Fig. 6.3). The correlation with hardness stems from the fact that hardness is a measure of local yield stress (Y), which for many materials obeys the simple relation: $Y = H/C$, where C is referred to as the constraint factor, and is typically 2.5 to 3 (22,25). This relation also holds for yield stress under shock, e.g., ballistic impact conditions, with the same or similar value for C (22). [Where there are significant differences in the yield stress for deformation, i.e., slip or twinning systems, that do not allow general deformation, e.g., they do not represent five independent slip systems, then the values of C may be quite different (22), especially where stresses are limited, e.g., in tension. Also, in materials such as silicate glasses and polymers' where the ratio $E/Y < 133$, then $C = 0.28 + 0.60 \ln (0.7 \, E/Y)(25–27))$. The H-$\sigma_c$ relation is a direct result of the yield stress ~ H/3 being the upper limit to the compressive strength of dense crystalline ceramics (1,4,28), even though their failure is brittle, i.e., cracks are generated and grown by microplastic processes that culminate in brittle failure. While there are a variety of effects that reduce compressive strengths below this H/3 limit, they are mainly extrinsic, so in well-made, well-tested, dense ceramics, their compressive strengths correlate reasonably well with this limit. (Such correlation is shown by the grain-size dependence of compressive strength (1,4,28).) Thus, while porosity is an important factor in reducing compressive strengths, the H-P trend should be approximately the σ_c -P trend. Again, grain size changes must be accounted for in the porosity dependence of compressive strength since both H and σ_c depend on grain size (though in differing degrees and trends), and E and B do not.

Since wear and related properties also generally correlate with hardness, their porosity dependence should approximately reflect that of hardness and hence also approximately that of Young's modules. Again, since these properties also depend on grain size, resultant variations of it must be accounted for in comparing their P-dependence with that of Young's modulus; however, the grain-size dependence of hardness, wear, and other properties are often similar, aiding their correlations with hardness. Again, as with hardness, wear and related properties can be sensitive to pores or heterogeneities of porosity on the scale of the penetration of the surface being worn, eroded, etc.

Some models have been developed for the erosive wear rate of ceramics (W_e) from particle impact based on indentation-cracking models, giving (29):

$$W_e \propto v_0^d R^e \rho^f K_{IC}^g H^h \tag{6.7}$$

where v_0 = particle initial velocity, R = particle radius, ρ = particle density, K_{IC} = fracture toughness, and H = the hardness of the body being eroded. For two models the exponents have respective values of: d = 3.2 or 2.4, e = 3.7 (for both), f = 1.3 or 1.2, g = −1.3 (for both), and h = −1.25 or + 0.11. Thus, the P-dependence of W_e is reflected in the P-dependence of density, fracture toughness, and hardness. A review of these models revealed that while they generally fit the data, there were discrepancies (e.g., greater than predicted dependence on H and K_{IC}) which were attributed to microstructural aspects of erosion not being adequately accounted for in the models (29). The interrelated effects of crack and grain size on the P-dependence of K_{IC} (Chapter 4) can be important factors in this, in view of the orders of magnitude differences in crack sizes between those commonly used for measuring fracture toughness and those involved in wear, erosive, etc., and removal of material by fracture.

Similarly, the modeling of the introduction of cracks from sharp indentations is used as a basis of modeling particle-impact damage, and for modeling material-removal processes in abrasive machining. Resultant models for the crack size, c, give (30):

$$ c \propto \left[\left(\frac{F}{K_{IC}} \right) \left(\frac{E}{H} \right)^{1/2} \right]^{2/3} \tag{6.8} $$

where F = the load on the indentor or abrasive particle. Since the crack size is related to the rate of machining, the porosity dependence of machining rates will be determined by that of H, E, and especially K_{IC}. The porosity dependence of machining flaw size via that of H is often greater than or equal to that of E and K_{IC}, but overall approximately that of E; however, the K_{IC} pertinent to machining (and wear and erosion) will typically be that for a much smaller crack scale than for normal fracture-toughness testing, the larger scale of the latter being the main source of K_{IC}-P versus E-P differences (Chapter 4). Note an experimental correlation of machining resistance and hardness has been shown (31).

Thus there are correlations between properties involving substantial compressive plus some (usually local) tensile stress with those properties that are determined mainly, or exclusively, under tensile loading. Differences and variations due to heterogeneity and anisotropy of porosity (and microcracking) have not been addressed. There can also be differences due to inherently different effects of tensile versus compressive stressing on some microstructures, namely the tendencies to respectively separate or pull apart, or to consolidate or push together small cracks or narrow portions, e.g., cusps, of pores. Thus, different effects of tensile versus compressive stressing of some graphites, rocks (32), and other materials [e.g., sintered ZrO_2 beads (33) (Fig. 3.9A)] are noted, even for elastic moduli (Chapters 3, 8, and 9) where certain porosity, anisotropy, or microcracking are involved. Collapse and crushing of foams were noted earlier in this section, and possible broader occurrence of these will be noted later, along with more on bodies of sintered ZrO_2 balloons.

6.3   COMPARISON OF MODELS AND DATA

6.3.1   Hardness

The only hardness data involving pores approaching those of ideal models is that for glassy carbon [mainly from Hucke et al. (1,34–37)], where pores are typical at least approximately spherical (Fig. 1.7A). This data (Fig. 6.4) gives a b value of 3.2 (in the e^{-bP} dependence), in good agreement with that expected for spherical porosity, i.e., ~3 (and is also consistent with data for compressive strength of this same material (see Fig. 6.11).

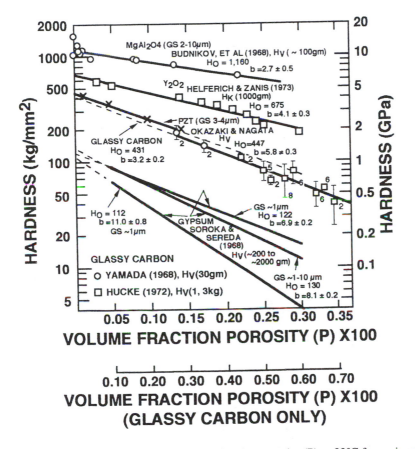

Figure 6.4   Hardness versus volume fraction porosity (P) at 22°C for various ceramics from a previous survey by Rice (1). Note: 1) the different scale for the glassy carbon from the other data and its close agreement with expectations for the MSA model for spherical pores; and 2) the similar trends for the different materials, including gypsum-based bodies. Published with permission of Academic Press.

Table 6.1 Summary of Ceramic Property Porosity Dependence for Different Processing[a]

Property	Extruded	"b" Values[b]			
		Cold (die) pressed	Iso-pressed	Colloidal	Hot pressed
Young's modulus (E)	2.3 (1)	3.6 ± 0.8 (25)	4±0 ± 1.2 (14)	3.9 ± 1.2 (4)	4.5 ± 0.7 (31)
Shear modulus (G)	2.2 (1)	3.2 ± 0.9 (8)	3.9 ± 1.3 (12)		4.3 ± 0.6 (17)
Bulk modulus		4.0 (1)			5.1 ± 0.8 (8)
Tensile strength (σ)	5.1 ± 2.1 (3)	4.0 ± 1.1 (3)	7 (1)	4 (1)	4.5 ± 1.2 (15)
Hardness (H)		4.1 ± 1.1 (5)			5.8 ± 1.4 (8)
Compressive strength					5.8 ± 1.5 (3)

Source: ref. (35); published with permission of J. Mat. Sci.

[a] Values shown are averages and ± 1 standard deviation where more than 1 value was available. Values in ( ) = number of studies averaged (35).

Data for other materials reflecting natural porosity from various processing from an earlier survey (1) (Fig. 6.4), a more recent survey (35) (summarized in Table 3.6, the pertinent portion of which is reproduced as Table 6.1), and additional individual data sets, some of which are not included in the surveys (Figs. 6.5–6.7) provide insight for such porosity dependence. This data is consistent overall with the trends expected based on MSA models and pore-processing correlations, i.e., higher b values for hot-pressed (e.g., ~ 4.5–6), and lower values (e.g., ~3–4) for die-pressed material. Also note the similar behavior, but

Table 6.2 Additional Hardness-Porosity Data at 22°C

Material	Fabrication[a]	Grain size (µm)	P	b (load)[b]	Investigator
TiC	HPHP	20	0.04–0.11	5.5 (not given)	Yamada (37)
TiC	RHP	44	0.03–0.10	1.9[d] (300 g ~ 3.3N)	LaSalvia et al. (38)
Si$_3$N$_4$	HIP	—	0–0.10	4.1 (500 g ~ 4N)	Greskovich and Yeh (39)
Si$_3$N$_4$	S	—	0–010	4.2 (10 N)	Mukhopadhyay et al. (23,24)[d]
Si$_3$N$_4$	GPS	—	0.05–0.35	6.3 (5 N)	Arato et al. (16) (Table 4.2)

[a] HPHP = high pressure hot pressing, RHP = reaction hot pressing (via SHS reactions), HIP = hot isostatic pressing, S = sintering (from various sources), and GPS = gas pressure sintering (at 1–2 MPa of N$_2$) of bodies at 290 MPa.

[b] 1 kg force ~ 9.81 N.

[c] The lower b value is consistent with much of the porosity being approximately spherical or polyhedral, hence not included with other hot pressing data.

[d] Includes data of Babini et al. (40).

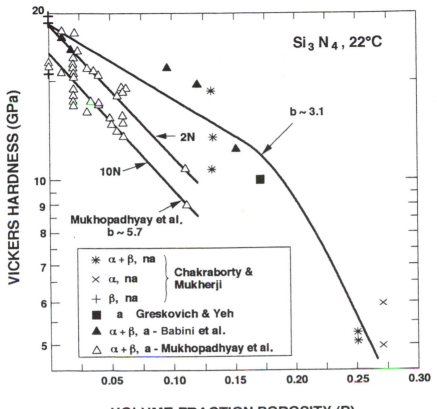

Figure 6.5 Hardness of various sintered, hot-pressed, or HIPed Si$_3$N$_4$ bodies at 22°C (16,21,23,24,39,40). Note: 1) identification of which phase was dominant, or if the phase content was mixed, and whether additives were (a) or were not (na) used (though limited and scattered, the latter mixed data suggests a lower b value for bodies made without additives), and 2) the rollover toward P$_C$ (see also Fig. 6.6).

somewhat higher b values, of gypsum-based materials (where mass increases as P decreases due to hydration) to other ceramics where mass does not change with P, e.g., for sintering. While such mass changes are likely to require some modifications of MSA models, evaluation of limited pertinent models (e.g., Fig. 8.2) and experimental data for gypsum, cements, and other materials (Chapter 8) indicate modest changes.

The earlier survey (1) reported higher b values for hardness (as well as for compressive strength and wear) versus those for elastic moduli and tensile strength, i.e., averages of ~6 ± 3 versus ~ 4 ± 2 respectively; however, this

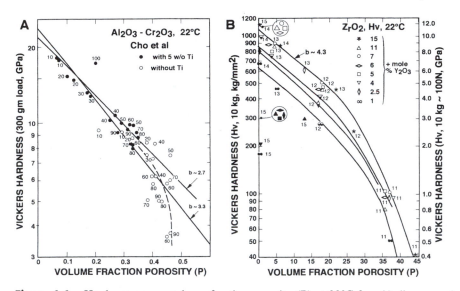

Figure 6.6 Hardness versus volume fraction porosity (P) at 22°C for: A) die-pressed and sintered Al_2O_3-Cr_2O_3 bodies (numbers next to data points are the percent Cr_2O_3) after Cho et al. (41), and B) die-pressed and sintered ZrO_2 with various degrees (1–15 m/o) of Y_2O_3 stabilization (numbers next to data points are the firing temperatures in 100°C, and half and fully filled symbols indicate respectively substantial and extensive monoclinic content in the as-fired samples) (42,43). Note: 1) both show at least the start of the roll over toward P_C values (see Fig. 6.5 for another case), and 2) revised evaluation of the curves in B) giving lower b values, based on more recent understanding (35,36) than in the original evaluation (43).

previous survey did not recognize the effects of processing and pore character on b values and had a substantially smaller data base than the later survey. Thus, nearly 3/4 of the available data in Table 6.1 is for hot-pressed material, which gives higher b values, hence driving the average b value toward the higher levels for hot pressing (36). Averaging the total available data, i.e., the 12 b values for hot-pressed and the 9 for sintered (mainly die-pressed) bodies give respectively 5.2 ± 1.5 and 4.2 ± 1.0. Subsequent broader evaluation corroborates higher b values for hot-pressing than for sintering, and again indicates that the overall higher average b value for hardness is due in substantial part to more of the data being from processes (hot-pressing in this case) giving higher porosity dependence, i.e., b values; however, there are some indications of higher b values for hardness than for Young's modulus for the same type of processing, e.g., by 10–25% instead of by ~50%.

The above trends are corroborated and additional aspects of the porosity dependence of hardness and its generally following MSA models are shown by

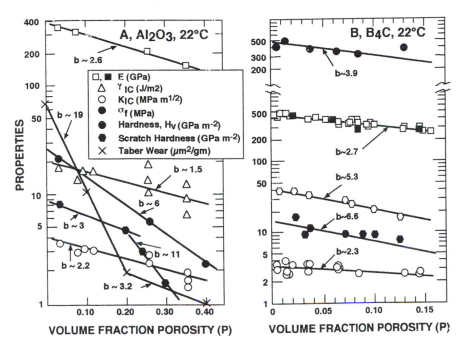

Figure 6.7   Hardness and other related mechanical properties of A) Al_2O_3 and B) B_4C versus volume fraction porosity (P) at 22°C after Wu and Rice (45).

considering a number of individual cases, e.g., Table 6.2 and Figs. 6.5–6.7. The slopes (b values) for Si_3N_4 at two different loads are nearly parallel (Fig. 6.5), implying limited, or no, effect of load on the P-dependence. However, Mukhopadhyay et al. (23) showed the Meyer parameter (the slope of a log-log plot of indent load versus indent diagonal) decreased from ~ 1.9 at P = 0 to ~ 1.8 at P = 0.1, i.e., the load dependence increased with increasing porosity, i.e., reflecting some of the common load dependence of hardness. This is approximately consistent with a value of ~ 1.4 from Thompson and Pratt (19) (mainly for P = 0.2–0.4). Thus, load can have some effect on H – P relations, but is not a dominant factor. Similarly, some aspects of the effects of phase content are illustrated in Fig. 6.5, e.g., providing some insight to the scatter of limited data, i.e., data for all α-Si_3N_4 generally lying above trend lines for data of mixed or all β material.

Consider now other aspects such as rollovers and P_C values. Except for one combined Si_3N_4 data set, no data set in Figs. 6.4 and 6.5 shows the rollover toward P_C values; however, these data sets do not extend to or beyond the limits of expected linearity on semilog plots, e.g., the glassy carbon data following the spherical pore model only extends to about the limit of expected linearity of

P ~ 0.7 (Figs. 2.11 and 3.1). Two studies of hardness clearly rolling over toward P_C values are shown in Fig. 6.6. The b values of ~ 2.5–4.5 prior to the rollovers (again generally consistent with those expected for die-pressing) are reasonably consistent with probable P_c values greater than or equal to 0.5. (These cases also again show that complications of changing phase composition are not dominant.) The ZrO_2-Y_2O_3 (42,43) data also reflects substantial decrease in hardness, and some possible increase in slope (b) and especially some decrease in P_C values with expected cracking from increased monoclinic content (especially in view of the larger grain size, mainly $\geq$ 20 μm) (43). In contrast to this note that data of Cottom and Mayo (44) for sintered TZP with P = 0 to 0.1 gives b ~ 6.2. What effect, if any, that the very fine microstructure (e.g., 55–390 nm grains) has on the higher b value is not known.

Two further points should be noted about H-P results. First, there has been at least one study measuring the porosity dependences of several properties on the same set of samples (Fig. 6.7). This shows general consistency of porosity dependences of hardness and most, but not all, other related mechanical properties. The differences are understandable as discussed below. Second, data from metals is consistent with data from ceramics, as discussed in more detail in Chapter 8. Thus, Salak et al. (46) showed that their 701 data points for sintered iron were fit well by the exponential relation, giving a b value of ~ 4.9. i.e., similar to, but higher than, the value of 4.3 they obtained for tensile strength (Section 5.3.1). Again note that this and other data presented earlier shows the overall H-P trends are consistent with MSA model trends, but may have somewhat higher b values than other related mechanical properties.

Finally, note the effects of porosity and especially of microcracking on the formation and character of indents in ceramics. Rice et al. (47) showed that secondary cracking, i.e., in addition or instead of the larger, nearly linear cracks extending from the sharp corners of the indent, often occurred around Vickers and Knoop indents in dense ceramics. Such cracking was most extensive when the indentation and grain size were about the same and was often of a spalling nature, commonly along grain boundaries. Such cracking appeared to correlate with boundary impurities and stresses from elastic and thermal expansion anisotropy.

More extensive observations and analysis have been made of Hertzian indents in ceramics, with particular pertinence to the present interests for recent observations of Lawn and colleagues (48–50) on effects of distributed microcracking associated with microstructural stresses on such indents. Such microcracking significantly aids deformation to accommodate the indent, hence limiting or eliminating formation of larger scale Hertzian cracks. Thus, while crystal transformation and twinning played important roles in MgO-PSZ, so did microcracking, with the relative roles of these shifting with heat treatment and microstructure (48). Similarly, microcracking plays a dominant role in some other bulk ceramics (e.g., larger grain Si_3N_4) and ceramic composites (e.g., crystallized glasses, especially those forming micatious crystallites) (49), as well as alumina plasma-sprayed coatings (50). While the modest porosity in the coatings plays a role in the indent deformation, microcracks associated with the laminar structure and the grain structure appear to be more important.

6.3.2 Compressive Strength

Turning to compressive strengths, there are six sets of data reflecting mainly pores close to idealized shapes. First is data of Weiss et al. (35,51) for Al_2O_3 bars (~ 2.5-cm lateral dimensions) with large (~ 0.25-cm diameter) cylindrical pores aligned with the compressive axis. Compressive strengths initially decrease more rapidly as the number of the axial, cylindrical pores increased from 0 to 7, then decreased more slowly with further increases in the number of the cylindrical pores, and hence so did P (Fig. 6.8). The latter decrease is close

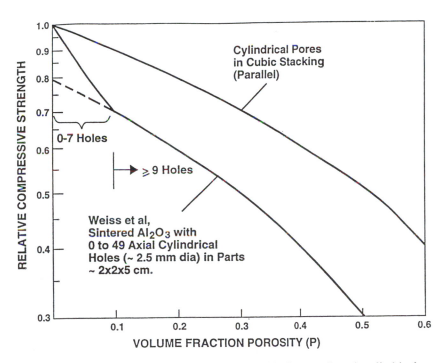

Figure 6.8 Compressive strength of Al_2O_3 with large, aligned cylindrical pores stressed parallel to their axes at 22°C (35,51). Note the faster initial decrease of compressive strength as the cylindrical pores are first introduced (besides the natural porosity of sintered bodies), then a slower decrease, nearly identical to that predicted by the pertinent MSA model (Figs. 2.10, 3.1). Published with permission of J. Mat. Sci. (35)

to that predicted by the MSA model (i.e., compare to Figs. 2.11 and 3.1), but somewhat greater, which is consistent with the presence of residual porosity of ~ 8 % from incomplete sintering. Further, note that the greater initial decrease indicates effects of heterogeneity, since even uniform distribution of a few cylindrical pores on a large scale does not behave like a homogeneous pore distribution on a finer scale; however, note that the initially higher slope (b ~ 3.7) reflects a reasonable average between that for stress parallel to aligned cylindrical pores (b ~ 1.4) and that for the residual porosity between incompletely sintered grains (b ~ 6). Failure from isolated cylindrical pores stressed normal to their axes is discussed below.

The second case is data of Trostel (52) for the (compressive) crush strength of Al_2O_3 and ZrO_2 sintered with 0 to 75 volume percent of Al_2O_3 and ZrO_2 respectively balloons (~ 2.5-mm diameter) to increase P. This data gives higher b values of 5 and 6.4 which is consistent with pores between sintering grains, but continue these values to higher P due to the increasing introduction of the balloons, consistent with the increasing mixture of pores from the balloons with those between the incompletely sintered grains (Fig. 6.9). The third case is data of Krauslin et al. (33) for bodies made of partially sintered ZrO_2 beads (20 to 200 μm diameter Fig. 3.9A). While not being adequately characterized to allow detailed quantitative evaluation, results agree semiquantitatively with expectations from combined MSA models for spherical pores and pores between sintering particles. Thus, they showed similar, initially higher values of Young's modulus and compressive strengths at smaller balloon diameters, but steeper decreases in these as the diameters (hence also P) increased compared to tensile loading. Also, as noted earlier they showed Young's modulus measured in compression being higher at smaller balloon size, but decreasing much faster with increasing balloon size than in tension. This may indicate some effects of increased contacts between smaller balloons in compression versus tension.

The fourth case is that of oriented spheroidal pores of controlled shape and orientation (53,54) in hydroxyapatite. Bodies were fabricated with dispersed plastic particles (0.42-mm diameter) that were subsequently burned out to give spherical pores or first deformed by warm pressing to give oriented pores on burn out. Pores from the plastic particles were 60–70% of the net porosity in the bodies. Specimens cut so the subsequent compressive stress axis was normal to the elongated pore axis were designated as oblate and those with the long axis and the stress axis parallel were designated as ellipsoidal. Their results were that the compressive strengths decreased with increasing pore size for all three pore geometries. The absolute strengths for a given porosity level were, in order of decreasing strength: 1) spherical, 2), ellipsoidal and 3) oblate (the latter two are elongated respectively parallel and perpendicular to the stress axis; Fig. 6.10), i.e., generally consistent with MSA concepts.

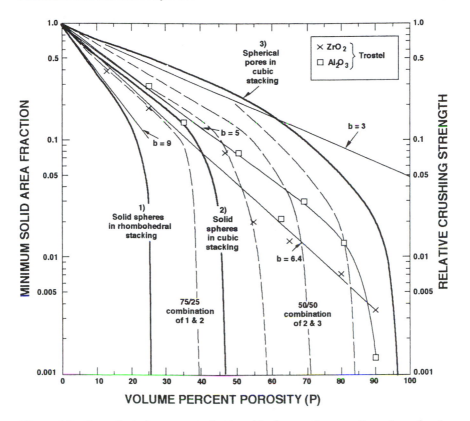

Figure 6.9   Log of relative compressive (crushing) strength versus the volume fraction porosity (P) of Al_2O_3 and ZrO_2 sintered with 0–75 volume percent admixtures of balloons of the corresponding composition used to increase P at 22°C after Trostel (35,52). Note that the increasing addition of balloons significantly extends the range of linear behavior on the semilog plot as expected from combination of the MSA model for spherical pores and for normal pores between the sintering powder particles. Published with permission of J. Mat. Sci. (35)

The fifth case of approximately ideal spherical pores is data of Nakashima et al. (55) for porous glass specimens made by leaching of phase-separated bodies with subsequent heat treatment. They showed that glasses with nominally 0.44 and 2.56 μm diameter pores both had the same compressive strength versus P, with the linear strength decrease on a semilog plot versus P giving a slope (b value) of 2.85 to a high degree of accuracy to the limits of their P (= 0.54), i.e., in excellent agreement with MSA expectations for spherical pores (which the pores should become with increased heat treatment). These investigators also

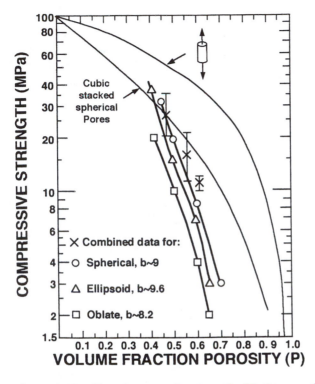

Figure 6.10 Plot of compressive strength of hydroxyapatite with substantial levels of spherical pores or spheroidal pores oriented with their long axes parallel (spheroidal) or perpendicular (oblate) to the compressive axis [after Liu (53,54)]. Combined data for all three pore types from separate measurements showing strengths for all three pore shapes decreasing as the pore diameter increased are also shown (x). Deviations below the two MSA models shown are attributed primarily to the substantial interparticle porosity (30–40% of the total P), and secondarily the lack of pore stacking characterization.

determined the Weibull modulus of their porous glass (P~ 0.5), obtaining m = 23.6 versus 2.56 for dense glass.

The sixth case of nearly ideally shaped pores is data for glassy carbon (Fig. 6.11). This material, where the porosity is expected to be predominantly spherical (Fig. 1.7A), gives a b value of ~ 3.3, modestly higher than for random packing of uniform spherical pores (b ~ 2.8), and in excellent agreement with the hardness data for the same material (Fig. 6.4).

Much of the limited compressive strength data for bodies with natural porosity is shown in Figures 6.11 and 6.12, along with data in Table 6.3. The earlier survey by Rice (1) indicated that compressive strength, like hardness and wear, had greater porosity dependence, e.g., b values of 6 ± 3 versus 4 ± 2 for elastic

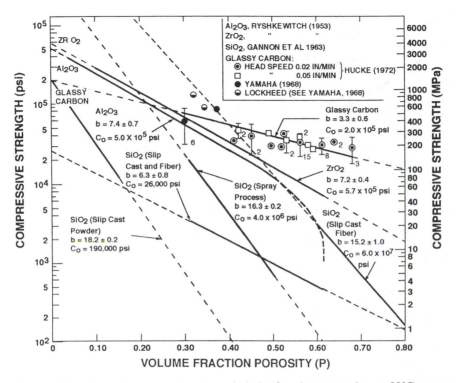

Figure 6.11 Log of compressive strength (σ_c) of various ceramics at 22°C versus volume fraction porosity (P), some from an earlier survey (1) ($C_0 = \sigma_c$ at P = 0). Note the: 1) glassy carbon data very similar to that for hardness (Fig. 6.4), 2) diversity of σ_c-P trends amongst these few materials, especially amongst the different SiO_2 bodies tested and fabricated (by differing processes from different raw materials) by the same investigators, again showing that processing and pore character is important, and 3) very high b values and P = 0 projections where data is only at higher P, indicating that the data reflects trends at and beyond the rollover toward P_C values.

moduli and tensile strength, but the more recent survey (36) (Table 6.1) questioned the extent of this, similar to the questioning of higher b values for H-P trends. Inclusion of more recently found data with identified processing brings the number of identified studies using hot pressing to five, giving an average b of 5.2 ± 1.6 and three sintered bodies giving an average b of 4.1 ± 2.4. Though there are uncertainties due to the limited number of tests, frequent lack of processing details, and substantial scatter (probably reflecting, in part, use of coarse starting raw materials and larger grains), these averages suggest similar results to those for hardness, namely that it generally follows MSA models but that there may be some increase in b values for compressive strength. This

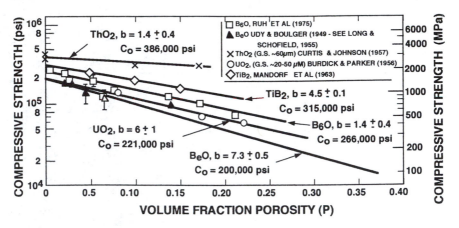

Figure 6.12 Log compressive strength for various ceramics versus P at 22°C. Published with permission of Academic Press (1).

implication is reinforced by the similar trends for hardness and the H-σ_c correlation, and especially by data for bodies with much of the pores being spherical giving results close to those for the corresponding MSA model and elastic behavior. Also note similar trends of cement materials that increase mass as P decreases follow similar trends as those for sintered ceramics which have constant mass as P varies (Fig. 8.12).

While high strength compressive failures, i.e., of dense samples, are often somewhat explosive, in properly conducted tests it typically occurs by progressive damage accumulation from a spatial distribution of sources, as opposed to failure in tension from a single flaw. Thus, the use of fractography is generally far more restricted for compressive failures. There are, however, cases of compressive loading that give insight into probable mechanisms in more general compressive loading. The first of these is the study of Sammis and Ashby (3). Although they did not confirm their complete model in limited tests, they corroborated the overall crack-growth behavior expected for the pores analyzed in tests of both glass and a brittle plastic (PMMA) with artificially introduced cylindrical holes with simple uniaxial compressive loading. Thus, while early crack growth was quite variable, it became much more systematic and consistent with their modeling as crack lengths increased. Such damage in compressive loading was corroborated in RSSC but not hot-pressed Si_3N_4 (6). The difference is probably due to the residual porosity and second-phase and larger microstructural scale of these and related features in the RSSC as compared to the much greater uniformity and finer microstructural scale of hot-pressed Si_3N_4. Other studies experimentally corroborate microcracking via acoustic emission as well as crack nucleation by slip and twinning (4,28,56).

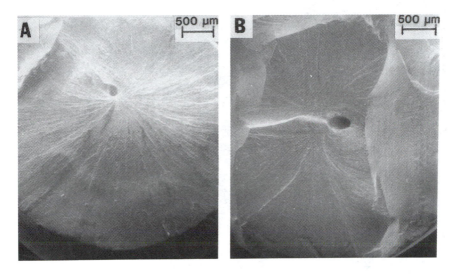

Figure 6.13 Examples of failure of small, experimental ceramic (ZrO_2-toughened alumina) balls from internal pores. The failures occurred during grinding them into bearing balls where multiple balls are simultaneously ground between parallel rotating plates, so failures are attributed to the approximate point impact, nominally axial loading. Note that the resultant tensile failure commonly occurred from one or two larger pores, often near the center of the isopressed balls, and is probably indicative of effects of pores in more general compressive loading.

Failure of individual spheres and cylinders under compressive loading normal (instead of parallel) to the cylinder axes are by themselves of significant technological importance, e.g., in ceramic bearings and milling balls. Figure 6.13 shows failure of small experimental ceramic ball bearings during grinding occurring from internal pores, often near the center of the balls. Such pores, when present, were very frequent fracture origins in such grinding which applies compressive loads between two rotating plates between which the balls are placed under some compressive loading. The stressing and failure is thus essentially that of diametral compression. Also, in rolling-contact fatigue testing for ball-bearing applications (of either balls or races) three steel balls are pressed to high radial-contact stresses about a central rod (~ 1.5-cm diameter) rotated at high speed. Good candidate materials fail at high loads after long times by limited, progressive spalling, as opposed to larger scale, catastrophic failure, i.e., allowing the start of failure to be detected by vibration and hence generally allowing the prevention of catastrophic failure. Microstructural heterogeneities are typically found as the origins of spalls, with isolated pores or pore clusters being common sources, especially in less developed materials (e.g., Fig. 6.14).

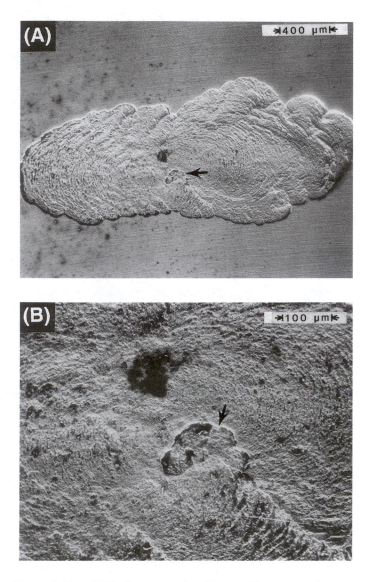

Figure 6.14 SEM micrograph of spall in a rolling-contact fatigue test (i.e., from rotation of the test rod while under a high radial compressive load applied by three balls symmetrically placed around a circumference of the rod) of a TZP: (A) complete spall; (B) higher magnification showing origin of spall from pore (approximately at the photo center).

Note that the substantial literature on powder compaction (57–59), although too extensive and specialized to be addressed in any detail here, provides additional information on compressive behavior of porous ceramics, as do such studies of hot pressing (briefly noted in Chapter 9). These include pressing to full density (57,58), even at room temperature [e.g., MgO at ~ 60,000 atm, ~ 6.2 GPa (58)]. Similarly studies of using unsintered or lightly sintered ceramic powders as high-pressure transmitting media can also be of value (e.g., ref. 59). Also note that the limited data clearly shows that pores can lead to locally enhanced slip when bodies containing them are loaded in compression or hydrostatically (60).

Next, consider the limited but useful data on compressive strength of highly porous foam or extruded cellular and related materials, starting with failure of individual, compressively loaded, sintered ceramic balloons. Chung and colleagues (12,13) corroborated the $(1 - P)^2$ dependence of Eq. 6.4, finding A to be ~ 0.252 for sintered alumina, i.e., about 1/3 of the value (~ 0.774) predicted from their finite element analysis. This difference in A values is attributed to variations in balloon geometry and quality. Tests of specimens of bonded balloons of sintered alumina or mullite, while showing some differences from their bilinear relation (Eq. 6.6), also provided some support for their relation. Again, note the work of Krasulin et al. (33) on ZrO_2 beads (Fig. 3.8A) and Liu (53,54) on apatite with spheroidal pores (Fig. 6.10), both showing strength decreasing with increasing pore size, and the former showing higher, but more rapidly decreasing strengths (and E) as pore size increased in compressive versus tensile loading.

Gulati and Schneider (10) reported axial compressive strengths of extruded square-cell cordierite honeycomb as a function of cell-wall P as e^{-5P} over the range considered (P~ 0.25–0.55). There is, however, some uncertainty in comparing this b value of five to other data since it is based on P starting at significantly >0. More generally they corroborated the equations of Fig. 6.2 by showing that σ_n versus σ_a and σ_d versus σ_n were straight lines (with respective slopes of 0.334 and 0.063). They also corroborated that the isostatic compressive strength was directly proportional to σ_n.

Dam et al. (61) studied compressive failure of sintered mullite open-cell foams (cell sizes ~ 1, 3.5, 4, and 4.5 mm; P ~ 0.83–0.88). Their results gave A and n of σ_c/σ_{c0} ~A $(1 - P)^n$ ~ 0.3 and 2.2 versus predicted values of 0.65 and 1.5, respectively. They argued that correction for the tubular pore in the struts (as done in similar studies of E and K_{IC}; see Sections 3.3 and 4.3) was appropriate, giving n ~ 1.7. i.e., much closer to the prediction of 1.5 of the model of Gibson and Ashby (7,8); however, they observed a modest but definite increase in both Young's modulus and compressive strength with increasing cell size, which is inconsistent with the model (and opposite the trends noted above for bonded ZrO_2 beads). They cited strut defects that are often found in these type of ceramic foams as a probable source of these differences. They also noted that the overall failure sequence of elastic deformation—crushing (e.g., layer by layer normal to the compressive stress), then crushed debris consolidation, i.e., densification, proposed by Maiti et al. (9)—was generally valid, but that there could be significant deviations in the specifics.

Brezny and Green (62) subsequently studied compressive and tensile behavior in a sintered alumina-mullite foam (P ~ 0.79 to 0.91, cell size ~ 2.25 ± 0.15 mm) and a glassy carbon foam (P = 0.96, cell size ~ 0.4 and 1.2 mm) using acoustic emission in addition to stress-strain and microscopic analysis. They showed that compressive strengths were substantially higher than tensile strengths, e.g., by a factor of 4–5 in the sintered foam, and that while tensile failure occurred by catastrophic propagation of a single flaw in tension, compressive failure was more distributed. In the sintered alumina-mullite foam the failure involved progressive crushing of layers due to random failure of struts and coalescence of the damage zone, then the consolidation of the debris, similar to the failure stages noted earlier. In the glassy carbon foam they found compressive failure occurred by the catastrophic collapse of a single layer of cells normal to the stress, i.e., suggesting similarities and differences from tensile failure. Thus, it appears that this material was attempting to follow the expected compressive failure mode, but was precluded from fully following this due to the much more rapid onset and propagation, and the greater localization, of compressive failure, i.e., in some respects a compromise between tensile and compressive failure modes. They suggested that this significant difference was due to a higher, narrower range of Weibull moduli of the struts in the glassy carbon that did not allow more gradual, progressive failure; however, other differences between the two materials must also be considered, namely the much smaller cell size (2–6 fold), lower relative density (2–5 fold), and lower fracture toughness of the struts of the glassy carbon. Results of Kurauchi et al. (63) showed the expected stages of elastic compression then the collapse of layers of cells, and then the consolidation of debris in two silicate glass foams to approach (to within 50–400%) and overlap with the glassy carbon foam in relative density and cell size respectively (and possibly in strut toughness); these results indicate that the extreme porosity of the glassy carbon foam may be a factor. Thus, at very high porosity (and possibly aided by lower toughness of struts), the progressive compressive failure mode, often layer by layer, may be precluded.

6.3.3 Wear and Related Resistance

Consider now data on the related topics of the resistance to wear and machining as a function of P, which have received very limited study, e.g., since porosity is typically quite detrimental to wear resistance. Available data is summarized in Table 6.3 and Figures 6.7 and 6.15. Two points should be noted. First, while

Table 6.3   Porosity Dependence of the Resistance of Ceramics to Wear and Machining

Material[a]	Grain size[b] (μm)	Porosity (P)	Test[c]	b[d]	Investigator
Al_2O_3	1	0.02–0.30	Scratch hardness	3 (11)[e]	Wu and Rice (45)
Al_2O_3 (85–97)	5	0.02–0.09	Milling ball (~3 cm) wear	20	Pearson et al. (64)
Al_2O_3	3	0.01–0.10	Milling ball (~3 cm) wear	23[f]	Pearson et al. (64)
Al_2O_3	1	0–0.16	Sawing	6 (0.80)	Rice and Speronello (31)[g]
Al_2O_3)	1	0–0.16	Grinding	4 (0.97)	Rice and Speronello (31)[g]
Al_2O_3	—	5 to 15	Abrasion	9	Hines (65)
Al_2O_3	1	0–0.40	Taber abrader	19 (3.2)	Wu and Rice (45)
MgO	1	0–0.14	Sawing	0.6 (0.14)	Rice and Speronello (31)[g]
B_4C		0–0.15	Scratch hardness	6.6[e]	Wu and Rice (45)
TiC		3–0.20	Unspecified wear	4.8	Artamonov and Bovkum (66)
ZrC		0–0.12	Unspecified wear	1.9	Artamonov and Bovkum (66)
WC		2–0.12	Unspecified wear	4.2	Artamonov and Bovkum (66)

[a] Material purity shown where known to be < 99%.
[b] Approximate average grain size.
[c] Scratch hardness test with diamond point; sawing and grinding with typical diamond tooling.
[d] Value of the exponential factor b.
[e] Numbers in ( ) > 1 are for the slope at higher P (see Fig. 6.7) and < 1 are correlation coefficients for the value preceding it; and superscript a designates the exponent determined by the originator of the data.
[f] Values determined by the original investigator.
[g] See Fig. 6.15.

there is substantial scatter that most likely reflects both porosity heterogeneity and probable greater sensitivity of wear, that the P-dependence, as measured by the b (slope) values, ranges from that for elastic moduli to much higher levels. (Note that the low b value for sawing MgO is quite uncertain and could be higher, e.g., ~ 4; Fig. 6.15.) Second, note that data of Wu and Rice (45) for both scratch hardness and abrasion testing of their hot-pressed Al_2O_3 bodies has two distinct slopes, with the slope changes occurring for both at P ~ 0.2 (Fig. 6.7). The second slope does not reflect the decrease toward P_C values since it occurs at too low a P and is inconsistent with other P data for other properties for the same bodies, and the change for the abrasion test is to a lower slope. Wu and Rice noted that the higher scratch-hardness slope, which commenced at P ~ 0.2

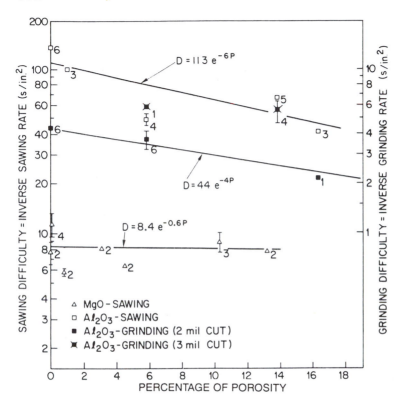

Figure 6.15 Sawing and grinding difficulty (i.e., the inverse of the rates) versus volume fraction porosity for hot-pressed Al_2O_3 and MgO [after Rice and Speronello (31)]. Least-squares lines are fitted, from top to bottom, at correlation coefficients of 0.80, 0.97, and 0.14, respectively. Published with permission of the Am. Cer. Soc.

was associated with the disappearance of most porosity immediately under the track of the conical diamond pin. The higher slope and hence increased scratch-track dimensions is attributed to greater penetration of the point into the sample, e.g., due to crushing, and hence consistent with the fact that the onset of the higher slope tended to decrease with load over the 0.4–2 kg range used. This is consistent with the absence of most porosity under the track and the absence of dislocation activity as well as more cracking on the track surface at higher P, i.e., limited plasticity. It is also consistent with the corresponding but opposite slope change in the abrasion test where the measure of wear is weight loss. Thus, crushing wear debris into the specimen surface porosity would reduce the weight loss at higher loads versus more weight loss by abrasion and no crushing.

Two important aspects of wear, as with other mechanical properties, are effects of low porosity and the interaction of porosity with the grain structure. Consider first the case of wear measurements at low porosity. He et al. (67) measured wear via a pin-on-plate test using a 4-mm diameter ball SiC recipro- cally moving unlubricated over a track length of 10 mm under a load of 8 N (858 MPa) with an average sliding velocity of 0.08 m/s in a dry nitrogen atmo- sphere. The plate sample was a very fine (180 nm) grain TZP made by sintering or sinter forging with a highly polished and annealed surface and volume frac- tion porosities of 1.5, 3, 5, and 7 %. Wear, as measured by weight loss, linearly increased fivefold over this range of porosity. Wear surface examination showed increased friction due to increased plastic deformation as P decreased. On the other hand, cracking transverse to the wear track increasingly occurred as P increased, increasing surface roughness. There was no transformation to monoclinic zirconia as a result of wear despite the differences in response as a function of porosity.

Consider next the changing intergranular to intragranular pore locations often associated with increased sintering due to grain growth as P decreases. This results in some pores being enveloped in the grains (becoming approxi- mately spherical or polyhedrahl; Fig. 1.4) and those remaining at grain bound- aries often becoming larger, but more cusp and often lenticular shaped (Fig. 1.3). These changes could reduce wear resistance since it typically decreases with increased grain size, and the increases boundary area of the intergranular pores may counteract probable benefits of small reductions in porosity and some possible wear reductions due to the intragranular pores; In practice, however, results are more complicated. For example many 99% commercial alumina products are sintered with MgO additions to control grain size (e.g., to ~ 8 μm) to give better mechanical performance, including wear resistance for various applications; however, for some wear applications better performance is found by leaving out the MgO, which about doubles the grain size from the same or similar firing with MgO additions. The resulting larger grain bodies of very similar density as those sintered with MgO have better wear resistance in some tests (e.g., rotating specimens in a water-sand slurry) and applications, but worse performance than its finer grain counterpart in other applications and tests, e.g., of grit blast resistance. A possible explanation for this better performance of the larger grain material may be along the lines suggested by Rice for wear (68) and earlier for benefits of machined strengths of larger grain MgO with limited intragranular pores (69). Thus, as sketched in Fig. 6.16, intragranular pores may limit the lengths of slip bands, twins, or cracks, and hence the extent of resultant cracking and spalling relative to those with only intergranular pores. Thus, if such effects occur, they may more than compensate for counter effects of limited increases of grain size; however, these results are a clear indication of

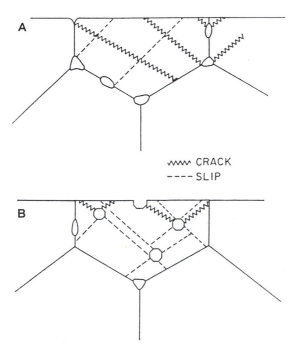

Figure 6.16 Schematic of possible effects of only intergranular pores on effects of machining and wear of cracking and surface spalling (A) versus possible effects with some intragranular pores replacing some intergranular pores (B). The latter can result in shorter slip (or twin) and distances, hence requiring higher stresses for resultant fracture initiation, along with smaller resultant spalls from such cracks as well as from cracks directly generated from the pores. Published with permission of J. Am. Cer. Soc. (69)

the complexities of microstructural interactions in wear, i.e., similar to those, but possibly even more complex than, for other properties (Chapters 4,5).

There apparently have not been any studies of erosion as a function of porosity, again in part because of the significant increase in erosion expected with porosity. Gulden's study (70) showing substantially higher (e.g., 5–10 fold) erosion of two RSSN bodies compared with other dense Si_3N_4 bodies indicates substantial effects of P ~ 0.2–0.3; however, porosity can provide beneficial protection of material surfaces from impact or erosion damage due to the above noted crushing action. Thus, Rice (71) showed that "fortified" Pyroceram® was far more resistant to surface damage and attendant strength loss than the same material with machined surfaces (Fig. 6.17). The fortification process is one of leaching much of the cristobalite out of the surface (71–73), leaving a substantial amount of fine pores (P ~ 0.3, 0.1–0.4 μm diameter) to substantial depths (e.g., ~ 0.6 mm). The resulting porous layer grades into, is well bonded to, and

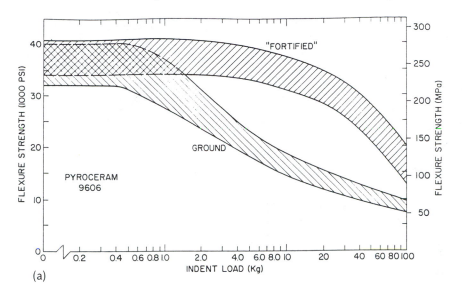

(a)

Figure 6.17   Summary of the effects of indentation damage on Pyroceram 9606® (a crystallized glass containing substantial coordierite) specimens with and without a porous surface layer, after Rice (71). a) Flexure strength at 22°C as a function of the load used to make a Vickers indent in the center of the tensile surface for material with a diamond-ground surface or for "fortified" surfaces (i.e., leached to remove much of the residual cristobalite). Note the much greater damage resistance of the latter surface due to crushing of the porous surface protecting it. b–d) Fracture cross-sections of specimens. b) A machined, unfortified surface showing fracture initiation from a median crack introduced by the indent (load = 50 kg, ~ 500 N). c) A fortified, unindented surface showing the lack of a crack in the unleached bulk material beneath the "fortified" i.e., unleached, surface. d) A fortified, indented surface showing that the failure initiation in the unfortified material is not directly above the indent, showing limited effect of the indent. I = indent, F designates "fortified" (lighter) layer. Published with permission of the Am. Cer. Soc.

quite compatible with the bulk material. Concentrated loads, e.g., from indents, on the surface simply result in crushing of the porous surface with little or no subsurface damage unless the porous layer is nearly, or fully, penetrated (Fig. 6.17). Similarly, Brennan (74) showed that the failure from ballistic impact (of a hardened chrome-steel pellet, 4.4 mm diameter, 0.34 g, fired from a modified air pellet gun) of dense Si_3N_4 was increased by of the order of fivefold from about 1.9–2.8 joules (1.4–2.1 ft-lbs) at both 22 and 1375°C by forming a layer of RSSN ~ 1 mm thick on the surface to be impacted. By controlling the processing-microstructure of the RSSN layer, only limited reduction of the flexural strength of coated samples occurred. Proper preparation of the porous coating also improved the Charpy impact resistance two- to threefold at both test

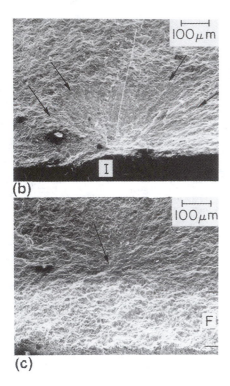

Figure 6.17   Continued

temperatures. (Note that porous surface layers can also favorably influence thermal stress performance, e.g., of Al_2O_3 (75); see Section 9.3.2). While crack branching and bridging in these porous layers may be a factor in the protection of the material underneath these layers, distribution of energy and load via crushing is probably much more important.

Whitney (76) measured the maximum cutting speed allowable for a tool life of 10 minutes in cutting 4340 steel with sintered 99+% alumina cutting tools. Their limited data (P ~ 0.07–0.22) shows an initial approximately linear decrease on a semilog plot (b ~ 1.6), then decreasing more rapidly above ~ P = 0.18.

There is also only limited data for the resistance of ceramics to ballistic impact. Ferguson and Rice (77,78) reported on the ballistic limit velocity (i.e., the projectile velocity above which the projectile will start to penetrate the target) of 22 caliber bullets used as fragment simulators in testing comercial and laboratory-prepared alumina bodies. They showed that this velocity decreased substantially, but linearly, as P increased from ~ 0 to ~ 0.05. Results were similar for two to three different scales of testing and whether done in one or different laboratories. Data ranged between two bounds about 200–400 ft/sec apart, apparently depending on purity and grain size factors.

6.4 DISCUSSION

Since compressive strength typically involves failure from cumulative damage, it is less susceptible to effects of heterogeneities than tensile strength, but it is not free from them. Again, porosity location in polycrystalline bodies, which is interrelated with grain size and growth, can also be a factor, as with fracture toughness, tensile strength, and hardness. Thus, pores in grains (mainly at moderate-to-large grain size) will often play a limited role in compressive failure if there is also substantial intergranular porosity, as is typically the case. Furthermore, grain boundary porosity is an increasing factor in intergranular fracture as temperature increases (Chapter 9). Finally, note that crushing and then consolidation occurs in compressive loading of more porous materials, e.g., lightly sintered powder masses and foams which is consistent with similar phenomena seen in wear and hardness tests.

While not studied, hardness would appear to be quite susceptible to effects of pores, or porosity heterogeneities, especially on a scale similar to that of the indent. Thus, indents adjacent to or over pores, or clusters of them whose sizes are a reasonable fraction of the indent size are likely to be irregular. Such effects can also be impacted by grain structure and pore location, i.e., the combination of grain-boundary pores and larger grains can be an important source of spalling around indents that also commonly results in their being discarded as "bad" indents. Since such areas often represent areas of higher P in the body, excluding such data is likely to bias results to a somewhat lower P-dependence. While numerical results from such irregular indents are uncertain (techniques to evaluate them should be investigated), at the minimum they can be used as a probe of porosity heterogeneity; however, the normal nonlinearity of H-P–dependence commonly has more impact on the porosity dependence of hardness than discarding "bad" indents. This and crushing of the underlying porous material may enhance the H decrease over that of E as P increases.

While the diversity of wear and erosion mechanisms and the sparse pore characterization and data are a limitation, it seems clear that the scale of the asperities, impact particles, and craters relative to the sizes of pores or pore clusters can be important in these phenomena. This is in part supported by some wear-P trends and direct evidence indicating crushing under indentations of the surface by scratch hardness points or asperities on mating wear surfaces. Similarly, grain-boundary pores and larger grains or other microstructural features should also often be factors as for spalling with hardness indents and compressive failure.

The limited data means that detailed comparison of model fits to data is restricted; however, it is clear that general MSA models as well as the related foam models of Gibson and Ashby are both useful, but both require modification to more accurately fit data. Currently modifications are empirical, but could ultimately be more basic. A critical factor is accounting for the variability of pore geometry and material nonuniformity. The indication that the coefficients for the models for foams and bonded ceramic balloons are often high by a factor of ~3 as a general correction factor (e.g., possibly reflecting statistics of defects in foam struts and balloon walls) is suggestive, but requires more evaluation.

Next, note some key aspects of self-consistency, evaluation of which, though important, has been very limited. Correlations of Young's modulus and hardness (and hence also of compressive strength, wear, and related behavior) were corroborated, showing generally consistent trends as well as some deviations. An important source of the deviations are effects of crushing indicated by property-P trends and corroborated by experimental observations in some hardness, wear, and compression tests (the latter in foams). The occurrence of such phenomena in different tests and materials (e.g., Fig. 6.7), including porous coatings, is another indication of self-consistency, but much more such evaluation is needed.

While there is broad overall similarity in trends of hardness, compressive strength, and wear as a function of porosity to those of elastic moduli and tensile strength, there are also differences. As noted in Section 6.2 compressive loading can entail more pushing together and less separation, while tensile loading entails the reverse. The various crushing and compaction noted above, as well as consolidation of powders are all manifestations of this. A key question is whether these are factors in indicated higher b values for hardness, compressive strength, and wear versus those in tension, e.g., for Young's modulus and tensile (or flexure) strength; however, as noted earlier, much of the earlier ~ 50% higher b average for compressive versus tensile loading is now seen as being due to more hot-pressed bodies evaluated for compressive versus tensile properties and the (then unrecognized) higher b values for hot-pressed material. Furthermore, the same or similar b values predicted irrespective of stress direction are observed, e.g., for glassy carbon (Figs. 6.4, 6.11) and silicate glasses with bubbles or balloons (Chapters 3 and 5); however, other cases of higher b values are seen in compressive versus tensile stressing, e.g., as reflected in averages of Table 6.1. The two known cases of tensile and compressive strengths measured on the same bodies by the same investigator give mixed results. Thus, Schiller (1,79) studies of plaster (P ~ 0.42–0.59), which showed compressive strengths ~ 2.5 times flexure strengths, and the latter ~ 3.5 times those in tension, with all three extrapolating to P_C ~ 0.87, gave parallel semilog plots versus P with b ~ 4. In contrast to the same b values in tension and compression,

Knudsen (80) obtained b values for ThO_2 at 22°C of 4.2 in flexure and 6.6 in compression (and b = 6.6 in flexure at 1000°C). Both these specific cases and more general data thus indicate that compressive loading can give higher b values than tensile loading, but that this difference depends on the specifics of the material, possibly the test, and especially the microstructures, in an as yet not completely understood fashions.

Finally, while there is little or no data on effects of microcracks on the properties considered in this chapter, they can be important. It is clear that extensive local microcracking can greatly reduce or eliminate the formation of larger, generally more serious, Hertzian cracks, and thus provide possibly useful protection against such damage; however, the effects of preexisting or stress-generated microcracks on other properties is a serious concern, e.g., one reason for lack of data on microcracks on properties such as wear and erosion is that they can be very detrimental to such properties.

6.5   SUMMARY

The porosity dependences of hardness, compressive strength, and wear and related behavior such as erosion and machining clearly depends on both the amount and character of the porosity, as for other mechanical properties. This dependence on pore character is generally consistent with MSA and related foam models, but further improvements are needed. A key need is to begin to account for variations in the pore structure and body quality, and differences that can occur between tensile and compressive loading for some microstructures (and possibly materials and tests). Some of the self-consistency expected from E-H, H-σ_c, H-W, and so on, correlations is found, but deviations and variations in these correlations often occur, probably generally increasing in the order listed. Some of these deviations and variations may reflect differing contributions of pore shape-stress concentrations, adding to and hence increasing the porosity dependence; however, much of these are due to varying interactions with porosity heterogeneities or with larger pores. Thus, distorted or chipped hardness, i.e., "bad" indents, due to interactions with larger pores or pore clusters will commonly be discarded rather than used, e.g., as a sign or measure of the porosity heterogeneity. Plotting of the remaining data, i.e., of "good" indents, will thus often reflect a less than average P giving a lower apparent P dependence; however, this will often be more than counteracted by greater impact of "good" indents in more porous areas due to increasing nonlinear decreases of hardness with increasing P. Effects of heterogeneous porosity on compressive strength are less evident, but are indicated by preferred failure of small balls under compressive loading (i.e., essentially under diametrical tensile loading), and by rolling-contact fatigue spalls often originating from pores or pore clusters (e.g., Fig. 6.14).

Wear, machining, and erosion show greater variation in part due to the diversity of mechanisms involved and probably greater effects of heterogeneity of the porosity. Mechanisms such as crushing and plowing, either locally or on a broader scale, are indicated as important factors, especially; 1) at medium to higher P, and 2) with larger pores or pore clusters relative to the sizes of surface penetrations. Crack branching or bridging may also be factors, but are likely to be restricted by the limited scale of cracks and their propagation, and would appear to decrease the P-dependence (i.e., as for γ and K_{IC} in Chapter 4), which appears to be contrary to greater P-dependence for most of these properties relative to other properties.

Use of well-bonded porous surface layers, preferably graded into the bulk, e.g., as from leaching a phase from the surface, can be very beneficial in increasing resistance to failure from gross or localized (i.e., particle) impact. This appears to result mainly from crushing of the surface layer, typically hundreds of microns thick with finer porosity, absorbing some of the energy and distributing the impact load.

Data on microcrack effects is limited, but it is clear that they can have complex beneficial, and especially detrimental effects that need much more documentation and understanding.

All of these variations and the diversity of phenomena and mechanisms present significant challenges to modeling, as well as to designing definitive experiments. Again, experiments comparing a number of related properties on the same set of materials would be of much greater value than the typical single-property studies. Finally, again note that pore character is important and thus cannot be neglected as is often the case. A particular example of this is the broad variety of compressive behavior of SiO_2 bodies made via different routes (Fig. 6.11).

REFERENCES

1. R. W. Rice, "Microstructural Dependance of Mechanical Behavior of Ceramics," Treatise on Materials Science and Technology, vol. 11: Properties and Microstructure (R. K. Mac Crone, Ed.), Academic Press, New York, pp. 191–381, 1977.
2. L. F. Nielsen, "Strength and Stiffness of Porous Materials," J. Am. Cer. Soc., **73**(9), pp. 2684–89, 1990.
3. C. G. Sammis and M. F. Ashby, "The Failure of Brittle Porous Solids Under Compressive Stress States," Acta Metall., **34**(3), pp. 511–26, 1986.
4. R. W. Rice, "The Compressive Strength Of Ceramics," Ceramics in Severe Environments (W. W. Kriegel and H. Palmour III, Eds.), Plenum Press, New York, pp. 195–227, 1971. Materials Science Research, vol. 5.

5. M. Adams and G. Sines, "Crack Extension from Flaws in a Brittle Material Subjected to Compression," Tectonophy., **49**, pp. 97–118, 1978.
6. G. Sines and T. Taira, "Tensile Strength Degradation of Ceramics from Prior Compression," J. Am. Cer. Soc., **72**(3), pp. 502–5, 1989.
7. L. J. Gibson and M. F. Ashby, "Cellular Solids, Structure & Properties," Pergamon Press, New York, 1988.
8. L. J. Gibson and M. F. Ashby, "The Mechanics of Three-Dimensional Cellular Materials," Proc. Roy. Soc. Lond. A **382**, pp. 43–59, 1982.
9. S. K. Matai, L. J. Gibson, and M. F. Ashby, "Deformation and Energy Absorption Diagrams for Cellular Solids," Acta. Metall, **32**, pp. 1963–75, 1984.
10. S. T. Gulati and G. Schneider, "Mechanical Strength of Cellular Ceramic Substrates," Proc. Ceramics for Environmental Protection, Enviceram '88, Cologne, Germany, 12/7–9/1988, Deutsche Keramische Gesellschaft e. V., Köln, Germany.
11. Y. Hiramatsu and Y. Oka, "Determination of the Tensile Strength of Rock by a Compression Test of an Irregular Test Piece:, Int. J. Rock Mech. Min., **3**, pp. 89–99, 1966.
12. J. H. Chung, J. K. Cochran, and K. J. Lee, "Compressive Mechanical Behavior of Hollow Ceramic Spheres," Mat. Res. Soc. Symp. Proc., **372**, pp. 179–86, 1995.
13. J. H. Chung, "Compressive Mechanical Behavior of Hollow Ceramic Spheres and Bonded-Sphere Foams," Ph.D. Thesis, Georgia Institute of Technology, Atlanta, GA, 6/1992.
14. P. W. Bratt, J. P. Cunnion, and B. D. Spivack, "Mechanical Testing of Glass Hollow Microspheres," Advances in Materials Characterization (D. R. Rossington, R. A. Condrate, and R. L. Snyder, Eds.), Plenum Press, New York, pp. 441–48, 1983.
15. S. R. Swanson and R. A. Cutler, "Fracture Analysis of Ceramic Proppants," J. Ener. Res. Tech., **105**, pp. 185–232, 1949.
16. P. Arato, E. Besenyei, A. Kele, and F. Weber, "Mechanical Properties in the Initial Stage of Sintering," J. Mat. Sci., **30**, pp. 1863–71, 1995.
17. I. J. McColm, "Ceramic Hardness," Plenum Press, New York, 1990.
18. J. H. Westbrook and H. Conrad (Eds.), "The Science of Hardness Testing and Its Research Applications," Am. Soc. for Metals, Metals Park, OH, 1973.
19. D. S. Thompson and P. L. Pratt, "The Mechanical Properties of Reaction-Sintered Silicon Nitride," Proc. Brit. Cer. Soc., **6**, pp. 37–47, 6/1966.
20. D. Chakraborty and J. Murherji, "Characterization of Silicon Nitride Single Crystals and Polycrystalline Reaction Sintered Silicon Nitride by Microhardness Measurements," J. Mat. Sci., **15**, pp. 3051–56, 1980.
21. C. Greskovich and G. E. Gazza, "Hardness of Dense α- and β-Si_3N_4 Ceramics," J. Mat. Sci. Letters, **4**, pp. 195–96, 1985.
22. J. J. Gilman, "Hardness-A Strength Probe," The Science of Hardness Testing and Its Research Applications (J. H. Westbrook and H. Conrad, Eds.), Am. Soc. for Metals, Metals Park, OH, pp. 51–74, 1973.
23. A. K. Mukhopadhyay, S. K. Dutta, and D. Chakrborty, "On the Microhardness of Silicon Nitride and Sialon Ceramics," J. Eur. Cer. Soc., **6**, pp. 303–11, 1990.
24. A. K. Mukhopadhyay, S. K. Dutta, and D. Chakrborty, "Hardness of Silicon Nitride and Sialon," Ceramics Intl., **17**, pp. 121–27, 1991
25. M. C. Shaw, "The Fundamental Basis of the Hardness Test," The Science of Hardness Testing and Its Research Applications (J. H. Westbrook and H. Conrad, Eds.), Am. Soc. for Metals, Metals Park, OH, pp. 1–11, 1973.

26. R. W. Rice, "Discussion of 'The Fundamental Basis of the Hardness Test'," The Science of Hardness Testing and Its Research Applications (J. H. Westbrook and H. Conrad, Eds.), Am. Soc. for Metals, Metals Park, OH, pp. 12–13, 1973.

27. D. M. Marsh, "Plastic Flow in Glass," Proc. Roy. Soc., A **279**, pp. 420–35, 1964.

28. J. Lankford, "The Compressive Strength of Strong Ceramics: Microplasticity Versus Microfracture," J. Hard Mat., **2**(1–2), pp. 55–77, 1991.

29. S. M. Wiederhorn and B. J. Hockey, "Effect of Material Parameters on the Erosion Resistance of Brittle Materials:," J. Mat. Sci., **18**, pp. 166–80, 1983.

30. D. B. Marshall, "Failure from Surface Flaws," Fracture in Ceramic Materials-Toughening Mechanisms, Machining Damage, Shock, (A. G. Evans, Ed.), Noyes Publications, Park Ridge, NJ, pp. 190–220, 1984.

31. R. W. Rice and B. K. Speronello, "Effect of Microstructure on Rate of Machining of Ceramics," J. Am. Cer. Soc., **59**(7–8), pp. 330–33, 1976.

32. E. D. Case, "The Effect of Microcracking upon the Poission's Ratio for Brittle Materials," J. Mat. Sci., **19**, pp. 3702–12, 1884

33. Y. L. Krasulin, V. N. Timofeev, S. M. Barinov, and A. B. Ivanov, "Strength and Fracture of Porous Ceramic Sintered from Spherical Particles," J. Mat. Sci., **15**, pp. 1402–1406, 1980.

34. E. E. Hucke, "Glassy Carbon," Univ. Michigan Rep. for Adv. Res. Projects Agency Order No. 1824, 1972.

35. R. W. Rice, "Evaluation and Extension of Physical Property-Porosity Models Based on Minimum Solid Area," J. Mat. Sci., **31**, pp. 102–18, 1996.

36. R. W. Rice, "Comparison of Physical Property vs. Porosity Behavior with Minimum Solid Area Models," J. Mat. Sci., **31**, pp. 1509–28, 1996.

37. O. Yamada, "High-Pressure Self-Combustion Sintering of Titanium Carbide," J. Am. Cer. Soc., **70**(9), pp. C-206–8, 1987.

38. J. C. LaSalavia, L. W. Meyer, and M. A. Meyers, "Densification of Reaction-Synthesized Titanium-Carbide by High-Velocity Forging," J. Am. Cer. Soc., **75**(3), pp. 592–602, 1992.

39. C. Greskovich and H. C. Yeh, "Hardness of Dense β-Si_3N_4," J. Mat. Sci. Letters, **2**, pp. 657–59, 1983.

40. G. N Babini, A. Bellosi, and C. Galassi, "Characterization of Hot Pressed Silicon Nitride Based Materials by Microhardness Measurements," J. Mat. Sci., **22**, pp. 1687–93, 1987.

41. S. Cho, F. J. Arenas, and J. Ochoa, "Densification and Hardness of Al_2O_3–Cr_2O_3 System With and Without Ti Addition," Ceramics Intl., **16**, pp. 301–309, 1990.

42. R. Ramadass, S. C. Mohan, and S. R. Reddy, "Studies on the Metastable Phase Retention and Hardness in Zirconia Ceramics," Mat. Sci. & Eng., **73**, pp. 215–17, 1985.

43. R. W. Rice, "Comment on 'Studies on the Metastable Phase Retention and Hardness in Zirconia Ceramics'," Mat. Sci. & Eng., **60**, pp. 65–72, 1983.

44. B. A. Cottom and M. J. Mayo, "Fracture Toughness of Nanocrystalline ZrO_2–3mol% Y_2O_3 Determined by Vickers Indentation, Scripta Metal-et Mat., **34**(5), pp. 809–14, 1996.

45. C. C. Wu and R. W. Rice, "Porosity Dependance of Wear and Other Mechanical Properties on Fine-Grain Al_2O_3 and B_4C," Cer. Eng. Sci. Proc., **6**(7–8), pp. 977–94, 1985.

46. A. Salak, V. Miskovic, E. Dudrova, and E. Rudnayova, "The Dependence of Mechanical Properties of Sintered Iron Compacts Upon Porosity," Pwd. Met. Mat., **6**(3), pp. 128–132, 1974.

47. R. W. Rice, "Hardness-Grain-Size Relations in Ceramics," J. Am. Cer. Soc., **77**(10), pp. 2539–53, 1994.

48. A. Pajares, F. Guiberteau, and B. R. Lawn, "Hertzian Contact Damage in Magnesia-Partially Stabalized Zirconia," J. Am. Cer. Soc., **78**(4), pp. 2609–18, 1996.

49. A. C. Fischer-Cripps and B. R. Lawn, "Stress Analysis of Contact Deformation in Quasi-Plastic Ceramics," J. Am. Cer. Soc., **79**(10), pp. 11083–86, 1995.

50. A. Pajares, L. Wei, and B. R. Lawn, "Contact Damage in Plasma-Sprayed Alumina-Based Coatings" J. Am. Cer. Soc., **79**(7), pp. 1907–14, 1996.

51. D. Weiss, P. Kurtz, and W. J. Knapp, "Strength of Specimens Containing Parallel Cylindrical Holes," Am. Cer. Soc. Bull., **45**(8), pp. 695–697, 1966.

52. L. J. Trostel, Jr., "Strength and Structure of Refractories as a Function of Pore Content," J. Am. Cer. Soc., **45**(11), pp. 563–564, 1962.

53. D. M. Liu, "Porous Hydroxyapatite Bioceramics," Porous Ceramic Materials, Fabrication, Characterization, Applications (D. M. Liu, Ed.), Trans Tech Publications, Switzerland, pp. 209–31, 1996.

54. D. M. Liu, "Control of Pore Geometry on Influencing the Mechanical Property of Porous Porous Hydroxyapatite Bioceramics," J. Mat. Sci. Let., **15**(5), pp. 419–21, 1996.

55. T. Nakashima, M. Shimizu, and M. Kukizaki, "Mechanical Strength and Thermal Resistance of Porous Glass," J. Cer. Soc. Jap., Intl. Ed., **100**, pp. 1389–93, 1992.

56. J. Lankford, "Uniaxial Compressive Damage in <$E alpha>-SiC at Low Homologous Temperature," J. Am. Cer. Soc., **62**(506), pp. 310–12, 1979.

57. E. Y. Gutmanas, A. Rabinkin, and M. Roitbarg, "On Cold Sintering Under High Pressure," High Pressure Science and Technology, Proc. VIIth Intl. AIRAPT Conf., vol. 1 (B. Vodar and P. Marteau, Eds.), Pergamon Press, New York, p. 312, 1980.

58. P. W. Montgomery, H. Stromberg, and G. Jura, "Solid Surfaces and the Gas-Solid Interface," Am. Chem Soc: Adv. In Chemistry Series **33**, p. 18, 1961.

59. A. Onodera, K. Suito, and N. Kawai, "Semisintered Oxides for Pressure-Transmitting Media," J. Appl. Phy., **51**(1), pp. 315–18, 1980.

60. W. Li-Chung and A. S. Wronski, "Slip Generation and Distortion of Cubical Cavities in LiF Single Crystals By Pressuration," High Pressure Science and Technology, Proc. VIIth Intl. AIRAPT Conf., vol. 2, Pergamon Press, New York, p. 286, 1980.

61. C. O. Dam, R. Brezny, and D. J. Green, "Compressive Behavior and Deformation-Mode Map of an Open Cell Alumina," J. Mat. Res., **5**(1), pp. 163–71, 1990.

62. R. Brezny and D. J. Green, "Uniaxial Strength Behavior of Brittle Materials," J. Am. Cer. Soc., **76**(9), pp. 2185–92, 1993.

63. T. Kurauchi, N. Sato, O. Kamigaito, and N. Komatsu, "Mechanism of High Energy Absorption by Foamed Materials-Foamed Rigid Polyurethane and Foamed Glass," J. Mat. Sci., **19**, pp. 871–80, 1984.

64. A. Pearson, J. E. Marhanka, G. MacZura, and L. D. Hart, "Dense, Abrasion-Resistant, 99.8% Alumina Ceramic," Aluminum Co. Am. Report, Alsoa, TN, 5/1967.

65. J. E. Hines, Jr., "Abrasive Wear of High Density Alumina," Ph.D. Thesis, Pennsylvania State University, University Park, PA, 1974.

66. A. Y. Artamonov and G. A. Bovkun, "Some Aspects of Abrasive Wear of Transition-Metal Carbides," Refractory Carbides, Studies in Soviet Science (G. V. Samsonov, Ed.; N. B. Vaughan, translator), Consultants Bureau, New York, pp. 371–76, 1974.

67. Y. He, L. Winnubst, A. J. Burggraaf, H. Verweij, P. G. Th. Van der Varst, and B. de With, "Influence of Porosity on Friction and Wear of Tetragonal Zorconia Polycrystal," J. Am. Cer. Soc., **80**(2), pp. 377–80, 1997.

68. R. W. Rice, "Micromechanics of Microstructural Aspects of Ceramic Wear," Cer. Eng. & Sci., **6**(7–8), pp. 940–58, 1985.

69. R. W. Rice, "Machining, Surface Work Hardening, and Strength of MgO," J. Am. Cer. Soc., **56**(10), pp. 536–41, 1973.

70. M. E, Gulden, "Study of Erosion Mechanisms of Engineering Ceramics, Static Fatigue of Ceramics," Solar Turbins Intl. Final report for 1/4/1973–1/1/1980 for ONR contract N00014–73–C-0401. 7/1980.

71. R. W. Rice, "Damage Resistance of Fortified Surfaces of a Coordierite-Based Crystallized Glass," J. Am. Cer. Soc. **65**(7), pp. C-106–7, 1982.

72. D. Lewis III, "Effect of Surface Treatment on the Strength and Thermal Shock Behavior of a Commercial Glass-Ceramic," Am. Cer. Soc. Bul., **58**(6), pp. 599–605, 1979.

73. D. B. Grossman, "Fortification of Radome Glass-Ceramics," Proc. 14th Symp. on Electromagnetic Windows, Georgia Institute of Technology, Atlanta, GA, 6/21–23/1978.

74. J. J. Brennan, "Development of Silicon Nitride of Improved Toughness," United Technologies Research Center Final Report for NASA Lewis Res. Cen. Contract NAS3–21375, 10/2/79.

75. M. F. Gruninger, J. B. Wachtman, Jr., and R. A. Haber, "Thermal Shock Protection of Dense Alumina Substrates by Porous Alumina Sol-Gel Coatings," Cer. Eng. Sci. Proc., **8**(7–8), pp. 596–601, 1987.

76. E. D. Whitney, "Microstructural Engineering of Ceramic Cutting Tools," Am. Cer. Soc. Bull., **67**(4), p. 1010, 1988.

77. W. J. Ferguson and R. W. Rice, "Effect of Microstructure on the Ballistic Performance of Alumina," Ceramics in Severe Environments (W. W. Kriegel and H. Palmour, III, Eds.), Plenum Press, New York, pp. 261–70, 1971. Material Science Research, vol. 5.

78. W. J. Ferguson and R. W. Rice, "Effect of Microstructure on the Ballistic Performance of Alumina," NRL (classified) Memo Rept. 2421, Washington, DC, 1972.

79. K. K. Schiller, "Porosity & Strength of Brittle Solids (with Particular reference to Gypsum)," Mechanical Properties of Non-Metallic Brittle Materials (W. H. Warson, Ed.), Interscience Pub. Inc., New York, pp. 35–49, 1958.

80. F. P. Knudsen, "Dependence of Mechanical Strength of Brittle Polycrystalline Specimens on Porosity and Grain Size," J. Am. Cer. Soc., **42**(8), pp. 376–388 (1959).

7

THE POROSITY AND MICROCRACK DEPENDENCE OF THERMAL AND ELECTRICAL CONDUCTIVITIES AND OTHER NONMECHANICAL PROPERTIES

KEY CHAPTER GOALS

1. Demonstrate the diversity of the porosity dependences of nonmechanical properties, ranging from dielectric constant depending only on P, to conductivities depending on MSA, to pore-size effects on some properties, including electrical breakdown, which also depends on oriented pore chains.
2. Introduce basic aspects of surface area and permeability porosity dependences basic to key applications of bodies of designed porosities.
3. Show similarities of the porosity and microcrack dependences of thermal conductivity and that microcracking can result in some porosity and grain-size dependence of thermal expansion.

7.1 INTRODUCTION

As noted in Chapter 2, some properties will not depend on porosity, some will depend on only the amount of porosity, and some will depend on both the amount and one or more aspects of pore character. While examples of other dependences were given, mechanical properties extensively discussed in Chapters 2–6, 8, and 9 primarily depend on both the amount of porosity and aspects of its geometry, e.g. as measured by the minimum solid area (MSA) of the body. This chapter addresses a broader range of porosity dependence for nonmechanical instead of mechanical properties.

Nonmechanical properties considered include electrical and thermal conductivities, which commonly have the same or similar porosity dependence as mechanical properties, but may deviate from this, e.g. when the conducted flux is no longer restricted to the solid phase. The limited but important occurrence of effects of porosity and grain size on thermal expansion via microcracking effects are noted. Dielectric constant porosity dependence is addressed, which generally does not follow a MSA or similar dependence, and frequently approaches a rule of mixtures dependence—i.e., there is no dependence on pore character, only on the amount of porosity. Electrical breakdown and some piezoelectric and magnetic properties as well as optical scattering are also considered, which add to the diversity of porosity dependences, e.g., introducing pore size-shape-spacing parameters for some properties. Surface area and aspects of permeation and flow in porous bodies are also outlined. Major factors determining which properties and the extent that they are considered are: 1) their technological importance, 2) there being at least a few data sets sufficient to indicate the character of the porosity dependence, and 3) their utility in illustrating the diversity and character of the porosity dependence.

Models and theoretical considerations (mainly bounds) are discussed for each of the various properties selected in the next section, mainly in the context of the specific property; however, a broader analysis is emerging that relates the behavior of one property to that of another, thus providing guidance in the bounds of a lesser known property from a better known one, e.g., from elastic moduli to electrical (and presumably also thermal) conductivity (1). Work showing the transformability of many equations for a property into others of the same or differing property also can aid in this (2); however, the applicability of equations for one property to another is limited to the underlying physics being valid for both. While many of the bounds for porosity dependence are of limited use by themselves, such cross-property bounds may be of use for the porosity dependence of some properties for which there is little understanding or data, and for checking self-consistency.

7.2   MODELS AND OTHER THEORETICAL CONSIDERATIONS

7.2.1   Thermal Conductivity and Expansion

First consider the effects of porosity on thermal conductivity and effects of microcracks on thermal conductivity and thermal expansion. There are four factors that need to be considered in modeling the thermal conductivity of porous materials: two for the solid phase and two for the pore phase. First and foremost is the effective conductive path through the solid phase. Second is the issue of whether there are regions of lower conductivity (i.e., higher thermal resistivity) other than the pores in the solid conductive path, e.g., due to grain boundary or other interfacial phases. Possible effects of such phases which are often neglected, but can be important in some cases, especially for some refractories (where microcracks can also be important). The third and fourth factors are convective and radiative transfer, respectively, that occur exclusively and mostly via the pore phase. Convective effects are dominated by the pore size, generally being considered to not become a factor until pore sizes of about 1 to 4 mm are reached; however, the nature of the fluid in the pores is a factor in the pore size where convective effects start to be a factor, i.e., the sizes mentioned depend on the type of gas in the pores (3). Further, data presented later shows the type and pressure of gases in pores (and microcracks) often plays an important role in the thermal conductivity of ceramic refractories. Radiative transfer greatly increases in its impact as temperature increases, often becoming the dominant factor for ceramics at high temperatures, e.g., > 1000°C: Radiative transfer also depends on the material and pore character, primarily increasing with the radiative path length through the pore phase, and hence with the size of pore and their degree of openness. Thus, in quite open polyurethane foams, e.g., $P = 0.8$ to ~ 1, radiation transfer has been estimated to be 5–25% of the apparent conductivity at 22°C, with higher contribution at higher P (3). Some radiative transfer can also occur through the solid phase as the density of the material in the pore walls and its infrared (IR) transmission limit increases. For most transmitting, i.e., dielectric, ceramics, the transmission limits are in the range of IR wavelengths of 3 to 12 μm, usually 3–6 μm, with the cutoff wavelength increasing as either the anion or cation atomic weights increase; however, the wavelength of peak black-body emission, while ~ 10 μm at 22°C, requires temperatures of a few hundred degrees Celsius to get down to the 3–6 μm range, which is consistent with radiative transfer not becoming significant in ceramics until temperatures of several hundred degrees Celsius (4).

Effects of porosity on thermal conductivity are important in general, especially for ceramics (and polymers) for thermal insulation, usually at higher porosity levels (e.g., polymeric foams). Thus, many of the models for effects of porosity are obtained from models derived for polymeric foams or composites, in the latter case by setting the properties of the dispersed phase equal to those of pores, often zero, e.g., where convection and radiation can be neglected; however, as noted in Chapters 2–6, such adaptation of composite models presents problems, in particular of typically not reflecting the porosity at which the percolation limit of the solid phase is reached, i.e., P_C values, and giving very wide bounds. A few models have also been derived directly for porosity, some of them for ceramics (though the material should not matter provided the microstructure and property parameters are consistent with the systems considered).

Table 7.1 Models for the Porosity Dependence of Thermal Conductivity of Porous Solids at ~ 22°C[a]

Equation	Form[b]	Bases; ranking[c]	Source[d]
1.	$1 - P$	Mixture rule UB (LB = 0); 15,d	Collishaw and Evans (3), Chen et al. (12)
2.	$1 - 1.5P$	Pores between approximately spherical particles; 13	Budiansky (13)
3.	$(1 - P)^n$	Foams/pores between aligned cylindrical particles; 3	Progelhof et al. (14), Wagh (15)
4.	$(1 - P)^{3/2}$	Spherical pores; 1	Bruggeman (16)
5.	$(1/3)(1 - P^2)$	Cubic foam cells, UB; 4	Schuetz and Glicksman (6)
6.	$(1 - P)\left(1 + \dfrac{P}{\alpha}\right)^{-1}$ typ. α 1, or mostly 2	Various, including spherical pores (Maxwell); 7,c	Hashin and Strikman (UB) (5), Eucken (17), Maxwell (18), Doherty et al. (19)
7.	$\dfrac{(1 - P^{2/3})}{(1 - P^{2/3} + P)^{-1}}$	Uniform cubic pores in simple cubic stacking; 12,D	Russell (20)
8.	$(1 - P)[1 + f(P,P_C)]^{-1}$	Various shaped pores[a]	Nielsen (21)

[a] Based on conduction only through the remaining solid phase, i.e., not including any possible added heat transfer via convection or radiation.

[b] For Eq. 8 by Nielsen, $f(P,P_C) = (P/A)[1 + P(1 - P_C)P_C^{-2}]$, with definitions of A = 1 for spherical pores, A = 3 for irregular pores, and 5–15 for platelet pores (contrast with Table 3.4).

[c] UB = upper bound, LB = lower bound. Numbers are the approximate ranking of that equation (1 = best) in Collishaw and Evans' (3) evaluation against data for polyurethane foams for P = 0.8 to 1, and lower and upper case letters are the rankings (a, A = best) from evaluations respectively of Rhee (8) and Dawson and Briggs (9).

[d] Most equations are from the review by Collishaw and Evans (3), as indicated by the ranking noted above.

As with other properties discussed earlier, there are basically three sets of models, i.e., those that are: 1) more versatile in the character of the porosity considered, but less rigorous in the specifics of the conduction; 2) more restrictive in the specifics or range of porosity addressed, but more rigorous or detailed in the conduction specifics; and 3) strictly empirical, especially for curve-fitting purposes. The minimum solid area (MSA) models (Figs. 2.11, 3.1) clearly reflect the first type of model. As noted earlier (Figs. 2.7–2.10), the localization of stress contours primarily in the MSA regions is also representative of flux lines from thermal (or electrical) conduction in the solid, so such models should be directly applicable to these properties. Thus, the initial slope of the approximately linear portion of semilog plots of the relative or absolute thermal conductivity versus volume fraction porosity, P, i.e., the b value, and the associated P_C value are key factors in the conductivity. These values most clearly reflect the character of single-pore populations or the weighted average of mixed pore populations. Again bounds are of limited use since the lower bounds of the rule of mixtures and Hashin and Strikman (5) are both ~ 0. The upper bound for Schuetz and Glicksman (6) (Table 7.1) is 80% that of the rule of mixtures, which means it may underestimate some higher thermal conductivities.

   There have been several reviews, mainly of the second type of models noted above for either ceramic (7–10), polymeric (3,11), or general or other materials. Representative equations for the effects of porosity on thermal conductivity at, or near, room temperature from these reviews [especially Collishaw and Evans (3)] are given in Table 7.1 along with rankings of how well the equations considered in some of the reviews fit data. The forms of the equations in Table 7.1 assume no convective or radiative effects, and most were obtained by setting the conductivity of the second phase = 0 in composites, [Hale (22) notes in his review that equations for dielectric constant, emphasized in his review, are also applicable to thermal conductivity. While some are the same or similar for both properties, many are not, and identical equations for most thermal conductivity and dielectric constant dependence on porosity is questioned, as discussed later.] More general forms of the models of Euken (17), and Russell (20) in Table 7.1 and other models allow for convection or radiation effects. For example, Loeb (23) developed a more complex model accounting for radiation effects (and for pore cross-sectional areas parallel and perpendicular to the overall heat flux). Radiation effects on the temperature dependence of porosity effects are discussed in Chapter 8. More fundamentally, these expressions are similar, and several are identical (e.g., Eqs. 1 and 3 of Table 7.1), to those for elastic properties (Table 3.1). Further, almost all are of the same general forms as for elastic properties, i.e., of the form $(1 - A_1P)^n$, $1 - A_2P^n$, or $(1 - A_3P)(1 + A_4P)^{-1}$. Such close relation to equations for elastic moduli (and hence also other mechanical properties) further reinforces the similarity of underlying factors controlling all of these properties, e.g., as indicated by MSA models fitting them.

   Consider now more specific aspects of the nature of equations for thermal conductivity. Collectively they are substantial in number, as for dielectric constant, and especially elastic moduli. As for these other properties, some of the number and diversity of the equations reflects different character and levels of porosity dependence, but much reflects the complexity of the problem and the variety of differing explicit or implicit assumptions used to make modeling tractable. The diversity and differences of the models is illustrated by Collishaw and Evans' (3) evaluation showing differences between some models to often be

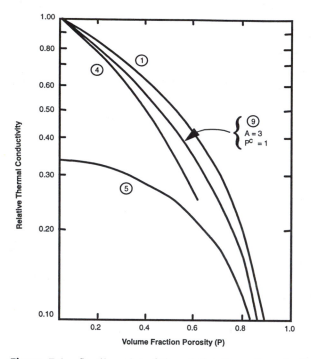

Figure 7.1 Semilog plot of the relative thermal conductivity versus volume fraction porosity (P) at 22°C for representative equations from Table 7.1. Note: 1) equation numbers are shown in circles next to curves, 2) all go to 0 at P = 1 whether they are valid there (including Eq. 9, which should go to 0 at P < 1), 3) all go to 1 at P = 0, except Eq. 5, which is for foams and not valid at low P, and 4) slopes (b values) of 1.2–1.5 for 1 and 1.5–2 for 4 for comparison to MSA models.

well over an order of magnitude in property values versus P. Plotting representative models from Table 7.1 shows extensive similarity for a few commonly used models (Fig. 7.1), except Eq. 5 at P < 0.6 where it is not valid, e.g., Eqs 6 and to some extent are similar numerically to Eqs. 8 and especially 4. All equations, except that of Nielsen (21) (Eq. 8), have no indication of P_C, usually reflecting their derivation for composites with the properties of the second phase being set to zero. Nielsen's equation, however, presents a problem at P = P_C, where it gives a conductivity of zero only if P_C = 1. Also, it does not vary much from its plot in Fig. 7.1 for other values of A and P_C (e.g., 1–10 and 1–0.2 respectively), questioning its utility for addressing a variety of porosity as might be implied by the presence of these parameters.

Litovsky and colleagues (24,25) reviewed models and data for conductivity of ceramic refractories that incorporate effects of both convection across a substantial range of temperatures and radiation (at higher temperatures) in pores and microcracks. They present an array of models, including simpler and more complex ones, that are applicable for various ranges and types of porosities and types of single or combined heat transfer mechanisms. These include models containing or solely consisting of some of the term forms given above, e.g., Eq. 3 in Table 7.1, and reflect various derivations and degrees of rigorness. Their review clearly demonstrates the need to match the proper model to the microstructures and mechanisms being dealt with, e.g., whether it includes microcracks, different bonding phases, and heat transfer mechanisms such as convective heat transfer of gases as a function of the type and amount (pressure) of the gas in pores or microcracks. They thus clearly illustrate the need for families of models.

As for elastic properties, models for the effect of cracks on thermal conductivity (hence also for thermal diffusivity) have been derived (26). These typically assume idealized arrays of planar cracks of simple, idealized shapes, e.g., discs, and negligible thickness (approximately true, reflecting the difficulty of determining crack contents experimentally). Representative results (Fig. 7.2 and Table 7.2) show significant effects of the geometry of the cracks, i.e., both their shape and how they are arrayed spatially in the body and relative to the conductive flux, i.e., analogous to the effect of pore shape and spatial arrangement. These models show thermal conductivity independent of crack content when the heat flow is parallel to the plane of the oriented cracks (reflecting the assumption of zero crack thickness), but large initial decreases in conductivity

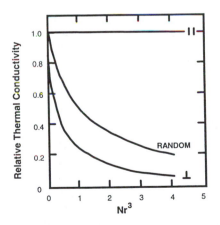

Figure 7.2 Thermal conductivity as a function of Nr^3, where N = the product of the number of cracks per unit volume and r = the radius of revolution of the idealized microcrack having orientations parallel, random, or perpendicular to the heat flow, after Hasselman and Singh (26). Published with permission of the Am. Cer. Soc.

Table 7.2   Examples of Models of the Effect of Cracks on Thermal Conductivity

Orientation of crack planes and heat flow	Resultant relative thermal conductivity (K/K_0)[a]
Crack planes and heat flow parallel	1
Crack planes and heat flow perpendicular	$(1 + Nr^3 8/9)^{-1}$
Crack planes random to heat flow	$(1 + Nr^3 8/3)^{-1}$

Source: From Hasselman and Singh (26).
[a] K = conductivity with cracks and K_0 without cracks; N = number of cracks per unit volume and r = crack radius.

as the crack content increases when they are oriented normal to the direction of heat flow. Randomly oriented cracks, which should be a reasonable approximation to microcracking due to phase transformations and thermal expansion anisotropy, show nearly as great an initial drop, then similarly reduced decrease in conductivity as crack densities respectively initially increase and then become substantial. Electrical conductivities are expected to follow similar trends.

Thermal expansion is normally independent of porosity, as noted in Section 2.1, and is also typically independent of grain size; however, microcracking can change this. As shown in Section 7.3.1, microcracking in highly anisotropic ceramics can lead to thermal expansion having substantial dependence on grain size and on porosity.

7.2.2   Electrical Conductivity and Electrical Breakdown

There are important similarities as well as differences of electrical and thermal conductivities of materials which impact their dependences on porosity. The basic similarity is that both represent a flux, usually mainly or exclusively, through the solid phase. The fundamental differences stem from the many more orders of magnitude covered by the range of electrical versus thermal conductivities of materials. Thus, for conductive particles in a poorly conducting matrix, percolation of the particles, i.e., connection of enough conductive particles to provide a continuous path through the body, often has a far greater effect on electrical than thermal conductivity. This means that resistive interfacial or grain-boundary phases can have much greater effects on electrical versus thermal conductivity, especially when a highly conductive phase is dispersed in a poorly conductive one. It also means that there can be greater problems in adapting models for thermal conductivity of composites to electrical conductivity of composites for effects of porosity by setting the conductivity of the dispersed phase equal to zero (27). This is in addition to problems of the solid phase by nature and definition always being a continuous phase regardless of whether the pore phase is discontinuous or also continuous in porous bodies.

Thus, setting the electrical conductivity of a dispersed phase in a model that deals with composites having highly conductive phases in poorer conducting matrices, and hence significantly impacted by percolation effects will not be representative of the effects of porosity. On the other hand, since electrical conductivity via gases in pores is often quite limited relative to that of the solid, such effects can often be more readily neglected in porous bodies where electrical, instead of thermal conductivity, is of interest. Thus, models for conductive composites may be applicable to porous bodies provided the matrix phase is conductive; however, the overall conclusions of Lux's review (27) of (mainly mechanistic) models for electrical conduction of composites are a useful guide, namely that: 1) no single model can account for the extensive impact of different processing, and hence pore character, and 2) because of this, fitting of a particular model to some data may be more an accident or favorable superposition of different mistakes in the derivation of specific equations or selecting parameters for them.

Since reduction of electrical conductivity by introduction of porosity is not of broad interest and adaptation of models for conductivity in composites is often of uncertain validity, there are few models for the effects of porosity on electrical conductivity; however, since the effect of pores in reducing electrical conductivity is typically not effected by the presence of boundary or interfacial phases and the flux paths are concentrated in the minimum solid areas as the stress contours are (Fig. 2.9), MSA models used for mechanical properties and thermal conductivity are again applicable. Thus, the slope of the initial approximately linear portion of the semilog plot, i.e., the b value, and the P_C value are key parameters reflecting effects of pore character.

A MSA-type model was derived by Yoshikawa et al. (28) assuming a composite having a simple cubic stacking of uniform conducting spherical particles in a nonconducting matrix and a contact resistance between the particles. They assumed that the: 1) resistance of the particles themselves could be taken as that of a cylinder whose length was the particle diameter (d), and whose volume was that of the spherical particle, and 2) contact resistance was $\propto (1/d)^2 R_p$, where R_p, the interparticle resistance, is a function of the surface character (e.g., roughness) and material resistivity. The resultant equation for the body (composite) resistivity, r is :

$$r \sim \left(\frac{1}{d}\right) R_p + 2r_0 \qquad (7.1)$$

where r_0 is the resistivity of the particle material. These authors noted that since R_p was a significant factor for the materials of interest to them (silver particles in silicone rubber) that the effects, and hence evaluation, of the effective conducting cross-section (which leads to the above factor of ~ 2 multiplying r_0)

was not very precise. Thus, results would not have been significantly different if the minimum solid area was used.

More recently Mizusaki et al. (29) presented a MSA model where they treated bodies of partially sintered particles as arrays of straight chains of uniform spherical particles. For mathematical purposes they treated both the main portion of each particle and the sintered neck regions as half-cycle sine waves of the same amplitude, but different frequencies.

Nielsen's model for thermal conductivity (Eq. 9 of Table 7.1) was also derived and proposed for electrical conductivity of composites (21); however, it thus has the same issues for electrical as it does for thermal conductivity, i.e., of setting the conductivity of the pore phase to zero and the equation being valid at P_C only if $P_C = 1$.

Next, consider electrical breakdown, i.e., the development of a, usually highly localized, conductive path (often accompanied by transient, highly nonlinear and localized melting and vaporization, as well as localized fracture). This occurs above a certain electrical field (applied voltage per thickness of material), and is of primary interest in bodies of very limited porosity and electrical conductivity since even modest levels of these greatly reduce high breakdown strengths commonly sought. Gerson and Marshall (30) developed a model for the effects of porosity on such breakdown assuming randomly distributed uniform spherical pores for any porosity level. They used techniques very similar to those for deriving MSA models, i.e., dividing the body to be modeled into cubic cells in simple cubic stacking such that each randomly distributed pore would just be contained in a random cubic cell. Electrical breakdown was assumed to occur in a column of such cubic cells (parallel to the uniaxial field) having sufficient pores (i.e., cells with pores) to reduce the amount of material in the column below the breakdown level of the dense solid. They calculated the probability of having such a column of cells occurring in a disc specimen, which obviously depends on both pore size and specimen dimensions and proportions (e.g., a cube versus a thin plate of the same volume). They gave the probability (p) of finding x cells with pores in a column of n cells of a body with a volume fraction of porosity, P, as:

$$p = (P)^n (1 - P)^{n-x} \qquad (7.2)$$

(n = specimen thickness divided by the pore diameter). Then they note that since the product of N (the total number of columns in the sample) and the probability for a column of cells with the maximum number of voids in it (over many samples), $p(x_m)$, is ~1, so:

$$p(x_m) \sim N^{-1} \qquad (7.3)$$

(Note that for a disc sample N is simply the square of the ratio of the specimen to pore diameters) and the probability of the particular sample having a column with x_m voids is, for large N (i.e., small void to specimen dimensions) $\sim 1 - e^{-1}$ ~ 0.63 where e = the naperian log base. From these relations they calculated breakdown as a function of (at least simple) specimen configurations and pore sizes and amounts. In view of the relation to e, it is not surprising that plots of their calculations are very close to straight lines on semilog plots over much of the range of specimen and pore dimensions encountered in their experimental evaluation (e.g., b values increasing from ~ 3.4 to ~ 7.8 as pore diameter increased from ~ 50 to 250 μm).

7.2.3 Dielectric Constant

An electrical/electromagnetic property of broad interest is the relative dielectric constant ($\in$, or relative permitivity), which is often referred to as simply dielectric constant, as will commonly be done here. The dielectric constant is basically a measure of the degree of polarization at the molecular or atomic level in a material (and hence its ability to develop an induced charge), as well as of the extent to which the field flux prefers, or is constrained to be in, the solid, i.e., the higher the dielectric constant, the more the flux is in the solid phase. The relative dielectric constant is the value of a given material to that of vacuum, whether measured with a static or varying electric field, or an electromagnetic field, although values can be frequency dependent. This constant is commonly measured by the capacitance of a simple parallel plate capacitor with the material of interest between the two plates versus the capacitance when the two plates have only vacuum between them (or only air since its relative dielectric constant is very close to that of vacuum, which is 1). Since $\in = 1$, not 0, for a vacuum, there is thus commonly much less difference in value from that of the solid phase, and hence commonly less uncertainty in adapting models for composites for porosity effects for $\in$ versus many other, e.g., mechanical, properties. The capacitor representation of $\in$ allows the two bounds for laminar composites to be calculated, giving the upper bound (the rule of mixtures) as:

$$\in = \in_0(1 - P) + P \tag{7.4}$$

and the lower bound is

$$\in = \in_0(1 - P + \in_0 P)^{-1} \tag{7.5}$$

where $\in_0$ is the value at P = 0. Note that the two bounds, whose separation increases from none at $\in_0 = 1$, as $\in_0$ increases are separated less than the bounds for mechanical properties since $\in_0$ for pores is 1 (unless they contain substantial

amounts of a high dielectric constant gas) instead of 0 for mechanical properties. These two bounding equations are special cases of a more general equation:

$$\in^n = \Sigma_i \phi_i \in_i^n \tag{7.6}$$

where ϕ_i = the volume fraction of each phase (e.g., $1 - P$ and P for porosity), and
$n = -1$ or $+1$, respectively, for the two above equations adding capacitances
respectively in series and in parallel, which are referred to respectively as the
series and parallel mixing rules. With $n = 0$ Eq. 7.6 is referred to as the
logarithmic mixing rule since it gives:

$$\log \in = \Sigma_i \phi_i \log \in_I \tag{7.7}$$

and is often used as an approximation to the average of the parallel and series
bounds (Fig. 7.3B).

There is substantial interest in effects of porosity on dielectric constant as for
a number of other properties, i.e., to minimize decreases in it where high values
are desired and to obtain large decreases in it where low values are desired.
Thus, a substantial number and variety of models have been derived for the
porosity dependence of dielectric constant, typically based on models for
composites (usually of simple, handleable phase geometry such as spherical
pores). As with other properties, use of composite models is subject to the
typical concerns of using such models for porosity, mainly that pore and solid
phases are not continuously interchangeable (as two solid phases are); however,
as noted earlier, there commonly is not the marked solid-pore property difference for $\in$ as for other properties. Models include those of classical physicists
such as Rayleigh (32) and Maxwell (18) to a variety of other early and more
recent models, e.g., of Bruggeman (16). Again, the diversity and number of
models precludes a detailed review here. Instead the reader is referred to other
reviews for details, e.g., Shin et al. (2) (who evaluated well-known equations
against a theoretical standard and considered transformation of equations to a
self-consistent form), as well as models (or bounds) briefly noted by Banno
(33), Payne and Cross (34), Wakino et al. (31), and Dunn and Taya (35), but
particularly by Hale (22), who devotes considerable attention to dielectric
constant of composites. [Note that while Hale's review is quite useful, his
assertion that equations for one property can be used extensively for other
properties, specifically those for dielectric constant for thermal conductivity or
magnetic permeability, which is common in the literature, must be questioned.
This is true only so long as the underlying physics is the same, which is
commonly not the case for dielectric constant and conductivity, since the
electric field flux is not totally confined to the solid phase, while the conductive
flux commonly is (in the absence of significant convection or radiation effects).

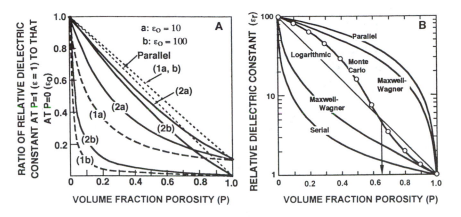

Figure 7.3 Plots comparing bounds and a few expressions for the ratio of the relative dielectric constant versus volume fraction porosity (P). A) Linear plot of the ratio of the relative dielectric constant at any P to that at P = 0 versus P (for ratios of ϵ of the matrix, at P = 0 of 10 and 100). Upper and lower bounds, 1 = rule of mixture/Wiener and 2 = Hashin and Shtrikman bounds [Bottcher's self-consistent system approximation for spheres in a continuous matrix lies between curve sets 2 but closer to the 2a set; after Hale (22)]. Published with permission of J. Mat. Sci. B) Semilog plot of ϵ versus P for the same or similar bounds and two approximations after Wakino et al. (31), published with permission of J. Am. Cer. Soc. Note: 1) the broad nature of the parallel-series bounds, the somewhat narrower separation of those of Hashin and Shtrikman (5) and Maxwell-Wagner (18), the good average of the logarithmic mixing rule, and that the narrowness of the bounds varies inversely with the level of the dielectric constant; 2) Wakino et al.'s Monte Carlo result, which is almost identical to their resultant empirical equation, is reasonably close to the logarithmic mix average and shows more parallel behavior at lower P and more series behavior at higher P (with the crossover point, V_0, at P ~ 0.65, which is a parameter in Wakino et al.'s model); and 3) greater compression of low ϵ values in A and of high ϵ values in B.

Furthermore, where conductive flux is not confined to the solid phase, e.g., due to convective or radiative transport, the behavior of such transport may not be the same as the distribution of electric field flux in pores. Thus, similar or identical equations for both properties must be viewed with caution, which is heightened by many equations for ϵ being different from those for thermal conductivity (compare Table 7.1 and Fig. 7.4).] Equations and resultant plots of the P dependence of dielectric constant selected from Hale's review are shown in Fig. 7.4, from which three factors should be noted. First, while Eqs. 1 and 2 for dilute solutions are not valid at P = 1, Eq. 1 is consistent with the limit there but Eq. 2 is not. Nor is the equation $\epsilon/\epsilon_0 = (1 - 1.5P)^{-1}$ presented by Payne and Cross (34) for dilute P, but the more general equation they presented:

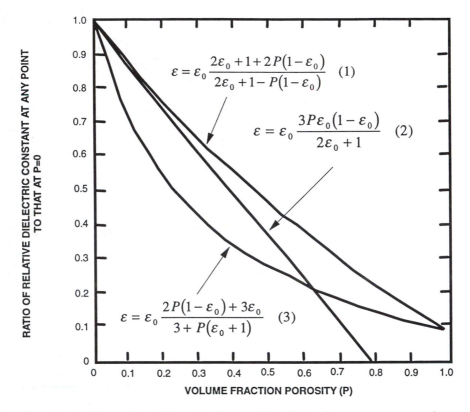

Figure 7.4 Linear plot of the ratio of the relative dielectric constant at any volume fraction porosity (P) to that at P = 0 versus P for three equations for composites from Hale's review (22) (obtained by setting the relative dielectric constant of the pores = 1 and that of the material = 10). Equations 1 (32) and 2 are for dilute concentrations of spherical particles (i.e., pores here), and Eq. 3 (16) is from the self-consistent scheme (SCS) where it is assumed that each second-phase particle is surrounded by material of the average properties of the composite. Although the latter results in the unknown composite ϵ on both sides of the equation, and hence commonly requires iterative methods of solution, the simplifications of setting ϵ = 1 for pores allows a simple closed form solution.

$\epsilon/\epsilon_0 = (1 - P)(2 + P)^{-1}$ (similar numerically to Eq.1, Fig. 7.4) is. Equations for dilute concentrations of other particle (i.e., pore) shapes, e.g., discs or rods, go to zero (i.e., below the P = 1 limit) faster than Eq. 2. Also, all equations give faster decreases as the dielectric constant of the material increases, some modestly, some substantially. Second, the equations vary from very close to a linear dependence to substantial nonlinear decrease with increasing P; however, third, the decrease of $\in$ with P is almost always substantially less than for many other, e.g., mechanical, properties (e.g., Figs. 2.11, 3.1).

Consider now models presented after Hale's review. Banno (33) summarized a model based on analyzing a composite of a simple cubic array of uniform rectangular parallelpiepids of the dispersed phase (i.e., pores for the case here). This model, which is thus based on closed pores, gives an approximately linear dependence on P for approximately isotropic pores (i.e., cubic pores in this case), but the dependence becomes progressively nonlinear as the closed pores become progressively elongated. Dunn and Tayas' model (35) is based on an isolated ellipsoidal pore (i.e., for limited quantities of closed pores) using Eshelby's method coupled to Mori-Tanaka's effective medium theory and computer computation. Their computations (for spherical pores) show dielectric constant decreasing essentially linearly with P to the limits of their computations of P = 0.5. Though their modeling is directed toward piezoelectric materials, as discussed below, it should be valid for other materials so long as the basic physics of the model, i.e., the nature of the pores and the interaction with the field is valid. Wakino et al. (31) have also used modeling based on computer simulation of electric flux concentration using a finite element (having arrays of 25×25 cells for each layer simulated) and a Monte Carlo approach. While their simulations were constrained to two dimensions, which means that their resulting exponential equation must be treated with some caution, their consideration of 1000 simulations for each case, and their consideration of multiple layers, mitigate many of the two-dimensional limitations. Their comparison of bounds and approximations (Fig. 7.2), along with some reviewed by Hale (22) is useful. Similarly, modeling and bound considerations of Miller and Jones (36) based on columns or layers of dielectric spheres embedded in a medium is useful. They noted that: 1) when such spheres are far apart (hence noninteracting) results follow Maxwell limits, but when the spheres touch in the columns or layers, the results approach the Wiener limits, 2) below the cubic packing limit for spheres (P < 0.52 for pores), the limits from their modeling were tighter than those of Wiener, and 3) ignoring interactions between layers or columns of such spheres introduces errors that increase with dielectric constant. Much dielectric constant dependence on porosity can often be expressed as a rule of mixtures of the pore ($\in \sim 1$) and solid phases ($\in_0$), Eq. 7.4, hence being dependent mainly or only on P. Deviations from the simple rule of mixtures may reflect charging of the pore surfaces, especially in materials of higher dielectric constant and with finer pores, making addressing internal charging effects, e.g., as in Bruggerman's model, important.

7.2.4 Other Electromagnetic and Magnetic Properties

There are few or no models for the porosity dependence of most other nonmechanical properties. Also, there has generally not been broader consideration of where there is dependence on both the amount and character of the porosity, and the extent to which many properties depend on the specifics of pore character is not well established. Nor has there been much attention to issues of self-consistency; however, there are some models that provide a useful starting point, guidance, or both.

Since the index of refraction is simply the square root of the relative dielectric constant, the above models should also in principle apply to the former; however, it must also be recognized that such application is valid only so long as other physical process are not significant. Thus, whenever electromagnetic wavelengths are beginning to be subject to measurable scattering as a result of the scale of pore dimensions versus the wavelength, models for dielectric constant/refractive index become less and less useful since scattering increasingly dominates the amount and distribution of the transmitted energy. Scattering, which is more commonly treated for dispersed particles rather than pores, has been treated theoretically, but involves a number of complexities, especially for multiple scattering. Suffice it to note here that scattering is a very peaked function of the ratio of the pore size to the wavelength, being a maximum when the pore size is ~ ¼–½ the wavelength, and for a fixed pore (or particle) size varies as the inverse fourth power of the wavelength. Thus, scattering from pores commonly found in ceramics is typically significant in the ultraviolet (UV), and especially the visible, and progressively less in the infrared (IR) and into the mm wavelength ranges. Such effects are illustrated by calculations of Erneta and Stockler (37) for PLZT showing the marked effects of small amounts of porosity (Fig. 7.5). An important challenge in more accurately addressing the issue of scattering is the effects of variable pore sizes, effects of which are illustrated in Fig. 7.5. Peelan and Metselaar (38) have calculated scattering coefficients for Al_2O_3 with different uniform size pores and for a lognormal size distribution of such pores based on Mie scattering.

Turning to other important, but more specialized electrical properties, consider those for piezoelectric materials (which also have applicability to ferroelectric materials). As noted above, models of both Banno (33) and Dunn and Taya (35) for dielectric constant were developed for piezoelectric materials (but are not restricted to them). Both applied their models to both elastic (Chapter 3) and electrical properties of these materials. Banno used the model of a simple cubic array of modest contents of uniformly sized and oriented rectangular parallelepiped second-phase particles (i.e., closed pores in this case) and using other reviewed results derived the porosity dependence of the piezoelectric constant, d_{31} (note that the hydrostatic piezoelectric stress constant, $d_h = d_{31} + d_{32} + d_{33}$, the latter being referred to as the piezoelectric stress constant), the electromechanical coupling factor (k_p), and the resonant impedance (Z_r). He showed the former decreasing and the latter two increasing, generally modestly and approximately linearly with increasing P (to the limits

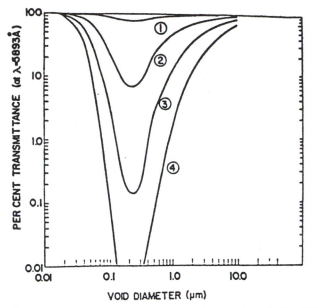

Figure 7.5   Calculated transmittance (at wavelength, λ = 5893 Å = 0.5893 μm) versus pore diameter for various percent porosities: 1) 0.01, 2) 0.1, 3) 0.25, and 4) 0.5%, in PLZT after Erneta and Stockler (37). Note the marked effects of even small amounts of P, e.g., ~ 0.01. Published with permission of the Am. Cer. Soc.

of their evaluation to, P = 0.5, which is clearly high for a closed pore model). Significant nonlinearity was calculated only as the aligned pores became significantly elongated.

Dunn and Taya (35) computer modeling of isolated (i.e., dilute, closed) spheroidal pores showed d_{33} decreasing a limited amount (e.g., by ~5–10%), and d_{31} even less, both in an approximately linear fashion over the range of P evaluated (again to P = 0.5, which is high for both dilute and closed pores). The specific acoustic impedance, Z, decreased more (e.g., by ~ 50%) with increasing P, but with limited nonlinearity, i.e., approximately linearly. They showed d_h increasing with P and to a greater degree (e.g., 100–200%) than the decrease of d_{33}, and even greater increases (e.g., 3–6-fold) and more nonlinearity for the piezoelectric voltage coefficients g_h and g_{33} (where $g_h = d_h/\epsilon_{33} = g_{31} + g_{32} + g_{33}$ as for the d's) as P increased. In all cases the extent of the change with P and the trend toward some nonlinearity showed greater increase as the pores increasingly became more oblate. More generally, one might expect models for the porosity dependence of piezoelectric properties analogous for those of magnetic materials discussed below based on pores inhibiting the motion of domain walls; however, effects of pores on domain behavior have apparently not been considered in the effects of pores on piezoelectric materials, except in a qualitative fashion for electrical fatigue effects discussed later.

Turning now to models for the porosity dependence of magnetic properties, Jain et al. (39) briefly reviewed modeling two properties, the initial permeability, μ_i, and the coercive force, H_c. They note that μ_i is mainly controlled by reversible motion of domain walls and adapted a model for the pinning of domain walls by grain boundaries to that for pores, i.e., that domain walls are pinned at pore boundaries until the magnetic field is high enough to unpin the domain wall. This gives μ_i as $\sim \propto \lambda$, where λ is the average wall-to-wall separation of the pores, and is quite dependent not only on the volume fraction porosity, P, but also on the size, shape, orientation, and spatial distribution of the pores. Thus, for a simple cubic array of uniform spherical pores of diameter :

$$\lambda = d\{[\pi(6P)^{-1}]^{1/3} - 1\}$$
(7.8)

(until the pores start to intersect at $P = \pi/6$)

while for a random array of pores the mean λ is given by Eq. 2.5, which is again directly proportional to the pore size, but with different dependence on P. (Note the differences in mean pore spacings between random and uniform pore stacking, Fig. 2.6.) Jain et al. (39) cited one model giving $H_c \propto P^{2/3}$, where the proportionality constant is linearly dependent on the ratio of the domain wall width to the pore diameter. They cited a second model giving a more complex power dependence on the diameter, spacing, and volume fraction of the porosity, and on the natural log (ln) of the pore spacing. Igarashi and Okazaki (40,41) adapted some theoretical concepts to obtain the functional dependence of initial permeability, μ_i, coercive force, H_c, and remnant magnetic flux density, B_r, on P. Their equation for μ_i, containing empirically determined parameters, gave a nonlinear decrease with increasing P that is greater for smaller μ_i of the dense material (which is also grain-size dependant). In contrast to Jain et al., they found H_c independent of P. For B_r they reported an initial approximately linear, then a much faster (nonlinear), decrease with increasing P. As will be discussed further later, separation of porosity and grain-size effects is an important challenge.

7.2.5 Surface Area

Surface area, especially specific surface area of porous bodies, as measured by gas absorption techniques, is a fundamental factor in many applications of highly porous materials such as sensors, and especially catalysts. It can also be a measure of porosity and densification (42), and hence offers another correlation with properties versus porosity. More recently, Martin and Rosen (43) have shown correlations between ultrasonic velocity and surface area versus volume fraction porosity (P) in three ZnO bodies sintered to varying degrees. They noted that since ultrasonic velocity can be measured at high temperatures it could thus be used as a real time monitor in sintering to specific surface areas and correlatable properties.

Surface areas of powders are typically modeled based on treating the particles as spheres, which readily shows the marked increase in their area with reduction in particle size; however, despite the importance of surface area, there has been limited general modeling of this as a function of the type and level of porosity. Jernot et al. (44) modeled specific surface area per unit volume and per unit mass (S_m) based on idealized sintering of uniform spherical particles of radius r, yielding:

$$S_m = 3(C_n - 1)(r\rho_0)^{-1}[(1 - P)^{-1} - 1] \tag{7.9}$$

where C_n = the particle coordination number and ρ_0 = the particle density (i.e., normally the theoretical density of the material).

This modeling of uniform spherical particles is quite useful, but it is only part of the picture since, as outlined in Chapter 2, a variety of pore shapes is needed to address general practical situations. There are two basic pore categories based on shape and processing origin, which clearly impact surface areas (Table 7.3). The first category is spherical and aligned cubic and cylindrical pores, representing intergranular pores larger than the grains or foam cell walls or struts defining them, e.g., respectively from: 1) bubbles or foaming, 2) fugitive impurity or added (e.g., sawdust) particles, and 3) extrusion. Such pore geometries also reflect, at least approximately, most intragranular pores (e.g., from grain or particle growth enveloping previously intergranular pores) which are thus smaller than the grains or particles. For this first type of pores, there is no difference in surface area between cubic and other stackings until the pores

Table 7.3 Relations for Simple Cubic Stacking of Selected Pore and Particle Geometries[a]

Pore geometry[b]	Volume fraction porosity, P	Minimum solid cross-sectional area (MSA)	Pore surface area, A
Cube (P = 1)	$(l/L)^3$	$1 - (l/L)^2$	$6l^2$
Sphere (P = $\pi/6 \sim 0.52$)	$(4\pi/3)(r/L)^3$	$L^2 - \pi r^2$	$4\pi r^2$
Cylinder (P = $\pi/4 \sim 0.78$)	$\pi (r/L)^2$	$L^2 - \pi r^2, L - r$[c]	$2\pi rL$
Pore between spherical particles (P ~ 0.036)	$1 - 4\pi/3 (r/L)^3$	$\pi/4 (R/r)^2$	$4\pi r^2 - 12\pi rh$

[a] See Figure 7.6A and text for notation and background.

[b] The P values for the open-closed pore transition are shown in ( ).

[c] Two expressions are needed to address stress or flux respectively parallel or perpendicular to the axes of the aligned cylindrical pores.

become open (which is a function of stacking, but does not occur until there is considerable porosity; Table 7.3). The second type of pores, those between packed particles and left by incomplete sintering, are inherently intergranular. Such pores are smaller than the grains if between individual grains, but larger than the typical grain structure if they are between agglomerates (e.g., from spray drying). The character and impact of such pores clearly depends on the particle stacking, as shown by Knudsen (45), who introduced such models for strength. Cubic stacking of particles, while being simpler to analyze, is also realistic since it is very close to random packing (46–48) and gives initial green densities (which are also the particle percolation limits for the body) consistent with those typically found for many bodies), i.e., $P \sim 0.5$.

The specific surface areas of such pore structures can be modeled via the same procedures used for MSA modeling (Chapter 2), which, gives a direct relation between S_m and MSA models. The computation methodology is to again consider the above idealized bodies as being made of identical cubic cells of the same simple cubic packing as that for the pores or particles (Fig. 2.1). Calculation of cell properties versus P was only done up to the $\sim$ P values where the first type of pores become open, i.e., until the pores intersect (Table 7.3). (Calculations can be extended to cover open, i.e., intersecting, pores as in open-cell foams, but are more complicated.) Though absolute cell sizes can be considered, as shown later, the cell size is determined by the ultimate pore size so for basic trends results are dependent only on the relative size of the pore and cell. Thus, the cell dimension, L, can be set at an arbitrary value, e.g., 1, with pore diameters ranging from 0 to 1 for calculations considered here. The resultant expressions for P and the surface area of the pore, A (along with the minimum solid cross-sectional area, MSA, for comparison to mechanical properties, specifically u_v and elastic moduli) are shown in Table 7.3 using the notation of Fig. 7.6A.

The methodology for treating pores between simple cubic stacking of spherical particles, i.e., as used in Section 2.4.3, is similar but not identical to that of treating stacked spherical pores. (These two cases reflect interchange of the solid and pore phases.) The initial cell size around each unsintered spherical particle is that in which the spherical particle just fits, i.e., is tangent at the two poles and at the four side cell-particle point contacts. Thus, the initial, maximum cell size is at the initial, maximum pore size, which is when the spherical particles just touch (i.e., the counterpart of spherical pores being at the open-closed pore transition); however, in contrast to stacked pores, which can increase their volume fraction beyond this point of contact by becoming open pores, the corresponding porosity of bodies of stacked particles cannot be similarly increased since this would entail eliminating contact between the particles, which cannot be done and maintain a solid body. Thus, while stacked pores can readily extend beyond the porosity of pore to pore contact, often to $P = 1$, ideal stacking of

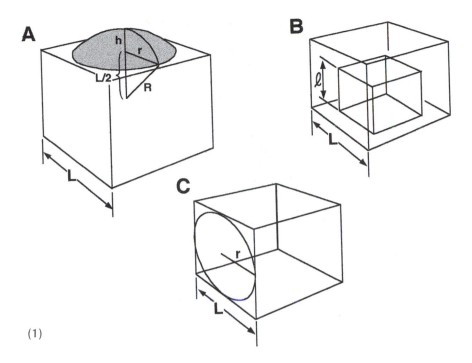

(1)

Figure 7.6 Modeling surface area using the same geometries as for MSA models (Figs. 2.1A, 2.10, 2.11. 3.1). 1) Schematic showing notation and treatment of A) pores between spherical particles, B) cubic pores, and C) cylindrical and spherical pores. Plot of the pore surface area, *A*, divided by 1 − P versus the volume fraction porosity P for the four pore geometries evaluated. Note that calculations do not continue to high P levels since they have no validity for pores between spherical particles, and increasing deviations occur for the simplified calculations used here at such P levels for cubic, cylindrical, and spherical pores.

spherical particles is limited to a maximum P determined by the contact (i.e., percolation limit) of the particles, i.e., at P ~ 0.48 for simple cubic stacking (and substantially less as the initial packing, hence green, density increases). Higher initial porosities, i.e., lower green densities, result only from inclusion of some additional pores of the first type via: 1) pores within the particles and thus of similar or smaller size than those between the particles, or 2) larger pores than those between the particles, e.g., resulting from particle bridging, bubbles, or fugitive material.

An essential difference for the second type of porosity is allowing for shrinkage of the stacked particles as sintering occurs via shrinkage of the cubic cell from its initial size (just enclosing the initial spherical particle). The six

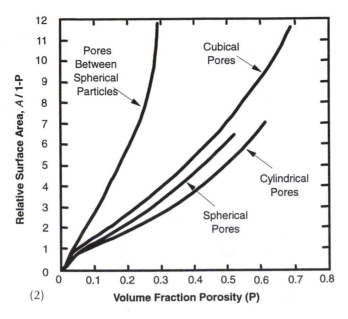

Figure 7.6 Continued

spherical segments that then begin to appear outside of the imaginary shrinking cubic cell walls are conceptually removed and added back into the truncated sphere inside the shrunken cube to preserve the original volume of the particles since mass is conserved in sintering. [Such modeling should also be applicable to forming bodies by reactions such as those involved in forming cementations or plaster bodies where the particle (and hence the body) mass increases as reaction increases and porosity decreases. Adaptation of models such as those of Rumpf (49), Harma and Satava (50), and Onada (51) should be useful in addressing such materials.] The end point, i.e., where the sintered body reaches theoretical density, is where the particle is now a cubical grain of the same volume as the starting spherical particle. Using the fact that the particle volume (and mass) are constant, the expressions for the curved surface area ($2\pi rh$) and volume ($h\pi/6$)[$h^2 + 3R_2$] of each of the six spherical segments removed, and the notation of Fig. 7.6 A, the expressions shown in Table 7.3 are obtained.

It is ultimately desired for some purposes, primarily specific comparisons or design, to obtain the specific surface area which is measured in meters squared per gram (m^2/g). [This can be obtained from the expressions for A in Table 7.3 by: 1) obtaining the absolute value of A by using the actual pore or particle dimensions, and 2) dividing this absolute value of A by the cell mass. For broader comparison, however, the density of the material at $P = 0$, ρ_0, can be neglected for relative comparison of bodies of the same material, or trends for different materials can be compared by multiplying them by the appropriate ratio of the densities.] The cell volume varies with pore geometry and size or particle size, but not widely for cases of primary interest (i.e., at intermediate and higher P levels, the values of are typically within a factor of ~ 2. Thus, by dividing A by $1 - P$, the resulting model trends can be compared on a general basis, which can be made more exact depending on the degree to which correction for differing cell dimensions are made.

The relative values of A divided by $1 - P$ for the selected pore geometries are plotted in Fig. 7.6 B based on relative pore to cell dimensions of 0 to ~ ½. Note first that the curves are all approximately linear, but not exactly; i.e., there is some concavity toward the upper left. (This is opposite to that of data, which is attributed to grain growth as P decreases as discussed later). Second, note the substantial higher surface area, but much more limited P range, for pores between stacked spherical particles, versus cubic, spherical or cylindrical pores. The much more limited P range arises from the percolation limit of stacked particles, i.e., at $P = 1 - \pi/6 \sim 0.48$ in this case (and at substantially lower values for other closer packed particle stackings) versus the solid percolation limit, i.e., P_C value, for cubic and spherical pores being theoretically 1. Aligned cylindrical pores theoretically break the body into "strands" of material above $P = \pi/4 \sim 0.785$, but for surface area purposes (as well as for some physical properties parallel with the aligned strands) this is not necessarily a major change. These differences are identical to those observed in using such models for mechanical and conductive properties, which is as it should be given their arising from basic geometry of the systems (Section 2.4.3).

Note three aspects of the results in Fig. 7.6B. First, a balance of porosity is often needed to achieve desired high surface area and performance goals with reasonable physical performance for important applications as catalysis, which requires a mixture of pore types to balance performance needs. Pores between particles give higher surface areas at a given level of P, but more rapid decreases in other key physical properties such as stiffness, strength, and erosion resistance, and lower permeability. Such porosity, by itself, can extend to at most P ~ 0.5, before reaching zero physical properties, which also limits the total surface area achievable with it alone. On the other hand, spherical, cubic, and especially tubular pores give greater permeability and much less decrease in mechanical properties for the same porosity level (Figs. 2.11, 3.1), and can reach substantial levels of surface area since they can extend to substantially higher P levels. Second, the calculations for cubic, cylindrical, and spherical pores are for when they are closed, and hence their surface areas are not measurable by normal techniques (again met in practice by a mix of pores, especially with pores between particles); however, the calculations are still valid whether the areas are readily accessible, hence measurable, or not. The trends of Fig. 7.6B would continue for a reasonable degree of openness (without correction for pore overlap), and in real bodies, variations of pore sizes, orientations, and secondary porosity (e.g., between partially sintered grains) would make many ideally closed pores open to various degrees. Third, as discussed later, the similarity of the curves (roughly parallel for absolute values) for different pores allows for comparison of surface areas from various pore types (or sizes) to be estimated by interpolation between curves for the various pores considered.

7.2.6 Permeation and Flow in Pore Structures

The pore structure has other important properties by itself, i.e., either independent of or in combination with those of the surrounding solid. These properties reflect the permeation or flow of fluids into and through porous bodies and their behavior in the body, e.g., as reflected in the conductive path length for conductive fluids in the pore structure. These represent a substantial and complex set of concepts, models, and measurements, whose comprehensive treatment is beyond the scope of this book, as noted earlier; however, it is of value to briefly outline these properties and key aspects of their dependence on the pore structure since they reflect important additional information about the pore structure and are critical to many important applications of porous materials.

The first of two basic factors that are specific to basic fluid behavior in porous bodies is the effective porosity, P_E, which simply reflects the fact that some porosity cannot participate in these processes. A major reason for lack of participation in such permeation and flow through bodies is that the pores are either totally closed, or are dead-end paths [e.g., in a glass bonded alumina, $P = 0.5$–0.7, 20–40% of the porosity may not contribute to permeability (52)]. Secondary reasons include factors such as the orientation of the pore or pore chain is such that there is no net pressure or chemical gradient across it. The labyrinth factor is defined as the ratio P_E/P, which in turn defines the structure factor, κ, as follows (52,53):

$$\frac{P_E}{P} = \kappa \, (1 + \kappa - P)^{-1} \tag{7.10}$$

The second factor is the tortuosity, τ, of the porosity, which is the ratio of the average path length for the fluid through the pore structure versus the actual thickness of the body in the direction of fluid flow.

Another important factor, although not specific to fluid behavior in porous bodies, is pore size; however, the aspects of pore size pertinent to fluid behavior in the pore structure may vary with the specific behavior of concern, and will typically be different from that pertinent to various physical properties of the porous solid. Thus, for example, smaller pore dimensions will typically play a more dominant role in limiting fluid behavior in porous solids versus average or larger dimensions often being more important in limiting physical properties of the porous solid.

Difinitive modeling and measurement of the behavior of fluids in porous solids is a large and complex subject, e.g., similar to that of the physical properties of the porous solid. Contrast of flow in the pore structure and the physical properties of the corresponding (porous) solid is important to the understanding of both, and is also important to the many applications that depend on both fluid behavior in the pores and the physical properties of the porous solid. The brief outline of diffusion and permeability in porous solids is presented below as a modest step in that direction. Permeation via diffusion in porous solids is very important, being a critical factor not only in the large and important field of catalysis, but also in construction materials, especially concretes (54). In the latter case diffusion of water in the processing and subsequent use of concrete is important, and diffusion of oxygen and chloride ions in the cement are important in determining the corrosion life of concrete.

Diffusive permeation in a porous body is governed by Fick's law giving the diffusion flux as the product of the concentration gradient and the effective diffusion coefficient, D_e. The latter is often given by (54):

$$D_e = (D_b)^{P/\tau} \tag{7.11}$$

where D_b = the bulk diffusion coefficient. Note that the tortuosity, τ, here may differ numerically from that for fluid flow through the same pore structure. Thus, for fluid flow in glass bonded alumina, $P \sim 0.5$–0.7, τ values were generally < 2 (52), but commonly 3–4 for alumina catalyst supports (54). Where other diffusion phenomena are involved, e.g., when the mean free path of the diffusing species is comparable to the pore size (i.e., Knudsen diffusion), this must be accounted for. This is often done by defining effective diffusion coefficients for these processes as before and combining them in parallel,

$$(D_e)^{-1} = (D_{eb})^{-1} + (D_{eK})^{-1} \tag{7.12}$$

When the diffusing species size is comparable to the pore size, the effective diffusion coefficient is replaced by the factor $(1 - \alpha)^4$ where α is the ratio of the species size to the pore size. Once D_e is estimated, various models can be used for estimating the rate of diffusion into the material.

Permeability, K, is the actual flow, by diffusive or other processes, through the material, which follows Darcy's law:

$$K = q\mu(ag)^{-1} \tag{7.13}$$

where q = the flow rate, μ = the viscosity, a = the cross-sectional area of the flow path (hence dependent on the character of the pore path), and g = the pressure gradient driving the permeation. The pressure may be externally applied, arise internally from capillary forces, or a combination. A basic, i.e., simple, model gives:

$$K = cP^3(\tau S_m^2)^{-1} \qquad (7.14)$$

where c = the Kozeny constant and S_m = the specific surface area; however, such models are commonly only approximate guides given the simplifying assumptions. Thus, Lukasiewicz and Reed (55) showed that such models did not describe well the behavior of porous alumina bodies they studied. Instead they showed that the specific permeability varied directly with the square of the mean entry pore radius (determined by mercury porosimetry) consistent with a capillary model they presented.

7.3 COMPARISON OF MODELS AND DATA

7.3.1 Thermal Conductivity and Expansion

Consider first the limited data for bodies having most pores reasonably close to ideal model shapes. Data of Francl and Kingery (7) for Al_2O_3 cubes (~ 2.5 cm on a side) with aligned cylindrical pores (~0.082 or 0.146 cm diameter) showed both good agreement with the relative difference between the conductivities parallel and perpendicular to the axis of the pores, as well as good absolute agreement with the MSA model (56) (Fig. 7.7). Minor deviations are attributed to the several percent residual porosity (e.g., 5–8%) in the Al_2O_3 and the fact that there were no cylindrical pores near the surface (hence giving a modest increase in conductivity both parallel and perpendicular to the pores). Francl and Kingery showed that Loeb's model (23) fit their data for aligned cylindrical pores well. This is not surprising since, as noted earlier, this model considers the cross-sectional area of the pores (and hence also of the solid) parallel and perpendicular to the overall heat flux. (The average and minimum solid areas for such pores are the same, so fitting of Loeb's model also supports fitting the MSA model for cylindrical pores.) Their data for Al_2O_3 sintered after burning out of naphthalene flakes added to generate porosity has only about half the slope (b value), i.e., ~ 1.3, expected for the reported "roughly spherical or ellipsoidal" pores of ~ 0.03 cm diameter; however, this lower slope is in good agreement with that expected for measurement perpendicular to the original pressing direction, and hence parallel to possibly aligned lenticular pores, as may well have occurred from the alinement of the naphthalene flakes in the slip

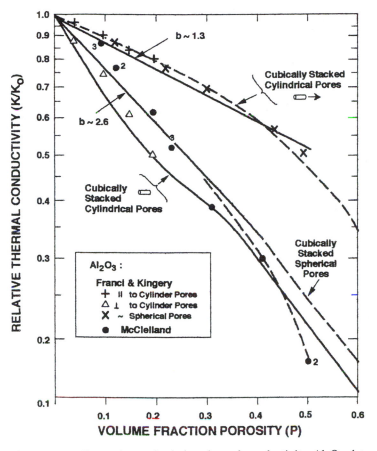

Figure 7.7   Comparison of relative thermal conductivity Al_2O_3 data of Francl and Kingery (7), Rice (56), and McClelland (57) at 22°C with MSA models for aligned cylindrical pores and spherical pores (numbers next to some data points reflect the number of tests where > 1). Note particularly good agreement between data and models for aligned cylindrical pores.

casting used to form these samples, which would lead to their forming lenticular pores (detailed characterization of their porosity was not provided).

Consider now data for bodies with more general, natural porosity. Data from previous evaluations (57–59) of ceramic conductive properties and their relation to MSA models is shown in Table 3.6, the pertinent portion of which is reproduced as Table 7.4. While the limited data is not sufficient to prove the validity of MSA models, it is consistent with the trends expected from these models. It also is consistent with the very similar, if not identical, behavior of

Table 7.4   Summary of Ceramic Property Porosity Dependence for Different Processing

Property	Extruded	Cold (die) pressed	Isopressed	Colloidal	Hot pressed
Young's modulus (E)	2.3 (1)	3.6 ± 0.8 (25)	4±0 ± 1.2 (14)	3.9 ± 1.2 (4)	4.5 ± 0.7 (31)
Shear modulus (G)	2.2 (1)	3.2 ± 0.9 (8)	3.9 ± 1.3 (12)		4.3 ± 0.6 (17)
Bulk modulus		4.0 (1)			5.1 ± 0.8 (8)
Thermal conductivity (K)	2.5 ± 1 (6)				4.4 ± 0.3 (3)
Electrical conductivity	1.9 ± 1 (4)	2.9 (1)			

Source: ref. (60); with permission of J. Mat. Sci.
[a] Note b values are for the porosity dependence e^{-bP}.

elastic moduli and thermal conductivity as a function of porosity, consistent with conductivity being in the solid phase as is stress transmission. Consideration of specific studies of ideal and other pore structures (mostly not included in Table 7.4) presented below substantially reinforces the consistency of thermal conductivity results and MSA model expectations.

The above overall trends are again supported by specific studies. First consider McClelland's alumina data (57,58), which approximately follows the spherical pore model (Fig. 7.7), but falling below this at higher P, consistent with a probable combination of pores between packed particles and larger (commonly approximately spherical) pores from particle bridging. The b value and initial roll over toward P_C are reasonably consistent. The overall trend of most data for the relative thermal conductivity of various rocks (60–64) and firebricks (10) at 22°C falls between the models for simple cubic packing of cylindrical, spherical, or cubic pores; (Fig. 7.8). This is consistent with these materials often having: 1) approximately spherical or more angular pores (e.g., as represented by cubic pores of various orientations whose average trend is similar to that for spherical pores; Figs. 2.11, 3.1), as is common in refractory bricks, and 2) a laminar character of some of the pores, which also falls in the stated range (Tables 3.2, 3.3). Also note that bodies with porosity extending to higher levels indicate rollovers toward P_C values as expected. Further, thermal conductivity data for die-pressed carbon bodies at both room (65,66) (Fig. 3.11A) and elevated temperatures (Fig. 9.11) is consistent with both the porosity and associated models, as well as with the elastic moduli and electrical conductivity, and rollovers of the data at higher P. Similarly data for sintered, hot-pressed, and single-crystal BeO (67) with a b value of ~ 4.4 (Fig. 7.9) is consistent with MSA model expectations, as is the parallelness of the data for different temperatures.

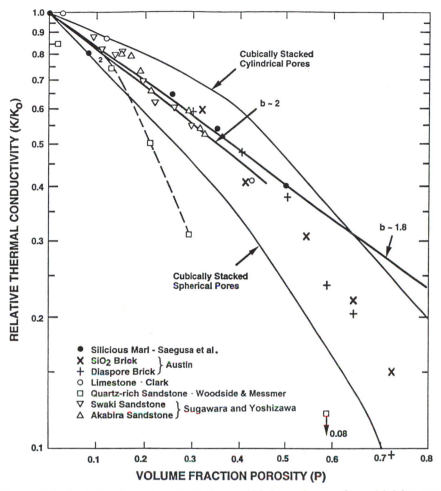

Figure 7.8 Relative thermal conductivity at 22°C for various rocks and bricks as a function of volume fraction porosity (P) from Rice (56), after Saegusa et al. (61), Clark (62), Woodside and Messmer (63), Sugawara and Yoshizawa (64), and Austin (10). Published with permission of J. Mat. Sci.

Turning to evaluations of non-MSA models, there have been three earlier but less extensive evaluations than that of Collishaw and Evans (3), results of which are summarized in Table 7.1. (Some better fitting equations in their evaluation, based on data for polyurethane foam, are not shown in Table 7.1 because of their complexity.) Francl and Kingery (7) found that equations of Loeb (23), Russell (20) (Eq. 8 of Table 7.1), and Eucken (17) (Eq. 6 of Table 7.1) were all reasonably close to their five data points (P = 0.12–0.49) for alumina with

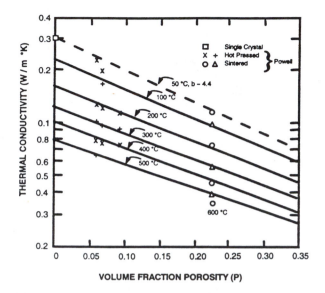

Figure 7.9 Thermal conductivity of single crystal and polycrystalline (sintered and hot-pressed) BeO near and modestly above room temperature versus volume fraction porosity, P, after Powell (67).

"isometric" pores at 200 and 800°C, but the closeness of fit decreased in the order listed. This order of fit also held for their thermal conductivity parallel with the aligned cylindrical pores in alumina, but Loeb's more complex equation was the only one agreeing with the conductivity perpendicular to the aligned cylindrical pores (since this model, like the corresponding MSA model, accounts for the cross-sectional area of pores parallel with the conduction direction). Dawson and Briggs' evaluation (9) of 10 equations based on their thermal conductivity for a porous (pressed) alumina (P = 0.45–0.62) resulted in the rankings shown in Table 7.1. The equations ranked second and third in their evaluation (not shown in Table 7.1) were not valid as P → 0 since they gave conductivities respectively of zero (for another equation of Russell) and 1/3 of the solid ceramic. (Note that three other equations they evaluated also gave zero total conductivity when the conductivity of the pores goes to zero, and one equation went to ∞, and two others gave values ≠ the conductivity of the solid at P = 0, i.e., only two equations were fully valid at P = 0.)

L. E. Nielsen's equation (Eq. 8 of Table 7.1, similar to that of L. F. Nielsen and his equation for elastic moduli, Eq. 11, Table 3.1) was proposed for both elastic moduli and conductive properties of composites; however, as noted earlier, despite incorporating P_C values, this equation continues to go to zero at P = 1 and does not vary extensively as it should to reflect the scope of pore character implied by the presence of pore parameters. Thus, while this equation has been used for composites (despite it going to infinite values as the percolation limit is reached, a problem that does not occur when applied to porosity), it is of doubtful utility. Rhee (8) evaluated four equations, three of which and their ranking are shown in Table 7.1, based on 10 sets of literature data (most of which are in Table 7.4). The best fitting equation in his evaluation was $K/K_0 = (1 - P) (1 + AP^2)^{-1}$, where A is an empirical fitting parameter, which gives values of 1 and 0 for $P = 0$ and 1 respectively for positive values of A. In between these bounds, it gives similar relative conductivity values as Eq. 4 of Table 7.1 and Fig. 7.1 over the range $A = 1$ to 10, i.e., falling modestly (typically ~ 10% above Eq. 4 for $A = 1$, being closer to (and crossing) Eq. 4 for $A = 5$, and falling below Eq. 4 for $A = 10$. Waugh (15) evaluated his equation (Eq. 3 of Table 7.1) against 10 data sets (most of which were the same as in Rhee's evaluation) obtaining exponential, q, values from 0.8 to 3.1. Such "testing" of non-MSA-based equations with a parameter has typically been essentially curve fitting since there has been no demonstrated (or considered) relation of the pore character to the parameter. Equations, with or without parameters, were often selected for commonly agreeing reasonably with much of the data whether or not they were derived and valid for the pore character and range they fit. A key test of models, however, is their ability to fit a variety of porosity-property data with equation parameters associated with the pore character, and thus makes them predictive where the pore character is known. Thus many non-MSA equations have failed in this key test, and hence require evaluation as to their ability to clearly associate parameter values with differing pore character.

Data to quantitatively evaluate effects of just microcracking on thermal conductivity or diffusivity are difficult to obtain since obtaining bodies with controlled microcrack populations that can be quantitatively characterized is at best difficult; however, thermal diffusivities or conductivities have been measured in graphite and other ceramics that often microcrack extensively due to highly anisotropic thermal expansion [depending on the degree of anisotropy, other properties, especially E (Chapter 3), and on grain size]. Such cracking should be approximately random (and on the scale of the grains), and is known to commonly close up on heating and to form or reform and open up on cooling, but with hysteresis. Study of the temperature dependence of the thermal diffusivity of a substantially anisotropic material such as Fe_2TiO_5 (68) qualitatively corroborates the random model of Table 7.2 (Fig. 7.2). Thus, with microcracks closing on heating and opening on cooling, the thermal diffusivity or conductivity, despite the normal decrease with increasing temperature, increases over most of the range of increasing temperature and decreases over most of the range of decreasing temperature (Figs. 9.3 and 9.14). These trends counter to the normal temperature dependence and reflect respectively the decrease of microcrack density and its increase. Thus, replotting these changes in diffusivity in terms of crack density, though not quantitatively possible, is qualitatively consistent with the random crack model, i.e., microcrack density decreases as temperature increases and vice versa. Data for less ($MgTi_2O_5$) and more (Al_2TiO_5) anisotropic materials of the same crystal structure show similar trends consistent with their relative anisotropies, except because of its greater anisotropy in the latter material it was not possible to get dense bodies of sufficiently fine grain size of the latter to have no microcracking.

Litovsky and colleagues (24,25) have extensively reviewed data and mechanisms for the net thermal conductivity of ceramic refractories. Their reviews address a diversity of single and combined mechanisms for a diversity of materials and microstructures, including mixtures of microcracks with various types and amounts of porosity. They treat both materials that are predominantly or exclusively crystalline or glassy, the former typically having higher conductivity which decreases with increasing temperature and the latter having lower conductivity which increases with increasing temperature. Within the context of these two basically different structure materials, the dominant microstructural effects on conductivity are the presence or absence of either grain-boundary pores or of microcracks (commonly at or crossing grain boundaries). Thus, in either case the type and especially amount (i.e., pressure) of the gas in the microcracks and grain-boundary pores (both of which are typically accessible to the gas environment of the refractory) has a substantial effect on conductivity. At and near room temperature the presence of grain-boundary pores, microcracks, or both commonly results in conductivities of bodies with such microstructures being a fraction, e.g., 0.2–0.5, under vacuum versus under an atmosphere of various test gases. The nature of the gas has a secondary effect on the conductivity versus its pressure. Such effects can persist at still substantial levels to low porosities (e.g., 2% open and 7% total) with limited microcracking. On the other hand, bodies with predominantly intragranular pores (hence not accessible by permeation or diffusion of external gassses) show no significant effect of the gas environment of the test atmosphere on the body conductivity.

As noted in Section 2.1, thermal expansion is normally independent of both porosity and grain size; however, as noted in Section 7.2.1 microcracking can change this. Thus, Manning (69) showed that the thermal expansion of Nb_2O_5 increased with increasing porosity (Fig. 7.10A). While this apparent porosity dependence probably results, at least in part, from decreasing grain size and thus decreasing microcracking as porosity increases, some direct effect of porosity may be occurring. The latter may occur due to pores reducing strain energy necessary for microcracking both on a global scale for all pores, as well as locally by intergranular pores reducing grain contact areas. Thus, some coupling of pore and microcrack effects probably occurs, and requires attention.

That grain size clearly plays a role in the thermal expansion of highly anisotropic ceramics was shown by Parker and Rice (70) in Al_2TiO_5-based

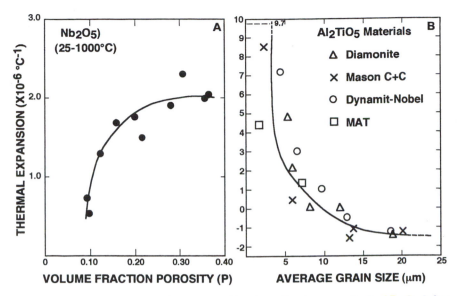

Figure 7.10   Thermal expansion versus: A) volume fraction porosity for Nb_2O_5 [after Manning (69)] and B) grain size for various Al_2TiO_5-based bodies [after Parker and Rice (70)]. Published with permission of U. Iowa (part A) and J. Am. Cer. Soc. (part B).

materials (Fig. 7.10B). Thus, as grain size increases above the critical size for the onset of microcracking (Sections 2.2.2, 10.2.5), the subsequent progressive increase in microcracking and attendant decrease in bulk thermal expansion with further increase in grain size continues until the microcracking saturates and the thermal expansion thus reaches its minimum. The decrease in bulk expansion of the body results from the microcracks, which simply open or close as temperature changes to accommodate the normal lattice expansion behavior of solid sections between the microcracks with reduced or no effect on the bulk expansion behavior of the body. Thus, as the body heats up, the microcracks close at the temperature of their formation for that body, so with further heating of the body exhibits the same bulk and lattice expansions. Similarly on cooling, the body exhibits the same lattice and bulk expansion (i.e., actually contraction on cooling) until microcracks reform or reopen, as discussed further in Section 9.3.1.

7.3.2   Electrical Conductivity and Breakdown

While the amount and scope of data for electrical conductivity is even less than that for thermal conductivity, it also agrees with the MSA models for the porosity expected from the processing method, where this is given. Thus, extruded

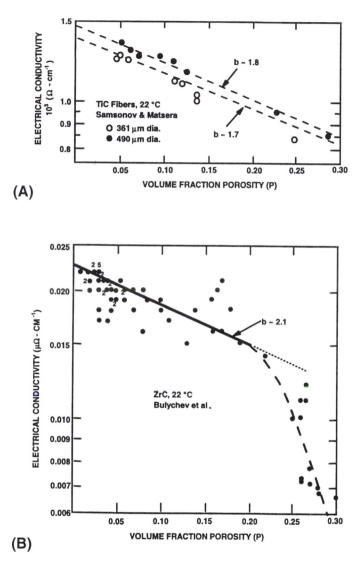

(A)

(B)

Figure 7.11 Electrical conductivity of extruded, sintered carbides. A) TiC fibers of two different diameters versus volume fraction porosity, P, at 22°C [after Samsonov and Matsera (72)]. B) Extruded rods of sintered ZrC (numbers next to some points reflect the number of values where >1) at 22°C after Bulychev et al. (73). Note the low b values consistent with tubular pores from extrusion for both materials and respectively a suggested start and a clear rollover toward values at $P_C > 0.3$.

carbon bodies show low b values, i.e., ~ 2 for conduction parallel to the extrusion axis (71), and somewhat higher for die-pressed bodies (~3) as expected (Table 3.6 and Fig. 3.11A). The lower b values for extruded bodies reflect a substantial tubular pore character with varying mixtures of other pores (mostly between grains), and the start of the rollover to P_C values and the b values are reasonably consistent with each other and MSA models (Figs. 2.11, 3.1, 7.11). Thus, note the low b values of 1.7 and 1.8 for extruded fibers of TiC (72) and of 2.1 for extruded rods of ZrC (73). The low b values (i.e., closer to the MSA model for conduction parallel with aligned cylindrical pores) of the fibers versus rods is consistent with greater tubular pores expected from the higher shear in extruding the finer fiber, versus coarser fibers and especially rods. Further, the indicated rollover toward the decrease to zero property values at P_C are also consistent, i.e., > P = 0.3 for the finer fibers, probably at P ~ 0.3 for the coarser fibers, and clearly at P ~ 0.2–0.25 for rods. Some tendency for lower projected P_C values is attributed to heterogeneous porosity, which is also indicated by greater data scatter. These trends are supported by data on other materials such as metals (e.g., Fig. 8. 5) and at elevated temperatures (74), e.g., Fig. 9.14.

All of these results are for materials with predominately electronic conductivity; however, Mizusaki et al. (29) showed that data for ZrO_2-Y_2O_3 (74), BaF_2 (75), and $La_{0.5}Sr_{0.5}CoO_3$ (76) materials with primarily ionic conductivity was consistent with their modified MSA model with values for the half wavelength for the particles from ~ 0.5 to 0.8 of the particle diameter. Since there is at present no way of determining whether these values are realistic for the specific bodies evaluated, it cannot be judged whether this is a true fit or simply curve fitting; however, the data evaluated is also consistent with expected conventional MSA models (Fig. 7.12).

Consider next the limited data for the effects of porosity on electrical breakdown of normally nonconducting materials. Gerson and Marshall (30) tested their model, based on the probability of pores aligned in a column through the sample thickness as discussed earlier, by sintering pressed PZT powders with dispersed plastic spheres, finding a good fit to the predictions of their model. While their model has both similarities and differences with MSA models for mechanical and conductive properties, it generally also gives straight lines on semilog plots of breakdown field versus P. Thus, the b values, although having a different relation to the porosity, are still of some possible value for comparison (i.e., it increases with pore size), but may also depend on other factors including specimen dimensions and configuration. Evaluation of their above data gives a b value of ~5.1 for spherical pores ~ 120 μm diameter.

Subsequent measurements were made by Beauchamp (77) on hot-pressed MgO with or without LiF additions to various residual porosities up to P = 0.12. Breakdown fields decreased from similar or somewhat lower than single-crystal

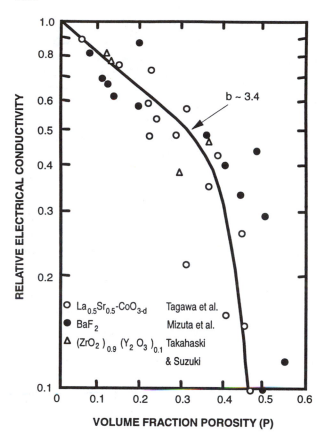

Figure 7.12 Relative electrical conductivity versus volume fraction porosity, P, at 22°C for three materials and investigations where conductivity is predominately ionic (versus electronic in Fig. 7.11) from Mizusaki et al. (29,74–76).

values with increasing P, with breakdown paths being mainly along grain boundaries, especially along triple lines in polycrystals and in <110> directions in single crystals; however, the decrease was initially more rapid, being nonlinear on semilog plots, e.g., with initially higher b values of 20 to more than 40, decreasing as P increased. Breakdown was also lower, as was the apparent rate of decrease, as grain sizes increased from 1–3, to 10–30, and 50–100 μm, respectively, when hot pressed or subsequently annealed. The greater, more rapid decrease of as-hot-pressed material and the apparent reduced porosity dependence as grain sizes increased are probably partly due to reduction in the residual fluoride additive content [mostly at grain boundaries (78)], and possibly some to the shift of pores from inter- to intra- granular locations; however, the effect of grain size was tentatively attributed to corresponding increases in intergranular pores and their being along preferred failure paths. Thus these results clearly support a significant effect of porosity on dielectric breakdown, and indicates grain boundaries and associated pores as a factor, but cannot be clearly compared with Gerson and Marshalls' model.

Tunkasiri and Rujijanagul (79) reported dielectric breakdown fields decreasing from 8.5 to 2.4 kV/mm as the grain size of sintered $BaTiO_3$ increased from ~ 3.5 to 25 μm. They attributed much or all of this decrease to failure occurring mainly along grain boundaries, especially with intergranular pores. Total porosity was ~ 0.03 in the finest grain bodies, and ~ 0.045–0.05 in the larger grain bodies., so larger grain bodies contained more porosity, part of which probably occurred as larger pores on some of the larger grain boundaries. Kishimoto et al. (80) reported similar results for (thin, ~ 200–270-μm thick film prepared) $BaTiO_3$ specimens with a similar grain size range (plus some larger grain bodies). While residual porosity was not specified, there must have been some, including along grain boundaries, that again probably played an important role in their failures.

Data of Morse and Hill (81) for electrical breakdown of alumina at various temperatures to P ~ 0.03–0.17 shows some similarities to that for MgO above, but indicate more consistency with Gerson and Marshalls' model. Thus their data was linear on a semilog plot with a lower b value consistent with finer pores (average ~ 2 μm) in their hot-pressed bodies and the aproximate extrapolation to values from single-crystal tests (Fig. 7.13). Extrapolation to somewhat below the single-crystal value at P = 0 and breakdown and resultant fracture along grain boundaries again indicate a role of grain boundaries in the breakdown process. Rupaal et al. (82) corroborated the model of Gerson and Marshall in pressed and sintered $MgAl_2O_4$ (~ 0.7 mm thick and ~ 7 mm dia.) with natural porosity from P = 0.062 to 0.17, as well as in single crystals. They showed that the larger pores (~ 40–80 μm) in the natural pore distributions dominate the breakdown and are thus the pore sizes to use in the model. Their results gave an average b value of ~ 9, which is higher than expected from pore size alone, but may reflect other factors.

Gerson and Marshalls' observations that, as specimen thickness increased, the breakdown field decreased (e.g., by ~ 50% as thickness increased from ~ 1 to 10 mm) and their theoretical calculations showing the porosity dependence increasing as the test sample diameter increases may seem counterintuitive to some; however, these specimen dimension dependences (which have been corroborated by others, e.g., Morse and Hill) are, at least in part, reflections of increased probability of finding greater approximate alignment of pores favorable to breakdown as specimen dimensions increase. Gerson and Marshall also noted surface space charge and heating effects as possible factors in the

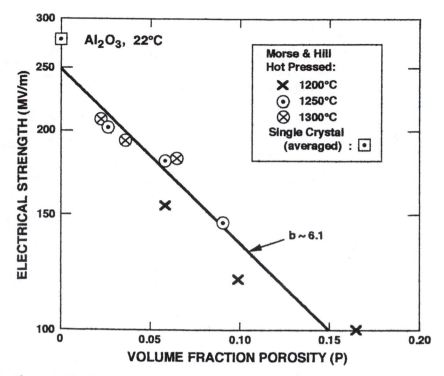

Figure 7.13 Electrical breakdown strength of hot-pressed alumina as a function of volume fraction porosity, P, at 22°C [after Morse and Hill (81)].

decrease with increased thickness; however, the decrease in breakdown field with increasing thickness may also reflect aspects of the processing-pore character. Natural porosity between particles (~5% and 5–10 μm diameter in their samples) is commonly reduced on surfaces of many samples, such as pressed bodies. This would reduce finer surface porosity to align with the larger, artificially introduced pores, giving some decrease in breakdown field as thickness increased, due to the increasing thickness progressively diluting the effects of a commonly denser surface. There may also have been some reduction of the opportunity for the large (~ 200 μm) plastic spheres used to introduce the bulk of the pores to occur near the surface in thinner samples. Furthermore, the general trend for less consolidation of, i.e., more (and possibly larger) pores in the central region, and for laminar pore character at higher pressures (and as a function of higher aspect ratio) of die-pressed bodies would also favor the above decreases in breakdown; however, tests at temperatures to 1400°C corroborate the decrease in breakdown with decreasing thickness below thicknesses of ~ 0.6

mm in both single crystal and polycrystals of alumina (83), thus indicating more basic causes which may be exacerbated by the processing porosity effects suggested above. These high temperature tests again indicate similar, but probably somewhat higher single-crystal values. Finally, note that heterogeneity of the porosity, i.e., of size, shape, and spatial distribution, probably plays a major role in electrical breakdown as indicated by Rupaal et al. (82). Both heterogeneity and surface effects become of increasing importance as dielectric insulators become thinner, especially for electronic systems where they become quite thin (e.g., approaching 1 µm for electronic applications).

An important question is the relation of electrical breakdown to mechanical failure under normal mechanical stressing since both result in mechanical failure and such failures commonly arise from pore clusters in both cases. Kishimoto et al. (84) first compared flexural and electrical failure of partially sintered TiO_2, showing nearly identical Weibull moduli for both failures, i.e., averaging respectively 6 ± 1.5 and 6 ± 1.9. Kishimoto et al. (85,86) have subsequently explored the possibility of using dielectric stressing to test for mechanical reliability. They first exposed porous TiO_2 (P~ 0.1) samples to fields causing electrical breakdown (failure) of 30% of the samples, then measuring the subsequent mechanical strengths of those samples not failing from the electric field exposure. They found that the exposure to high voltage truncated the lower end of the strength distribution of the subsequently flexure tested samples; however, the truncation was not complete, i.e., some samples surviving the voltage test (failing 30% of the samples) left samples failing at mechanical stresses that were below the 30% failure level in flexure testing of fresh samples. The correlation coefficient between electrical breakdown failure and flexure testing was 0.77. Thus, such voltage screening had some useful but imperfect benefits in removing weaker samples. These trends are consistent with their earlier results (84) showing different dependences of the two failures, i.e., failure fields increasing only ~ 30% as flexure strengths increased approximately sixfold as P decreased from ~ 0.9 to 0.4.

Kishimoto et al. (80) also compared electrical failure of their $BaTiO_3$ cited earlier with mechanical failure of the same type of thin samples. They found the Weibull moduli (typically ~ 5 ± 1) and distributions very similar for both types of failure, but did observe some definite differences for larger grain samples. These differences were attributed to grain sizes no longer being small relative to specimen thicknesses and probable different effects of surface flaws on the two failure mechanisms.

The similarities and differences between the statistical, e.g., Weibull, behavior of dielectric breakdown and flexure strength are consistent with the porosity dependence of the two properties having both similarities and differences. Use of true tension instead of flexure should improve the correlation (since the dielectric stress was more comparable to a uniform tensile stress distribution than the large stress gradients in flexure, e.g the latter emphasizing surface and near surface flaws); however, there are more fundamental differences that limit the degree of correlation. While both failures are impacted by larger pores and pore (or pore-flaw) clusters, they differ in their dependences on specifics of these, especially the geometry of their spatial distribution. Thus, dielectric failure depends substantially on chains of pores (or pore-crack combinations) nominally parallel with the field that substantially reduce the solid material barrier along an essentially tubular path of possible failure. Tensile failure may be more impacted by other flaws (especially surface flaws), but flaw-pore combinations and larger pores or pore clusters are also important; however, for all of these a more compact geometrical distributions of the flaws, pores, or combinations having larger cross-sectional area normal to the tensile stress are more serious sources of failure (Sections 5.3.3, 5.3.4) rather than their being in chains parallel with the stress or field.

7.3.3 Dielectric Constant and Piezoelectric Properties

Consider first data for dielectric constant versus P (31,33,87–100), much of which is presented in Fig. 7.14. Note first that all data decreases essentially linearly with increasing P, even at P in the 0.5–1 range, except for a probable exception of the $CaTiO_3$ data, which appears to be bilinear. Furthermore, note that data for materials with moderate relative dielectric constants (e.g., ~ 10) all follow or are close to a rule of mixtures relation, except for the limited Si_3N_4 data. Such a dependence is consistent with this property for these materials being dependent primarily, or exclusively, on only the volume fraction porosity (P), and thus being given by a rule of mixtures, i.e., ~ 1 − P dependence (not exactly since the dielectric constant is ~1, not zero; see Eqs. 2.1, 7.4). The linearity and approximate rule of mixture behavior of dielectric constant data was also shown in a previous survey (60), where the A values in 1 − AP averaged ~ 1 (Table 7.5). That dielectric constant generally has much less dependence on P than other, especially mechanical, properties such as indicated by work of Liu and Wilcox (96) on coordierite bodies sintered with increasing volume fractions of glass balloons as the main source of increasing P. The dielectric constant data for these bodies is very close to 1 − P, in contrast to the flexure strengths (b value of ~ 2) and is reasonably consistent with the MSA model for spherical pores.

The primary deviants from being near a simple rule of mixtures are materials with high dielectric constants, e.g., >1000, i.e., primarily $BaTiO_3$ and PZT. Earlier $BaTiO_3$ data evaluated by Payne and Cross (34) also showed both an approximate rule of mixtures behavior and deviations below it. Furthermore, much of this $BaTiO_3$ and PZT data suggests increasing deviations below the rule of mixtures as the size of the pores decreases, e.g., data of Biswas (88), Hikita et al. (99), and Bast and Wersing (100), having pores respectively of ~ 100–150,

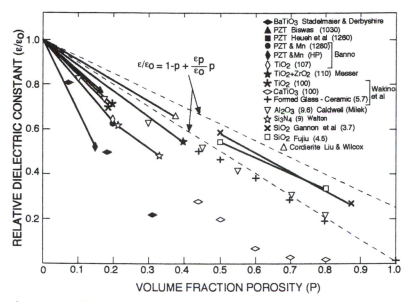

Figure 7.14 Linear plot of ratio of the relative dielectric constant at a some P divided by the value at P = 0 versus volume fraction porosity at 22°C for various ceramics (31,33,87–97). Note: 1) the relative dielectric constant of the dense material is given for the specific bodies in () wherever it was known, and 2) for simplicity, much data is identified by only a single symbol at both or one end of the line.

~ 100, and 20–50 μm (the first two are approximately spherical and the latter is the thin dimension of long, aligned, rectangular pores). These large pore sizes are in contrast to the known and expected much smaller, normal processing pores in the PZT of Banno (33), Okazaki and colleagues (41,101) and BaTiO$_3$ (87). Both the indicated increased deviations from a rule of mixtures trend with higher dielectric constant and finer pores suggests pore-charging effects as an important, possibly dominant, factor. Thus, combination of such effects with the rule of mixtures relation, or further development of models addressing charging, e.g., that of Bruggeman (16; also 22,34,95) should be pursued.

Yamai et al.'s (102) sintered Nb$_2$O$_5$, Nb$_2$TiO$_7$, and a K-Sr niobate all had rapid initial and then slower decreases from initial ε values of ~ 350, 175, and 3500, respectively, for P ~ 0.05–0.35. Introducing microcracks (~ 1 vol. %), apparently from grain growth from heat treatment (leaving P ~ 0.1 ± 0.05), gave ε of Nb$_2$TiO$_7$, ~10% below values for porosity only, indicating greater microcrack effects. The other two materials had greater ε reductions, 30–60% below those for porosity alone, indicating marked microcrack effects consistent with possible crack surface charging effects, especially at higher ε.

The first of two other probable aspects of porosity dielectric constant variations to be noted is porosity heterogeneity. While variations of P within the

Table 7.5 Porosity Dependence of Dielectric Constant

Material	Processing[a]	P[b]	Data points[c]	A[d]	Investigator
SiO_2	S	0.03–0.85	>18	0.84	Gannon et al. (94)
SiO_2	S	0.03–0.83	>14	0.9	Harris and Welsh (97)
SiO_2	S	0.08–0.16	65	0.9	Harris and Welsh (97)
SiO_2	Foamed gel	0.50–0.80	6	0.8	Fujiu et al. (95)
Borosilicate glass	S	0.05–0.43	14	0.9	Sacks et al. (98)
TiO_2 (+14%ZrO_2)	S	0.05–0.38	13	1.5	Messer (91)
PZT	S	0.03–0.16	8	1.4	Biswas (88)
PZT	S	0.06–0.25	16	1.6	Hsueh et al. (89)
Si_3N_4	RS	0.22–0.32	4	1.2	Walton (93)

Source: ref. (60); with permission of J. Mat. Sci.
[a] S = Sintering, RS = reaction sintering.
[b] Volume fraction porosity.
[c] Number of data points (> indicates some data points reflected more than 1 measurement).
[d] A is from the equation $\epsilon = \epsilon_0 (1 - AP)$ where P = volume fraction porosity; see Table 7.6 for additional data.

measuring area will not result in deviations where the property dependence is linear with P, variations will occur if the P of the sample is different from that in the measurement area, e.g., as could commonly be the case in die-pressed disks frequently used for measurement, since there are often some density gradients, especially near the specimen surfaces. Second is changes in pore structure that effect surface charging of pore surfaces or flux "leakage" into the pores (or other effects leading to the faster decrease of dielectric constant with increasing P than for the rule of mixtures). Both should occur as a function of pore shape, e.g., between spherical pores and the cusp-shaped areas of pores between particles. Such pore-shape effects can also be a factor in effects of heterogeneity where this entails finer pores or other pore changes that give deviations from rule of mixture behavior; however, such changes in pore character must also occur with much processing to achieve high P. Thus, as discussed in Chapter 2, the percolation limit for the solid particles in porous bodies made by consolidating powder particles where the pores are primarily those between the packed particles occurs at $P_C < \sim 0.5$. Therefore, obtaining bodies of the same material with higher P requires introduction of other pores which must be larger than those between packed particles, e.g., by foaming or use of larger burnout particles. Such changes in pore character to obtain higher P may reduce pore charging effects, hence reducing the greater decrease of dielectric constant versus that due to the rule of mixtures rate. They also appear to be those that are associated with the reduced rate of dielectric constant decrease with increasing

P seen with $CaTiO_3$ in Fig. 7.14. Thus, the character of the porosity and changes in it with increasing porosity need to be considered even for dielectric constant. The failure of existing models to recognize this limits their utility, not as much as for most other properties, especially mechanical ones, but still presents limitations.

Next consider piezoelectric properties, the porosity dependence of some of which are summarized in Table 7.6. Note that these are at least approximately linearly dependent on porosity, as is consistent with much of the analysis noted earlier, provided pores are not highly anisotropic in shape. Dunn and Taya's (35) evaluation of piezoelectric properties from the data of Bast and Wersing (100) for PZT with printed tubular pores showed that the somewhat scattered data was equally consistent with either the limited nonlinearity predicted from their modeling or the approximately linear dependence on P shown in Table 7.5; however, Dunn and Taya's fitting to their model involves an assumption of the aspect ratio of the assumed ellipsoidal pores, which appears to be mainly a fitting parameter with little or no justification in the actual microstructures. Thus, in view of the broader linearity of many of these properties in Table 7.5 and uncertainties in their modeling, linear dependence of many of them should be seriously considered. Also note that their analysis questions the proposed nonlinearity with elongated pore since the data they evaluated involved extreme, i.e., very long, aligned tubular pores. Additionally, data of Hikata et al. (99) on (coarse grain/agglomerated) sintered PZT with P = 0.25–0.7 showed d_{33} $\in$ and d_{33} varying linearly with P (with A values respectively of 1.4 and −0.5,

Table 7.6 Porosity Dependence (A Values)[a] of Some Piezoelectric Properties

Material	Processing[b]	Porosity[c] (%)	d_{31}[a,d]	K_p[a,e]	Z_r[a,f]	$\in$[a,g]	Investigator
PLZT	H	N, 0–8	2.4	4.0	13		Banno (33), Okazaki et al. (41,90)
PZT+Mn	H	N, 2–14	1.9	1.6	16	2.9	Banno (33), Okazaki et al. (41,90)
PZT+Nb	S	N, 5–19				1.9	Banno (33), Okazaki et al. (41,90)
PZT	S	PR,0–16			3.4	2.5	Bast and Wersing (100), Dunn and Taya (35)
PZT+Nb	S	N, 3–15	1			1.4	Biswas (88)

[a] In the porosity dependence of the relative property: (1 − AP).

[b] H – hot pressing, S = sintering.

[c] Porosity character: N = natural porosity, PR = printed rectangular (tubes; ~ 20–50 μm in the thickness direction), and amount in percent.

[d] Piezoelectric strain coefficient.

[e] Planar coupling coefficient.

[f] Resonance impedance.

[g] Relative dielectric constant (see also Table 7.4).

the latter minus sign meaning a linear increase with increasing P), and g_{33} increasing very nonlinearly with increasing P, e.g., the relative value being ~ $(1 - P)^{-2.4}$. (Note that this nonlinearity is such that there appears to be no significant linear section of plots of the log of g_{33} or its reciprocal versus P, i.e., indicating greater decreases than for MSA models.) Finally, note that differing dependences of the piezoelectric strain coefficients (e.g., d_{31} and d_{33}) and voltage output coefficients, e.g., g_{33}, especially the increase of some with P instead of decreasing with P means that the response of these materials to signals while under hydrostatic pressure can be improved, e.g., Nagata et al. (101). This has been used to design the pore structure of sonar detectors to greatly improve their sensitivity under hydrostatic pressure, e.g., for use at significant ocean depths (103,104), as discussed in Section 10.3.1.

An important factor in the operation of piezoelectric transducers to generate ultrasonic waves is electrical fatigue. Fatigue manifests itself by reduction in the polarization and resultant elastic strain from the applied alternating electric field as the number of cycles increases. Such fatigue, which is related to domain behavior, can occur in fully dense material (including single crystals) and can be affected by atmosphere (e.g., H_2O) or other surface species (105); however, it is also significantly affected by porosity as shown by Jiang and Cross (106). They demonstrated that the number of cycles for fatigue to occur in a PLZT (grain size 2–5 μm) and two PZT materials investigated decreased from > 10^9 at P<0.01 to ~ 10^4 at P ~ 0.08. Bodies without extensive fatigue could be restored to near or full original performance by annealing them above the Curie temperature. They proposed two fatigue mechanisms: 1) gradual domain stabilization (pinning) by space charges, and 2) surface deterioration caused by the large surface AC fields.

Next consider the limited data on effects of microcracking (and in some cases combined effects with porosity) on piezoelectric properties. Hiroshima et al. (107) reported complex interactions of composition-firing temperature-grain size-microcrack-porosity effects on relative permativity and Curie temperature in various $Pb_{1-x}Ba_xNb_2O_6$ (x = 0.3–0.5). Increasing porosity and microcracking both increased Curie temperatures in materials with higher internal (microstructural) stresses, which they attributed to relaxation of the internal stresses. Relative permativity increased with decreasing porosity, but was apparently increased by some increases in microcracking, but not greater extents of microcracking (which increased with grain size). Chung et al. (108) reported microcracking increasing with both DC polling fields and grain sizes (varying, respectively, from 36 and 11 to 53 and 17 μm) in both $BaTiO_3$ and PZT materials. They also noted that the microcracks were trans- rather than intergranular. These poling-grain size-microcracking results are generally consistent with earlier work of Chiang et al. (109) on PZT showing higher voltages and larger grain sizes leading to microcracking during poling. They also reported that the planar coupling coefficient was reduced some by both microcracking and slow crack growth, especially in larger grain samples.

7.3.4 Magnetic Properties and Optical Scattering

Consider now the limited data on the porosity dependence of magnetic properties. Jain et al. (39) sintered MnZn ferrite bodies to test the two models for the dependence of initial permeability, μ_i, and coercive force, H_c, noted earlier. They found that μ_i increased linearly with increasing spacing between intergranular pores, λ, as proposed, i.e., ~ a threefold increase as spacing of the intergranular pores increased from ~ 3 to ~ 9 μm; however the dependence of H_c, was ~ $\lambda^{-0.5}$ instead of λ^{-2} as predicted. They suggested that the dependence of μ_i as predicted reflected its dependence on small domain displacements from their pinned positions, while that of H_c involved considerable displacement (via bowing), that was not accurately reflected in the model.

Comparison of Jain et al.'s results with other limited data shows some similar, and some different, behavior (Fig. 7.15). Data of Igarashi and Okazaki (40,41) for two grain sizes of NiZn ferrite (~ 0.6 and 21 μm) with larger pores being respectively of the order of 1 and 10 μm is linear on a semilog plot of μ_i versus P. The resultant slopes (b values) of ~3 and ~ 5, respectively, for sintered and hot-pressed material are consistent with MSA models; however, this may be simply coincidence. Since pore spacing commonly decreases as P increases, the data is also at least qualitatively consistent with $\mu_i \propto \lambda$ as found by Jain et al. above. Earlier data of Economos (110) for $MgFe_2O_4$ is similar to the μ_i trend of Igarashi and Okazaki at very low P, but then decreases much faster with increasing P (possibly due to reduced grain size as discussed below). Igarashi and Okazaki showed H_c to be independent of P, while Economos showed it rapidly increasing with increasing P (Fig. 7.15), e.g., in a similar fashion as a $P^{2/3}$-dependence as given by one model noted by Jain et al. Smit and Wijn (111) also observed H_c increasing as P increased [which Igarashi and Okazaki (40,41) attributed to corresponding grain size changes]. H_c increasing with P is qualitatively consistent with it decreasing with λ as found by Jain et al. Even greater contrast was found for the P dependence of remnant magnetic flux density, B_r. Igarashi and Okazaki found it decreasing with increasing P similar to their μ_i data, but somewhat faster for the same body (i.e microstructure), while Economos found it increasing with increasing P.

Magnetic properties thus show a diversity of P-dependence, including opposite P-dependences for a given property in different bodies, as do piezoelectric properties. Some of these differences in behavior, even substantial ones, probably reflect effects of differing grain sizes (as indicated above) in view of grain size having substantial and often similar effects via the same mechanism, e.g., effecting domain behavior. Thus, much better grain characterization is needed along with much better porosity characterization. For the latter, possible effects

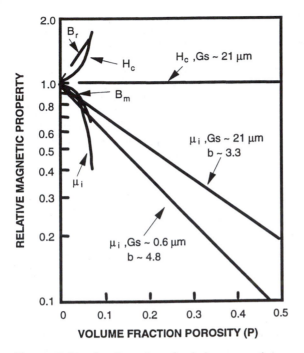

Figure 7.15 Semilog plot of relative magnetic properties versus volume fraction porosity, P, at 22°C. Data for the following relative properties: 1) coercive force (H_c), 2) initial permeability (μ_1), and 3) remnant magnetic flux density B_r for both: a) sintered $MgFe_2O_4$ (110) (including also data for maximum induction, B_m), and b) sintered (grain size ~ 21 μm) and hot-pressed (grain size ~ 0.06 μm) NiZn ferrite (40,41). Note: 1) the data for B_r follows a similar trend as for μ_i, especially the lower, hot-pressed data, and 2) the linear μ_i, semilog trend and b values, i.e., slopes. Note designation of grain size (Gs) and its effects on porosity dependence of properties.

of separations of intragranular pores on domain behavior, but not of intergranular pores (since their effect is likely to be overridden by that due to grain boundaries), need to be addressed; however, both the consistency and the diversity of results as well as the limited modeling both indicate that much of the diversity is real, namely, that it will be modified but not eliminated by better microstructural characterization and models. This expectation is supported by the similar diversity of piezoelectric dependence on P and the underlying similarity of mechanisms, e.g., via domain effects.

As noted earlier, scattering becomes important, and commonly totally dominant at shorter wavelengths, e.g., in the visible, when the pore size is a reasonable fraction of the wavelength, which is commonly the case for much residual porosity from sintering, e.g., Fig. 7.5. Limited data of Goodman (112) on the effects of porosity in porcelain, although complicated by the multiphase effects, illustrate this, showing translucency rapidly decreasing to ~ 70% and 40% of P = 0 values by ~ 5% contents of pores with respective mean diameters of 130 and 30 µm. Saturation of the translucency decreases occurred beyond ~ 5% porosity, probably due to multiple scattering from pores and second phases, was found. Clearer are results of Grimm et al. (113) for alumina lamp envelope material (~ 0.03–0.63% porosity, grain size ~ 27 µm) showing a linear decrease in transmission with increasing porosity (giving a b value of ~ 100 !). Tests of grain-size effects on transmission confirmed the gross predominance of porosity on scattering relative to scattering from grain boundaries (without gross second-phase content), which is also supported by observations of Hing (114).

7.3.5 Surface Area

A summary of representative S_m versus P data for various metal (44) and ceramic (43,115) pressed and sintered powder compacts is given in Fig. 7.16. The general character of these curves as well as absolute values are fairly consistent with the models (Fig. 7.6B), e.g., nearly linear curves (until P ~ 0.45–0.55) with some concaveness toward the plot origin. Martin and Rosen (43), based on their measurements of ZnO partially sintered powder compacts [on which they also measured elastic moduli (116)] showed S_m-u_v correlations demonstrating that monitoring u_v during sintering could be a useful tool for following densification. Such possible use is of added interest because of the direct correlation between u_v and elastic moduli (Chapter 3). Jernot et al. (44), in fitting their model to the metal data they surveyed (Fig. 7.16), showed that the co-ordination number of the metal particles generally increased from 5–6 to 6–7 as the surface area increased with increasing density over the range they covered.

The correlation of Martin and Rosen raises the issue of whether the body with the highest S_m has the lowest sound velocity at any P and vice versa. While an absolute explanation for this apparently intuitive contradiction cannot be given, the present modeling gives insight into two possible reasons for this opposite trend, and indicates that it is not intrinsic, but depends on pore parameters. First, it may arise in part due to differing pore geometries. Thus, as noted earlier, differing pore shapes lead to some variations in unit-cell dimensions that affect S_m differently than MSA, and hence ultrasonic velocity. Further, the presence of varying amounts of anisotropic pores is another possibility. Pores, other than spherical ones, are not completely symmetrical about an arbitrary axis, and hence can affect physical properties differently in different directions relative to their alignment. This may be pertinent to Martin and Rosen's three bodies die-pressed from three different powders at different pressures, with the sonic velocity measured in the axial direction of their pellets after sintering. Thus there could have been varying degrees of laminar pores (along with the normal porosity between individual particles or agglomerates). Laminar pores act similarly to aligned cylindrical pores which reduce physical properties such as

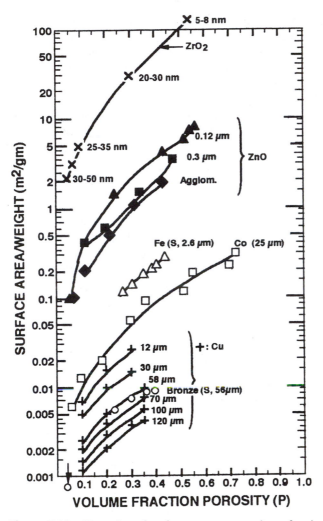

Figure 7.16   Examples of surface area versus volume fraction porosity (P) for various partially sintered metals (44) and ceramics (43,115).

u_v more in a direction normal to their alignment (i.e., parallel with the axis of pressing in this case), while the specific surface area has no directionality. Thus, varying amounts of laminar pores, which have effects similar to cylindrical pores, could give an inversion of specific surface area versus sonic velocity data as a function of P. Second and of probable broader applicability, is differing pore sizes. As shown by calculated absolute A or specific surface area values, differing pore sizes have a marked effect on specific surface area; however, they have limited or no effect on physical properties (provided that the P levels compared are the same and porosity is homogeneous and pore sizes are small in comparison to the body or flaw dimensions). Consider now extensions of the use of surface area modeling. An important extension of its use arises from the similar shape of the $A (1 - P)^{-1}$ versus P curves. This allows reasonable inter- polation between various curves, as is also the case for MSA modeling of physi- cal properties (Fig. 2.11). Thus, curves for various pore shapes or sizes or both can be calculated and used to give reasonable estimates of the SSA resulting from various mixtures of pore size, shape, or both. Other lesser extensions would be to correct for some of the idealizations commonly used. Thus other pore shapes can be calculated and some can be modified. An example of the latter is modeling of approximately spherical pores that are not large in comparison to the surrounding grains, e.g., those resulting from grain bridging in green-body forming. These typically have greater pore volume and especially surface area due to the cusps formed at the intersection of the grain boundaries and the pore than are reflected by the approximate radius of the resultant pore. Corrections for such factors can be included.

7.4 DISCUSSION

Evaluation of the porosity dependence of thermal conductivity shows four important results. First, for equations with no fitting parameters, i.e., Eqs. 1, 2, 4–8 of Table 7.1, there is a general trend for increasing agreement with data in the order listed. Thus, the linear equations (Eqs. 1 and 2 of Table 7.1) are gener- ally among the poorer fitting and are consistent with $1 - P$ being the upper bound consistent with its fitting data for conductivity parallel with aligned cylindrical pores. Data generally not fitting $1 - P$ corroborates the designation of thermal conductivity as a property dependent on not just P, but also pore character. Second, while some equations containing more P terms may fit data better, this is not always so, e.g., Eq. 8 of Table 7.1. Third, except for the MSA- based models and that of L. E. Nielsen, the occurrence and effects of P_C terms are absent, further questioning other models not reflecting effects of P_C values. Fourth, data generally agree with MSA models. More generally, complex microstructures require an array of models that need to be matched to the microstructure being treated, with grain-boundary pores and microcracks often playing a major role due to effects of gas type, and especially pressure, in them, e.g., due to convection.

The sparse data on the porosity dependence of electrical conductivity shows that it also follows MSA models and that, as expected, the solid percolation effects found in composites of conductive particles in a nonconducting matrix are not a factor; however, P_C values reflecting when the body is no longer a coherent solid and thus has no conductivity are still very pertinent. It should also be noted that the limited evaluation of the model of Yoshikawa et al. (28) for conductivity of composites of conductive spheres in an insulating matrix using a contact area intermediate between the average and the minimum solid area was not at all definitive. The conductivity was dominated by the interparticle resistivity, so the area effect could not be accurately evaluated.

Other electrical and magnetic properties show a diversity of behavior and known or implied mechanisms. Thus, electrical breakdown theory and data show a dependence on pore size and the statistics of pore size, shape, and spatial distribution to give a breakdown path with the minimum solid material. While, this minimum solid material is similar to the MSA for mechanical and conductive properties, it is also different. The minimum material refers to the collective content of material, i.e., thickness, along a potential breakdown path nominally parallel with the field, while MSA is the minimum cross-sectional area normal to the stress or flux between individual particles or pores. Thus, theoretical and experimental breakdown data having nearly straight line trends on semilog plots is due not to specific MSA mechanisms being operative as for mechanical and conductive properties, but to similarities, especially the exponential relation of the probability of forming a favorable breakdown path. Electrical breakdown is clearly biased toward the larger extremes of the pore size, orientation, and spatial distribution, especially as the dielectric layer gets thinner.

The dielectric constant shows some similarities and some differences in its porosity dependence relative to that of electrical breakdown. For moderate relative dielectric constants (e.g., < 10) data follows or is close to a rule of mixtures of air and the material, similar to some but not most models; however, some inverse dependence on pore size is indicated as is the porosity dependence of relative dielectric constants decreasing below the rule of mixtures trend as dielectric constant increases. Thus fine porosity in materials of high polarizability (e.g., piezoelectric materials with domains) and hence probable significant charging and interaction in pores appears to substantially alter the rule of mixtures trend. This effect of charging should increase with decreasing pore size, and be especially important for microcracks, as indicated by Yamai et al. (102). Again, while some models may provide guidance for such effects, many appear to be of, at best, limited use. It is clear that overall the porosity dependence of relative dielectric constant, differ significantly from that of mechanical, conductive, and electrical breakdown properties.

Much greater diversity of behavior is found for piezoelectric and magnetic property dependences on porosity. Thus, some properties may be independent of porosity, some follow or are close to a linear dependence on the volume fraction of porosity or pore size, some properties follow or are close to the nonlinear dependence on P of MSA-type behavior, and some properties have even more nonlinear dependences on porosity. Further diversity and complexity is shown by some properties increasing with increasing porosity in the above array of functional trends, and by some properties showing opposite trends with porosity depending on its (unidentified) character. Such variability shows the extreme need for much better porosity characterization and understanding of its effects. Such understanding is prompted not only by the basic scientific need, but also by the promise of important engineering payoffs, e.g., as indicated for piezoelectric transducers (see Section 10.3.1).

Consider next the issue of self-consistency, which has been widely neglected for these (as well as other) properties. While data is often particularly limited for the properties in this chapter, there are still opportunities to evaluate self-consistency. Thus, consistency of electrical and thermal conductivites, and these in turn with elastic moduli, i.e., all following consistent MSA models (e.g., Figs. 7.7–7.9, 7.11, and 7.12), is a clear demonstration of this. Deviations due to other effects such as convective or radiative thermal transport can be indicated by such comparisons. Evaluation of these collective properties as a function of temperature can similarly be valuable, i.e., the same functional dependence should be found as temperature increases (e.g., Fig. 7.9.) until such other conductive mechanisms are sufficiently activated. Comparison of other tests should also be valuable, e.g., electrical breakdown tests as a sensitive indicator of the heterogeneity of the porosity, especially as compared to conductive, elastic, and tensile strengths on the same set of samples, as well as to magnetic or piezoelectric properties. Unfortunately, there is a huge lack of such information, especially for the latter properties.

Finally, consider the effects of porosity heterogeniety, which clearly dominates electrical breakdown as noted above. For other properties the effects of heterogeneous porosity generally increase with the degree of nonlinearity of the dependence on porosity. In principle properties having linear dependence on P would not be effected by heterogeniety since property fluctuations from porosity variations within the test area should cancel; however, in tests such as those frequently used for dielectric constant data, scatter is likely due to gradients and other variations in porosity between the test sample and the specific area over which the dielectric constant is measured. Thus, die-pressed and sintered test disks can often have denser peripheries, which raise the disk density (especially at lower P) but not in the area measured. Also, heterogeneities in pore location, i.e., intra- versus intergranular locations are indicated for their differing effects on electrical breakdown and domain behavior in magnetic (and thus similarly implied for piezoelectric) materials. More generally, any property (where other aspects of the porosity than just P affect it) can vary due to heterogeniety of the distribution of these other pore characteristics that effect properties. Again, the issue of grain-size variation with changing levels of porosity and its heterogeniety are also an important issue. This is true especially where grain size directly affects properties, instead of indirectly via impacts on porosity, but is also a factor where this is not extensively the case. Thus, many of the properties considered in this chapter are clearly candidates for such effects of porosity and grain size and their interactions, which are probably important factors in some of their variability, e.g., electrical breakdown.

7.5 SUMMARY

Data supports minimum solid area models (MSA) models as being among the better fitting models for electrical and thermal conductivities where convection and radiation effects are not important, which is usually the case until one or more of higher porosities, larger pores, and higher temperatures occur. The self-consistency of MSA models fitting both elastic (and other mechanical) and conductive properties adds to its credibility as a useful modeling approach. Since MSA models are the only ones clearly reflecting pore character, and thus offering some predictive capabilities, they are recommended as a component of evaluation and for prediction. Complications found in composites due to percolation of more conductive particles or poorly conducting interfaces generally are not a factor in the porosity dependences of conductivity.

Data for anisotropic ceramics show thermal expansion clearly decreasing with increasing grain size due to the effects of grain size on the resultant microcracking. An increase in thermal expansion with increasing porosity was also shown in such materials. This probably reflects, at least in part, decreasing grain size and associated microcracking; however, it also indicates some possible basic coupling of porosity and microcracking effects, which deserve attention.

Electrical breakdown of dielectric materials under electric fields, while having some similarities, clearly also shows different porosity dependence from that of electrical (and hence also thermal) conductivity and mechanical properties. Breakdown is found theoretically and experimentally to be determined by statistical extremes of pore distribution giving breakdown paths nominally parallel with the applied electric field that have the minimum material content. This depends not only on the amount of the porosity and pore size and shape (although only spherical pores have been modeled so far), but also the specimen configuration and dimensions relative to those of the pores. Breakdown behavior typically is very close to a linear behavior on a semilog plot, e.g., with b values (slopes) of 3–10, with b values increasing with pore size; however, such linear semilog trends arise from the exponential character of the probabilities or the necessary pore clustering, and not from control of the process by MSA character, with which the minimum material has both some similarities and differences. (Note that breakdown also involves surface effects, some of which may be associated with pore distributions that are process dependent, but also other surface factors, e.g., probably surface flaws from machining, that lead to the breakdown field strength decreasing as sample thickness increases). The statistics of dielectric breakdown and flexure strength have been shown to correlate, but imperfectly. Thus, while some high voltage exposure can be used to screen out mechanically weaker samples, it is imperfect in doing so. This is attributed to intrinsic differences in the specifics of the role of pores on dielectric versus mechanical failure, especially the geometry of their spatial distribution. Thus, mechanical failure from pores is impacted by single or closely spaced, clustered pores (or pore-flaw combinations) with larger cross-sections normal to the stress axis (Sections 5.3.3, 5.3.4). On the other hand, dielectric failure is dominated by chains of pores (or pore-flaw combinations) that minimize the solid material in a path parallel to the field.

In contrast to electrical breakdown, the dielectric constant generally shows substantially less dependence on the amount and character of the porosity, and related specimen geometry (though some possible effects of specimen preparation are noted). Thus, much dielectric constant data follows, at least approximately, a rule of mixtures relation between the solid and pore phase (relative dielectric constant ~ 1), i.e., a linear dependence on P. Bodies with a higher dielectric constant fall below the linear rule of mixtures trend, generally increasingly so as the dielectric constant and the amount of porosity increases, and as the pore size decreases. While some models reflect at least part of this behavior, there are serious questions of pore-shape effects on surface charging and flux "leakage" in the pores that are not addressed by current models. Charging effects are a possible explanation for indicated greater effects of microcracks versus pores. Thus, such models should be used with caution and substantial cross-correlation of models and trends for different materials and porosity should be made.

Significantly extending the trend away from similar porosity dependence of nonmechanical properties from that of mechanical and conductive properties are the more diverse and sometimes complex porosity dependences of piezo-electric and magnetic properties. These range from no porosity dependence to linear, and nonlinear dependences on porosity. The nonlinearity may be similar to that typically found for MSA control of properties (which may be related to underlying mechanisms, but may more commonly be coincidences arising from other sources) or be substantially greater. Furthermore, while many properties decrease with increasing porosity, some properties show the same array of linear to nonlinear change, but instead increase with increasing porosity; these opposite trends with P occur in bodies of the same material with different processing, and hence different microstructures. While much of this porosity dependence is not well understood, both theoretical and empirical results indicate that much of this diversity of behavior results from various combinations of pore size, shape, and spatial distribution factors such as pore location, i.e., intra- versus inter-granular, and spacing of the former, being factors in differing properties. Thus, for example, spacing of intragranular pores can effect domain behavior that can affect different properties in different fashions, but intergranular pores have much less or no effect since the grain boundaries on which they are located have similar or greater effect. Furthermore, many of the properties of interest are affected by grain size, often directly and substantially, or indirectly via effects on pore character on a global or local scale or both. While, extending models for the porosity dependence of other related, e.g., elastic, properties via computer evaluations of the property tensors can be useful, much broader study is needed, with more and better data and characterization being critical needs. Evaluation of the self-consistency of results via comparison of several related as well as different properties, while important for any property, is particularly important for magnetic and piezoelectric properties because of their diversity and complexity. An important driving force for such understanding is the technological possibilities indicated earlier (and in Section 10.3.1) of improving hydrostatic performance of sonar transducers by engineering their pore structure.

Finally, note that idealized pore models such as those used in MSA models for mechanical and conductive properties have also been used successfully for modeling surface area versus the amount and character of porosity. Such calculations agree with and provide theoretical insight to interpreting the limited available data for specific surface area as a function of volume fraction porosity (P). The models show near but not full linear dependence on semilog plots of pore area over the predominant porosity range encountered, e.g., to P ~ 0.6. The similar, simple nature of the curves for various pore geometries means that the various models can be combined to predict approximate surface areas of various combinations of pore types, sizes, or both by interpolating between the different curves representing the pertinent porosities. The models also give insight into differences, e.g., inversions, in the trends of surface area versus ultrasonic velocity (hence also other properties such as elastic moduli) as a function of P. The contrast in the surface area and MSA versus P trends again shows the diversity that can occur in the P dependence of differing properties, even models of the same pore structures for properties dependent on similar, but not identical, aspects of the pore cell geometry. Models for specific surface area versus porosity are also consistent with expected increases in particle coordination number, C_n, again reinforcing its use and the need to address changing pore character with changing levels (and hence character of) porosity.

REFERENCES

1. L. V. Gibiansky and S. Torquato, "Link Between Conductivity and Elastic Moduli of Composite Materials," Phy. Rev. Let., **71**(18), pp. 1927–30, 1993.
2. F. G. Shin, W. L. Tsui, and Y. Y. Yeung, "Dielectric Constant of Binary Mixtures," J. Mat. Sci. Let., **8**, pp. 1383–85, 1989.
3. P. G. Collishaw and J. R. G. Evans, "An Assessment of Expressions for the Apparent Thermal Conductivity of Cellular Materials," J. Mat. Sci., **29**, pp. 486–98, 1994.
4. G. E. Youngblood, R. W. Rice, and R. P. Ingel, "Thermal Diffusivity of Partially and Fully Stabilized Zirconia Single Crystals," J. Am. Cer. Soc., **71**(4), pp. 255–60, 1988.

5. Z. Hashin and S. Shtrickman, "A Variational Approach to the Theory of the Mechanical Behavior of Multiphase Materials," J. Mech. Phys. Solids, 11, pp. 127–40, 1963.

6. M. A. Schuetz and Glicksman, "A Basic Study of Heat Transfer Through Foam Insulation," J. Cell. Plast., 20, pp. 114–21, 1984.

7. J. Francl and W. D. Kingery, "Thermal Conductivity: IX, Experimental Investigation of Effect of Porosity on Thermal Conductivity," J. Am. Cer. Soc., 37(2), pp. 99–106, 1954.

8. S. K. Rhee, "Porosity-Thermal Conductivity Correlations for Ceramic Materials," Mat. Sci, & Eng., 20, pp. 89–93, 1975.

9. D. M, Dawson and A. Briggs, "Prediction of the Thermal Conductivity of Insulating Materials," J. Mat. Sci., 16, pp. 3346–56, 1981.

10. J. B. Austin, "Factors Influencing the Thermal Conductivity of Non-Metallic Materials," Intl. Symp. Thermal Insulating Matl., pp. 3–67, ASTM, Philadelphia, PA, 1936.

11. D. M. Bigg, "Thermally Conductive Polymer Compositions," Polymer Comp., 7(3), pp. 125–40, 1986.

12. F. C. Chen, C. L. Choy, and K. Young, "A Theory of the Thermal Conductivity of Composite Materials," J. Phys. D: Appl. Phys., 10, pp. 571–86, 1977.

13. B. Budiansky, "Thermal and Thermoelastic Properties of Isotropic Composites," J. Comp. Mat., 4, pp. 286–95, 1970.

14. R. C. Progelhof, J. L. Throne, and R. R. Ruetsch, "Methods for Predicting Thermal Conductivity of Composite Systems: A Review," Polymer, Eng. Sci., 16, pp. 615–25, 1976

15. A. S. Waugh, "Porosity Dependence of Thermal Conductivity of Ceramics and Sedimentary Rocks," J. Mat. Sci., 28, pp. 3715–21, 1993.

16. D. A. G. Bruggeman, "Brechnung Verschiedener Physikalischer Konstanten von Hetergener Substanzen: I. Dielektrizitatskonstanten und Leitfahigkeiten der Mischkorper aus Isotropen Substanzen," Ann. der Phys., 5(24), pp. 636–79, 1935.

17. A. Eucken, "Thermal Conductivity of Ceramic Refractory Materials; Calculation from Thermal Conductivity of Constituents," Forsh. Gebiete Ingenieurw., B3, Forschunggsheft 353, p. 16, 1932.

18. J. C. Maxwell, "A Treatise on Electricity and Magnetism, vol. 1," Clarendon Press, Oxford, p. 440, 1892.

19. D. J. Doherty, R. Hurd, and G. R. Lester, Chem. Ind., p. 1340, 1962.

20. H. W. Russell, "Principles of Heat Flow in Porous Insulators," J. Am. Cer. Soc., 18, pp. 1–5, 1935.

21. L. E. Nielsen, "Thermal Conductivity of Particulate-Filled Polymers," J. Appl. Polymer Sci., 17, pp. 3819–20, 1973.

22. D. K. Hale, "Review: The Physical Properties of Composite Materials," J. Mat. Sci., 11, pp. 2105–41, 1976.

23. A. Loeb, "Thermal Conductivity: VIII. A Theory of Thermal Conductivity of Porous Materials," J. Am. Cer. Soc., 37(2), pp. 96–99, 1954.

24. E. Litovsky and M. Shapiro, "Gas Pressure and Temperature Dependences of Thermal Conductivity of Porous Ceramic Materials: Part 1, Refractories and Ceramics with Porosity Below 30%," J. Am. Cer. Soc., 75, pp. 3425–29, 1992.

25. E. Litovsky, M. Shapiro, and A. Shavit, "Gas Pressure and Temperature Dependences of Thermal Conductivity of Porous Ceramic Materials: Part 2, Refractories and Ceramics with Porosity Exceeding 30%," J. Am. Cer. Soc., 79, pp. 1366–76, 1996.

26. D. P. H. Hasselman and J. P. Singh, "Analysis of Thermal Stress Resistance of Microcracked Brittle Ceramics," Am. Cer. Soc. Bul., **58**(9), pp. 856–60, 1979.
27. F. Lux, "Review: Models Proposed to Explain the Electrical Conductivity of Mixtures Made of Conductive and Insulating Materials," J. Mat. Sci., **28**, pp. 285–301, 1993.
28. S. Yoshikawa, G. R. Ruschau, and R. E. Newnham, "Conductor-Polymer Composite for Electronic Connectors," Advanced Composite Materials, Processing, Microstructures, Bulk and Interfacial Properties, Characterization Methods, and Applications (M. D. Sacks, Ed.), American Ceramics Society, Westerville, OH, pp. 373–79, 1991. Ceramic Transactions, vol. 19.
29. J. Mizusaki, S. Tsuchiya, K. Waragai, H. Tagawa, Y. Arai, and Y. Kuwayama, "Simple Mathematical Model for the Electrical Conductivity of Highly Porous Ceramics," J. Am. Cer. Soc., **79**(1), pp. 109–13, 1996.
30. R. Gerson and T. C. Marshall, "Dielectric Breakdown of Porous Ceramics," J. Appl. Phy., **30**(11), pp. 1650–53, 1959.
31. K. Wakino, T. Okada, N. Yoshida, and K. Tomono, "A New Equation for Predicting the Dielectric Constant of a Mixture," J. Am. Cer. Soc., **76**(10), pp. 2588–94, 1993.
32. J. W. Rayleigh, "On the Influence of Obstacles Arranged in Rectangular Order upon the Properties of a Medium," Phil Mag., **34**, pp. 481–502, 1892.
33. H. Banno, "Effects of Shape and Volume Fraction of Closed Pores on Dielectric, Elastic, and Electromechanical Properties of Dielectric and Piezoelectric Ceramics—A Theoretical Approach," Am. Cer. Soc. Bull., **66**(9), pp. 1332–37, 1987.
34. D. A. Payne and L. E. Cross, "Microstructure-Property Relations for Dielectric Ceramics: I. Mixing of Isotropic Homogeneous Linear Dielectrics," Ceramic Microstructures, vol. 76 (R. M. Fulrath and J. A. Pask, Eds.), Westview Press, Bolder, CO, pp. 584–97, 1977.
35. M. L. Dunn and M. Taya, "Electromechanical Properties of Porous Piezoelectric Ceramics," J. Am. Cer. Soc., **76**(7), pp. 1697–706, 1993.
36. R. D. Miller and T. B. Jones, "On the Effective Dielectric Constant of Columns or Layers of Dielectric Spheres," J. Phys. Appl. Phys., **21**, pp. 527–32, 1988.
37. M. Erneta and H. A. Stockler, "Light Scattering by Pores in Ceramic (Pb, La) (Zr, Ti)O_3," J. Am. Cer. Soc., **56**(7), pp. 394–95, 1973.
38. J. G. J. Peelen and R. Metselaar, "Light Scattering by Pores in Polycrystalline Materials: Transmission Properties of Alumina," J. Appl. Phy., **45**(1), pp. 216–220, 1974.
39. G. C. Jain, B. K. Das, R. S. Khanduja, and S. C. Gupta, "Effect of Intergranular Porosity of Initial Permeability and Coercive Force in a Manganese Zinc Ferrite," J. Mat. Sci., **11**, pp. 1335–38, 1976.
40. H. Igarashi and K. Okazaki, "Effects of Porosity and Grain Size on the Magnetic Properties of NiZn Ferrite," J. Am. Cer. Soc., **60**(1–2), pp. 51–54, 1977.
41. K. Okazaki and H. Igarashi, "Importance of Microstructure in Electronic Ceramics," Ceramic Microstructures '76 (R. M. Fulrath and J. A. Pask, Eds.), Westview Press, Bolder, CO, pp. 564–83, 1977.
42. F. N. Rhines, R. T. Dehoff, and R. A. Rummel, "Rate of Densification in the Sintering of Uncompacted Metal Powders," Agglomeration (W. A. Knepper, Ed.), Interscience Pub., New York, pp. 351–78, 1962.
43. L. P. Martin and M. Rosen, "Specific Surface Area and Ultrasonic Velocity in Porous Zinc Oxide," J. Am. Cer. Soc., **80**(4), pp. 839–846, 1997.

44. J. P. Jernot, M. Coster and J. L. Chermant, "Model of Variation of the Specific Surface Area During Sintering," Powder Technology, **30**, pp. 21–29, 1981.

45. F. P. Knudsen, "Dependence of Mechanical Strength of Brittle Polycrystalline Specimens on Porosity and Grain Size," J. Am. Cer. Soc., **42** [8] 376–88 (1959).

46. W. O. Smith, P. D. Foote, and P. F. Busang, "Packing of Homogeneous Spheres," Phys. Rev., **34**, pp. 1271–74, 1929.

47. M. J. Powell, "Computer-Similuated Random Packing of Spheres," Powd. Tech., **25**, pp. 45–52, 1980.

48. J. S. Reed, "Introduction to the Principles of Ceramic Processing," John Wiley & Sons, New York, p. 188, 1988.

49. H. Rumpf, "The Strength of Granules and Agglomerates" Agglomeration (W. A. Knepper, Ed.), Intrescience Pub., Inc., New York, pp. 379–418, 1962.

50. P. Harma and V. Satava, "Model for Strength of Brittle Porous Materials," J. Am. Cer. Soc., **57**(2), pp. 71–73, 1974.

51. G. Y. Onada, Jr., "Theoretical Strength of Dried Green Bodies with Organic Binders," J. Am. Cer. Soc., 59(5–6), pp. 236–39, 1976.

52. S. K. Roy, "Characterization of Porosity in Porcelain-Bonded Porous Alumina Ceramics," J. Am. Cer. Soc., **52**(10), pp. 543–548, 1969.

53. E. Manegold and K. Solf, "Capillary Systems: XIX-C," Kolloid-Z., **81**, pp. 36–40, 1937.

54. C. J. Pereira, R. W. Rice, and J. P. Skalny, "Pore Structure and Its Relationship to Properties of Materials," Pore Structure and Permability of Cementatious Materials (J. P. Skalny, Ed.), Mat. Res. Soc. Symp. Proc., **137**, pp. 3–21, 1969.

55. S. J. Luckasiewicz and J. R. Reed, "Specific Permeability of Porous Compacts as Described by a Capillary Model," J. Am. Cer. Soc., **71**(11), pp. 1008–14, 1988.

56. R. W. Rice, "Evaluation and Extension of Physical Property-Porosity Models Based on Minimum Solid Area," J. Mat. Sci., **31**, pp. 102–18, 1996.

57. J. D. McClelland and L. O. Petersen, "The Effect of Porosity on the Thermal Conductivity of Alumina," Atomics Intl. Report contract: At(11–1)-GEN-8, Oct. 1961.

58. J. E. Hove and W. C. Riley (Eds.), "Introduction: Ceramics for Advanced Technologies," John Wiley & Sons, Inc., New York, p. 5, 1965.

59. J. H. Enloe, R. W. Rice, J. W, Lau, R. Kumar, and S. Y. Lee, "Microstructural Effects on the Thermal conductivity of Polycrystalline Aluminum Nitride," J. Am. Cer. Soc., **74**(9), pp. 2214–19, 1991.

60. R. W. Rice, "Comparison of Physical Property-Porosity Behavior with Minimum Solid Area Models," J. Mat. Sci., **31**, pp. 1509–28, 1996.

61. T. Saegusa, K. Kamata, Y. Iida, and N. Wakao, "Thermal Conductivities of Porous Solids," Kagaku Kogaku, **37**, pp. 811–814, 1973.

62. H. Clark, "The Effects of Simple Compression and Wetting on the Thermal Conductivity of Rocks," Am. Geophy. Union, Reports and Papers, Tectonophysics, pp. 543–44, 1941.

63. W. Woodside and J. H. Messmer, "Thermal Conductivity of Porous Media, II: Consolidated Rocks," J. Appl. Phy., **32**(9), pp. 1699–1706, 1961.

64. A. Sugawara and Y. Yoshizawa, "An Experimental Investigation on the Thermal Conductivity of Consolidated Porous Materials," J. Appl. Phy., **33**(10), pp. 3135–3138, 1962.

65. P. Wagner, J. A. O'Rourke, and P. E. Armstrong, "Porosity Effects in Polycrystalline Graphite," J. Am. Cer. Soc., **55**(4), pp. 214–219, 1972.

66. S. K. Rhee, "Discussion of Porosity Effects in Polycrystalline Graphite," J. Am. Cer. Soc., **55**(11), pp. 580–81, 1972.

67. R. W. Powell, "The Thermal Conductivity of Beryllia," Trans. Brit. Cer. Soc., **53**, pp. 389–97, 1954.

68. H. J. Siebeneck, J. J. Cleveland, D.P. H. Hasselman, and R. C. Bradt, "Thermal Diffusivity of Microcracked Ceramic Materials," Ceramic Microstructure, 76 With Emphasis on Energy Applications (R. M. Fulrath and J. A. Pask, Eds.), Westview Press, Bolder, CO., pp. 743–62, 1977.

69. W. R. Manning, "Anomalous Elastic Behavior of Polycrystalline Nb2O5," Ph.D. Thesis, Iowa State Univ., Ames, IA, 5/1971.

70. F. J. Parker and R. W. Rice, "Correlation Between Grain Size and Thermal Expansion for Aluminum Titanate Materials," J. Am. Cer. Soc., **72**(12), pp. 2364–66, 1989.

71. E. A. Belskaya and A. S. Tarabanov, "Experimental Studies Concerning the Electrical Conductivity of High-Porosity Carbon-Graphitic Materials," Institute of High Temperatures, Academy of Sciences of the USSR, Tr. from Inzhenerno-Fizicheskii Zhurnal, **20**(4), pp. 654–659, 4/1971.

72. G. V. Samsonov and V. E. Matsera, "Manufacture and Properties of Powder Metallurgical Refractory Compound Fibers and Porous Fiber Materials," Sov. Pwd. Met. Mat. Cer., **11**(9), pp. 719–731 (1972).

73. V. P. Bulychev, R. A. Andrievskii, and L. B. Nezhevenko, "Theory and Technology of Sintering, Thermal, and Chemicothermal Treatment Processes, The Sintering of Zirconium Carbide," Tr. from Poroshkovaya Metallurgiya, **4**(172), pp. 38–42, 1977.

74. T. Takahashi and Y. Suzuki, "Effect of Porosity on Electrical Conductivity of Sintered $(ZrO_2)_{0.9}$ $(Y_2O_3)_{0.1}$," Denki Kagaku, **37**, pp. 782–86, 1969.

75. S. Mizuta, K. Shirasawa, and H. Yanagida, "The Effect of Water and Porosity on Ionic Conduction of Sintered BaF_2," Denki Kagaku, **41**, pp. 913–18, 1973.

76. H. Tagawa, J. Mizusaki, Y. Arai, Y. Kuwayama, S. Tsuchiya, T. Takada, and S. Sekido, "Sinterability and Electrical Conductivity of Variously Prepared Perovskite-Type Oxide, $La_{0.5}Sr_{0.5}CoO_3$," Denki Kagaku, **58**, pp. 512–19, 1990.

77. E. K. Beauchamp, "Effect of Microstructure on Pulse Electrical Strength of MgO," J. Am. Cer. Soc., **54**(10), pp. 484–87, 1971.

78. W. C. Johnson, D. F. Stein, and R. W. Rice, "Analysis of Grain-Boundary Impurities and Fluoride Additives in Hot-Pressed Oxides," J. Am. Cer. Soc., **57**(8), pp. 342–44, 1973.

79. T. Tunkasiri and G. Rujijanagul, "Dielectric Strength of Fine Grained Barium Titanate Ceramics," J. Mat. Sci. Let., **15**, pp. 767–69, 1996.

80. A. Kishimoto, K. Koumoto, and H. Yanagida, "Mechanical and Dielectric Failure of $BaTiO_3$ Ceramics," J. Mat. Sci., **24**, pp. 698–702, 1989.

81. C. T. Morse and G. J. Hill, "The Electric Strength of Alumina: The Effect of Porosity," Proc. Brit. Cer. Soc., **18**, pp. 23–35, 6/1970.

82. A. S. Rupaal, J. E. Garnier, and J. L. Bates, "Dielectric Breakdown of Porous $MgAl_2O_4$," J. Am. Cer. Soc., **64**(7), pp. C-100–101, 1981.

83. M. Yoshimura and H. K. Bowen, "Electrical Breakdown Strength of Alumina at High Temperatures," J. Am. Cer. Soc., **64**(7), pp. 404–10, 1981.

84. A. Kisimoto, K. Koumoto, H. Yanagida, and M. Nameki, "Microstructure Dependence of Mechanical and Dielectric Strengths—I. Porosity," Eng. Fract. Mech., **40**(4/5), pp. 927–30, 1991.

85. A. Kishimoto, K. Endo, Y. Nakamura, N. Motohira, and K. Sugi, "Effect of High-Voltage Screening on Strength Distribution for Titanium Dioxide Ceramics," J. Am. Cer. Soc., **78**(8), pp. 2248–50, 1995.

86. A. Kishimoto, K. Endo, N. Motohira, Y. Nakamura, H. Yanagida, and M. Miyayama, "Strength Distribution of Titania Ceramics After High-Voltage Screening," J. Mat. Sci., **31**, pp. 3419–25, 1996.

87. H. H. Stadelmaier and S. W. Derbyshire, "Variation of Dielectric Constant and Spontaneous Strain With Density in Polycrystalline Barium Titanate," The Role of Grain Boundaries and Surfaces in Ceramics (W. W. Kriegel and H. Palmour, III, Eds.), Plenum Press, New York, pp. 101–110, 1966. Materials Science Research, vol. 3.

88. D. R. Biswas, "Electrical Properties of Porous PZT Ceramics," J. Am. Cer. Soc., **61**(9–10), pp. 461–462, 1978.

89. C-C. Hsueh, M. L. Mecartney, W. B. Harrison, M. R. B. Hanson, and B. G. Koepke, "Microstructure and Electrical Properties of Fast-Fired Lead Zirconate-Titanate Ceramics," J. Mat. Sci. Letters, **8**, pp. 1209–1216, 1989.

90. K. Okazaki and K. Nagata, "Effects of Grain Size and Porosity on Electrical and Optical Properties of PLZT Ceramics," J. Am. Cer. Soc., **56**(2), pp. 82–86, 1973.

91. P. F. Messer, "Ceramic Processing, A Systematic Approach," T. & J. Brit. Cer. Soc., **82**, pp. 190–192, 1983.

92. J. T. Milek, "Aluminum Oxide-Data Sheets," Hughes Aircraft Co. (Culver City) Report AD 434 173, p. 85, 3/1964.

93. J. D. Walton, Jr., "Reaction Sintered Silicon Nitride for High Temperature Radome Applications," Am. Cer. Soc. Bul., **53**(3), pp. 255–258, 1974.

94. R. E. Gannon, G. M. Harris, and T. Vasilos, "Effect of Porosity on Mechanical, Thermal, and Dielectric Properties of Fused Silica," Am. Cer. Soc. Bull., **44**(5), pp. 460–462, 5/1965.

95. T. Fujiu, G. L. Messing, and W. Huebner, "Processing and Properties of Cellular Silica Synthesized by Foaming Sol-Gels" J., Am. Cer. Soc., **73**(1), pp. 85–90, 1990.

96. J. G. Liu and D. L. Wilcox, Sr., "A Layered Sphere Bruggeman Dielectric Mixture Model for a Hollow Ceramic Sphere Composite," J. Appl. Phy., **77**(12), pp. 6456–60, 1995.

97. J. N. Harris and E. A. Welsh, "Fused Silica Design Manual, I," NSC Special Publication, Georgia Institute of Technology, Atlanta, GA, report for Naval Ordinance Systems Command Contract N 00017-72-C-4434, 5/1973.

98. M. D. Sacks, M. S. Randall, G. W. Scheiffele, R. Raghunathan, and J. H. Simmons, "Processing of Silicate Glass/Silicon Nitride Composites with Controlled Microporosity," Ceramic Trans., **19**, Advanced Composite Materials, M. D. Sacks, ed., Am. Cer. Soc., Westerville, OH, pp. 407–420, 1991.

99. K. Hikita, K. Yamada, and M. Nishioka, "Effect of Porous Structure on Piezoelectric Properties of PZT Ceramics," Jap. J. Appl. Phy., **22**, Sup. 22–2, pp. 64–66, 1983.

100. U. Bast and W. Wersing, "The Influence of Internal Voids with 3–1 Connectivity on the Properties of Piezoelectric Ceramics Prepared by a New Planar Process," Ferroelectrics, **94**, pp. 229–42, 1989.

101. K. Nagata, H. Igarashi, K. Okazaki, and R. C. Bradt, "Properties of an Interconnected Porous Pb(Zr, Ti)O_3 Ceramics," Jap. J. Appl. Phy., **19**(1), pp. L37–40, 1980.

102. I. Yamai, T. Ota, and J. Takahashi, "Effects of Microcracks on Dielectric Property," Ceramics Today–Tomorrow's Ceramics, P. Vincenzini, ed., Elsevier Science Publishers, NY, pp. 2039–45, 1991.

103. M. Kahn, "Acoustic and Elastic Properties of PZT Ceramics with Anisotropic Pores," J. Am. Cer. Soc., **68**(11), pp. 623–28, 1985.

104. M. Kahn, A. Dalzell, and B. Kovel, "PZT Ceramic-Air Composites for Hydrostatic Sensing," Adv. Cer. Mat., **2**(4), pp. 836–40, 1987.

105. Q. Jiang, W. Cao, and L. E. Cross, "Electric Fatigue in Lead Zirconate Titanate Ceramics," J. Am. Cer. Soc., **77**(1), pp. 211–15, 1994.

106. Q. Y. Jiang and L. E. Cross, "Effects of Porosity on Electric Fatigue Behavior in PLZT and PZT Ferroelectric Ceramics," J. Mat. Sci., **28**, pp. 4536–43, 1993.

107. T. Hiroshima, K. Tanaka, and T. Kimura, "Effects of Mocrostructure and Composition on the Curie Temperature of Lead Barium Niobate Solid Solutions," J. Am. Cer. Soc., **79**(12), pp. 3235–42, 1996.

108. H. T. Chung, B. C. Shin, and H. G. Kim, "Grain Size Dependence of Electrically Induced Microcracking in Ferroelectric Ceramics," J. Am. Cer. Soc., **72**(2), pp. 327–29, 1989.

109. S. S. Chiang, R. M. Fulrath, and J. A. Pask, "Influence of Microcracking and Slow Crack Growth on the Planar Coupling Coefficient in PZT," J. Am. Cer. Soc., **64**(10), pp. C-141–43, 1981.

110. G. Economos, "Effect of Microstructure on the Electrical and Magnetic Properties of Ceramics," Ceramic Fabrication Processes (W. D. Kingery, Ed.), John Wiley & Sons, Inc., New York, pp. 201–12, 1958.

111. J. Smit and H. P. J. Wijn, "Ferrites," Wiley & Sons, New York, 1965.

112. G. Goodman, "Relation of Microstructure to Translucency of Porcelain Bodies," J. Am. Cer. Soc., **33**(2), pp. 66–72, 1950.

113. N. Grimm, G. E. Scott, and J. D. Sibold, "Infrared Transmission Properties of High Density Alumina," Am. Cer. Soc. Bul., **50**(12), pp. 962–65, 1971.

114. P. Hing, "The Influence of Some Processing Parameters on the Optical and Microstructural Properties of Sintered Alumina," Science of Ceramics, **8**, Brit. Cer. Soc., pp. 159–72, 1976.

115. A. J. Allen, S. Krueger, G. Skandan, G. G. Long, H. Hahn, H. M. Kerch, J. C. Parker, and M. N. Ali, "Microstructural Evolution During the Sintering of Nanostructured Ceramic Oxides," J. Am. Cer. Soc., **79**(5), pp. 1201–12, 1996.

116. L. P. Martin, D. Dadon, and M. Rosen, "Evaluation of Ultrasonically Determined Elasticity-Porosity Relations in Zinc Oxide," J. Am. Cer. Soc., **79**(5), pp. 1281–89, 1996.

8

POROSITY AND MICROCRACK DEPENDENCE OF MECHANICAL PROPERTIES OF CERAMIC COMPOSITES, OTHER CERAMICS, AND OTHER MATERIALS AT 22°C

KEY CHAPTER GOALS

1. Demonstrate the broad similarities and some differences of primarily the porosity and secondarily the microcrack dependences of the properties of monolithic ceramics relative to those of plastics, metals, plasters, cements, and traditional and designed ceramic composites.
2. Show that the location of pores in one or the other of, or between, the phases of a composite can be important in the porosity effects on properties.
3. Show that (1) much more study and characterization of microcracks is needed, especially to differentiate between those that are preexisting versus those generated with macrocrack propagation and their (not necessarily identical) increases in fracture toughness, and (2) toughnesses and Weibull moduli are often, and strengths are usually if not always, less in microcracked than nonmicrocracked bodies.

8.1 INTRODUCTION

Previous chapters addressed the porosity dependence and, where data was available, the microcrack dependence of various physical, especially mechanical, properties in nominally single-phase ceramics, i.e., those with little or no second solid phase (which when present is mainly a grain boundary or other limited impurity phase). This chapter similarly addresses the porosity and some micro-crack dependence of other ceramics and materials. The literature on micro-cracking and related effects such as crack bridging has become substantial, much of it not directly related to effects of microcracks analogous or related to those of porosity. The latter is the focus here, making the discussion of micro-cracking limited. Other, nonceramic materials addressed are plastics and metals, for which some data on porosity dependence was noted earlier in conjunction with evaluation of the porosity dependences of specific properties; however, some plastic and substantial metal data is presented here for a better perspective on the common trends of their physical property-porosity dependence for comparison with the corresponding trends for ceramics.

Other ceramics addressed are those having substantial amounts of at least a second solid phase, and hence are basically one of two types of composites. The first type includes a substantial portion of traditional ceramics such as white-wares and porcelains, as well as glasses of varying degrees of crystallization, including "glass-ceramics" (though there is typically very little or no porosity in the latter, and hence very little data on the porosity dependence of their proper-ties, but there is some on microcracking effects). The second type of ceramic composites are those designed with two or more desired phases and resultant microstructures to achieve some desired level or combination of physical prop-erties. Such materials have been developed mostly in recent years particularly for mechanical properties and applications, but some important composite developments have also been directed at various nonmechanical, e.g., various magnetic and especially electrical applications. Many of the composites consist of particles of one phase dispersed in a matrix, but other dispersed forms, including fibers, exist. Such designed composites are commonly sought as dense bodies for their best performance, but some limited data on porosity effects on them is available to provide useful guidance to porosity effects in them and comparison with other ceramics. A number of these materials also exhibit microcracking, frequently sought for toughening. Although microcracking in such composites is a specialized subject more appropriate for treatment in a comprehensive evaluation of composites, some models and data that provide insight to microcracking are presented.

Other related materials addressed are rocks and materials made by reactions with water, i.e., plaster, and especially cement-based materials, which are also typically composite materials. Rocks contain varying porosity and plasters and cement materials typically contain substantial porosity; this plays an important role in their properties. Understanding their porosity dependence is important for the development and use of these materials, but again is also of value to compare to that of ceramics and ceramic composites. Since there is substantial data on the porosity dependence of these materials, much of it intimately related to specifics of the particular material/field, the treatment here must be broader, much of it based on surveys of these materials.

Before addressing data for the above material groups, theoretical and modeling considerations are addressed. Again, because of the diversity and frequently specialized nature of the literature on most of these materials and of composites in general, the treatment here is broad. Basic concepts, needs, and approaches are emphasized, rather than detailed evaluation of specific models. Then the data for the porosity and microcrack dependence of each of the groups of above materials is summarized. For each material group, to the extent that the data allows, the properties are sequentially treated in the order of the preceding chapters—elastic, crack propagation, tensile strength, hardness, compressive strength, and related properties; and nonmechanical properties.

8.2 THEORETICAL CONSIDERATIONS AND MODELS

8.2.1 General Considerations

Modeling of single-phase, nonceramic materials, i.e., of metals and plastics, commonly follows the same routes as for ceramics. In fact, many of the models applied to ceramics were derived either explicitly for metals or effectively for a generic, i.e., nonspecific, material provided that it met certain, usually very broad, behavior requirements. While macroscale plastic deformation may occur to varying extents in some of these materials, porosity commonly severely restricts this, so most observations and models for ceramics are applicable to metals and plastics and vice versa. Thus, both minimum solid area (MSA, and related foam)- and mechanics-based models have been applied to these materials, and some of these models were originally derived for porous (e.g., cast or sintered) metals. Exceptions to this are elastomeric materials, higher porosity foams of some metals and plastics, and some porous metals and plastics at higher temperatures. Thus, some foam models have been modified to account for plastic deformation of cell walls or struts (or plastic buckling of these).

Three points should be noted from porosity studies of metals. First, the exponential relation has been widely used on both an empirical and an analytical basis. The former is seen in much of the data to be reviewed later, and the latter stems substantially from Exner and Pohl (1). They derived this relation based on random pores close to the fracture plane impacting the MSA. Furthermore, Navara and Bengtsson (2) showed analytically that use of the exponential MSA approximation and treating the fracturing ligaments as notched cylinders gave fracture toughness going through a broad maximum (e.g., at $P \sim 0.2$) as a function of P. A novel concept for strengthening metals via the inclusion of a population of small (typically submicron, inert-gas filled) pores has been put forward by Louat et al. (3) that could also be applicable to some other materials. They note that the equilibrium pressure of an insoluble gas in pore is simply $2s (r)^{-1}$, where s is the surface energy of the metal and r is the pore radius. For submicron pores, pressures can be a significant fraction of the shear modulus, and thus result in significant pore-dislocation interaction and thus possible strengthening greater than the loss of strength due to the material missing due to the pores, e.g., reduction due to reduced elastic moduli. They further note that such inert-gas filled pores should be stable at elevated temperatures, thus providing strengthening at higher temperatures. Such dislocation as well as possible related crack effects could be a factor in the mechanical behavior in ceramics, e.g., Fig. 6.16.

Modeling of the porosity dependence of composites, as required for many materials addressed in this chapter, fundamentally has the same requirements as modeling porosity dependence of single solid phase materials, namely the key geometrical aspects of the pores and hence also the solid phase; however, composites pose two additional requirements that can be challenging, namely the key geometrical and association aspects of: 1) the two (or more) solid phases with each other, and 2) the pores with one or more of the solid phases. Both of these geometrical and association aspects can vary widely, driven mainly by fabrication methods, so pores may be associated predominantly with one phase or another, or have mixed (and variable) phase association. The solid phases may have varying associations with one another, e.g., ranging from one phase substantially encompassing grains of the other solid phase (s) to random distribution of particles of the other solid phase (s). At dilute distributions of a dispersed solid phase in a matrix, both of these situations pertain, i.e., the dispersed solid phase is typically randomly distributed in the other solid, matrix phase, but also encompassed by the matrix phase; however, as the volume fraction of the dispersed phase increases the situation may follow different paths of varying structure and complexity. As the content of the dispersed phase increases its contiguity increases, ultimately becoming the matrix phase in a "mirror image composite," i.e., where the two solid phases exchange matrix and dispersed-phase roles, as is typically the case in consolidating particulate mixtures of two randomly mixed phases. In processing where reactions, significant addition of other material, or both occur, a variety of microstructures may evolve. A simpler form of the microstructure that can evolve from some processes may be essentially grains bonded together by a continuous grain-boundary phase, but many more complex structures can result.

To put the above solid microstructures and their relation to porosity in perspective, consider some of the specific processing methods applicable to materials of interest in this chapter. One that is widely used, especially for composites of interest for mechanical properties, is the consolidation of matrix particles around dispersed particles of varying morphology. While this is

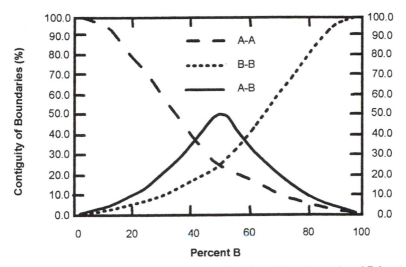

Figure 8.1   Schematic of the contiguity of grains of the phases A and B in a two-phase composite of uniform, ideal particles of uniform or random stacking. This relates to evaluating porosity effects in composites since the contiguity of the phases determines the potential of having intergranular pores on A-A, A-B, and B-B grain boundaries. These curves will vary as the particle size, shape, and stacking of each phase changes, especially for real particles and significnt differences in particle character of phases A and B.

conceptually the simplest process and microstructure, the microstructural aspects of the solid and associated pore phase present modeling challenges. These are best seen by considering the contiguity of the two solid phases with themselves and with each other (Fig. 8.1). The frequency of contiguity of the matrix phase grains or particles with themselves decreases continuously as more dispersed particles are added, and contiguity of dispersed-phase and matrix-phase particles first increases, reaches a maximum, then decreases. Contiguity of the dispersed phase is already significant at its percolation limit and that of the matrix phase is still significant at its percolation limit with increasing volume fraction of the dispersed phase (which essentially becomes the matrix phase when it constitutes more than 50 vol. % of the body). Both percolation limits are dependent on the particle or grain sizes and morphologies of both phases.

Consider first the case where the only porosity in the above structure arises from incomplete sintering of the particles of the two (A and B) phases, and remains as intergranular porosity. Even in this simplest pore structure and having (and maintaining) essentially equiaxed particles there is a changing mix of pores at A-A, A-B, and B-B grain boundaries as the volume fraction of the dispersed phase (B) increases. For composite properties of interest that do not differ significantly between phases A and B and that the presence of the pores does not significantly change the mechanism (s) determining the property, the changing mix of pore-phase association does not have much effect. Elastic properties are a common example of this, so models applicable to single-phase bodies should be applicable to such composite bodies. On the other hand, many other, real, cases are more complex, especially if the property differences are substantial. In this case the effect of pores within or between all A or all B grains, while having the same relative effect on elastic moduli, have differing absolute effects on the composite properties. Thus, for example, a 10% decrease in properties of phase A or B which differ substantially, e.g., by fivefold, has differing effects on the composite depending on whether the pores causing the reduction are primarily in one phase or the other. The situation becomes more complicated where the particulates are not equiaxed, or where the presence of pores changes the mechanisms determining the properties. Thus, whisker (or short fiber) and platelet composites will also commonly affect the morphology of the pores, and commonly their orientation to the degree that these added phases have preferred orientations. Again, if the properties of the matrix and dispersed phases are similar, then the orientation of pores can be addressed (if noted and characterized) by methods outlined in previous chapters; however, combining significant property differences with preferred pore morphologies and orientations clearly is a more complex problem. Note that the same issues of microcrack association with grains of the same or differing phases exists as discussed below.

Consider next the common and more complex cases of composite bodies formed by in situ reactions. These encompass such diverse materials as plasters and cements, along with many other reaction-processed ceramics. The latter includes bodies infiltrated by a reactant phase (e.g., molten Si for processing reaction-sintered or bonded SiC, or oxygen or nitrogen in Lanxide-type processes) and bodies whose final chemistry and microstructures are formed by reactions that occur fully or partly simultaneously with consolidation-densification processes. Different reactions can result in varying simple and complex microstructures, mixes, and extents of porosity from intrinsic density changes from reaction and extrinsic effects (e.g., outgassing). Thus, besides reflecting a diversity of grain structures, these processes commonly result in a diversity of pore structures, often with more than one pore structure in a given body, commonly with none of these characterized, even if recognized.

A common tendency for such reactions is that they characteristically proceed from the surface of individual particles inward, whether they entail macroscopic, moving reaction fronts or only individual, adjacent particles reacting. There are three complications to these reactions, mainly on the particle scale: 1) varying degrees of interaction between the particle scale reaction front and that of the macroreaction front (if the latter exists), 2) the extent of the reaction of individual particles, i.e., part or all the way through individual particles, and 3) the relation of the reaction products to the reacting particle and the infiltrated or other particle reactants. The first of two important ramifications of the latter is that the reaction product often does not form a simple conformal layer on the original particle surface or its unreacted core. For example, reaction products of some hydration reactions initially form as whiskers in an essentially radial fashion from the particle surface or the particulate reaction sites on it. The whiskers subsequently impinge and constrain each others growth. Second, both irregular reaction-product morphology and intrinsic changes from reactant to product densities can result in varying amounts and character of porosities associated with the reaction.

8.2.2 Porosity Modeling Approaches

Modeling of materials considered in this chapter has often been via use of general models applied to single solid phase materials whether they were single or multiphase. Some nominally single solid phase models have also been derived for materials of this chapter. Thus, Helmuth and Turk (4) used an idealization of cements as made up of cubes of laminar composites such that random orientation of the laminar cube structure gave an isotropic body to model their elastic moduli. Pores were created by elimination of composite cubes, and treated on a load-bearing basis. For an idealized single cube layer this gives the relative modulus as $(1 - P)^n$, with $n = 1$, consistent with MSA models for such pores since their single layer is essentially a MSA model; however, they argue that n becomes 3, based on probabilities of association of cubical pores in adjacent layers above and below any layer under consideration.

There are two basic modeling approaches that are, in principle, applicable and have been used for bodies of two or more solid phases and pores. The first approach is use of models for multiphase, composite bodies. Those of Kerner for shear and bulk moduli (5) and for conductivities (6) and those of Budiansky for shear and bulk moduli as well as Poission's ratio (7), as well as for thermal expansion, heat capacity, thermal conductivity, and Gruneisen constant (8) are particular examples. These models can be used for composites with pores, recognizing the basic problems that the use of composite models for porous bodies present. Thus, such models typically cannot reflect some critical geometrical aspects of the pore structure such as pore stacking and P_C—i.e., the percolation limit of the solid phase(s). Based on evaluations in previous chapters, Budiansky's models, which are simple in form, may be more useful for at least some bodies.

The second basic modeling approach is to adapt existing models for porosity or two-phase composites to bodies of at least two solid phases plus porosity.

This can frequently be done by treating the body as a combination of two composites: first one of the two solid phases and then one of this solid composite as the matrix phase for the pore phase. Thus, a pore or composite model can be applied to one part of the microstructure with these results then taken as input into a second application of the same, or another, model (9). Such repeated application of models must be done with considerable caution and should entail two steps. The first step is evaluation of the microstructures and the applicability of the model or models used to such structures. The second step is to compare results from use of different models. More complex models requiring iterative solutions for just two phases may be difficult to apply to such situations, but deserve consideration.

There are three types of microstructural situations in which the above combination, or repeated application, of models should be simpler and more accurate. One is where the pore structure is larger than the composite microstructure, e.g., as in a closed- or open-cell foams, so the solid, composite phase can be treated as essentially a homogeneous solid phase containing and defining the pores. In this case, composite modeling can be used to calculate the properties of the composite solid phase, which then becomes input for the solid properties in a model of porosity dependence for the porous composite being sought. The second case is where the porous composite consists essentially of a mixture of grains of the two solid phases with the pores being between these grains. Again, the mixed grains of the two solids may be treated approximately as a composite with the resulting properties then being used as input for a porosity model. The accuracy of such approximations is likely to decrease substantially as the difference in properties of the two solid phases increases, and it should be at least fairly sensitive to the character of the intergranular porosity. This character commonly will be significantly impacted by fabrication methods and parameters, as discussed in earlier chapters, but may be modified by the nature and interaction of the two solid phases. The third case is where the porosity is essentially all contained in one solid phase; here the properties of the solid-plus-pore phase can be calculated from appropriate porosity models, then these properties used as input for the pore-containing phase in appropriate composite models. In all three of these cases, MSA models may play a useful and appropriate role.

Finally, consider possible modeling approaches for at least some porous composites involving reactions of an infiltrated phase was particles, with cements and plasters being important examples. Bache (10) introduced a MSA-type model for the strength of concrete based on bonding of the aggregate (idealized as spherical particles of diameter D) by a thin layer of cement paste assuming that the critical flaw size for brittle failure is D. His model gives the failure strength (σ) from the following proportionality:

$$\sigma \propto \left(\frac{d}{D}\right)^{3/2}(\gamma E D^{-1})^{1/2} \qquad (8.1)$$

where d = the diameter of the sphere to sphere contact area, and the proportionality depends on shape factors (unspecified). He then used this model, which is most appropriate for tensile failure, for compressive strength, and used the empirical relation of $\sigma_c = \sigma_c (1 - P)^n$ (finding n ~6), claiming that it fitted his (limited) experimental data. While this model provides some guidance, it needs more scrutiny. Harma and Satava (11) subsequently presented an essentially MSA model, i.e., assuming that strength is proportional to the ratio of the area of contacts between particles to that at P_C, allowing for growth of the particles (due to reaction) obtaining;

$$\sigma \propto f - f_C \qquad (8.2)$$

where $f = (1 - P)^{4/3}P^{-1/3}$, f_C has the same form with P replaced by P_C, and the proportionality constant reflects a fairly complex function of several shape and geometrical factors (which generally cannot be calculated for real systems). The function $f - f_C$ is very similar to those of Fig. 8.2, i.e., going to ∞ at P = 0 and being an approximately straight line at intermediate P values on a semilog plot, except it does not reach values less than 1 till P ~ 0.1–0.3 and gives similar to somewhat higher b values (e.g., 4–8), consistent with the same or lower P_C values. Harma and Satava also plotted Eq. 8.2 on the alternate semilog plot, i.e., versus log P, giving a straight line, hence being similar to, and thus providing some analytical support for, Shiller's equation (12):

$$\sigma \propto \log\left(\frac{P_C}{P}\right) \qquad (8.3)$$

However, this equation also goes to infinity at P = 0. (Although many investigators of materials such as plasters which typically do not reach P = 0, have not been concerned with such behavior at P = 0, this can be important for relating extrapolations and special processing giving low P values to better understand overall behavior.)

Onada (13) developed two related MSA-type models [based on a similar model for randomly packed spherical particles by Rumpf (14)] for the tensile strength of green ceramic bodies due to the amount and wetting of binder on the ceramic particles. Although the latter were assumed to be uniform spheres of random packing, these models may have applicability to some reaction-processed composite bodies, e.g., some formed by hydration. Both versions of the model treat spherical particles bonded to each other at contact points. One

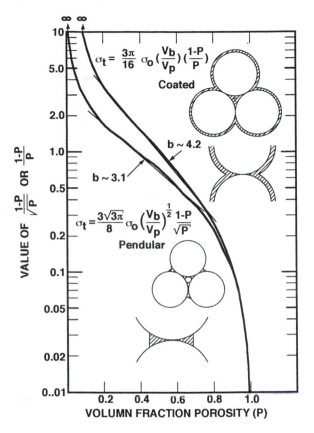

Figure 8.2 Plot of the functional dependence of the porosity dependent terms in Onada's two models (13) versus volume fraction (P). These models, derived for green body strengths due to binders, are suggested here as possible models (or as least as a starting point for broader modeling) of some reaction, e.g., cementatious, materials. Note: 1) the two different model versions (inserts and discussed in the text), where σ_t = tensile strength, σ_0 = the strength of the bonding phase, V_b = the total volume of the binder (reacted) phase, and V_p = the total volume of unreacted particles in the body, 2) they are not valid at low P since they go to infinity at P = 0, but then show behavior typical of MSA models (e.g., with b values of ~ 3–4), and 3) the numerical parameters are ~ 0.59 (coated) and ~ 1.15 (pendular).

version treats a bonding phase only immediately around the contact area, referred to as the pendular state (Fig. 8.2), which could reflect early stages of reaction bonding. In the other version, the bonding phase completely coats the bonded particles, referred to as the coated state (Fig. 8.2). This version could represent a more advanced (but still intermediate stage) of reaction, but not of bonding, or a concrete material with a limited cementatious phase around the aggregate. Besides these limitations, the models clearly are not valid below P ~ 0.2 since they go to infinity at P = 0, but have behavior typical of other MSA models at higher P, i.e., going to zero at P = 1, thus not accounting for particle stacking.

Kendall and Birchall (15) proposed accounting for porosity effects on strengths of cements via the porosity dependences of E and fracture energy or toughness and the size of larger pores as the flaw size in the Griffith equation as done in Chapter 5 for general ceramics. Their selection of the two different functional forms for the porosity dependences of E and γ was apparently arbitrarily based on modest empirical data, and they did not address the possible dependence of either of these functional forms on pore character. Nonetheless, their approach is basically valid, reflecting the dependence of strength on both porosity and flaw size; the latter dependence being explicitly demonstrated by them. While their statement that the porosity dependence of strength was fortuitous is in principle valid, it neglects extensive data showing such dependence for cements, related materials, other ceramics, as well as other materials as reviewed in these chapters.

8.2.3 Microcrack Modeling

Microcracking, not only occurs but is commonly sought in ceramic composites. Much of the background on such microcracking is given in Chapters 1–7, and 9–10, an important aspect of which is the difficulty of characterizing the microcrack content, i.e., size and especially spatial distribution. Ceramic composites aid in such characterization by providing control of the microcracks via the composition, size, and volume fraction of the dispersed particles, but still leave uncertainties, e.g., of the fraction of particles with microcracks and their scale. There is some modeling of elastic moduli and fracture toughness as well as of thermal conductivity, but little or none for strengths and other mechanical properties of microcracking composites.

Modeling of elastic properties of microcracked composites can further aid in characterization of the microcracking since elastic properties reflect mainly preexisting, as opposed to stress-induced, microcracks. Thus, Pan et al.'s (16) recent modeling (via the effective medium approach) of elastic moduli of particulate composites in terms of the fraction of particles microcracked and the extent of cracking of the particles provides some guidance. As discussed later, comparison with one more comprehensive sets of data yields reasonable estimates of the extent of microcracking and indicates a linear dependence of moduli on microcrack extent, i.e., number and size; however, while the nature of particulate composites can aid microcrack characterization, they also present challenges, especially when one or both phases are anisotropic. In such cases microcracks can occur not only with (between or within) abutting grains/particles of the two phases, but also with those of the same phase if it is sufficiently anisotropic relative to the grain/particle scale. Singh et al. (17) presented analysis to explain experimental observations showing that maxima of microcracking in dense BeO-SiC composites occurred at ~ 30 and 70 volume percent SiC, not at ~ 50% as might be intuitively expected.

The substantial literature on crack bridging and related phenomena as well as resultant R-curve effects has some relevance; however, this is a large subject, with the specific roles of microcracking undefined, so the treatment here is cursory. Thus, for example, while crack bridging clearly involves forming microcracks, whether they form ahead of the main crack or behind it in the forming of grain bridges is not known. Finally, there is no modeling of interactions of pores and microcracks, which though commonly neglected, can occur, e.g., Fig. 7.10A.

8.3 EVALUATION OF DATA AND COMPARISON WITH MODELS

8.3.1 Porosity Effects on Plastic and Metal Properties

Data for the porosity dependence of properties of plastics is limited since they are normally used and made with little or no porosity, except at high porosities of foams, which are of substantial interest and use; however, Young's modulus data for an epoxy (18) and a polyester (19) body with approximately spherical pores (bubbles) was shown to be consistent with the spherical pore MSA model (20) (Fig. 8.3). Thus, the more extensive data clearly shows a rollover at $P \sim 0.5$ toward a $P_C > 0.8$, consistent with a b value of ~ 2.3, which is low but in the range expected for spherical pore MSA models. The other data for lower P gives a b value of ~ 3.5, again in the range expected for spherical pores from MSA models. These variations in b values probably reflect variations in pore size and stacking, and hence spacing, while their average is in very good agreement with the value for simple cubic (hence also random) stacking of uniform spherical pores.

The substantial data for mainly polymeric foams reviewed by Gibson (21), and Gibson and Ashby (22), commonly presented as a log-log plot of the absolute or relative property versus the relative density $(1 - P)$, generally agree with their model dependence of $(1 - P)^2$; however, it also agrees overall about as well with MSA models, as shown by a semilog plot of the relative E versus P (Fig. 8.4). This shows that their average trend falls between MSA models for simple cubically (or random) stacked spherical or cylindrical (aligned) pores over most of the P range as expected; however, for $P > 0.8$ the average data trend crosses the lower curve for cubically stacked spherical pores. This is also expected since deviation from perfect stacking of these (or other) pores would decrease P_C values and hence the rollover toward P_C. The very limited G data

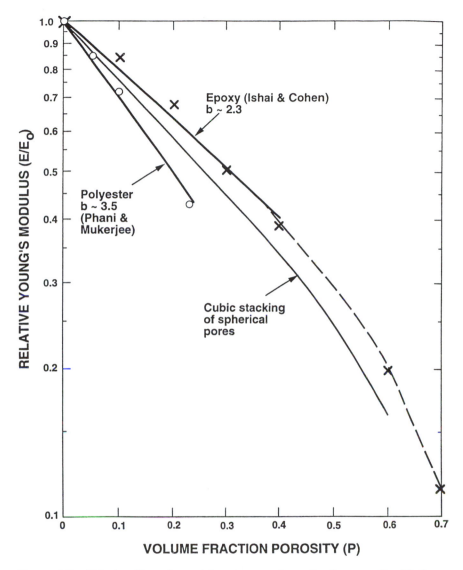

Figure 8.3 Relative Young's modulus versus volume fraction porosity (P) for an epoxy (18) and a polyester (19) polymer with mostly spherical pores (bubbles). Published with permission of J. Mat. Sci. (20)

also approximately agrees with their proposed $(1 - P)^2$ dependence, but since it is normalized by E_0, not G_0, and the E/G ratio will vary some with different materials, such normalization is uncertain in comparing to other materials such as metals and ceramics. Their data summary for Poisson's ratio showed it being independent of P, though there is some suggestion of a decrease at $P > 0.99$ (insert, Fig. 8.4). Comparison of trends for other properties is complicated by limited data (especially for crushing, i.e., compressive, strength and to some extent for K_{IC}), different mechanisms (such as plastic deformation of cell walls or struts), or both.

There is substantial data for metals, although often limited in any one study in terms of porosity parameters (especially porosity characterization) and different properties measured on the same body. Data for the porosity dependence of elastic properties is summarized in Table 8.1 and for fracture toughness, tensile strength, and hardness in Table 8.2.

Data in Table 8.1 is consistent with ceramic data of Chapter 3 and with MSA models in at least six ways. First, differing fabrication/processing clearly changes the porosity dependence, generally in the direction expected. Thus, not only are the changing processing conditions used by Goetzel (23) the primary source of substantial data variation, the trend is clearly for higher b values as the consolidating pressure increases, consistent with increased particle packing density, i.e., C_n. The scatter of results from each processing condition generally give too limited a P range to give difinitive quantitative results by itself, but clear qualitative shifts as noted. The unequivocal different b values of Pohl (27) for different processing show a strong dependence on processing and related pore character (although not specifically identified). Though less extreme, this is also the case for the HIPed Cu where higher HIP pressures presumably reduced closed porosity (31). Finally, note that the one case of hot pressing [for at least some Be bodies (28)] is at least approximately consistent with such results for processing of ceramics.

A second consistency with ceramic results and MSA models is that the average b value for all tests of E is ~ 4 as expected. Third, the limited data for G is consistent with its b value being similar to, but generally somewhat less than that for E, and the two values for ν are consistent with these being of the order of twice the difference between the values for E and G. Also, despite the variations in data of Haynes and Egediege (26), their data clearly showed a higher b value for B than for E and G. Fourth, the limited P-dependence data for ultrasonic wave velocities is also reasonably consistent with expectations for the b value of the modulus being twice that of the corresponding velocity plus 1, e.g., respective b values for Be of ~ 1.5 and ~ 3.5 (28). Similar consistency was shown for the ultrasonic and resulting modulus data of Panakkal et al. (30) by Miatra and Phani (32). Fifth, where Poisson's ratio was determined as a function of P, the b values were generally consistent with those for ceramics and MSA

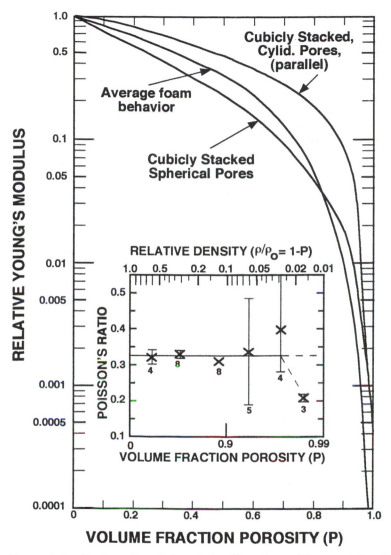

Figure 8.4 Semilog plot of the relative Young's modulus trend for plastic foams versus volume fraction porosity (P) at 22°C. Mean data trend shown with MSA models for simple cubic stacking of uniform spherical and (aligned) cylindrical pores, showing expected agreement as discussed in the text. Also note the insert showing replotted Poisson's ratio versus P, which shows it independent of P, except for a possible decrease at high P [after Gibson (21)].

Table 8.1   Porosity Dependence of Elastic Properties of Metals at 22°C

Metal	Pro- cessing[a]	P range[b]	Data points[c]	Approx. b values[d] E	G	v	Investigator
Fe	S[e]	0.01–0.26	45	4±2			Goetzel (23)
Fe[f]	S	0.02–0.61	150	4±0.5			McAdam (24)
Fe	S	0.07–0.29	8	4	3.8		Artusio et al. (25)
Cu-Sn	S	0.08–0.26	4	3.8	3.8		Artusio et al. (25)
Fe	S	0.04–0.27	~60	3–4	3–4	1–2	Haynes and Egediege (26)
Fe	S[g]	0.13–0.24	20	3.1			Pohl (27)
Fe	GCI[g]	0.08–0.10	≥30	7.3			Pohl (27)
Fe	NCI[g]	0.08–0.10	≥20	1.6			Pohl (27)
Be	HP[h]	0.02–0.22	0–0.22	3.7			Beasley and Cooper (28)
Bronze	S[i]	0.2–0.6	14	4			Jernot et al. (29)
Co	S	0.06–0.63	19	3.5			Jernot et al. (29)
Ni	S	0.05–0.36	14	4.0			Jernot et al. (29)
Fe	S[j]	0–0.22	20	3.8	3.5	0.8	Panakkal et al. (30)
Cu	HIP	0.13–0.49	12	3.3–3.8[k]			Takata and Ishizaki (31)

[a] S = sintering, GCI = gray cast iron, NCI = nodular cast iron, HP = hot pressed, HIP = hot isostatically pressed (HIPed).
[b] Range of volume fraction porosity (P).
[c] Number of data points over the P range.
[d] Approximate slopes for the approximately initial, linear portion of a semilog plot of elastic property versus P, i.e., the MSA slope for E = Young's modulus and G = shear modulus, as well as v = Poission's ratio.
[e] Reflects various process parameters which clearly varied E for a given P, hence giving considerable scatter for the collective data.
[f] Survey of various iron materials.
[g] All three b values are based on and consistent with a P = 0 value of E = 150 GPa.
[h] Some, possibly all, bodies appear to have been hot pressed.
[i] Spherical particles.
[j] P = 0 values from HIPed material.
[k] Values decreased from 3.8 to 3.3 as the HIP pressure increased from 0.1 to 200 MPa.

model expectations. Sixth, where data has been measured to sufficiently high P levels relative to the pertinent P_C values, rollovers of the moduli dependence to greater rates of decrease with increasing P are seen. Thus, the E data of Takata and Ishizaki (27) for Cu and McAdam's (24) survey for Fe show rollovers at ≥ P = 0.4 and 0.3, respectively, which are reasonably consistent with the inverse trend expected for the respective b values of ~ 3.5 and 4. More data on rollovers, especially extensive ones, and agreement with MSA model trends is shown in Fig. 8.5A. Ultrasonic velocity versus P data for hot-pressed W (33),

Table 8.2 Porosity Dependence of Tensile Strength (σ_t) and Hardness for Metals at 22°C

Metal	Processing[a]	P range[b]	Data points[c]	Approx. b values[d] σ_t	H	Investigator
Fe	S[e]	0.01–0.26	45	5±2	4±2 (B)	Goetzel (23)
Fe	HP	0.01–0.2	12	3–5	3–5 (B)	Henry and Cordiano (39)
Fe[f]	S	0.02–0.7	>45	3.2		Eudier (40)
Fe	S[e]	0.13–0.29	8	5–7		Gallina and Monnone (41)
Cu	S[g]	0.01–0.37	28	5	5 (R)	DeHoff and Gillard (42)
Fe[f]	S	0.01–0.35	>700	4.3	4.3 (B)	Salak et al. (43)
Be	HP[h]	0–0.9	≥15	5–7		Beasley and Cooper (28)
Brass	S	0.01–0.15	10	4±0.6[i]		Rostoker and Liu (44)
Fe	S	0.03–0.22	30	4.8		Jernot and Chermont (45)
Fe	S		12	6–10[h]		Kuroki and Tokunaga (46)
Steel	S	0.1–0.45	4	3		Duwez and Martens (47)
Steel	S	0.1–0.45	4	5		Duwez and Martens (47)
Various		0.08–0.24		4–13		German (48)

[a] S = sintering, HP = hot pressed.

[b] Range of volume fraction porosity (P).

[c] Number of data points over the P range.

[d] Approximate slopes for the initial approximately linear portion of a semilog plot of the property versus P, i.e., the MSA slope. B = Brinell hardness, and R = Rockwell hardness.

[e] Reflects various process parameters which clearly varied E for a given P, hence giving considerable scatter for the collective data.

[f] Survey of various iron materials.

[g] Spherical particles.

[h] Some, possibly all, bodies appear to have been hot pressed.

[i] Yield stress for approximately spherical porosity.

although only for P ≥ 0.2, gave b values of 0.7 to 2.2, which are reasonably consistent with higher values expected for hot pressing. More important, this study clearly showed distinct differences in the porosity dependences of bodies made by the same processing for similar, but not identical, powders. Similarly, Matikas et al. (34) showed b values for shear and longitudinal sound velocities

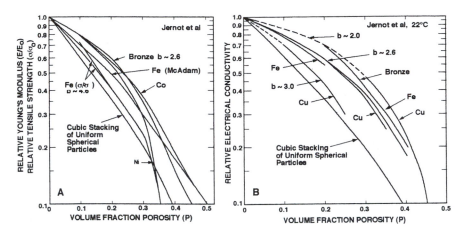

Figure 8.5   Representative examples of metal property data over a substantial P range at 22°C. A) Young's modulus and tensile strength from Jernot et al. (29) and McAdam (24), and B) electrical conductivity from Jernot et al. (29). Note the similarity of the P-dependence for all of these properties and their general agreement with MSA models.

versus P in hot-pressed TiAl as ~ 2 and ~3, respectively, and giving b values for shear and Young's moduli of ~ 5 and 8, respectively. These higher values are consistent with hot pressing typically giving denser particle packing, i.e., higher C_n and b values.

As with ceramics, there is very limited data on the porosity dependence of fracture toughness of metals; however, the limited data shows that there is a similar trend as for elastic properties (and tensile strength) as well as the variations found for ceramics. Thus, Cooper (35) reported four sets of data of others for a Ni-Mo steel (P = 0–0.15) tested at 200 and 295°K with two different tempering conditions, giving b values of 3–7. The primary differences in b values are due to variations of ≥ 50% in the values at P = 0. Whether these arise due to differences in grain size, residual stresses, changes in failure-causing flaws, or ductility at P = 0 is not known, but the similarity to effects in ceramics (e.g., Fig. 5.6) should be noted. At and beyond P = 0.04 the b values were all ~ 4, i.e, consistent with the average trend for elastic moduli. On the other hand, he also presented data for sintered and plasma sprayed Be which showed a definite maxima at P ~ 0.06 (Fig. 8.6), which is qualitatively but not necessarily quantitatively consistent with the model of Navara and Bengtsson (2). Studies of Clarke and Queeney (36) on stainless steel (P~ 0.05–0.12) gave b ~ 6–8 and those of Drachinskii et al. (35) (P~ 0–0.1) gave b ~ 5–6 for each of three levels of intergranular fracture (which varied the K_{IC} level). Finally, although not directly measuring K_{IC} versus P, Barnby et al. (38) showed a linear correlation of K_{IC} and yield stress for a range of porous bodies of a sintered steel covering a

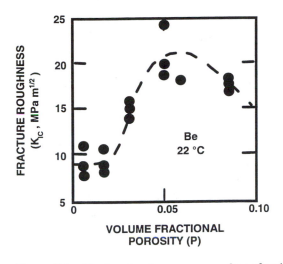

Figure 8.6   Fracture toughness versus volume fraction porosity, P, for plasma-sprayed and sintered Be (35). Note the clear maxima. Published with permission of Pergamon Press.

severalfold change in each property. Finally, as noted in Chapter 4, Bompard et al. (49) have shown substantial crack branching in porous Ni.

Summary data for the porosity dependence of tensile strength and hardness (Table 8.2) also shows the same trends amongst these properties and between various mechanical properties as found for ceramics. Thus, both the overall average b values for strength and hardness are the same or higher than for elastic moduli, as also shown by the direct comparisons of these values with those for Young's modules on the same bodies (23,28). Also, where properties have been measured to higher P values relative to expected P_C values, rollovers to greater decrease with increasing P occur at higher P for lower b values, i.e., at $\geq P = 0.5$, 0.3, and 0.3, respectively, for Eudier's survey of Fe strengths (40), Duwez and Martens (47) of sintered Ni-Mo steel (but clearly at higher P for their stainless steel with b ~ 5), and DeHoff and Gillards' (42) measurements of strength and hardness of sintered spherical Cu particles. Furthermore, demonstration of rollovers and of the general agreement with MSA model trends is again shown in Fig. 8.5. DeHoff and Gillard also showed their Cu results were independent of the particle size over the range of 10–200 μm diameter. More fundamentally they showed experimentally that the strength of the copper directly correlated with the minimum solid area, as measured on the fracture surfaces. Also note that Ul`yanov et al. (50) showed both Brinell and Rockwell hardnesses of two sintered steels having b values of 4–8 over P ~ 0.05–0.1.

While the exponential relation is widely used for and fits metal data, as shown above and noted earlier, other expressions have been used. The dependence of mechanical properties on P via $(1 - P)^n$ is one of the other expressions used. Thus, McAdam (24) fitted his survey of E for sintered Fe bodies with n = 3.4 (b ~ 4). Maitra and Phani (32) fitted the shear and logitudinal wave velocities and resultant shear and Young's moduli data of Panakkal et al. (30) with n = 1.1, 1.4, 3.2, and 3.9, which are close to the b values (Table 8.1). German's (48) substantial log strength versus log $1 - P$ data shows n values typically in the 4 to 10 range and are consistent with the higher b values for his data (which may reflect some data not being at P < 0.1–0.2).

Provenzano et al. (51) have conducted initial experiments to test concepts of strengthening metals by dispersion of fine (submicron) pores as suggested by Louat et al. (3). Copper and nickel bodies hot pressed or HIPed from fine (e.g., ~ 1 micron) commercial powders were reported to have strengths from about those for dense bodies to four times higher with modest amounts of porosity, though the pore sizes were typically larger than desired. Some of the expected thermal stability of inert gas-filled pores at higher temperatures was also shown, but possible benefits in damping capability of the metals due to pores were not demonstrated.

As with other materials, heterogeneity of the porosity is an underlying but usually unknown factor that is probably frequently significant. However, Dixon et al. (52) study of sintered nickel steel did address heterogeneity. Although, they used only a limited range of porosity, their results clearly showed greater nonlinear decrease of strength with increasing P as heterogeneity increased (Fig. 8.7). This trend is logical since heterogeneities clearly have a greater impact on tensile strength. They found that both strength and hardness varied approximately linearly with P in homogeneous bodies, although this is probably due in part to the modest P range covered.

A few measurements have been made of nonmechanical properties of metals as a function of porosity. Thus, Ul'yanov et al. (50) showed saturation magnetization and coercive force generally decreasing modestly. This data gave b values of ~1 generally over the range P ~ 0.05–0.1; however, different heat treatments reduced the decreases in coercive force so it was approximately independent of P, or showed small but definite increases with increasing P. Beasley and Cooper (28) showed very limited decreases and substantial anisotropy of thermal conductivity of hot-pressed Be with increasing P (~ 0–0.3), e.g., b values of 0.3–0.5 at 800°K (and substantially lower b values and conductivity levels at 1249°K). The lower conductivity (e.g., by ~ 30%) and higher porosity dependence normal to the pressing direction versus parallel to it probably reflects orientation of both the pores and the grain crystallography from the hot pressing. Finally, semilog plots of the electrical conductivity data

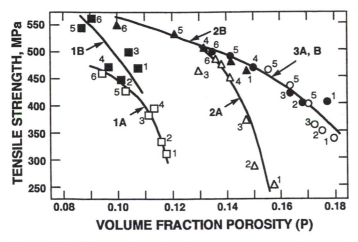

Figure 8.7 Effects of porosity heterogeneity on tensile strength–volume fraction porosity (P) for sintered nickel steel at 22°C, after Dixon et al. (52). A nickel steel was sintered from each of three powder constituent preparations: 1) mechanical blended, 2) plated, or 3) prealloyed powders. The compositional (as well as qualitatively the porosity) homogeneity increased in the order listed. Sintering was at: A) 1150 and B) 1300°C, with increased homogeneity for B, especially for powders 1 and 2. Note greater strengths at a given P and lesser decrease at higher P with greater homogeneity. Published with permission of Powd. Met.

for sintered metals of Jernot et al. (53) clearly shows overall trends consistent with MSA models and with elastic moduli and tensile strength (Fig. 8.5B).

8.3.2 Porosity and Microcrack Property Dependence of Rocks

While the physical properties of rocks have been widely studied, data on their porosity dependence is restricted by several factors. First, there are significant limitations on obtaining samples where the primary variable is porosity covering a reasonable range, and even then there are often problems of anisotropy, gradients, and heterogeneity. Second, understanding effects of other important variables such as water and especially confining pressure on properties is also needed for rock mechanics. Thus, while data on dry, unconfined rocks is basic to much modeling (and of primary interest here), much other data (beyond the scope of this section and book) is needed, limiting efforts on obtaining just basic rock properties. Nonetheless, there is reasonable data available, either as general, qualitative or semiquantitative observations, but also some quantitative data of specific properties as a function of porosity.

An important type of semiquantitative evaluation of porosity effects is in the form of property correlations. Since the range of properties correlated is typically significantly extended by the presence of porosity, these correlations provide some indication of the porosity dependence, and corroborate the interrelations of properties discussed in earlier chapters. Thus, Judd and Huber (54), using the Bureau of Mines data base on Mine Rock, did a statistical analysis of the correlation of various mechanical properties with factors such as apparent porosity and especially apparent specific gravity. (The term apparent, i.e., open, porosity reflects the common problem for many materials in this chapter of not knowing their true theoretical density, and hence not being able to readily determine their total porosity from a density measurement.) They showed the generally expected decreases of longitudinal wave velocity (v_1), G, E, σ_t, σ_c, hardness (sclerescope), as well as decreases of impact toughness and specific damping capacity, and some possible decrease of Poission's ratio with increasing apparent porosity. Cross-correlations of properties, e.g., of σ_c with v_1, G, E, and σ_t were also shown, which were consistent with the correlations with porosity and density.

More extensive correlations over a narrower range of properties, namely between Young's modulus (E) and compressive strength (σ), was shown by Stagg and Zienkiewicz (55) for a variety of rocks, all covering a moderate to substantial range of moduli and strengths (Fig. 8.8). They showed granites, diabase, basalt and other flow rocks, gneiss, quartzite, and most limestone and dolimite typically correlated in the range of E/σ_c ratios of 200–500 to 1, i.e., averaging ~ 350 to 1. Marble and some schists correlated at substantially higher ratios, e.g., averaging ~ 750 to 1, while sandstone and other schists averaged lower, e.g., an average of ~ 200 to 1, and shales even lower, e.g., averaging ~ 100 to 1.

Varying qualitative and quantitative effects of porosity on crack propagation and fracture energies in some sedimentary rocks (sandstones) were shown by Hoagland et al. (56) (Fig. 4.5). This study also showed the significant anisotropy of fracture toughness, i.e., values as much as 2–5 times higher for crack propagation normal to the bedding planes versus parallel to the bedding planes. They also showed measurable anisotropy, e.g., by factors of 1.2–2.2 for the two planes normal to each other and the bedding planes, with the anisotropies being different in both magnitude and direction between the limestone and sandstone used. Peck et al. (57) provided qualitative observations on crack propagation and fracture energy of similar and other rocks. They concluded that porosity (and weak bonding of interfaces) gave lower fracture energies. Higher fracture energies and greater distances of crack propagation for steady-state fracture were found when there was a preexisting network of interconnecting microcracks (or when the rock texture provides multiple, incipient fracture surfaces). The highest fracture energies were found in crystalline rocks without intercon-

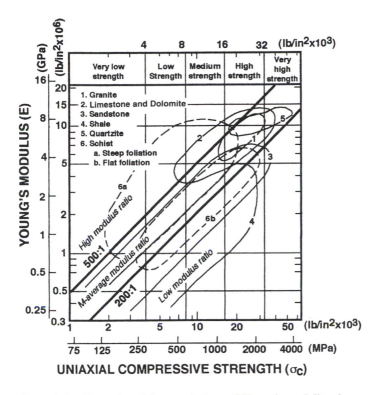

Figure 8.8   Examples of the correlations of Young's moduli and compressive strengths of rocks having a substantial range of these properties at 22°C after Stagg and Zienkiewicz (55). Published with permission of John Wiley & Sons.

nected microcracks. Their comparisons of their observations with those of other investigators on similar rocks is also useful.

Of the limited, detailed quantitative studies of several properties (of unconfined rocks) versus porosity, those of Price and colleagues (58,59) are perhaps the most comprehensive (Fig. 8.9). They conducted two studies of Yucca Mountain tuff at 22°C, covering a substantial P range, usually from 0.05–0.1 to 0.4–0.65 (although with limited other pore characterization). Their earlier study on cylindrical samples (~ 2.5 cm diameter, by ~ 5.1 cm long; saturated with water) were of E, B, and compressive strength. These gave respective b values of ~ 6, 5, and 7–8.5 when the volume fraction of clay in the samples was included as part of the porosity (defined as functional porosity) which gave P values to 0.64. Data points at this maximum P value (consisting of a single measurement for each property) clearly indicated rollovers for E and B at

P ~ 0.5, but not for compressive strength (although this may be due to there being only a single measurement) at the highest porosity level. Their second, more recent and extensive study of the same rock material was conducted on cylinders of twice the dimensions of the first test (specimens were saturated with water, then drained). This study, which used only the actual porosity (i.e., not including the clay as part of the porosity, so P levels extend to only ~ 0.54) showed a modest b value of ~ 1.5 for ultrasonic velocity (Fig. 8.9, < half the b values for E of 5.5–7 and that for B of ~ 5 in their earlier study). Other b values are ~ 5 for tensile strength and 6–9.5 for compressive strength. Rollovers to greater property decreases were well defined by several multiple sample data points for tensile strength (P ~ 0.4–0.45), and fairly well indicated for E and B (at P ~ 0.5), as in the study having single data points at higher P (0.64). The b values in this second study tend to be somewhat higher than in the first study due to the inclusion of clay content as part of the volume fraction porosity in the first but not the second study.

The results of Price and colleagues are consistent with several expectations based on MSA models, but present questions for other expectations. Thus they are qualitatively consistent with lower b values expected for ultrasonic velocity, but the difference from b values for E is greater than expected. The b value for tensile strength is approximately that for B and the lower end of the range for E, consistent with b values for tensile strength normally being about the same as that for E, but not consistent with it frequently being somewhat greater than for E, instead of possibly the other way around as indicated by this data. The b value for v (~1) is consistent with b values of ~ 4–5 for E, again consistent with the lower range of b values, but uncertain for the higher b values for E. The high b values (6–9.5) for compressive strength range from being modestly higher than for tensile strength and the lower end of the range for E as commonly found to substantially higher than for E and especially tensile strength.

Much, and quite likely all, of the above differences and uncertainties are due to two factors. The first is data over a P range that was often probably marginal to clearly define the b values and rollovers (a typical large challenge in rock studies, and still an issue in the more comprehensive studies considered here). The second is the substantial data scatter and the substantial heterogeneity it must reflect, along with possible anisotropies and gradients it may reflect (again particularly important challenges to the study of rocks, but also pertinent to other materials). The scatter can be put in quantitative perspective by calculating the Weibull moduli (m) for different properties, as discussed in Sections 2.3, 3.5, and 10.5.2. This gives m values of 10–50 for sonic velocity, 3–4 for elastic properties, and 2 for tensile and compressive strength. These are all substantially lower than for ceramic materials of similar porosity levels and character, e.g., by an order of magnitude. Thus, the scatter and no or limited

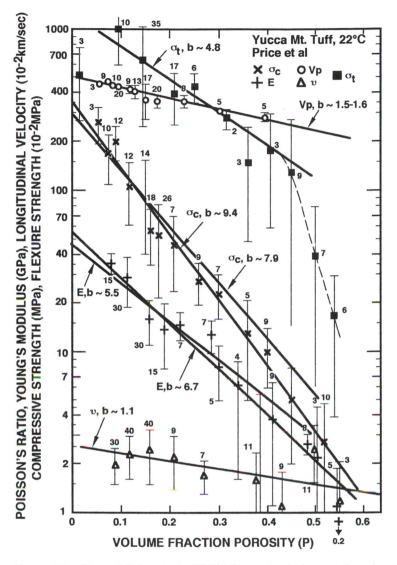

Figure 8.9 Data of Price et al. (58,59) for mechanical properties of wet, drained, unconfined samples of Yucca Mountain tuff. Note clear evidence of a rollover of strength, and some evidence of this for Poisson's ratio.

data at P < 0.1 and above P = 0.4–0.54 will typically give higher b values (e.g., see Fig. 3.18), and obscure rollovers. Therefore correction for these probable problems would bring the rock data from agreeing on overall trends and on some quantitative expectations, but failing other quantitative expectations, to much better, probably overall agreement.

Data on the porosity dependence of thermal conductivity of rocks was shown earlier to be reasonably consistent with MSA models which are expected to be pertinent (Fig. 7.8).

8.3.3 Porosity Dependence of Plaster Properties

Consider now plaster materials, which represent simpler hydration reactions in that they generally involve a single material, most or all of which is consumed in the reaction. These materials thus generally represent an intermediate level of complexity between normally sintered ceramics and reaction-processed materials such as cements. Valuable sets of experiments were conducted by Soroka and Sereda (60) and by Schiller (12). Soroka and Sereda investigated five different gypsum bodies: 1) cast, hydrating powder-water mixtures, 2) cast, hydrated bodies compacted, 3) compacted, unhydrated powder, 4) unhydrated material compacted then hydrated, and 5) hydrated powder subsequently compacted. Their measurements of both Young's modulus and Vickers microhardness (Fig. 8.10) presented for bodies 1, 2, 4, and 5 versus P show three important trends. First, both properties of the different bodies measured at modest P extrapolated to essentially a single value for the property at P = 0. Second, properties of all bodies were essentially linear on semilog plots over a substantial P range (Fig. 8.10, and Table 8.3 for b values), except data for body 1 appears to reflect a rollover towards a P_C value (e.g., > 0.7). This is also true for their hardness data, since the data for body 1 (not plotted in Fig. 8.10) is very similar to the extension of data for body 4 for E. Also, data for body 5 (hydrated then compacted) may reflect a start of a rollover toward P_C > 0.3 at P ~ 0.2. This would make its extrapolation to P = 0 more consistent with those of bodies 2 and 4, and reduce its b value to 8–9 from ≥ 11. The former is more consistent with trends expected for consolidation of powders, and the increased b value for the body hydrated in situ, then compacted (body 2) is also consistent with this. Effects of compaction on the hydration reaction and also possibly via deformation of hydrated products are also expected. Thus, their results indicate substantial similarity to MSA-model trends, which is logical given the common role of compaction in the gypsum materials and pertinent MSA models.

Schiller's plot of tensile, flexure, and compressive strengths on a linear scale versus the log of the porosity (P ~ 0.42–0.6) for gypsum at 22°C was linear, with all three projecting to P_C ~ 0.87 (9,12). Analysis of the data showed all three

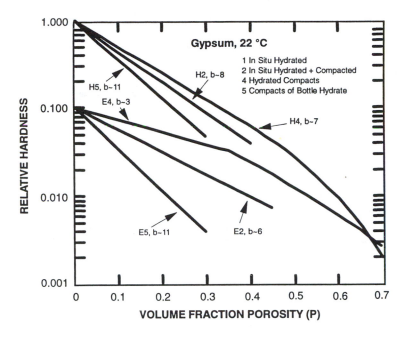

Figure 8.10   Relative Vickers hardness (H, loads adjusted to give indent depths of 35–50 μm) and relative Young's modulus of various gypsum bodies after Soroka and Sereda (60). Note: 1) clear differences in behavior with different processing and hence different microstructures, i.e., porosities; 2) distinct rollovers (at P = 0.34–0.45) where data extends well past this, and possible starts of rollovers at lower P with higher b values where data has not been obtained significantly beyond the possible rollover; and 3) that the b values are typically the same or higher for H versus E for the same processing and microstructure.

strengths gave parallel plots of the log of the strength versus a linear scale for P, all with b values of ~ 4. Such values are in the range of those for normal hydration (without consolidation) from Soroka and Sereda (3–7, Table 8.3). Note that the range of the latter can not be attributed to grain size variations since the highest and lowest b values were for bodies with grains of ~ 1 μm, again indicating that pore character (shape, stacking, etc.) is a factor. Besides showing some overall similarities to MSA-model trends, as well as some specific ones (i.e., higher b values for compacted bodies), there are also similarities and variations for different properties for the same bodies. Thus note the same b values for compressive and tensile strengths (the latter in both true tension and flexure), as well as some higher b values for hardness versus E (Table 8.3).

Table 8.3 Comparison of b Values for Various Properties of Gypsum Bodies at 22°C

Investigator	Processing[a]	P range[b]	Data points[c]	E	σ_t	σ_c	H_V
Soroka and Sereda (60)	Hydrated, cast (4)	0.07–0.23	10	3			7
Soroka and Sereda (60)	Hydrated, cast, then pressed (2)	0.11–0.45	18	6			8
Soroka and Sereda (60)	Pressed hydrate (5)	0.04–0.30	9	~11[d]			11
Schiller (9,12)	Hydrated, cast	0.43–0.59	4		4	4	

[a] Numbers in ( ) correspond to property curves in Fig. 8.10.
[b] Range of volume fraction porosity (P).
[c] Number of data points over the P range.
[d] This value may be 8 to 9 if there is a rollover toward a lower P_C value as discussed in the text.

8.3.4 Porosity Dependence of Properties of Cements

Cements generally represent a greater degree of complexity, involving more than one reactive constituent, and broader variation in the degree of reaction. These, especially Portland cements, are a step toward concrete, i.e., a composite of cement with dispersed aggregate. Because of the extent, complexity, and specialization of these materials, only the basic trends for cements pertinent to broader understanding of porosity dependence of materials in general are addressed. The cement literature understandably has focused extensively on composition, processing, and their complex interactions, which have been extensively studied (61–65). The reactions involve substantial, approximately radial, whisker growths that become entangled. There has also been substantial study of porosity in cements since this obviously has a complex, detailed relation to the reaction processes and the extent of their progress. Porosity in cements is generally divided into: 1) gel pores (micopores) which are an intrinsic part of the hydrated products and independent of the water/cement or solids (w/c or w/s) ratio; and 2) capillary (meso-, i.e., finer-) pores which are directly effected by the w/c ratio. The latter are commonly 50–70% of the total porosity, thus giving some overall correlations of the porosity level with the w/c ratio (61–67); however, porosity values vary with both the processing and the measurements and calculations used to obtain them, making porosity characterization a challenge, especially at moderate to lower porosities. Thus, for example, the two common methods of measuring (open) porosity, mercury porosimetry and helium pycnometry, begin to differ significantly below porosities of about 25% (67). This is due to accessibility of pores, as well as possible damage to the pore structure in determining it by mercury intrusion as noted in

Table 8.4 Comparison of b Values for Various Properties for Cement-Paste Bodies at 22°C

Investigator	P range[a]	Data points[b]	Approximate b Values for					
			E	G	B	ν	$K_{IC}(\gamma)$	σ_c
Kendall and Birchall (15)	0.01–0.15	5	4–5				(7–8)	
Ahmed and Struble (69)	0.19–0.50	19					3–4	4–7
Roy and Gouda (79)	0.02–0.2	30						6
Verbech and Helmuth (80)	0.08–0.30	4						3–5
Helmuth and Turk (4)	0.3–0.56	44	2.5–3	2.2–2.3	3	~0		
Alford et al. (76)	0–0.08	12						6
Marsh and Day (68)	0.27–0.43	>30						7–10
Taylor (65)	0.28–0.44	5						4
Taylor (65)	0.01–0.41	22						5
Taylor (65)	0.06–0.50	14						11?[c]

[a] Range of volume fraction porosity (P).

[b] Number of data points over the P range.

[c] This value reflects 1 data point for one body at P = 0.06 and the remaining points for other bodies with P = 0.24–0.50; however, the former may or may not represent a valid extrapolation of the latter, which may instead mostly reflect the initiation and a substantial portion of a rollover toward a P_C of > 0.5 from a b value of < 6.

Chapter 1. Nevertheless, there is considerable data on porosity effects on (mainly mechanical) properties, especially compressive strengths, as outlined below and in Table 8.4. Although much of this data is over limited porosity ranges or combines bodies with possible different microstructures (other than that directly related to the changing porosity), some trends are indicated.

Unfortunately, limited detailed study has been made of the porosity dependence of elastic properties; however, Helmuth and Turk (4) carried out a fairly comprehensive study of ultrasonically measured elastic properties versus porosity for a tricalcium cement and two Portland cements, for various curing times from 6 to 24 months. They showed that their data for E and G fit the model they presented where the P-dependence is given by $(1 - P)^n$ with n = 3 for both the capillary porosity of their two cements and varying curing times, as well as their total porosity; however, semilog plots of their data (Fig. 8.11) shows that the data is clearly bilinear, so extrapolation of their data to P = 0 via $(1 - P)^3$ gives values that are high by 20–40%. Whether the change in slope at P ~ 0.3 is the start of rollovers to P_C, or other changes in the composite structure is not certain, but clearly serves as a warning about assuming consistency of behavior over a larger range of porosity on log-log plots. Note also that the relative trends for b values to decrease modestly in the order B, E, and G are shown by their data as generally found for most ceramics. Kendall and Birchall (15) showed Young's modulus of burnt-out and rehydrated macrodefect-free (MDF) cements

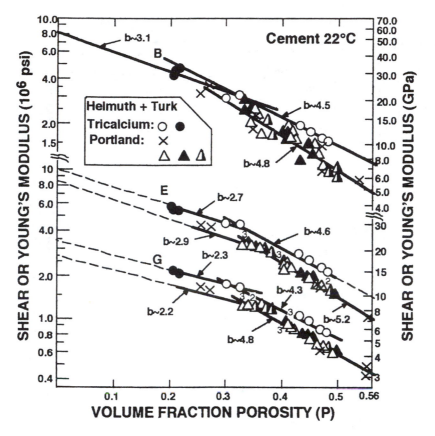

Figure 8.11   Semilog plot of bulk (B), Young's (E), and shear (G) modulus of a trical-cium and two Portland cements for curing times of 6 to 24 months versus volume fraction porosity (P) at 22°C [after Helmuth and Turk (4)]. In each set of curves the upper curves are for the tricalcium cement and the lower curves for the two Portland cements. Note the definite bilinear character of the curves, and that data for various cure times all lie along the same curve for the given material. Bulk moduli values, though more scattered, showed the same trends, i.e., bilinear curves and relative b values as shown in the figure and discussed in the text.

significantly deviated at low P above the relation $E = E_0 (1 - P)^3$ that they and others have often used for cements. (Analysis of this data shows that changing the exponent from a value of 3 does not solve the fitting problem, but that the data is approximately fitted by the exponential relation with a b value of ~ 4–5.)

Kendall and Birchall (15) showed that these same bodies they evaluated for E versus P reasonably fitted the exponential relation for fracture energy versus P data for these bodies, giving a b value of 8 (Table 8.4, analysis shows this b value could possibly be as low as ~7). Ahmed and Struble (69) measured K_{IC} of several cement bodies as a function of P, showing a linear trend for the collective data versus P; however, a semilog plot of this data shows a distinct rollover at just under P = 0.4, giving a b value of ~ 3 below this P level (Table 8.4, Fig. 8.12). Although very limited, these results indicate similar trends as for MSA models, but with some variations, as for sintered ceramics, but possibly over a wider range, as suggested by b = 8. As with other ceramics there are also other variations. Thus, Eden and Bailey (70) reported that de-airing Portland cement paste and removing large pores (> 100 μm diameter), increased strength

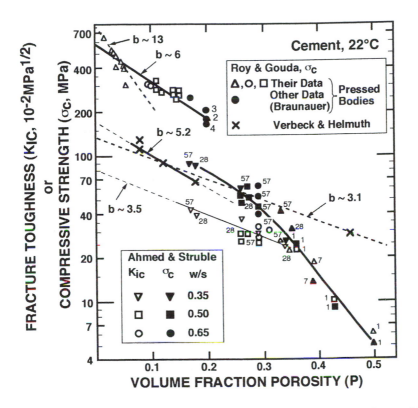

Figure 8.12 Semilog plot of compressive strength and fracture toughness of Portland cement samples versus volume fraction porosity (P) at 22°C. Data of Roy and Gouda (79) and others for consolidated cement samples (note the possibly higher slope for the samples consolidated at > 22°C, giving the lowest P values) showing distinctly higher strengths of compacted versus uncompacted samples (80). Numbers next to some points represent number of tests where > 1. Data of Ahmed and Struble (69) for both fracture toughness and compressive strength showed the same overall trend for various processing (including shown water to solids ratios, and aging 1 to 57 days). Both properties show a clear rollover to greater decreases with increasing P, but at lower P for compressive strength.

(modestly, as discussed below), but made a small decrease in K_{IC}. The latter may reflect some of the temporary reversals of K_{IC} decreasing with increasing porosity with added, larger pores, as with similar RSSN (Fig. 4.3). More generally R-curve effects, i.e., a temporary increase in K_{IC} with increasing crack propagation and hence length, are observed in cement pastes. Igarashi and Kawamura (71) reported such R-curve effects were associated with a microcracking zone (as revealed by dye penetration examination) around the macrocrack. The width of this zone decreased from 100–200 μm on either side of the macrocrack for a water/cement (w/c) ratio of 0.55 to about half of this at a w/c ratio of 0.35 (and even more with the addition of fume silica), i.e., it increased with P. These zone dimensions varied roughly inversely with the level of the initial or subsequent K_{IC}, i.e., the K_{IC} levels for the lower w/c ratios (and hence lower P) were approximately twice that for the lower w/c ratio, opposite from the zone width trends. On the other hand, the addition of fume silica gave the lowest K_{IC} levels and zone widths.

Consider next tensile (and flexure) strength, which has received some, but not extensive attention for cements. It is clear that such strength decreases with increasing P (15), and that strengths at P = 0 are not intrinsic, but in fact depends on the character, especially the size, of other flaws that are independent of porosity, probably more than for other ceramics (Fig. 5.6). Also, while fractography is not nearly as extensively or as successfully used for cements as for other ceramics, it also shows similar failure from isolated larger pores (72,73). More extensive in recent years has been the application of some fractography and fracture mechanics (74–77) showing strengths (for a given w/c ratio and hence approximately constant P) decreasing as the maximum pore size increases (70,74–77). There are variations in the strength increases obtained eliminating larger pores, e.g., from > 100% (76–78) to only 15% (70), which must reflect other sources of failure, e.g., clusters of smaller pores, unreacted particles, combinations, etc. Although not extensively studied, limited measurement of the Weibull modulus also supports a multiplicity of flaw sources by the low modulus values, e.g., ~ 3 (78).

Most extensive studies of effects of porosity on compressive strengths show substantial dependence on porosity, as indicated by the higher b values, which are as high, or higher than, for other ceramics (Table 8.4). Some of the higher values may reflect inappropriate combination of bodies reflecting different processing and resulting different pore structures, e.g., as noted in Table 8.4; however, such combinations may also give lower b values, e.g., while the overall b value is 6 from the data of Roy and Gouda (79) for their hot-pressed cement that they combined with other data for consolidated cements, the b value for the hot-pressed material alone could be ~ 13. Such higher b values are indicated by other data in Table 8.4, as well as more recent data of Lu and Young (81) giving values of ~ 10 and 12 for two cements. Again there is also some limited but clear evidence of a rollover to greater compressive strength decrease at higher P. Thus, Ahmed and Struble's (69) compressive data (with a higher b value for compressive strength than for K_{IC}) shows a clear rollover at just under P = 0.4.

Regardless of the above uncertainties, the limited data for the porosity dependence of mechanical properties of cements and for gypsum show similar trends and reinforce each other. Thus, both show similar trends to those for MSA models, i.e.,: 1) higher b values with greater mechanical compaction (hence usually higher C_n), 2) variable, but often lower b values for fracture toughness, 3) variable, but often somewhat higher b values for compressive strengths versus elastic moduli and fracture toughness, and 4) evidence of rollovers to greater property decreases with increasing P (i.e., toward P_C values), but apparently commonly at greater P values. The latter may well reflect effects of hydration on pore structure-property relations.

The subject of the porosity dependence of concretes, i.e., of cement matrices with dispersed material, especially aggregate, is large and complex, in part because specific attention and control of porosity is limited; however, one observation is pertinent, namely that fracture of concretes is typically through the cement phase or along the cement-aggregate interfaces (82,83). Thus, the porosity in the cement phase (which is typically much greater than in the aggregate) is a major factor in properties involving fracture. There are also correlations between different mechanical properties, e.g., Ahmad and Shah (84) compiled substantial data showing broad, nonlinear correlations with varying scatter of the elastic modulus, tensile strength, and flexure strength with compressive strength (from ~2 to nearly 100 MPa). The nonlinearity of the elastic modulus versus compressive strength appears to be broad, possibly reflecting the differing impacts of cement and aggregate properties on the two properties, including the differing effects of porosities noted above. The nonlinearities in the strength correlations are mainly or exclusively at lower strengths (e.g., to tensile strengths of 2–3 MPa and flexure strengths of 3–4 MPa, i.e., about 1/3 of their range), where isolated defects are likely to be more extreme and significant in differentiating compressive and tensile failure. Roy et al. (85) have also shown a linear correlation between Vickers microhardness and compressive strength of high-strength mortars with varying amounts and types of sands. The linear correlations again indicate similar porosity dependences of the correlating properties in view of the fact that porosity remains an important factor in the ranges of properties correlated.

8.3.5 Porosity and Microcrack Property Dependence of Traditional Ceramics

Two types of related traditional ceramics for which there is some reasonable data on porosity effects are porcelains and whitewares. Consider first porcelains for which there is both more total data and a few more comprehensive studies of porosity dependence. Porcelains are a collection of bodies typically made from varying proportions of clays (commonly ~50%), a feldspar (e.g., Lucite, $K_2O \cdot Al_2O_3 \cdot 4SiO_2$) and quartz (often in roughly approximately equal proportions), and frequently modest additions of other materials such as Al_2O_3. The resulting in situ reactions during firing yield a silicate-glass matrix with various amounts and distributions of quartz grains (due to incomplete dissolution) and mullite (from precipitation or prior reaction and incomplete dissolution of ingredients). These materials are another good example of difficulties in accurately determining total porosity by only density measurements since over the range of firing temperatures of $\leq 1100°C$ to $\geq 1300°C$ the density goes through a maximum and apparent porosity through a minimum (86,87) (Fig. 8.13). These changes reflect the countereffects of sintering, reducing porosity and increasing density, forming more glass phase from crystalline phases, and the resultant reduction in density and increased porosity. Both careful quantitative phase identification and especially quantitative stereology along with direct measurement of apparent (open) porosity can give reasonable values for total porosity.

Wayne et al. (86) showed flexural strength correlating fairly well with density, i.e., reaching a peak versus firing temperature similar to that of density (e.g., Fig. 8.13). More recent studies of Kobayashi et al. (88) showed a similar correlation between density and Young's modulus as well as a somewhat similar strength-density correlation; however, deviations in the latter appear to be due to a considerable extent to the forming of a denser surface layer on test bars fired at < 1200°C. Williams et al. (87) corroborated the strength-density corre-

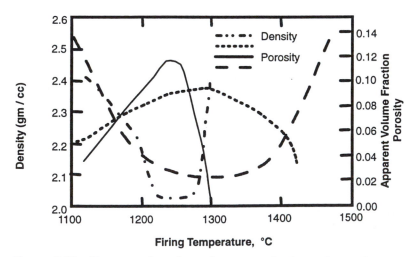

Figure 8.13 Representative plots of apparent density and porosity versus firing temperatures for electrical porcelains [after Wayne et al. (86) and Williams et al. (87)].

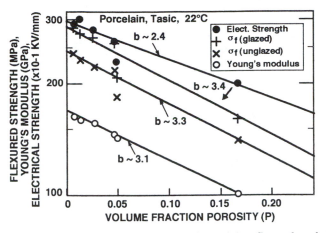

Figure 8.14   Semilog plot of Young's modulus, flexural, and electrical strength versus volume fraction porosity at 22°C [after Tasic (90)]. Note the linear trends over the modest range of estimated total porosity (~ 0.01 to 0.17) and associated b values reasonably consistent with mainly approximately spherical pores.

lation, showing a high linear correlation of flexural strength increasing with increasing density (from ~ 2.10 to ~ 2.35 g/cc). Estimating the total porosity as a function of density (giving P ~ 0.04–0.15) yields a b value of ~ 2.5–3 for the porosity dependence of flexure strength. This range would be consistent with most porosity being approximately spherical, which is likely since much of the porosity occurs in the glassy phase. The latter, as well as useful fracture observations, was shown by Kobayashi et al.'s (89) observation that pores between the glass and crystalline phases were fracture origins. A useful data set for the porosity dependence of porcelain properties is that of Tasic (90) (Fig. 8.14). This shows b ~ 3–3.5 for both E and flexure strength (the latter both with and without glazed surfaces), consistent with the estimate from the data of Williams et al.. These studies thus show very similar porosity dependence of strength and E, as was also indicated in earlier work of Roy (91) showing flexure strength being a close linear ratio of E, i.e., flexures strength = 1.63 E times 10^{-3}, for porcelain-bonded alumina with P 0.52–0.67 over the approximately whole and half order of magnitude range of each property.

A complication in porcelains can be variable microcracking, mainly or exclusively from quartz particles. The occurrence of this has been confirmed by acoustic emission showing microcracks from larger and smaller quartz particles occurring respectively earlier and later in cooling cycles (92).

Kalnin (93) measured both static and dynamic Young's modulus and flexural strength of whiteware bodies prepared from varying clay and feldspar components giving P 0.04–0.28 with 10–40 wt.% mullite in an alkali-alumino-silicate glass, i.e., similar to the above porcelain bodies. He showed a relatively tight 1:1 correlation of the two moduli measurements and a similar correlation of strength to Young's modulus times 10^{-3}. Considering the ranges of the modulus, strength, and porosity, these measurements indicate a b value in the range of ~3 to 7.5 in the extreme, and most likely in the range of 3 to 4.5. The latter range is consistent with b values for porcelains discussed above, as well as the glass contents and expected, substantial, approximately spherical porosity.

Turning to nonmechanical properties, Tasic (90) also measured electrical breakdown strength of his porcelains, giving a b value of ~ 2.4 (Fig. 8.13). Such a value appears qualitatively consistent with probable modest pore sizes. Also, as noted earlier, Goodman (94) showed translucency of porcelains bodies decreasing rapidly with initial increases in porosity, with a tendency for either or both to show greater or more rapid drops as pore size increases, as might be intuitively expected; however, by ~ P 0.05 the decrease in translucency with increasing P was saturated (to at least P = 0.2–0.3), with the saturation level of translucency tending to be lower with finer pore sizes, again as might be intuitively expected.

"Glass-ceramics," i.e., extensively crystallized glasses, are another important but mostly more modern class of ceramic composites. They are complicated by varying degrees and characters of crystalline grain structure as well as stresses from both the phase change (s) and the resultant phases. Another complication is varying extents of microcracking from the microstructural stresses. Thompson et al. (95) demonstrated by direct SEM examination (showing considerable transgranular fracture) as well as by Young's modulus changes that lithium disilicate bodies with large crystallized grains had microcracks, while those with finer grains did not, i.e., E ~ 72 and ~ 137 GPa, respectively, for grain sizes of ~ 150 and 15 μm. Fracture from indent flaws gave fracture toughnesses of 1.3 and 3.0 MPa m$^{1/2}$, respectively, for the microcracked and nonmicrocracked bodies. Greater fracture roughness and fractal dimensions were found for the tougher, finer grain, nonmicrocracked material.

8.3.6 Porosity Effects in Ceramic Composites

Modern ceramic composites with selected dispersions of ceramic particulates, platelets, whiskers, or continuous fibers, have been of substantial interest, especially for structural applications due to hoped for improved resistance to failure, mainly as a result of higher toughness; however, data as a function of porosity of these materials is quite limited since the performance of such composites is usually optimum at P = 0. Because of this and frequent difficulties in densification of such composites by sintering, many composites are hot pressed to near zero porosity. Data for other composites made by processes that do not give full density have often been reported only at the higher densities achieved; however, the modest data available—some covering a reasonable P range in a well-behaved fashion and much that is more limited, scattered, or both—is summarized in Table 8.5 and discussed below. Again, there is little or no characterization of the porosity, so evaluation is limited to the dependence on the volume fraction porosity P, and attempting to identify some dependence on aspects of the pore character implied from fabrication/processing. The latter is far more uncertain in composites because of the uncertainty or lack of models and pore character trends.

Although the composite data is limited, the overall results in Table 8.5 generally show similar trends as for other ceramics and other materials, with one important exception (for some fiber composites, discussed below). Consider three similarities. First, the b values are typically in the same range as for monolithic, i.e., essentially single (solid) phase, ceramics, although they are possibly on average shifted to higher levels, e.g., average b values of ~ 6 versus ~4 for monolithic ceramics. In particular, similar b values are observed for the CVI fiber composites where the mass conservation of processing on which current MSA models are based is not valid. Furthermore, this is also true where dispersed silicide particles undergo some degree of reaction with the polymer-derived matrix material, and thus again violates the conservation of solid density and volume. These observations indicate that while some changes of MSA models for such cases are needed, the results will often be modifications, commonly modest ones, from current MSA models. Second, in the two cases where rollovers to faster property decreases with increasing P are observed, the P ranges where this occurred varied inversely with the b value as expected, i.e., at P ~ 0.6 with b ~ 4.6 versus at P ~ 0.25–0.3 with b ~ 7–8. Note that the former rollover is shown not only by Fisher et al.'s (103) data (showing $P_C \geq 0.93$, but also supported by other data (104,105) (as is data at lower P (106)). Third, since the E-P and σ_f -P data are for the same investigator (103) and samples, they support the b value for the latter commonly being at least as large as that for elastic properties.

Two other observations on the composite data are also useful. First, in cases where there is limited or no difference in E of the matrix and the dispersed phase of, for example: 1) Si_3N_4 with TiN or Al_2O_3 and 2) Al_2O_3-TiB_2 and SiC-TiC), there should be limited or no effect of pore location relative to the two phases on Young's modules, other than effects on pore shape. There can of course be more effect on other properties such as fracture toughness and flexure strength; however, the Al_2O_3-TiB_2 (98) and the SiC-TiC (99) data (where the latter properties were determined in bodies with no, very limited, and substantial amounts of dispersed phase) do not show large differences in the P dependence. Second, the limited data indicates significant decreases in b values, e.g., by a factor of ~ 2 at high temperatures, This is shown by the fracture toughness and the strength results for SiC-TiC (99) at 1500°C.

Table 8.5   Effect of Porosity on Mechanical Properties of Ceramic Particulate, Whisker, and Fiber Composites at 22°C [a]

Composition	Pro-cessing[b]	P range[c]	Data points[d]	E	K_{IC}	σ_f	H	Investigator
A. Particulate composites								
Si_3N_4-0.4 $MoSi_2$	PP-RP	0.05–0.15	7			3		Greil (96)
Si_3N_4-0.4 $CrSi_2$	PP-RP	0.04–0.21	5			4		Greil (96)
Si_3N_4- 0.2 to 0.7Al_2O_3	RSSN	0.13–0.3	5			7–8		Yasutomi and Sobue (97)
Si_3N_4- 0.2 to 0.7TiN	RSSN	0.12–0.35	6			4–5		
Al_2O_3- 0 to 0.37 TiB_2	IP, S	0.02–0.14	25			4–12		Stadlbauer et al. (98)
SiC- 0 to 0.33 TiC	HP	0.01–0.17	6		3–5			Lin et al. (99)
SiC- 0.5 TiC	HP	0–0.19	4			5.5		Endo et al. (100)
SiC- 0 to 0.9 TiC	HP	0.01–0.26	10			4–6		Endo et al. (100)
B. Whisker composites								
SiO_2-Al_2O_3-Cr_2O_3 -0.04 SiC	S-G	0.12–0.35	6				3	Lee and Park (101)
Al_2O_3- mullite	HP	0.01–0.18	6			4		Tamari et al. (102)
C. Continuous fiber composites								
SiC-SiC	CVI	0.2–0.93	C	3		4.6		Fisher et al. (103) and others (104–107)
SiC-SiC	PP	0.12–0.42	7			1–14[f]		Jamet et al. (108)
Mullite-SiC	HP	0.01–0.33	5			2–4[f]		Coblenz et al. (109)
MgO-SiC	HP	0.04–0.20	3			3–4		
ZrO_2-SiC	HP	0.02–0.23	25			4 ± 2[f]		
SiO_2-SiC	HP	0–0.3	7			1–5[f]		

Header spanning note: "~ b values for [e]" spans the E, K_{IC}, σ_f, H columns.

[a] See also refs. (110) and (111) for summaries of much of the data in A and B.

[b] PP = polymer pyrolysis, RP = reaction processed, RSSN = reaction sintered silicon nitride, IP = isopressed, S = sintered, HP = hot pressed, S-G sol-gel, CVI = chemical vapor infiltration.

[c] Range of the volume fraction porosity (P).

[d] Number of data points at different P levels; C = compilation of data from different sources (see Fig. 8.15).

[e] Slope of the initial, linear portion of semilog plots of the property vs. P, i.e., b value, for E = Young's modulus, K_{IC} = fracture toughness, σ_f = flexure strength, and H = hardness.

[f] Indicates clear or probable bilinear values as discussed in the text.

While the limited available data on the porosity dependence shows overall trends similar to other ceramics and materials, there is considerable scatter. This is not surprising for composites given the pore character-location variations discussed earlier; however, further evaluation of existing fiber composite data indicates that such composites are an important, probably extreme, example of bodies in which the association of the pores with the different phases as a function of processing can be quite important since fibers typically have limited or no porosity. While there has been no detailed characterization or study of porous ceramic fiber composites, the data indicates probable processing differentiation as follows. The CVI composites show considerable similarity and normal scatter amongst themselves and in comparison to other materials (Fig. 8.15). It is suggested that this reflects most of the deposition occurring on the fibers, so much of the pore character and location are respectively somewhat

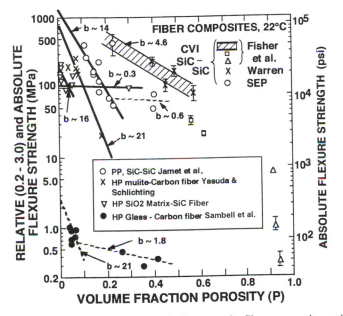

Figure 8.15 Flexural strength for ceramic fiber composites with matrices obtained from CVI (103–107) [other data at P = 0.88–0.94 (105) also agrees with data shown], polymer pyrolysis (PP) (108), and hot pressing (HP) (109,113,114). [One of the latter is in terms of relative values, which exceed 1 due to using "theoretical" strengths to allow use of data of differing fiber contents (112,113).] Note the somewhat higher b value for the CVI composites and the clear overall consistency with the trends for MSA models (including a high P_C value consistent with the modest b values), and the bilinear character of the PP (108) and the HP-SiO_2 glass matrix composites (109), and the similarity of these to other data as discussed in the text.

tubular and in the matrix, away from the fiber. This would be consistent with limited scatter, modest b values, and a high P_C. On the other hand, fiber composites where the matrices and densification are obtained by polymer pyrolysis or hot pressing of powders commonly show much more scatter. Further evaluation found that data for the latter two processes showed a bilinear trend on semilog plots (Fig. 8.15), including the original ceramic (glass) matrix-carbon fiber composite work of Phillips and colleagues (112,113). At low to moderate P (i.e., to P = 0.06–0.25) there is a rapid decrease in strength with increasing P, i.e., b values of ~ 15–25, but this transitions to a much lower rate of strength decrease with further increase in P, e.g., b values of ~1 to 2. Of the seven body-data sets available, four show this and the other three are partly or fully consistent with this bilinear trend. Thus, the MgO-SiC fiber data (109) consists of only three data points that are not inconsistent with, but are insufficient to establish the above trends. Mullite-C fiber (114) and coordierite-SiC fiber (112,113) data both showed high b values (~ 20) over their limited P ranges (respectively 0.04–0.17 and 0,01–0.13), which may be at or below the transition to low decreases with further P increases (i.e., lower b values). (Also, the more extended decrease of the hot-pressed mullite matrix-C fiber composite to its more modest P level may reflect extension of interfacial pores, in this case due to reaction of the mullite and C, since this lowest strength was from the highest temperature hot pressing, 1800°C, where reaction is likely.)

While there are probably other factors, such as grain growth and fiber degradation, that complicate these trends, i.e., contributing to the scatter, it is suggested that pore-phase association is the major factor. In both polymer pyrolysis and hot pressing (or sintering) of the matrix, essentially all of the porosity occurs in the matrix, as with CVI; however, in contrast to most pores in CVI matrices being well removed from the immediate vicinity of the fibers and a common simple rounded tubular shape, both polymer pyrolysis and hot pressing yield substantial porosity throughout the matrix and a reasonable amount at the fiber matrix interface. Like grain-boundary porosity, interfacial porosity is expected to result in more serious property decreases, possibly much more so given the necessity of stress transfer across fiber-matrix boundaries for good composite mechanical performance. Furthermore, interfacial pores are probably as slow, if not more so, in being eliminated, i.e., much earlier densification is probably in the matrix, which by itself contributes only a limited amount to strength increases. On the other hand, decreasing interfacial porosity, which is expected to occur more as processing is approaching limited porosity, should increase strengths faster. Thus, the combination of these factors are probable reasons for low b values at higher P and high b values at low P. Such effects are likely to be quite variable, e.g., depending on fiber size, grain structure and size, and processing, and hence a source of both variation and scatter. For example, the lower b values for both branches of the bilinear curves for the polymer pyrolysis matrix (Fig 8.15) would be consistent with its porosity in the bulk of the matrix being generally spherical, as is also suggested in the two sets of SiO_2 matrix data, all generally being amorphous matrices.

Hsueh and Becher (115) reported fracture toughness and R-curve effects in composites of Al_2O_3 with SiC whiskers with P = 0.006, 0.049, and 0.115. Their data shows both the initial and the saturation toughness decreasing with increasing porosity, but with decreases at the higher porosity being less than for the intermediate porosity. Modeling of the R-curve effects was presented, but relied on curve fitting for each porosity level, thus providing little insight into the role of the porosity. No detailed characterization of the porosity is available, so while more pores might be expected at the whisker-matrix interfaces, especially at lower porosity, there is no verification of this.

Brooks and Winter (105) also measured strengths and Weibull moduli of a variety of commercial ceramic foams including some made by sintering powders (as well as by CVI of SiC fibers, Fig. 8.15). While their data by itself is not sufficient to determine the P-dependence of their properties, it is useful in conjunction with other data above. Further, their Weibull modulus data also suggests an overall trend to decrease with increasing P (Fig. 5.7, focusing on the average trend, since their use of only 5–7 specimens is too few to indicate anything other than overall trends). Their CVI-SiC fiber composite data is also consistent with a previous compilation of data for similar bodies (107), and reinforces indications of a continuous variation of strengths across a broad P range, e.g., of P ~ 0.2–0.95. This trend is consistent with those of MSA models (Figs. 2.11, 3.1), e.g., with an initial slope or b value of ~ 4.6, although the existing models were derived for bodies in which changes in P are achieved without changes in mass, while CVI bodies clearly change mass with P; however, such application of MSA models is expected and is consistent with data of cements and plasters (Figs. 8.10–8.12), but requires model refinement to account for mass changes to be fully predictive. Replotting this CVI composite data as log strength versus log 1 – P gives n ~ 2.5 for $(1 - P)^n$, which is different from exponents for foams. Such differences are logical given the probable different pore character, but again emphasizes that n values can vary, and pore character is an important factor.

Finally, note that thermal conductivity has been measured on Al_2O_3-SiC, MgO- SiC, and BeO-SiC composites as a function of porosity (P ~ 0.005 to ~ 0.09, 0.09, and 0.03, respectively) (116,117). These showed relative conductivity rapidly dropping to 0.4–0.6 of the P = 0 value by P = 0.01–0.02, then very small (or no) decreases to relative values of 0.4–0.5 of the P = 0 value at the maximum P studied. These trends varied little or none over the range of SiC additions of from ~ 0.02 to over 0.8, indicating that the presence of the SiC rapidly dominates the conductivity as P increases. An important variation that can occur in some composites is microcracking due to the second phase, which can have substantial effect on properties, including thermal conductivity, which is discussed below.

Effects of microstructure on thermal conductivity of ceramic refractories, which are commonly composite materials, are summarized at the end of the following section since they often include combined effects of porosity and microcracks.

8.3.7 Microcracking Effects in Ceramic Composites

An important observation on effects of porosity on microcracking and its effects on crack propagation and fracture toughness was made by Lutz, Claussen, and Swain in duplex TZP-alumina composites (118–120). They observed crack propagation in these bodies having a dispersion of particles of mixtures of monoclinic ZrO_2 + Al_2O_3 in TZP matrices depended significantly on limited quantities of porosity left from sintering. Thus, they showed that sintering, which leaves some porosity in the dispersed particles (~ 20–60 μm diameter), gives substantial microcracking zones (revealed by dye penetration) up to 1–3 mm wide about the mean macrocrack path, along with substantial R-curve effects. In contrast to this, little or no microcracking and R-curve effects occurred in bodies with little or no porosity in the dispersed particles (i.e., in HIPed bodies). Residual sintered porosity allowed compositionally dependent microcracking to occur. As microcracking increased, so did R-curve effects, fracture toughness, and residual strengths after serious thermal shocks, but initial strengths decreased substantially, e.g., from 1300–1700 MPa to 100–800 MPa (120).

Although porosity can play an important role in microcracking, microcracking clearly occurs in composites without the presence of pores. A key issue in microcrack effects with or without interaction with pores is whether they are predominately preexisting or whether substantial microcrack generation occurs as a result of stressing. The latter occurs mainly or exclusively with crack propagation, as discussed in Chapter 4, while nondestructive measurements such as for elastic moduli primarily reflect preexisting microcracks. Pan et al. (16) reported the elastic moduli (E, G, and B) of a dense, commercial SiC-15 vol. % TiB$_2$ composite (Hexoloy ST) being a few percent below values for the SiC matrix, and about 2/3 those of the TiB$_2$; the difference was logically attributed to preexisting microcracking. Application of their effective medium model predicted similar values as those measured, but about 15% higher than actual values, and indicated that microcracking probably occurred with 60–100% of the TiB$_2$ with cracks extending respectively 90 and 60 degrees around the particles.

While microcrack generation by propagating cracks is fairly well accepted in composites as well as some rocks (Fig. 4.5) and anisotropic ceramics (Fig. 4.7), there are still significant uncertainties. Issues of the relative roles of preexisting versus generation of microcracks were examined by Faber et al. (121) using measurements and a survey of composites of silicate glass matrices with dispersed alumina particles (typically ~ 30 μm diameter). They found that the improvements in fracture toughnesses of most of these composites was due to preexisting microcracks, with the increase being linear with the volume fraction of alumina to at least ~ 0.4, increasing by ~ 100% for an increase in alumina volume fraction of ~ 0.2. The few systems found to have microcrack generation during crack propagation (shown by careful volumetric measurements, e.g., similar to dilatometer measurement of Al_2TiO_5 in Chapter 4) showed greater rises in toughness with increasing volume fraction of alumina particle content to a maximum at 0.3. Toughness decreased to or slightly below the level with only preexisting microcracks by ~ 0.4 (and presumably lower with further increases in alumina particle contents); however, the key factors determining whether or not microcrack generation and additional toughening occured could not be determined. Singh et al. (17) reported that strengths and toughnesses of hot-pressed BeO-SiC composites had maxima (levels not quantified) at ~ 20 and 75 vol. % SiC, i.e., similar to but shifted somewhat from the minima for thermal conductivity. They attributed these maxima and the minima of toughness and strength, respectively, and of thermal conductivity to both be consistent with microcracking and the shifts for the mechanical properties to reflect effects of microcracks generated by macrocrack propagation.

Turning to the effects of microcracks on strength, they almost always if not universally decrease strengths. Although, there are occasional claims, usually not quantified as for BeO-SiC noted above, for increased strengths with micro-cracking, detailed strength measurements of microcracked materials are more limited than those of toughness. Much of this limitation apparently arises from the assumption that strengths should follow toughness values; however, there is substantial strength data showing that strengths commonly decrease in composites as toughness increases (122,123). The opposite trends of strengths and toughness are attributed to similar crack-scale effects as for pores (Chapter 4). Thus, the large scale of cracks and their extent of propagation for most toughness tests versus those typically controlling strengths allows substantial crack branching, bridging, etc., to occur in most toughness tests. On the other hand, there is commonly little or no opportunity for such toughness effects to come into play with the smaller cracks and their more limited propagation to catastrophic failure and hence strength control. An important exception to this is where substantial crack propagation occurs, as in severe thermal shock.

Limited data on Weibull moduli supports the argument that much of the toughening with large cracks in most toughness measurements does not have much or any effect on normal strength behavior. The premise of increasing tougheness is that it should increase reliability, and hence Weibull moduli, since they reduce the dependence of strength on local material/microstructure variations; however, except in limited cases, Weibull moduli of ceramic particulate composites are the same or less than for bodies of the individual constituents with comparable processing, microstructures, and surface finishes (122). The only established exceptions are in self-reinforced bodies, mainly Si_3N_4, where larger, elongated grains are generated in situ; however, these grains, or clusters of them, commonly become the fracture origins and thus may raise the Weibull moduli by being a more uniform source of failure sources (123).

Finally, consider effects of microcracking on other properties, specifically thermal conductivity. As noted earlier, minima of the thermal conductivity were noted at about 30 and 80 vol. % SiC in BeO-SiC composites (17). Some quantitative agreement with microcrack models was presented for these bodies. Similarly, Buykx (124) showed that the dominant effect of U_4O_9 second-phase particles on the conductivity of UO_2 occurred when they lead to microcracks (mainly along grain boundaries, e.g., of grains ~ 50 μm). These results are qualitatively consistent with expected microcrack effects.

Ceramic refractories, which are commonly composite materials and frequently involve combined effects of various pore and microcrack structures, show a complex diversity of thermal conductivity behavior. This behavior and the mechanisms and modeling have been reviewed by Litvosky and colleagues (125,126), as summarized in Sections 7.2.1, 7.3.1, 9.2.3, and 9.3.2.

8.4   DISCUSSION AND SUMMARY

Extensive similarities are found for the porosity dependence of the other materials considered and those of nominally single phase ceramics. Thus modest data for plastics (Fig. 8.3) and more extensive metals data (Tables 8.1, 8.2; Fig. 8.5) are consistent with each other and general ceramic trends, including the same or similar consistency with MSA models. They also show the occurrence of a maxima in the fracture energy–P-dependence of Be (Fig. 8.6) and clearer demonstration of greater decreases in properties due to heterogeneity of the character, and especially the spatial distribution of porosity (Fig. 8.7). DeHoff and Gillards' (42) measurements on fractures of sintered Cu particles showed a direct correlation with minimum solid area. Data for traditional ceramics, such as whitewares, porcelains and composites of glass and crystalline phases, are generally quite consistent with nominally single-phase ceramics. In particular they often reflect effects of approximately spherical pores consistent with expectations of much of the porosity being in the glass phases. Some variations may occur due to microcracks associated with some phases, e.g., quartz in porcelains.

Other materials, while also showing overall similarities to nominally single-phase ceramics, also show more deviations. Thus, rocks, although more constrained in the extent of data, especially over a substantial P range, show similar trends, including observed or indicated rollovers (Fig. 8.9). They tend to give higher b values, which in part reflect both more limited P ranges and not accounting for possible or probable starts of rollovers; however, a major factor in the trend for higher b values and greater scatter for rock data is probably heterogeneity of the microstructure, especially the porosity. Rocks are thus an added and important reason for addressing issues of heterogeneity and their modeling, as discussed in Chapter 10.

Cements and related materials, specifically gypsum, also show overall similarities to other ceramics, including the overall shape of the property-P trends (Figs. 8.10–8.12). This includes definite rollovers to faster property decreases where data covers a sufficient P range, and probable rollovers where data probably extends only part way into the rollover. It also includes a trend for higher b values for hardness versus Young's moduli. Also, there is a trend, as for other ceramics, for higher b values. This again may often in part reflect data over too limited a P range (both at lower and higher P), and probably also varying aspects of heterogeneity; however, a probable factor in the higher b values is the interparticle bond character resulting from the reaction process that generate the product material. Thus, while models for strengths of green bodies as a function of two idealized binder distributions and their adaptation to cements and related reaction-derived materials have significant uncertainties, they have some application and indicate higher b values (Fig. 8.2). Further modeling improvements are needed, as discussed in Chapter 10.

Designed ceramic composites, which cover a broad range of behavior, also show overall similarities with much of the porosity dependence of other ceramics and other materials, as well as some important differences. Overall similarities include the initial linear decrease of properties on semilog plots versus P giving b values ranging from those for other ceramics and other materials (e.g., Table 8.5). Important similarities are specifically seen for CVI fiber composites, which include overall consistency with MSA-model trends, and low-to-moderate b values (higher for strength than the modules, as is common) consistent with high P_C values. These are in turn consistent with deposition on the fibers leaving most or all porosity away from the fiber-matrix interface, and generally of tubular character. There is also a frequent trend for higher b values, which in some cases reflects a similarity due to heterogeneity effects that can be more serious in composites, especially when there are substantial differences in properties of the phases; however, there are also intrinsic differences indicated for some types of composites and processing, especially certain fiber composites.

Two important, interrelated, differences in composites associated with their two- or multiple-phase character are indicated; both are related to pore location and the resultant character. As shown in Fig. 8.1, there are intrinsic differences and changes of these with regard to pore-phase association, even for idealized composites. In particular, interfacial pores between the different phases increasingly can occur as a 50:50 composition is approached. This is probably a factor in the greater scatter and higher b values in particulate (and whisker) composites; however, essentially all of the limited data for fiber composites made by polymer pyrolysis or hot pressing shows (or is consistent with) a bilinear trend for semilog plots of properties versus P that is attributed to pore location. Thus, at low porosity, very rapid decreases in properties of such fiber composites are found (e.g., b values of ~15–20 to P ~ 0.05–0.15) and are followed by very low rates of decrease (e.g., b values of <1 to ~ 3; Fig. 8.15). The very high b values are suggested to be due to interfacial pores between the fibers and the matrix generally being among the last to be eliminated and having by far the most effect on mechanical properties. On the other hand, much of the earlier porosity eliminated in processing is that in the matrix, which has less, often much less, effect on properties, hence lower, possibly much lower, b values. Within this bilinear trend, there is some indication of expected effects of pore character, i.e., lower b values in both branches for more spherical pores in the matrix. Progressively lower degrees of such interfacial pore effects in platelet, whisker, and particulate composites need to be considered, along with other microstructural factors such as the relative sizes and morphologies of the matrix and dispersed phases.

Finally, microcracking occurs in many ceramic composite materials, e.g., in some porcelains and crystallized glasses, and particularly in many more modern composites, which are often selected to obtain some microcrack toughening. While composite microstructures can aid in defining the amount and character of microcracking, they also introduce complexities such as microcrack formation in single-phase areas of anisotropic grains as well as with abutting grains of different phases. Another complexity is the nature of the microcracks: while intergranular ones are often likely and almost universally assumed, they can often have considerable transgranular character. Models for elastic moduli, which are determined mainly by preexisting microcracks, show promise of aiding microcrack characterization along with giving guidance to the moduli as a function of composite composition. A further complexity is the extent to which microcracks are preexisting or form in association with propagation of a larger crack and the effects of crack size and extent of propagation. Comparison of fracture properties such as strength and fracture toughness with elastic moduli as a function of composition and microstructure and careful measurements of sample volume or thermal expansion as a function of stressing can be good indicators of stress-crack induced microcracking. Some comparison of preexisting versus crack propagation-induced microcracking has been made showing some similarities and some differences in toughness changes; however, the key issue of the effects of crack scale and character of propagation on fracture toughness versus strength has not been adequately addressed. The common assumption that strength behavior follows that of fracture toughness, which unfortunately has lead to a paucity of strength measurements, is seriously questioned. While more study is needed, strengths should be measured and their values or trends not assumed from toughness values. This is critical since some strength behavior has opposite microstructural trends from that of fracture toughness, i.e., that strength decreases as toughness increases. Furthermore, increased toughness may decrease, not increase, the Weibull modulus (m), and where m does increase in composites, it may arise from other sources than the increased toughness. The increased toughness in composites is of greater use in improving resistance to serious thermal shock damage, i.e., analogous to effects of porosity. Porosity and microcrack effects can often be independent, as commonly assumed; however, there are clearly important cases where the two have had significant interactive effects (beneficial for thermal shock), so such interactions must be considered.

REFERENCES

1. H. E. Exner and D. Pohl, "Fracture Behavior of Sintered Iron," Pwd. Met. Intl., **10**(4), pp. 193–96, 1978.
2. E. Navara and B. Bengtsson, "Fracture Toughness of P/M Steels," Intl. J. Pwd. Met. Pwd. Tech., **20**(1), pp. 33–43, 1984.
3. N. P. Louat, M. S. Duesbery, M. A. Iman, V. Provenzano, and K. Sandanada, "On Dispersion of Voids as Sources of Strength," Phil. Mag. Let., **63**(3), pp. 159–63, 1991.
4. R. A. Helmuth and D. H. Turk, "Elastic Moduli of Hardened Cement Pastes," Symp. On Structure of Portland Cement Pastes and Concrete, Highway Res. Bd., Spec. Rept. No. 90, Highway Research Board, Washington, D.C., 1966, pp. 135–44.
5. E. H. Kerner, "The Elastic and Thermo-Elastic Properties of Composite Media," Proc. Roy. Soc. (London) Ser. B, **69**, pp. 808–13, 1952.
6. E. H. Kerner, "The Electrical Conductivity of Composite Media", Proc. Roy. Soc. (London) Ser. B, **69**, pp. 802–7, 1956.
7. B. Budiansky, "On the Elastic Moduli of Some Heterogeneous Materials," Mech. Phys. Solids, **13**, pp. 223–27, 1965.
8. B. Budiansky, "Thermal and Thermoelastic Properties of Isotropic Composites," J. Comp. Mat., **4**, pp. 286–95, 1970.
9. R. W. Rice, "Microstructural Dependence of Mechanical Behavior of Ceramics," Treatise on Materials Science and Technology, vol. 11, (R. McCrone, Ed.), Academic Press, New York, pp. 199–381, 1977.
10. H. H. Bache, "Model for Strength of Brittle Materials Built up of Particles Joined at Points of Contact," J. Am. Cer. Soc., **53**(12), pp. 654–58, 1970.
11. P. Harma and V. Satava, "Model for Strength of Brittle Porous Materials," J. Am. Cer. Soc., **57**(2), pp. 71–73, 1974.

12. K. K. Schiller, "Porosity and Strength of Brittle Solids, with Particular Reference to Gypsum," Mechanical Properties of Non-Metallic Brittle Materials (W. H. Warson, Ed.), Interscience Pub. Inc., New York, pp. 35–49, 1958.

13. G. Y. Onada, Jr., "Theoretical Strength of Dried Green Bodies with Organic Binders," J. Am. Cer. Soc., 59(5–6), pp. 236–39, 1976.

14. H. Rumpf, "The Strength of Granules and Agglomerates" Agglomeration (W. A. Knepper, Ed.), Intrescience Pub., Inc., New York, pp. 379–418, 1962.

15. K. Kendall and J. D. Birchall, "Porosity and Its Relationship to the Strength of Hydraulic Cement Pastes," Mat. Res. Soc. Symp. Proc. (J. F. Young, Ed.), Materials Research Society, Pittsburgh, PA, pp. 143–48, 1985. Very High Strength Cement-Based Materials, vol. 42.

16. M. J. Pan, D. J. Green, and J. R. Hellmann, "Influence of Interfacial Microcracks on the Elastic Properties of Composites," J. Mat. Sci., 31, pp. 3179–84, 1996.

17. J. P. Singh, D. P. H. Hasselman, W. M. Su, J. R. Rubin, and R. Palika, "Observations on the Nature of Micro-Cracking in Brittle Composites," J. Mat. Sci., 16, pp. 141–50, 1981.

18. O. Ishai and L. J. Cohen, "Elastic Properties of Filled and Porous Epoxy Composites," Intl. J. Mech. Sci., 9, 539–46, 1967.

19. K. K. Phani and R. N. Murkerjee, "Elastic Properties of Porous Thermosetting Polymers," J. Mat. Sci., 22, pp. 3453–59, 1987.

20. R. W. Rice, "Evaluation and Extension of Physical Property-Porosity Models Based on Minimum Solid Area," J. Mat. Sci., 31, pp. 102–18, 1996.

21. L. J. Gibson, "Modeling the Mechanical Behavior of Cellular Materials," Mats. Sci. & Eng., A110, pp. 1–36, 1989.

22. L. J. Gibson, and M. F. Ashby, "Cellular Solids, Structure and Properties," Pergamon Press, New York, 1988.

23. C. G. Goetzel, "Sintered, Forged and Rolled Iron Powders," Iron Age, 150(14), pp. 82–92, 1942.

24. G. D. McAdam, "Some Relations of Powder Characteristics to the Elastic Modulus and Shrinkage of Sintered Ferrous Compacts," J. Iron and Steel Institute, 168, pp. 346–358 (8/1951).

25. G. Artusio, V. Gallina, G. Mannone, and E. Scambetterra, "Effect of Porosity and Pore Size on the Elastic Moduli of Sintered Iron and Copper-Tin," Pwd. Met., 9, pp. 89–100, 1966.

26. R. Haynes and J. T. Egediege, "Effect of Porosity and Sintering Conditions on Elastic Constants of Sintered Iron," Pwd. Met., 32(1), pp. 47–52, 1989.

27. D. Pohl, "On the Fatigue Strength of Sintered Iron," Pwd. Met. Intl., 1(1), pp. 26–28, 1969.

28. D. Beasley and R. E. Cooper, "The Effects of Porosity on the Mechanical and Physical Properties of Beryllium," presented at the Be 77 Conf. 10/1977.

29. J. P. Jernot, M. Coster and J. L. Chermant, "Model to Describe the Elastic Modulus of Sintered Materials," Phys. Sta. Sol. (A), 74, pp. 325–32, 1982.

30. J. P. Panakkal, H. Willems, and W. Arnold, "Nondestructive Evaluation of Elastic Parameters of Sintered Iron Powder Compacts," J. Mat. Sci., 25, pp. 1397–1402, 1990.

31. A. Takata and K. Ishizaki, "Mechanical Properties of Hiped Porous Copper," Porous Materials (K. Ishizaki, L. Shepard, S. Okada, T. Hamasaki, and B.

Huybrechts, Eds.), Am. Cer. Soc., Westerville, OH, pp. 233–42, 1993. Ceramic Transactions, vol. 31.

32. A. K. Maitra and K. K. Phani, "Ultrasonic Evaluation of Elastic Parameters of Sintered Powder Compacts," J. Mat. Sci., **29**, pp. 4415–19, 1994.

33. J. T. Smith and S. A. LoPilato, "The Correlation of Density of Porous Tungsten Billets and Ultrasonic-Wave Velocity," Trans. Met. Soc. AIME, **236**, pp. 597–98, 1966.

34. T. E. Matikas, P. Karpur, and S. Shamasundar, "Measurement of the Dynamic Elastic Moduli of Porous Compacts," J. Mat. Sci., **32**, pp. 1099–1103, 1997.

35. R. E. Cooper, "Toughness-Porosity Phenomena," ICF 4, Advances in Research on the Strength and Fracture of Materials, vol. 3B (D. M. R. Taplin, Ed.), Pergamon Press, New York, pp. 809–18, 1978.

36. G. A. Clarke and R. A. Queeney, "Fracture Toughness and Density in Sintered 316L Stainless Steel," Intl. J. Pwd. Met., **8**(20), pp. 81–87, 1972.

37. A. S. Drachinskii, A. V. Krainikov, A. E. Kushchevskii, and Yu. N. Podrezov, "Correlation of Fracture Toughness with the Intergranular fracture in Powder Iron Specimens," Trans. Poroshkovaya Met., **277**(1), pp. 43–45, 1986.

38. J. T. Barnby, D. C. Ghosh, and K. Dinsdale, "The Fracture-Resistance of a Range of Sintered Steels," Pwd. Met. **16**(31), pp. 55–71, 1973.

39. O. H. Henry and J. J. Cordiano, "Hot-Pressing of Iron Powders," Trans. Am. Inst. Mining Met. Eng., **166**, pp. 520–32, 1946.

40. M. Eudier, "The Mechanical Properties of Sintered Low-Alloy Steels," Pwd. Met., **9**, pp. 278–90 (1962).

41. V. Gallina and G. Mannone, "Effect of Porosity and Particle Size on the Mechanical Strength of Sintered Iron," Pwd. Met., **11**(21), pp. 73–82, 1968.

42. R. T. DeHoff and J. P. Gillard, "Relationship Between Microstructure and Mechanical Properties in Sintered Copper," Modern Developments in Powder Metallurgy, vol. 4 (H. H. Hausner, Ed.), Plenum Press, New York, pp. 281–90, 1971.

43. A. Salak, V. Miskovic, E. Dudrova, and E. Rudnayova, "The Dependence of Mechanical Properties of Sintered Iron Compacts Upon Porosity," Pwd. Met. Intl. **6**(3), pp. 128–32, 1974.

44. W. Rostoker and S. Y. K. Liu, "The Influence of Porosity on The Ductility of Sintered Brass," J. Mat. Sci., **5**, pp. 605–17, 1970.

45. J. P. Jernot and J. L. Chermont, "Model to Predict the Tensile Strength of Sintered Materials," Phys. Sta. Sol. (A), **74**, pp. 579–86, 1982.

46. H. Kuroki and Y. Tokunaga, "Effect of Density and Pore Shape on Impact Properties of Sintered Iron," Intl. J. Pwd. Met. & Pwd. Tech., **21**(2), pp. 131–37, 1985.

47. P. Duwez and H. Martens, "The Powder Metallurgy of Porous Metals and Alloys Having a Controlled Porosity," Trans. AIME, **175**, pp. 848–77, 1948.

48. R. M. German, "Strength Dependence on Porosity for P/M Compacts," Intl. J. Pwd. Met. & Pwd. Tech., **13**(4), pp. 259–71.

49. P. Bompard, D. Wei, T. Guennouni, and D. Francois, "Mechanical and Fracture Behavior of Porous Materials," Eng. Fract. Mech., **28**(5/6), pp. 627–42, 1987.

50. A. I. Ul'yanov, V. S. Korobeinikova, G. V. Sterkhov, and N. A. Sidorov, "Effect of Porosity on the Reliability of Magnetic Quality Control of the Heat Treatment of Sintered Steels," Trans. Poroshkovaya Met., **295**(7), pp. 45–49, 1987.

51. V. Provenzano, M. S. Duesbery, N. P. Louat, and K. S. Kumar, "Damping and Strength Properties of Voided Copper," Proc. Micromechanics of Advanced Materials (S. N. G. Chu, et al., Eds.), Minerals, Metals & Materials Soc., Warrendale, PA, pp. 213–220, 1995. **AU: OK?**

52. H. Dixon, A. J. Fletcher, and R. T. Cunhill, "Relationship Between Degree of Homogeneity and Physical and Mechanical Properties of a Sintered Nickel Steel," Pwd. Met., **21**(3), pp. 131–42, 1978.

53. J. P. Jernot, J. L. Chermant, and M. Coster, "A New Model to Describe the Variation of Electrical Conductivity in Materials Sintered in the Solid Phase," Phy. Stat. Sol. (A), **74**, pp. 475–83, 1982.

54. W. R. Judd and C. Huber, "Correlation of Rock Properties by Statistical Methods," International Symposium on Mining Research, vol. 2 (G. B. Clark, Ed.), Pergamon Press, New York, pp. 621–48, 1962.

55. K. G. Stagg and O. C. Zienkiewicz, "Rock Mechanics in Engineering Practice," John Wiley & Sons, New York, pp. 5–11, 1969.

56. R. G. Hoagland, G. T. Hahn, and A. R. Rosenfeld, "Influence of Microstructure on Fracture Propagation in Rock," Rock Mech., **5**, pp. 77–106, 1973.

57. L. Peck. C. C. Barton, and R. B. Gordon, "Microstructure and the Resistance of Rock to Tensile Fracture," J. Geophy. Res., **90**(B13), pp. 11, 533–46, 1985.

58. R. H. Price, "Analysis of the Elastic and Strength Properties of Yucca Mountain Tuff, Nevada," Res. and Eng. Appl. of Rock Masses, **1**, pp. 89–96, 1985. Proceedings of the 26th U.S. Symposium on Rock Mechanics (E. Ashworth, Ed.).

59. R. H. Price, R. J. Martin III, P. J. Boyd, and G. N. Boitnott, "Mechanical and Bulk Properties of Intact Rock Collected in the Laboratory in Support of the Yucca Mountain Site Characterization Project," Presentation Workshop on Rock Mechanics Issues in Repository Design and Performance Assessment, Rockville, MD, 9/19–20/1994.

60. I. Soroka and P. J. Sereda, "Interrelation of Hardness, Modulus of Elasticity, and Porosity in Various Gypsum Systems," J. Am. Cer. Soc., **51**(6), pp. 337–40, 1968.

61. H. F. W. Taylor, "Mineralogy, Microstructure and Mechanical Properties of Cements," Proc. Brit. Cer. Soc., **28**, pp. 147–63, 6/1979.

62. J. E. Bailey and D. Chescoe, "Microstructure Development During The Hydration of Portland Cement," Proc. Brit. Cer. Soc., **28**, pp. 165–77, 6/1979.

63. P. L. Pratt and H. M. Jennings, "The Microchemistry and Microstructure of Portland Cement," Ann. Rev. Mat. Sci., **11**, pp. 123–49, 1981.

64. H. M. Jennings, "The Developing Microstructure in Portland Cement," Advances in Cement Technology (S. N. Ghosh, Ed.), Pergamon Press, New York, pp. 349–96, 1983.

65. H. F. W. Taylor, "Cement Chemistry," Academic Press, New York, 1990.

66. S. Mindess, "Relationship Between Strength and Microstructure for Cement-Based Materials: An Overview," Materials Research Society Symposium Proceedings (J. F. Young, Ed.), Materials Research Society, Pittsburgh, PA, pp. 53–68, 1985. Very High Strength Cement-Based Materials, vol. 42.

67. S. Diamond and M. E. Leeman, "Pore Size Distributions in Hardened Cement Paste By SEM Image Analysis," Microstructure of Cement-Based Systems/Bonding and Interfaces in Cementatious Materials (S. Diamond, et al., Eds.), Materials Research Society, Pittsburgh, PA, pp. 217–26, 1995. Materials Research Society Proceedings, vol. 370.

68. B. K. Marsh and R. L. Day, "Some Difficulties in the Assessment of Pore-Structure of High Performance Blended Cement Pastes," Materials Research Society Symposium Proceedings (J. F. Young, Ed.), Materials Research Society, Pittsburgh, PA, pp. 113–21, 1985. Very High Strength Cement-Based Materials, **42**

69. A. Ahmed and L. Struble, "Effects of Microstructure on Fracture Behavior of Hardened Cement Paste," Microstructure of Cement-Based Systems/Bonding and Interfaces in Cementatious Materials (S. Diamond, et al., Eds.), Materials Research Society, Pittsburgh, PA, pp. 99–106, 1995. Materials Research Society Proceedings, vol. 370.

70. N. B. Eden and J. E. Bailey, "On the Factors Affecting Strength of Portland Cement," J. Mat. Sci., **19**, pp. 150–58, 1984.

71. S. Igarashi and M. Kawamura, "Detection of Failures Around Main Cracks and Their Relation to the Fracture Toughness of Cement Paste," J. Am. Cer. Soc., **78**(7), pp. 1715–18, 1995.

72. S. J. Hanna, Ph.D. Thesis, School of Civil Engineering, Perdue U., Dissertation Abstr. Part B 29 No. 11, Univ. Microfilms No. 69–7454, p. 4157 B, 1968.

73. S. Diamond, "Very High Strength Cement-Based Materials-A Prospective," Materials Research Society Proceedings (J. F. Young, Ed.), Materials Research Society, Pittsburgh, PA, pp. 233–43, 1985. Very High Strength Cement-Based Materials, vol. 42.

74. J. D. Birchall, A. J. Howard, and K. Kendall, "Strong Hydraulic Cements," Proc. Brit. Cer. Soc., **32**, pp. 25–32, March 1982.

75. K. Kendall, A. J. Howard, and J. D. Birchall, "The Relation Between Porosity, Microstructure and Strength, and the Approach to Advanced Cement-Based Materials," Phil. Trans. R. Soc. Lond., A **310**, pp. 139–53, 1983.

76. N. M. Alford, J. D. Birchall, A. J. Howard, and K. Kendall, "Comments on 'The Factors Affecting Strength of Portland Cement'," J. Mat. Sci. Disc., **20**, pp. 1134–36, 1985.

77. R. Baggott and A. Sarandily, "High Strength Autoclaved Mortars," Materials Research Society Proceedings (J. F. Young, Ed.), Materials Research Society, Pittsburgh, PA, pp. 69–77, 1985. Very High Strength Cement-Based Materials, vol. 42.

78. P. Kittl and R. Aldunate, "Compressive Fracture Statistics of Compacted Cement Cylinders," J. Mat. Sci., **18**, pp. 2947–50, 1983.

79. D. M. Roy and G. R. Gouda, "Porosity-Strength Relations in Cementatious Materials with Very High Strengths," J. Am. Cer. Soc., **56**(10), pp. 549–50, 1973.

80. G. J. Verbeck and R. A. Helmuth, Proc. Fifth Intl. Symp. Chem. of Cement, Tokyo, 1968, vol. III, (The Organ. Com. For V-ISCC and The Cement Assn. Japan, Eds.), pp. 1–32, 1969.

81. P. Lu and J. F. Young, "Slag-Portland Cement Based DSP Paste," J. Am. Cer. Soc., **76**(5), pp. 1329–34, 1993.

82. S. Mindess and S. Diamond, "Fracture Surfaces of Cement Pastes, Rocks, and Cement/Rock Interfaces," Microstructure of Cement-Based Systems/Bonding and Interfaces in Cementatious Materials (S. Diamond, et al., Eds.), Materials Research Society Proceedings, vol. 370, pp. 295–307, 1995.

83. D. Walsh, M. A. Otooni, M. E. Taylor, Jr., and M. J. Marcinkowski, "Study of Portland Cement Fracture Surfaces by Scanning Electron Microscopy Techniques," J. Mat. Sci., **9**, pp. 423–29, 1974.

84. S. H. Ahmad and S. P. Shah, "Properties of High Strength Concrete for Structural Design," Materials Research Society Proceedings (J. F. Young, Ed.), Materials Research Society, Pittsburgh, PA, pp. 169–81, 1985. Very High Strength Cement-Based Materials, vol. 42.

85. D. M. Roy, Z. E. Nakagawa, B. E. Scheetz, and E. L. White, "Optimized High Strength Mortars: Effects of Chemistry, Particle Packing, and Interface Bonding," Materials Research Society Proceedings (J. F. Young, Ed.), Materials Research Society, Pittsburgh, PA, pp. 245–52, 1985. Very High Strength Cement-Based Materials, vol. 42.

86. B. E. Wayne, B. Gibson, B. Hales, and G. James, "On the Vitrification and Fired Properties of an Electrical Porcelain Body," Trans. Brit. Cer. Soc., **62**, pp. 421–41, 1963.

87. E. C. Williams, R. C. Reid-Jones, and D. T. Dorril, "The Physical Properties and Behavior in Kiln of Hard Electrical Porcelain as Interpreted from Density Measurements," Trans. Brit. Cer. Soc., **62**, pp. 405–20, 1963.

88. Y. Kobayashi, O. Ohira, and E. Kato, "Effect of Firing Temperature on Bending Strength of Porcelains for Tableware," J. Am. Cer. Soc., **75**(7), pp. 1801–6, 1992.

89. Y. Kobayashi, O. Ohira, and E. Kato, "Bending Strength and Microstructures of Porcelains for Tableware," J. Jap. Cer. Soc., **99**, pp. 495–501, 1991.

90. Z. D. Tasic, "Improving the Microstructure and Physical Properties of Alumina Electrical Porcelain with Cr_2O_3, MnO_2, and ZnO Additives," J. Mat. Sci., **28**, pp. 5693–701, 1993.

91. S. K. Roy, "Characterization of Porosity in Porcelain-Bonded Porous Alumina Ceramics," J. Am. Cer. Soc., **52**(10), pp. 543–48, 1969.

92. G. Kirchoff, W. Pompe, and H. A. Bahr, "Structure Dependence of Thermally Induced Microcracking in Porcelain Studied by Acoustic Emission," J. Mat. Sci., **17**, pp. 2809–16, 1982.

93. I. L. Kalnin, "Strength and Elasticity of Whitewares: Part I, Relation Between Flexural Strength and Elasticity," Am. Cer. Soc. Bull., **46**(12), pp. 1174–77, 1967.

94. G. Goodman, "Relation of Microstructure to Translucency of Porcelain Bodies," J. Am. Cer. Soc., **33**(2), pp. 66–72, 1950.

95. J. Y. Thompson, K. J. Anusavice, and B. Balasubramaniam, "Effect of Micro-cracking on the Fracture Toughness and Fracture Surface Fractal Dimension of Lithia-Based Glass-Ceramics," J. Am. Cer. Soc., **78**(11), pp. 3045–49, 1995.

96. P. Greil, "Active-Filler-Controlled Pyrolysis of Preceramic Polymers," J. Am. Cer. Soc., **78**(4), pp. 835–48, 1995.

97. Y. Yasutomi and M. Sobue, "Development of Reaction-Bonded Electro-Conductive Titanium Nitride-Silicion Nitride and Resistive Alumina-Silicon Nitride Composites," Cer. Eng. Sci. Proc., **11**(7–8), pp. 857–67, 1990.

98. W. Stadlbauer, W. Kladnig, and G. Gritzner, "Al_2O_3–TiB_2 Composite Ceramics," J. Mat. Sci. Let., **8**, pp. 1217–20, 1989.

99. B. W. Lin, T. Yano, and T. Iseki, "High-Temperature Toughening Mechanism in SiC/TiC Composites," J. Cer. Soc. Jpn., **100**(4), pp. 509–13, 1992.

100. H. Endo, M. Ueki, and H. Kubo, "Hot Pressing of SiC-TiC Composites," J. Mat. Sci., **25**(5), pp. 2503–6, 1990.

101. B. I. Lee and S. Y. Park, "Sol-Gel Processing of SiC-Whisker-Reinforced Silica-Based Ceramic Composites," J. Am. Cer. Soc., **72**(12), pp. 2381–5, 1989.

102. N. Tamari, I. Kondoh, T. Tanaka, and H. Katsuki, "Mechanical Properties of Alumina-Mullite Whisker Composites," J. Cer. Soc. Jpn., **101**(6), pp. 721–4, 1993.

103. R. E. Fisher, C. Burkland and W. E. Bustamante, "Ceramic Composite Thermal Protection Systems," Cer. Eng. & Sci. Proc., **6**(7–8), pp. 806–19, 1985.

104. J. W. Warren, "Fiber and Grain-Reinforced Chemical Vapor Infiltrated (CVI) Silicon Carbide Matrix Composites," Cer. Eng. & Sci. Proc., **6**(7–8), pp. 684–93, 1985.

105. D. L. Brooks and E. M. Winter, "Material Selection of Cellular Ceramics for a High Temperature Furnace," Columbia Gas System Service Corp., Columbus, OH, Report, 1989.

106. P. J. Lamicq, G. A. Bernhart, M. M. Dauchier, and J. G. Mace, "SiC/ SiC Composite Ceramics," Am. Cer. Soc. Bul., **65**(2), pp. 336–38, 1986.

107. R. W. Rice and D. Lewis, III, "Ceramic Fiber Composites Based upon Refractory Polycrystalline Ceramic Matrices," Reference Book for Composites Technology, vol. 1 (S. M. Lee, Ed.), Technomic Press, Lancaster, PA, pp. 117–42, 1989.

108. J. Jamet, J. R. Spann, R. W. Rice, D. Lewis II, and W. S. Coblenz, "Ceramic-Fiber Composites Via Polymer-Filler Matrices," Cer. Eng. & Sci. Proc., **5**(7–8), pp. 443–74, 1984

109. W. S. Coblenz, R. W. Rice, D. Lewis III, D. Shadwell, B. Bender, and C. C. Wu, "Progress in Ceramic Refractory Fiber Composites," Metal Matrix, Carbon, and Ceramic Matrix Composites (J. D. Buckley, Ed.), NASA Conf. Pub. 23567, pp. 191–216, 1984.

110. C. X. Campbell and S. K. El-Rahaiby, "Databook on Mechanical and Thermal-physical Properties of Particulate-Reinforced Ceramic Matrix Composites," Ceramic Information Analysis Center, West Lafayette, IN, and American Ceramics Society, Westerville, OH, 1995.

111. C. X. Campbell and S. K. El-Rahaiby, "Databook on Mechanical and Thermal-physical Properties of Whisker-Reinforced Ceramic Matrix Composites," Ceramic Information Analysis Center, West Lafayette, IN, and American Ceramics Society, Westerville, OH, 1995.

112. R. A. J. Sambell, D. C. Phillips, and D. H. Bowen, "The Technology of Carbon-Fiber-reinforced Glasses and Ceramics," Carbon Fibers; Their Place in Modern Technology, pp. 105–13, 1974. Proc. Intl. Conf., Plastics Inst., London.

113. D. C. Phillips, "Fibre Reinforced Ceramics," Handbook of Composites, vol. 4 (A. Kelly and Yu. N. Rabothnov, Eds.), North Holland, New York, pp. 373–428, 1983. Chapter VII.

114. E. Yasuda and J. Schlichting, "Kohlenstoff-Faserverstarktes Al2O3 und Mullit," Z. Werkstofftech, **9**, pp. 310–15, 1978.

115. C. H. Hseuh and P. F. Becher, "Evaluation of Bridging Stresses from R-Curve Behavior for Nontransforming Ceramics," J. Am. Cer. Soc., **71**(5), pp. C-234–37, 1988.

116. D. P.H. Hasselman and K.Y. Donaldson, "Thermal Conductivity of Whisker-and Particulate-Reinforced Ceramic Matrix Composites," Handbook on Discontinuously Reinforced Ceramic Matrix Composites (K. J. Bowman, S. K. El-Rahaiby, and J. B. Wachtman, Jr., Eds.), Ceramic Information Analysis Center, West Lafayette, IN, and American Ceramics Society, Westerville, OH, pp. 357–406, 1995.

117. l. D. Bentsen and D. P. H. Hasselman, "Role of Porosity in the Effect of Micro-cracking on the Thermal Diffusivity of Brittle Matrix Composites," (T. Ashworth and D. R. Smith, Eds.), Plenum Press, New York, pp. 485–98, 1985. Thermal Conductivity, vol. 18.

118. E. H. Lutz, N. Claussen, and M. V. Swain, "K^R-Curve Behavior of Duplex Ceramics," J. Am. Cer. Soc., **74**(1), pp. 11–18, 1991.

119. E. H. Lutz and N. Claussen, "Duplex Ceramics: II, Strength and Toughness," J. Eur. Cer. Soc., **7**, pp. 219–26, 1991.

120. E. H. Lutz, M. V. Swain, and N. Claussen, "Thermal Shock Behavior of Duplex Ceramics," J. Am. Cer. Soc., **74**(1), pp. 19–24, 1991.

121. K. T. Faber, T. Iwagoshi, and A. Ghosh, "Toughening by Stress-Induced Micro-cracking in Two-Phase Ceramics," J. Am. Cer. Soc., **71**(9), pp. C-399–401, 1988.

122. R. W. Rice, "Toughening in Ceramic Particulate and Whisker Composites," Cer. Eng. Sci. Proc., **11**(7–8), pp. 667–94, 1990.

123. R. W. Rice, "Microstructural Dependance of Fracture Energy and Toughness of Ceramics and Ceramic Composites Versus that of Their Tensile Strengths at 22°C," J. Mat. Sci, **31**, pp. 4503–19, 1996.

124. W. J. Buykx, "The Effect of Microstructure and Microcracking on the Thermal Conductivity of UO_2–U_4O_9," J. Am. Cer. Soc., **62**(7–8), pp. 326–31, 1979.

125. E. Litovsky and M. Shapiro, "Gas Pressure and Temperature Dependences of Thermal Conductivity of Porous Ceramic Materials: Part 1, Refractories and Ceramics with Porosity Below 30%," J. Am. Cer. Soc., **75**, pp. 3425–29, 1992.

126. E. Litovsky, M. Shapiro, and A. Shavit, "Gas Pressure and Temperature Depen-dences of Thermal Conductivity of Porous Ceramic Materials: Part 2, Refractories and Ceramics with Porosity Exceeding 30%," J. Am. Cer. Soc., **79**, pp. 1366–76, 1996.

9

POROSITY AND MICROCRACK EFFECTS ON THERMAL STRESS AND SHOCK FRACTURE AND PROPERTIES AT ELEVATED TEMPERATURES

KEY CHAPTER GOALS

1. Show that most porosity dependence of properties at 22°C continues to substantial temperatures, i.e., properties scale with their inherent temperature dependences (e.g., of single crystals), until substantial diffusion, creep, or intergranular sliding and fracture occur to alter mechanical properties or radiation transfer to increase thermal conductivity.

2. Demonstrate that microcracks commonly close and at least partially heal (or form less) as temperature increases, temporarily reducing or reversing the normal property decrease with increasing temperature and vice versa as temperature decreases, with important exceptions indicated in larger grain HfO_2 showing the need for further research.

3. Show that thermal shock resistance is commonly significantly increased by porosity, probably partly due to crack branching or bridging, but more research is needed to determine the specifics.

9.1 INTRODUCTION

This chapter summarizes models, concepts, and data concerning the temperature dependence of primarily porosity and secondarily microcracks effects on ceramic properties considered in earlier chapters. For most properties there are very limited models and limited data, so the discussion is focused more on the known or expected changes in mechanisms. Changes in effects of porosity on strength due to creep and stress rupture from various sources, especially grain-boundary sliding, which are extensive and complex topics themselves, are only outlined here.

A key factor in addressing the effects of porosity or microcracks on properties as temperature changes is sorting out property changes strictly due to temperature changes versus those due to temperature effects on the porosity or microcrack dependence. The fundamental porosity dependence of many of the properties of interest should not necessarily change as temperature increases, or may only change in degree but not character; however, some porosity dependence may also change in character due to introduction of new mechanisms, e.g., from significant changes in other factors such as diffusion, and particularly grain-boundary behavior, which can lead to changes in other properties, their porosity dependence, or both. Thus, grain-boundary sliding becomes a major factor in high temperature mechanical properties, especially compressive, and particularly tensile strength, but also affects hardness and may affect (especially improperly measured) elastic moduli. Nonmechanical properties can also change substantially with temperature, e.g., thermal conductivity due to substantial changes in convective and especially radiative transport across pores. In contrast to often limited and gradual change of porosity effects with increasing temperature, effects of increased temperature on microcrack effects on properties are typically more pronounced and different due to common microcrack closure and possibly healing, greatly reducing or eliminating their effects on properties. Note that closure refers to the crack opening being reduced to approximately atomic lattice dimensions. While closure is a necessary condition for crack healing, i.e., forming of most or all normal atomic bonds across the previous crack faces (e.g., with modest diffusion) it is not a guarantee for healing since poor registrary or contamination of crack surfaces may inhibit or prevent healing.

Properties are generally treated in the order that their porosity dependence at 22°C was treated in preceding chapters. Thus, mechanical properties are treated in the order of elastic moduli; crack propagation and fracture energy and toughness, tensile strength, hardness, and compressive strength and related properties are covered; then nonmechanical properties are covered. One important exception to this sequence is thermal stress and shock failure, for which there is modest modeling and data. This important topic is treated following treatment of tensile strength.

9.2 MODELS AND CONCEPTS

9.2.1 Mechanical Properties Other Than Thermal Stress and Shock Failure

Elastic moduli typically decrease substantially with increasing temperature due to their reflecting basic atomic bonding and the resultant decreases of this with temperature. Decreases of 1–2% per 100°C temperature rise are common as shown later. Anderson (1) showed that Wachtman's empirical equation (2) for Young's modules (E) at any temperature (T):

$$E = E_0 - ATe^{-(T_0/T)} \qquad (9.1)$$

where E_0 = Young's modules at absolute zero temperature (T_0), and A a constant, was theoretically correct, i.e., it could be derived from the Mie-Gruneisen equation of state if Young's modulus was replaced by bulk modulus. The equation was also theoretically correct for Young's modulus provided that the temperature dependence of Poission's ratio is small, which is true for some ceramics but not others. Others have corroborated and extended these results, (e.g., ref. 3); however, such temperature dependence does not mean there is any effect of temperature on the porosity dependence, and in fact there should not necessarily be any. Such intrinsic changes in elastic moduli with temperature are an important factor in the temperature changes of many other properties that scale with elastic moduli.

Some greater decrease of elastic moduli than predicted by the above equation is commonly seen in polycrystalline bodies at elevated temperatures that is associated with grain-boundary sliding (e.g., as also indicated by internal friction). Grain-boundary porosity may affect such changes, but it must be seriously questioned whether such effects are intrinsic or an artifact of the measurement. The latter seems likely since if the stressing period is sufficiently long relative to the rate of creep such that some measurable nonelastic strain occurs during the stress cycle, a decrease in elastic moduli will be indicated; however, such a strain would be a measurement artifact and would be incorrectly used as a modulus change in modeling creep behavior, since this may be essentially including the effects of creep twice.

Other property changes with temperature may cause changes in the temperature dependence of elastic properties. A prime example of this is reduction in microcracking from thermal expansion anisotropy (TEA) between grains (or other particles). Such cracks typically form as stresses increase with decreasing temperature (and hence also increasing elastic moduli), occurring at characteristic grain sizes and temperatures for a given material, and hence TEA; however, such cracks also generally decrease in size, number, or both with many closing and partly or fully healing as temperature substantially increases. Depending on the extent and nature of such microcrack changes, the normal decrease of elastic (and other mechanical) properties with increasing temperature may be temporally reduced or reversed due to such closure-healing. With further temperature increase the normal decrease in stiffness (and other mechanical properties) with increasing temperature resumes. The reverse trend occurs on cooling.

Like elastic properties, the initial temperature effects on the porosity or microcrack dependence of crack propagation, fracture energy and toughness are primarily determined by effects on other properties, primarily Young's modulus; however, as changes in temperature become more substantial (e.g., to 500–1000°C) two other factors can become significant. First, in microcracked ceramics, the reduced formation or closure of such cracks, while reducing the decrease of Young's modulus or possibly temporarily reversing it as temperature increases (e.g., see similar effects on thermal conductivity, see Fig. 9.3), may be effected by other direct effects of the reduction and ultimate loss of the microcracks on these properties. Thus various combinations of crack deflection, branching, and bridging may decrease then disappear as temperature increases; however, when temperature changes alter porosity effects, these may complicate changes in microcracking and its effects on properties. A more general and often more significant change should occur due to intergranular pores enhancing intergranular crack propagation at elevated temperatures, especially at the lower crack velocities commonly used for determining crack propagation characteristics. Unfortunately, there has been no experimental study of such effects.

Turning to tensile stress response and strength, there are two regimes to consider. The first is brittle fracture due to various combinations of more moderate temperatures (e.g., < 1200°C), higher strain rates, or more refractory materials. As with brittle fracture at low temperatures, there are basically two aspects of pore effects on brittle fracture to consider. One is fracture initiation from a single pore or a cluster of pores, with this conceptually becoming even more of an issue with intergranular pores in polycrystals as temperatures increase and strain rates decrease. Unfortunately, both the limited overall application of fractography and the frequent reduction in clarity of fracture markings and hence in identification of fracture origins with predominantly intergranular fracture leaves specific demonstrations of this in polycrystalline bodies, at best, limited; however, while tests and fractography of single crystals are even more limited, fracture initiation from intragranular pores at elevated temperatures in sapphire filaments will be shown later (see Fig. 9.5).

Consider now the other more general effect of porosity on higher temperature brittle tensile failure in polycrystals. Enhanced intergranular fracture occurs (4), playing a larger role in fracture, increasing the porosity dependence, especially of tensile strength; however, this depends on how much of the porosity is intergranular and the extent of intergranular crack propagation in failure. Possible effects of intergranular pores in enhancing crack deflection, branching, and bridging may be limited or precluded because strength is usually determined by smaller cracks than in typical crack propagation and fracture energy and toughness tests, as is generally the case at lower temperatures; however, the degree of reduction of fracture toughnesses and strengths increases with increasing temperatures can increase the extent of crack propagation to failure, hence the opportunity for large crack effects on tensile strength.

The second tensile stress regime is where nonelastic, i.e., creep, or other deformation processes becomes significant. Porosity plays an important role in tensile creep and creep rupture via a variety of mechanisms including reduced stiffness and strength, and impacting deformation processes. Intergranular pores play a particularly important role in enhancing grain-boundary sliding and failure in polycrystalline bodies. Models for the effects of porosity on creep and stress rupture are thus far only obtained via the porosity dependence of the individual parameters in models for creep and stress rupture, hence being limited by the cumulative uncertainties in each parameter. Since there is often some to substantial uncertainty in the accuracy of these models, this gives considerable uncertainty in modeling the effects of porosity via such models. This is compounded by the expected increasing effects of intergranular pores enhancing intergranular fracture and possible effects of such fracture. The generally very limited pore characterization, including the amount of inter- versus intragranular pores compounds these problems, although frequently fractography on failed samples can help substantially (but is rarely done). For reference and guidance, the following representative equation for the strain rate ($d\epsilon/dt$) in creep is:

$$\frac{d\epsilon}{dt} = AG\,(kT)^{-1} \left(\frac{b}{D}\right)^m \left(\frac{\sigma}{G}\right)^n \tag{9.2}$$

where A = a constant, G = the shear modulus, k = Boltzman's constant, T = absolute temperature, b = the Burgers vector, D = grain size, σ = stress, and m and n are constants for a given body and condition, which are typically 0–3 and 1–5, respectively. Thus, to account for porosity effects on creep, its effect on both G and σ must be accounted for. The challenges of having two properties dependent on porosity is compounded by the frequent substantial exponents magnifying uncertainties in such corrections. This challenge is further compounded by pores being generated, changed, or both, in tensile creep and are often eliminated, changed, or both in compressive creep. Again, the impact of grain size and its interaction with porosity, e.g., via pore location, can be important.

Turning to compressive strength, the effects of pores on creep and crack propagation are again important as in tensile stressing in both similar and different fashions. The major similarity is a reduction of the load-carrying ability due to porosity, with the extent of this increasing at higher temperatures and with more intergranular pores as shown later. Major differences arise due to compressive stresses driving densification and typically being a process of cumulative failure steps. Thus, with compressive stressing, typically the more

porosity there is, the more it may be reduced by compressive stressing, especially as temperatures increase (although there can be some pore generation in
compressive stressing in bodies, especially those initially having low or no
porosity). Higher strain rates and more intragranular pores will limit these
processes, and lower strain rates and more intergranular pores will enhance
them. There is considerable literature on HIPing, press forging, and hot pressing
that is pertinent to much but not all of these phenomena, and there is limited
direct study of just compressive creep of bodies with substantial porosity.

Hardness, wear, and erosion will show some of the porosity effects seen in
tension and some seen in compression, as well as other changes, with these
differences being greatest for wear and erosion. Thus, both the amount and
character of strain rate-dependent crushing-consolidation is expected to occur or
be enhanced under compressive stresses as temperature and the amount of intergranular porosity increase. Increased plasticity at higher temperatures and lower
strain rates both change the character and increase the extent of consolidation,
which can make major changes, especially in wear and erosion. On the one
hand, increased plasticity will reduce local fracture which will typically reduce
wear and erosion; however, increased plasticity also means reduced hardness
and hence more penetration of asperities into mating wear surfaces, thus
increasing wear. Increased plasticity in particulates causing erosion, or in the
surfaces they impact can also make complicating changes in erosion, e.g., shifting of the angle for maximum erosion from 90 degrees (i.e., normal) to the
eroding surface for brittle processes to 30 degrees for ductile erosion processes
(5,6). At higher temperatures increased chemical reaction can also become an
important factor in erosion and especially wear.

All pores can effect deformation due to stress concentration effects on stress-
dependent diffusion, with such effects varying with the amount, shape (and
orientation), and size, as well as the location of the pores. Pore size, while not
effecting the stress concentration of an isolated pore, affects pore spacing (as
does heterogeneity), and hence the interaction of stress concentrations between
pores. Increasing the size of pores or pore clusters also enhances opportunities
for more penetration from hardness indentors, asperities on wear surfaces, and
from impacting particles. Location of pores (which also affects pore shape, e.g.,
Figs. 1.2–1.4) becomes much more important at temperatures where grain-
boundary sliding becomes significant, which is significantly impacted by grain
size and grain-boundary chemistry. The introduction or increased importance of
these interrelated variables greatly complicates these properties, their measurement, and especially their modeling.

9.2.2 Thermal Stress, Shock, and Fatigue Failure

The classical way of addressing thermal stress and shock failure was to solve the
boundary value problems, or more commonly use the existing compilations of

solutions (e.g., ref. 7). Such solutions almost invariably assumed isotropy of the body and that properties are temperature independent, since to do otherwise substantially complicates the problem. Application of these solutions, which are typically quite dependent on component geometry and boundary conditions, is by comparing the resulting thermal stresses to the known or expected failure stresses of the material, e.g., via use of Weibull statistics of failure (8). Such solutions and experimental results yielded a number of thermal-stress failure criteria depending on the material parameters and the thermal environment (9–13). These basically fall in two broad categories of thermal stress failure resistance, i.e., where: 1) no significant crack generation or growth occurs so there is no loss of strength, or 2) thermal stresses are so extreme that crack generation or more commonly growth cannot be avoided, but can sometimes be limited so some reasonable level of strength remains.

The maximum tensile stress σ_t from a rapid temperature change is

$$\sigma_t = E\alpha \, (\Delta T)(1 - v)^{-1} \tag{9.3}$$

Where E = Young's modulus of the body, α = its thermal expansion, ΔT = the temperature difference (typically in quenching), and v = Poission's ratio. Various modifications of this are made, e.g., to account for time dependence of establishing this stress. Thus, a factor reflecting the ratio of the time-dependent maximum stress to the theoretical maximum stress, i.e., of ≤ 1, is often multiplied into the right side of Eq. 9.3 to account for this (14). Similarly, the right side of Eq. 9.3 may be multiplied by the thermal conductivity to reflect the higher temperature gradients from lower conduction in the body. Such equations are readily solved for the critical ΔT, i.e., ΔT_C, from quenching that causes measurable strength degradation. Thus, although there are uncertainties or refinements such as specifics of the heat transfer from the specimen to the fluid (15) and effects of the bath temperature on this (16), such quench tests are widely used to compare materials in standardized quench tests to rank materials. Various versions of Eq. 9.3 can be used to address the porosity dependence by incorporating the porosity dependence of E, σ_t, and v, e.g., as recently done by Arnold et al. (17); however, as discussed briefly in Sections 4.3–4.4 and in more detail in Section 9.3.2, this can be in serious error if crack branching and related phenomena occur.

Modeling of thermal stress and shock effects was advanced by analysis of bodies with arrays of cracks based on previous mechanical analysis of such bodies. Models based on such crack arrays were developed, applied, and extended by Hasselman (9–13), including showing their consistency with fracture mechanics formulation (13). This in turn focused attention on effects of crack arrays (18), especially of microcracks (19,20), including pore-crack combinations, e.g., Bowie (21). Hasselman also noted that thermal stresses could cause failure of function due to excessive buckling or fracture from bending stresses (22), with criteria for resistance to these being respectively: 1) low thermal expansion and a low aspect ratio of the body, and 2) a high body aspect ratio and value of $\sigma^2 (\alpha E^2)^{-1}$. Some analysis of thermal fatigue has also been made (23); however, with the exception of microcracks and pore-crack combinations, all of the above developments introduce porosity dependence only via that of the constituent properties. Another exception is analysis showing that some improvements in thermal stress resistance can be achieved by introduction of reduced thermal conductivity in the surface region (24,25).

Gibson and Ashby (26) applied the analysis of Eq. 9.3 to foam materials using their models for the porosity dependence of E and σ_t, giving:

$$\Delta T_{cf} \sim 0.65 (\Delta T_{cs})(1 - P)^{-0.5} \tag{9.4}$$

where ΔT_{cf} = the critical ΔT for the foam, and ΔT_{cs} = the critical ΔT for the solid struts.

A major development in analysis of thermal stress failure has been computer analysis, especially via finite element analysis. There are a variety of codes to address a variety of thermal stress environments, with some able to handle the challenge of temperature-dependent properties, which can be important in some cases, e.g., high temperature applications.

9.2.3 Temperature Dependence of Nonmechanical Properties

The effects of porosity can change significantly as temperature increases due to changing transport across the pores to some extent for electrical and especially for thermal conductivity i.e., due to convective and especially radiative transport. Both of these can vary considerably with pore size, shape (and orientation), and interconnection (also varying with heterogeneity). Because of the substantial added complications that these effects contribute to the behavior, many simplifications are again necessary. An important example is separating out convection and radiation effects based on their relative temperature dependence. Convection can be more significant at lower temperatures, while radiation typically becomes an important and often dominant factor at higher temperatures. A significant challenge is to determine which simplifications are reasonable, and which are not, and over what temperature ranges.

Models for the porosity effects on thermal conductivity that allow for convective, radiative, or either or both effects were noted in Sections 7.2.1, 7.3.1, and 8.3.6, particularly for ceramic refractories, which are often composite materials. A key material effect is whether the material, without pores or microcracks, is crystalline or glassy. Crystalline materials have initially higher conductivities that decreases with increasing temperature, while glasses start from lower conductivities that increase with increasing temperatures. The two commonly approach similar conductivities at higher temperatures. The presence of grain-boundary pores, microcracks, or both lowers the conductivities of both types of materials and introduces considerable dependence on the type and especially pressure of the gas atmosphere of the test, and hence also in much of such porosity and microcracks.

9.3 COMPARISON OF MODELS, THEORETICAL EXPECTATIONS, AND OBSERVATIONS

9.3.1 Effects of Temperature on the Mechanical Properties Other Than Thermal Shock and Stress Resistance

POROSITY DEPENDENCE OF ELASTIC PROPERTIES

Earlier studies of refractory ceramics, e.g., of Wachtman and Lam (27), showed that Young's modules typically decreases by 1–2% per 100°C to at least 1000°C. Beyond this, single-crystal values continue to decrease at the same rate till temperatures of 1500°C or more; however, polycrystalline materials commonly showed changes to substantially higher rates of decrease at 1000–1200°C, and sometimes substantially lower temperatures. Subsequent work, mainly on rare earth oxides, corroborated these trends for both Young's and shear moduli, but with greater decreases generally occurring at higher temperatures, e.g., at ≥ 1400°C (28–30). Such increased rates of decrease are attributed to effects of grain boundaries based not only on their absence in single crystals, but also on the onset of increased grain-boundary sliding (28,29). [Lower temperatures for such increased moduli decreases appear to be related to the material character, e.g., lower absolute melting temperature, and greater impurities, as expected. Also, in evaluating specifics of increased internal friction at such temperatures in polycrystals, effects of changing grain sizes on changes in internal friction (28) must also be considered.] The extent to which such increased reductions represent intrinsic decreases, however, or simply artifacts of the measurements due to the strain rates of the test being less than the relaxation times for grain-boundary sliding processes is uncertain. This leaves an important issue of whether using such "elastic" data in modeling creep effects may count such effects through both decreases in elastic properties and creep mechanisms being modeled.

There is a reasonable amount of data for Young's and shear moduli of bodies of different porosity as a function of temperature (28–36). These show the temperature dependences of different bodies of differing porosities from the same material and processing (typically differing only in the degree of sintering) being parallel to each other or the relative modulus of bodies of different

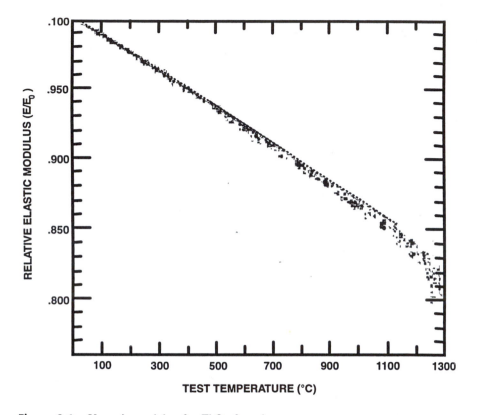

Figure 9.1 Young's modulus for ThO$_2$ from bar resonance versus test temperature normalized to the value at P = 0 for each temperature representing about 200 or more measurements. Thus, initially there is nearly perfect coincidence of normalized moduli giving a band only slightly wider than the line, with gradual, limited broadening as temperature increases showing limited change in the porosity dependence until ~ 1200°C. Plot after Spinner et al. (36).

porosity versus temperature falling on top of each other (Fig. 9.1), e.g., to the limits of volume fraction porosity tested, P ~ 0.4 (36). Such parallelness, i.e the same or similar b values (Table 9.1), means that there is no change in the porosity dependence over the range investigated, as expected, and that plots of the data as relative modulus versus P would all give a common curve. Also, note that the b values for G generally remain somewhat less than for E as at 22°C (Table 9.1). The degree of consistency of such parallelness of modulus versus P curves, or curves for relative modulus versus P being identical, for bodies of differing levels of porosity can thus be used as an indication of homogeneity of the porosity (along with data scatter).

Table 9.1 Summary of Room and Elevated Temperature Elastic Moduli Dependence on Porosity

Material	Processing[a]	GS[b] (μm)	P (#)[c]	T (#)[d]	b (E)[e]	b (G)[f]	Investigator
Al_2O_3	SC	23	0.5 (9)	1200 (8–10)	3.3	2.6	Coble and Kingery (32)
Al_2O_3	DP?	5–60[g]	0.18 (~4)	1000 (8)	3.2	—	Neuber and Wimmer (33)
BeO	Ex	3–70[g]	0.16 (5)	1000 (6)	2.5[h]	2.5[h]	Chandler et al. (34), Fryxell and Chandler (35)
Gd_2O_s	D, IP	—	0.37 (3)	1400 (26)	3.0[i]	2.7	Haglund and Hunter (31)
Sm_2O_3	D, IP	—	0.37 (6)	1400 (15)	3.5	2.7	Hunter et al. (29)
Sc_2O_3	D, IP	—	0.31 (16)	1400 (21)	5.6	4.2	Dole et al. (30)
Tm_2O_3	D, IP	—	0.24 (8)	1400 (21)	4.3	3.9	Dole et al. (30)
Y_2O_3	D, IP	0.2–220	0.37 (8)	1660 (16)	4.4	—	Marlowe and Wilder (28)
ThO_2	IP	~ 10[j]	0.39	1300 (10)[k]	4–6	6	Spinner et al. (36)

[a] Processing (of green bodies prior to sintering): SC = slip casting, DP = die pressed (? means this designation is suggested but not verified), D, IP = die- followed by isopressing.

[b] GS = grain size.

[c] P = upper limit of volume fraction porosity, figures in ( ) are the number of porosity steps covering the range tested.

[d] Upper limit of temperature for tests. Numbers in ( ) are the number of temperatures at which tests were made from 22°C for all except for Neuber and Wimmer (33), which includes tests at −200 and −78°C.

[e] b value (i.e., slope on a semilog plot of Young's modulus, E, versus P).

[f] b value (i.e., slope on a semilog plot of shear modulus, G, versus P).

[g] Measurements analyzed for different sets of specimens having similar average grain sizes, not for all specimens together over the entire range of grain sizes.

[h] Values were the same or very similar for specimens with or without preferred orientation.

[i] Similar trends shown for bulk modulus as a function of porosity and temperature.

[j] Grain sizes varied from the approximate value shown to values about 2–5 times larger due to use of increasing particle sizes to increase P.

[k] Porosity varied from natural porosity between particles of changing size and from various admixtures of organic beads of > 40 μm size.

MICROCRACK DEPENDENCE OF ELASTIC PROPERTIES

Consider next effects of microcracks on elastic properties as temperatures increase. The predominant source of microcracks in ceramics is microstructural stresses from crystal structure transformation, and especially thermal expansion anisotropy, both of which are a function of temperature. Such microstructural stresses may by themselves generate microcracks, i.e., result in microcracks that are preexisting for subsequent testing or use of the specimen or component. Alternatively, or additionally, microcracks may be generated as a result of the superposition of applied mechanical or thermal stresses and microstructural stresses that by themselves had not generated microcracks. In the latter case the microcracks may form throughout the body or, more commonly, only locally due to stress concentrations from cracks and possibly pores. Where microcracks form in conjunction with macrostresses, increasing temperatures reduce the microstructural stresses and hence the resultant microcracking. Where microcracks have already formed, they typically close and often heal up as temperature increases so their effects decrease or disappear at or below the transformation or original processing temperatures, but then reform or open to decrease properties as the body cools. Such microcrack formation typically begins and increases above a property- (hence material-) dependent grain size (Sections 1.3.2 and 10.2.5). Microcrack effects are reduced or disappear on heating and appear and increase on cooling due to partial or complete closure and healing, respectively, and reforming and opening. Temperature cycling typically shows substantial hysteresis of such property changes. Microcracking could also conceivably form on heating above processing temperatures due to continuing thermal expansion anisotropy or other transformations (if not precluded by stress relief). When such microcrack healing and reforming occurs, the resultant changes in elastic moduli from the microcrack changes can be opposite from those due to temperature alone in the absence of microcracking, as discussed earlier. Thus the decreases of properties such as elastic moduli with increasing temperature and their increases with decreasing temperature are reduced or actually reversed.

The most extensive data showing these effects are in graphite, along with some data for isomorphous hexagonal BN, both of which have high anisotropy in thermal expansion, about an order of magnitude higher in the c direction (perpendicular to the hexagonal planar structure) than in the plane of the hexagonal structure (37) (referred to as the a direction or the a-b plane). Again, specific correlation with microcrack parameters such as size, shape, and orientation is not possible, and quantitative analysis is complicated by the significant anisotropy in grain shape (typically platelet grains with their plane being parallel to the a-b plane of the hexagonal structure) and the resultant significant variation in degree of preferred orientation of grains, with different processing. There is almost always extensive (but not perfect) preferred orientation in pyrolitically deposited (i.e., CVD) graphite (PG) and BN (PBN), and less and variable (from approximately none to substantial) preferred orientation of these materials, depending on fabrication method and parameters for their processing via consolidation from particulates. There is also typically limited grain or porosity characterization, except for P ($\sim$ 0 in CVD and $\geq$ 0.2 for much partic-

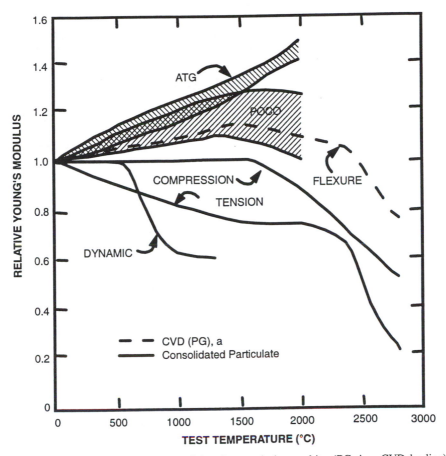

Figure 9.2   Relative Young's modulus for pyrolytic graphite (PG, i.e., CVD bodies) and some representative particulate-derived graphites versus test temperature. Note differences in PG values measured in the a direction in flexure, compression, tension, and dynamically (i.e., by resonance methods), and both the similarities and differences in the values and trends.

ulate-based processing, where there is often considerable anisotropy of shape and preferred orientation of pores depending on fabrication methods and parameters, adding to the anisotropy of bodies). However, comparison of bodies of different fabrication and generally known preferred orientations, and tests of these in different directions relative to the processing direction (particularly the temperature dependence of their properties) clearly show a major role of microcracking in their moduli, as well as other properties as shown below.

Data summarized from the commercial production literature, independent studies (38,39), and other reviews (40,41) show, often substantial effects of microcracking on elastic moduli (Fig. 9.2). Measurements of E versus temperature for PG parallel with the basal plane orientation (i.e in the a direction, also referred to as with grain) shows some variability in both values and trends with temperature with different measuring stress conditions, i.e., from flexure, compression, tensile, and sonic testing. (Evaluation of such differences can be valuable in further understanding behavior, e.g., as discussed later in conjunction with compressive strengths.) Overall, this group of measurements on the same pyrolitic graphite (PG) initially shows, on average, approximately no decrease until temperatures of ~ 500–1500°C, and substantial decreases primarily above 1500–2500°C.

Marlowe's studies of particulate derived graphites of either substantial or approximately zero preferred orientation (i.e., the latter are approximately isotropic), shows similar to greater increases in G and E than seen in some PG measurements as temperature increases (Fig. 9.2). Thus, these graphites increased E by ~5 to > 40% as temperature increases (38). These materials, of course, start from (and remain) at substantially lower E values than for the a direction in PG. The lower E values for the molded graphite reflects a substantially less-preferred grain orientation and hence less of the much stiffer a direction and more in the more compliant c direction along the stress axes than for PG measurements that are approximately parallel with the a-b plane. (Note that anisotropic, particulate graphites commonly show substantial anisotropy of E due to preferred orientation, e.g., ~ 5 and ~ 13 GPa, respectively, across and with the grain of an extruded graphite and ~ 6 and 10 GPa for a molded graphite (41).) Similar trends were reported for the shear modulus, but with increases being from ~ 0 to < 10% below those for E. In both cases the two grades of isotropic graphite reached maxima at between ~ 1300–1800°C, while the anisotropic graphite (ATJ) had not reached its peak at the limit of testing (2000°C). Furthermore, the lower the temperature of the peak in E or G, the lower the peak increase in moduli was relative to the other graphites. Finally, measurements by Sato et al. (39) of three reactor (molded) graphites that were near isotropic behavior and an anisotropic (extruded) one all showed only a few to several percent decrease in E with increasing temperatures up to ~ 1000°C. Beyond this E increased, reaching modest maxima of 0 to a few percent below room temperature values at ~ 2000°C, then decreasing more rapidly to the limits of their measurements (2400°C). Again, the absolute values were substantially less than for PG, e.g., averaging ~ 10 GPa, and the lowest value (~ 7 GPa) was for the extruded material (measured normal to the extrusion direction, the others were measured parallel with the molding direction).

These results are all basically self-consistent (and with other properties discussed below). Thus, the lower temperature and level of the maxima of the isotropic graphites is basically consistent with their typically finer grain sizes and smaller, and commonly fewer, microcracks. The extent of microcracking and hence potential moduli increases with temperature roughly correlating inversely with the degree of preferred crystallographic orientation, i.e., both are less in PG versus various particulate derived graphites. These results and evaluations are also supported by the limited measurements showing some hysteresis in the moduli and thermal expansion (42) as temperature is cycled, as well as other behavioral aspects addressed in other reviews (37), especially that of Brocklehurst (40). The latter includes observations on acoustic emission, some nonlinearity of stress-strain relations at higher stress, and reductions of E by prestressing; however, more detailed, i.e., quantitative, evaluation is limited by incomplete characterization, as well as uncertainty in the extent of coupling between porosity and microcracking. While it is commonly assumed as a first approximation that the effects of pores and microcracks are independent, e.g., as done by Hasselman (43), this is not always so as shown earlier in Nb_2O_5 (44) (Fig. 7.10A) and Al_2O_3- ZrO_2 duplex composites (45) (Section 9.3.2). A key question is the extent to which microcracks are already formed versus the extent they are generated during stressing (and hence a function of such stressing). Direct crack propagation studies of graphite, effects of prestressing, the occurrence of acoustic emission, limited hysteresis, and little or no effect on thermal conductivity (Section 9.3.3) indicate that substantial microcrack generation in graphite can occur during stressing.

That similar effects should occur in other microcracking ceramics is corroborated by Manning's measurements on Nb_2O_5 (44). He showed that both E and G for hot-pressed bodies decreased at 22°C with increasing P (to ~ 0.2, with b values respectively of ~ 4.7 and 4.6), while moduli values for sintered bodies at the same P were about half those of the hot-pressed bodies and were nearly independent of porosity (Fig. 3.16A). (Note, for the hot-pressed and sintered bodies the grain sizes were respectively ~ 8 μm and ~ 20 μm). The hot-pressed bodies showed the normal decrease and increase in moduli with increasing and decreasing temperature to the limits tested (~1100°C), i.e., ~ 17%, Fig. 9.3). On the other hand, the moduli of the sintered samples, which were only 20–30% of the hot-pressed values at 22°C, increased on heating and decreased on cooling. The increase began at temperatures of ~ 600°C and saturated at ~ 800°C at ~ 70–80% of the hot-pressed modulus, i.e., reasonably consistent with the reduction expected for the porosity alone in the sintered versus hot-pressed samples. Limited data for further heating indicate that the modulus of sintered samples then decreases with further temperature increase. On cooling the modulus of sintered samples increased less than for hot-pressed samples until temperatures of ~ 400°C, then decreased rapidly with further temperature decreases to 22°C. This substantial hysteresis in moduli is consistent with the microcracks closing and at least partially healing on heating, and then opening or reforming on cooling. This, and the effect of heating rate on the hysteresis, is supported by Manning's observation that while holding at a given temperature (above the

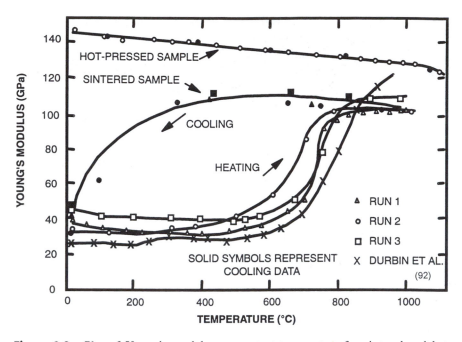

Figure 9.3 Plot of Young's modulus versus test temperature for sintered and hot-pressed Nb_2O_5 after Manning (44). Note that the hot-pressed material, which had too fine a grain size for microcracking, shows the normal temperature dependence (1–2% change per 100°C) on both heating and cooling. The sintered material clearly exhibited microcracking and a resultant decrease in $\bar{E}$ on cooling and increasing $\bar{E}$ on heating, to the point of the closure or reforming of the microcracks with substantial hysteresis.

minimum for modulus increase on heating) results in the modulus increasing with time at such fixed temperature. These modulus increases at fixed temperatures on heating increased in extent with time. The saturation of such increases at a fixed temperature resulted in a greater net increase as the level of the hold temperature increased. For example the increase at ~ 700°C was of the order of 8% and is near saturation after about 60 min., while at 800°C the increase is about 30% and had not saturated after 90 min.

These effects of microcracking on Young's modulus of Nb_2O_5 are corroborated by other studies. Ohya et al. (46) showed Young's modulus of Al_2TiO_5 increasing from ~ 30 GPa at 22°C to ~ 100 GPa at 1100°C with ~ 80% of the increase occurring between 700–1000°C, which is very similar to data of Manning (Fig. 9.3). The Al_2TiO_5 increases were corroborated by increased changes and hysteresis of thermal expansion over the latter temperature range (i.e., confirming microcracking) and by flexural strengths following nearly identical but somewhat greater increases with increasing temperature (~ 50 to

270 MPa) from 22–1100°C, again almost all of this between 700 and 1000°C. Similarly, Dole et al. (47) showed microcracking commencing in HfO_2 (P~ 0.05) with grain sizes > 2 μm and that microcracking progressively increased above this grain size with corresponding decreases in Young's and shear moduli. They also showed that specimens without microcracks (i.e with grains at or below 2 μm) had normal (~ 1.3%/100°C) changes of E and G until ~ 1200°C, and somewhat greater changes between 1200 and 1600°C under both heating and cooling; however, increasing grain size resulted in increasing changes and the onset and increase of hysteresis in E and G due to closing and opening (and probably some respective healing and reforming) of microcracks like in other microcracking materials such as Nb_2O_5 and Al_2TiO_5. Dole et al. (47) also observed a change in microcracking character with further increased grain size (e.g., to ~ 17 μm) and microcracking. The change entailed the onset of what they called "unstable" microcracks which were substantially less amenable to closure-healing and associated increases in properties, presumably due to their larger size, greater complexity, or both (Section 3.3.4). The absence of property recovery at higher temperatures and expected microcrack closure or healing was shown in another study of HfO_2 (48). Repeated heated and cooling of HfO_2-1 mol% Er_2O_3 to 1600°C leads to progressive decreases in E to the point of losing the ultrasonic signal in specimens that were above the micro-cracking threshold. Whether the occurrence of microcracks that do not close or heal on heating is due to just larger microcracks (i.e., scaling with grain size) or whether they become more complex, e.g., by becoming larger than one grain facet in size, due to transformation effects, is not known. Finally, more recent data of Davidge (49) for a dense, large grain (~290 μm), pure Al_2O_3 body showed clear hysteresis (20–1450°C), but substantially less than the above materials, consistant with its substantially lower TEA.

POROSITY AND MICROCRACK EFFECTS ON CRACK PROPAGATION, FRACTURE ENERGY, AND TOUGHNESS

There is limited data for crack propagation and fracture energy and toughness at elevated temperatures for any bodies, and even less for porous bodies. There is some data for graphites, often to quite high temperatures, but less data than for elastic moduli. For more conventional ceramics, there is quite limited data, mostly at temperatures < 1000°C, and, with two minor exceptions discussed below, none address the effects of porosity. This leaves the expected mechanisms and effects based on this limited data, other behavior, and general principles as discussed in Section 9.2 as the primary guides, along with the associated uncertainties of interactive mechanisms briefly outlined there.

Of the few studies of crack propagation and fracture energy and toughness in graphite, one examined crack propagation as a function of temperature and environment (e.g., atmospheres of H_2O, CO, or He). Freiman and Mecholsky

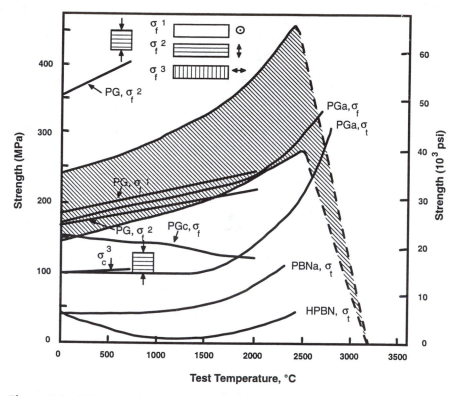

Figure 9.4 Relative and absolute strengths of various graphite and some BN bodies from manufacturer's and other literature (40,41). The cross-hatched region is the range of strengths of several particulate derived graphites reported by Riley (41). Note differences in tensile (σ_t), flexure (σ_f), and compressive (σ_c) strengths and as a function of the orientation of the stress to the deposition plane and c-axis direction for PG [e.g., for σ_f in the vertical direction and the c axis being: 1) normal to the plane of the plot, 2) in the plot plane and vertical, and 3) in the plot plane and horizontal, and 2) and 3) for σ_c].

(50) showed that both an isotropic (POCO-AXF-5Q) and an anisotropic (ATJ-S) graphite exhibited stable crack propagation to respectively ~ 800 and 1600°C in H_2O, CO, or He atmospheres. Above these temperatures crack propagation was catastrophic. They attributed the stable crack propagation to stress corrosion due to H_2O in the pores (hence explaining no influence of the external test atmosphere, and therefore consistent with strength results discussed below). They showed fracture energies in both materials increasing by of the order of 50% and fracture toughnesses by about 20% from 22 to 1400–1600°C. Similarly, Sato et al. (39) showed fracture toughnesses of three approximately isotropic (molded) and one anisotropic (extruded) graphites increasing by respectively

~50 (for the anisotropic, extruded material) and > 80% to maxima at ~ 2100°C or at, or beyond, their maximum test temperature of 2600°C. These changes are substantially more than they found for Young's modules of these same graphites discussed earlier (and about the same or intermediate for tensile strength, as discussed below), but were similar in overall trend. Thus the extruded material again had the lowest toughness and the least increase, consistent with its measurement being with the oriented grains versus those of the molded graphites being across the grain orientation. Subsequent graphite studies have further extended work on requirements for measuring toughness, associated acoustic emission, and documentation of R-curve effects and grain bridging resulting from microcracking (e.g., refs. 51–53).

Turning to the first of two limited fracture energy tests versus temperature for porous oxide ceramics, Baddi et al. (54) tested fine grain (0.3–0.7 µm) MgO they made by reactive hot pressing of Mg (OH)$_2$, then annealing at 1200°C. Their limited data for P = 0.07, 0.12, and 0.18 shows fracture energy (NB) increasing slightly from 22–800°C, but with the b value essentially constant at 10–11 (neglecting the systematic increase of grain size as P decreased). Above 800°C fracture energies started increasing rapidly as temperature increased; fracture energies of the highest P, finest grain body increased faster and so equaled that of the lowest P, largest grain body before 1100°C. This higher temperature behavior was attributed to plastic deformation, e.g., superplasticity, but illustrates probable serious complications due to neglecting grain-size effects. Unfortunately, there is insufficient information to determine such effects, which may not be a simple extrapolation of behavior at larger grain size (which indicates little or no grain-size dependence of fracture energy of MgO). The other data of Goretta et al. (55) for an open-cell Al$_2$O$_3$ foam (P ~ 0.76–0.91, cell size 1.4–1.6 mm) showed K$_{IC}$ first increasing with temperature, reaching a maximum of about twice the 22°C value, then decreasing to about 80 % of the 22°C level by 1400–1500°C.

MICROCRACK DEPENDENCE OF TENSILE STRENGTH

There are few direct studies on the effects of porosity on strength of graphite at high temperature, and especially of pore initiated failure. Riley (41) reported a substantial, e.g., twofold, increase in tensile strength as density increased from 1.56 to ~ 1.78 g/cc (i.e., for P decreasing from ~ 0, 3 to ~ 0.2) at 1200°C; however, there is substantial evidence to qualitatively show substantial effects of microcracks (37,40,41). Thus, it is widely known that most or all graphites, especially particulate-derived ones, tested in the a direction typically show substantial, e.g., ~ 100%, increases in tensile strength as temperature increases (i.e., at strain rates limiting creep, Fig. 9.4). They typically reach strength maxima at temperatures of the order of 2400°C, i.e., similar to their E and K$_{IC}$ trends discussed earlier. [Note that data of Sato et al. (39) is one of the few cases where all three properties were measured on the same sets of material.] Thus, while there are issues of deformation contributing to these maxima (and the subsequent significant strength losses at temperatures beyond the maxima), the major factor in these strength maxima is taken to be the closure or lack of formation of microcracks. This is also consistent with the observation of clear mechanical fatigue in graphites. Again, while some of this may result from deformation processes, much of it is attributed to microcracks based both on the fatigue, as well as other (e.g., elastic) behavior. Also, as with fracture toughness studies, evacuation of the gas in the pores results in increased strength and reduction or elimination of slow crack growth. Finally, there are various effects due to different aspects of preferred orientation and stress, including significant anisotropies and shifts between flexure, tensile, and compressive testing (Fig. 9.4.) that can be valuable in further defining the microstructural dependence of behavior.

There is much less information on the effects of temperature on the strength of other microcracked ceramics. Ohya et al. (46) showed a very similar but greater relative strength increase in Al_2TiO_5 with temperature (e.g., 50–270 MPa) from 22–1100°C, with the great majority of this increase in the 700–1000°C range. Again, the increased strength with increased temperature is attributed to closure, healing, or both of microcracks; however, the fact that these changes closely parallel the changes in Young's modules shows that most or all of the strength increases reflect changes in existing microcrack populations and little or no effect of microcrack generation in conjunction with the propagation of the failure-causing flaw.

Very similar effects should occur in BN and, in fact, do in the few cases where they are not overridden by other, processing-related phenomena, namely the nearly universal use of densification aids for all but pyrolytic (i.e., CVD) BN. Such densification aides, usually oxides such as CaO or MgO (combined with natural or added B_2O_3), both limit TEA stresses, and hence microcracking, and result in grain-boundary sliding that limits high temperature performance, especially strength. Thus, note the similar increase in tensile strength of pyrolitic BN (PBN) stressed in the a direction as temperature increases, as well as the initial decrease, then increase, in strength of hot-pressed BN (with limited grain-boundary phase, Fig. 9.4).

POROSITY DEPENDENCE OF TENSILE STRENGTH AND OTHER
MECHANICAL PROPERTIES

Limited data on the porosity dependence of tensile strength at elevated temperatures is summarized in Table 9.2, all of which fortunately involves similar testing at 22°C for direct comparison. The bodies made from coarser Al_2O_3 (fused) grain by Coble and Kingery (32) gave identical strength-P plots for tests at 22 and 750°C. The more limited data of Neuber and Wimmer (33), which is

Table 9.2   Summary of the Data for the Porosity Dependence of Flexure Strength of Ceramics at Room and Elevated Temperatures

Material	Processing[a]	GS[b] (μm)	P[c]	b (22°C)[d]	b (HT)[e]	Investigator
Al_2O_3	SC	23	0.04–0.55	~4	~4 (750)	Coble and Kingery (32)
Al_2O_3	DP?	~20	0.04–0.18	3+	3+ (1000)	Neuber and Wimmer (33)
Al_2O_3	HP	2.2	0.01–0.66	9.35	9.29 (1200)	Passmore et al. (56)
Al_2O_3	HP	30	0.3–0.66	0.34	0.33 (1200)	Passmore et al. (56)
BeO[g]	Ex	25–30	0.02–0.17	2.2–2.7[f]	4.2 (1200)	Chandler et al. (34), Fryxell and Chandler (35)
BeO[h]	Ex	25–30	0.02–0.17	2.5–3.7[f]	1.4–4[f] (1200)	Chandler et al. (34), Fryxell and Chandler (35)
MgO	RHP	0.03–0.07	0.07–0.18	~6.1	~ 4.4	Baddi et al. (54)
ThO_2	DP?	6–43	0.07–0.31	4.2	6.6 (1000)	Knudsen (57)
UO_2	D, IP	~20	0.05–0.25	3.2 ± 0.8	6.8 ± 0.6 (1000)	Knudsen et al. (58)
UO_2	D, IP	35–60	0.05–0.25	1.7 ± 0.7	1.8 ± 1.6 (1000)	Burdick and Parker (59)
ZrB_2	HP	6–11	0–0.16	3.2 ± 1.5	0.4 ± 1.8 (800)	Clougherty et al. (60)
ZrB_2	HP	6–11	0–0.16	3.2 ± 1.5	2.8 ± 0.8 (1400)	Clougherty et al. (60)
ZrB_2	HP	6–11	0–0.16	3.2 ± 1.5	1 ± 1 (1800)	Clougherty et al. (60)
SiC	DP		0.01–0.08	~4	~4 (1370)	Seshadri et al. (61)

[a] Processing (of green bodies prior to sintering): SC = slip casting, DP = die pressed (? means this designation is suggested but not verified), HP = hot pressing, Ex = extruded, RHP = reaction hot-pressed from $Mg(OH)_2$.

[b] GS = grain size.

[c] P = range of volume fraction porosity tested.

[d] The slope on semilog plots of strength versus P at 22°C.

[e] The slope on semilog plots of strength versus P at upper temperature shown in (°C).

[f] Range of values indicates differences in accounting for grain-size effects between this author and the original investigators.

[g] Specimens had some preferred grain orientation (which increased with firing and grain size) due to some crystallographic powder morphology character of the UOX powder.

[h] Specimens had no preferred grain orientation from the essentially equiaxed AOX powder.

complicated by a broader grain-size range (and probable mixed intra- and inter-granular porosity, especially at larger grain sizes), indicates the same or higher b value as for E (Table 9.1) at both 22 and 1000°C, though with some possible increase in b at the higher temperature and some reduction of b values at larger grain size. The BeO data of Fryxell and Chandler (35) for samples that had a preferred orientation showed a definite increase in b values with increasing temperature (tests were also made at intervening temperatures of 200, 400, 600, and 800°C). These changes were found by the original authors using a larger data set by normalizing values for differing grain sizes, as well as by this author for individual data sets of approximately constant grain size. Values were higher for strength than for Young's modulus. On the other hand, the BeO without preferred grain orientation showed a decrease in b values as evaluated by the original authors by normalizing the effects of grain size so a larger data base of 100–300 specimens could be used for each b value. These resulting values were respectively similar to and below those for E at 22 and 1000°C; however, evaluation of this author using the substantially more limited number of samples with similar grain sizes but differing porosity gives higher b values than those of Fryxell and Chandler and limited or no increase with temperature. These latter values were higher than those for Young's modules (Table 9.1) and closer to the values for the oriented BeO found by both methods.

Data of Baddi et al. (54) for their three porous reaction, hot-pressed, and annealed MgO bodies clearly showed the b value for strength decreased starting by ~ 700°C similar to that for their K_{IC} values discussed earlier. Both the level of the b values for strength and their changes arise due to greater decreases of the larger grain size, less porous, bodies versus greater increases in the fracture energy of the finest grain, most porous body. Knudsen's (57) ThO_2 data clearly showed a substantial increase in b value as temperature increased, which appears to correlate with the greater decrease of E at similar or the same temperatures for very similar bodies (Fig. 9.1). The two UO_2 bodies (of essentially identical raw materials and processing) both showed an increase in b values at the 1000°C-temperature test, but the larger grain-size range clearly showed significantly lower b values and substantial uncertainty in the increase in the b value at 1000°C. ZrB_2 results showed limited or uncertain to definite decreases in b values at higher temperature from the b value at 22°C (which is low for hot pressing, but is consistent with the b for E at 22°C). The scatter and limited range of the SiC data precludes detailed comparison other than that there is not a substantial difference between b values at 22 and 1370°C.

While there is some data on the effects of porosity on elevated temperature tensile strength, there is little or no data on failure from individual pores or pore clusters at elevated temperatures. Such pores surely continue to be preferred fracture origins as temperatures increase, but factographic features to clearly identify such origins decrease as the degree of intergranular fracture increases, so the evidence of their being origins becomes more uncertain. This greatly constrains or precludes determining whether the propensity of pores acting as fracture origins decreases at higher temperatures because of blunting of possible fracture from the pores due to enhanced grain-boundary sliding around the pores due to stress concentrations.

One important exception to the absence of observations of pores as fracture origins at elevated temperatures is Newcomb's thesis (62), subsequent publications with Tressler (63,64) and further analysis (65) providing useful data on the tensile fracture of c-axis (Saphikon) sapphire filaments, mostly from 800–1500°C at strain rates of 4.8×10^{-1} to 9.6×10^{-5}. These showed that internal pores, singly or as pairs closely spaced or partially joined, were about half of the sources of failure identified fractographically (62) (feasible primarily from ~ 800°C and above). Fracture features over the mirror region on both fracture halves, rather than being mirror images of each other, matched each other, indicating propagation of a sharp crack at all temperatures, and did not reflect changing to a proposed finger-crack growth mechanism (66). Until ~ 1400°C (or lower temperatures at higher strain rates) pores at origins were the only discernable feature there with dimensions consistent with the flaw size, indicating the pores acted as sharp flaws. Above this strain rate-dependent transition temperature range, peripheral cracks around the pores were found, whose size was consistent with the flaw size for brittle fracture. At lower strain rates and higher temperatures crack growth reached ~ 4–6 times the size of the pore from which the crack originated (Fig. 9.5A). Thus, the resultant failure causing crack size had varying relation to the size of the pore from which it originated (Fig.9.5B).

In view of pores being essentially sharp flaws in single- and polycrystal bodies, (but often not in glass bodies) (Sections 5.2, 5.3.3, and 5.3.4), it is useful to briefly consider differences between such bodies that could cause their differences in pores acting as sources of failure in them. Clearly pores in polycrystalline bodies have grain boundaries and associated grooving and anisotropies of properties from differing crystallographic orientations in adjoining grains, as well as more frequent associated (usually smaller) pores differentiating such failure from pores from those in glasses and single crystals, as previously discussed (67). Single crystals and glasses lack grain boundaries and associated grooving and property mismatches (and generally have fewer associated pores); however, crystals have preferred fracture (usually cleavage) planes and related crystalline and property anisotropy and resultant pore faceting and surface steps (roughening), all of which are possible factors that should be evaluated in differentiating the effect of pores acting as fracture origins in crystals versus glasses. In glasses, their isotropy gives no pore faceting, roughening, etc., and no preferred fracture planes, making fracture initiation from pores more difficult in them unless assisted by the presence of associated (typically machining) cracks as observed (67).

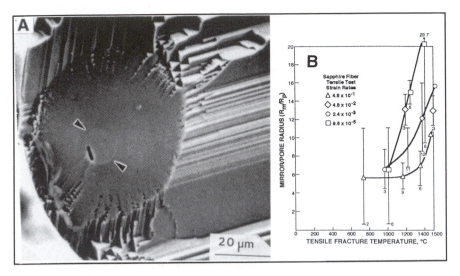

Figure 9.5 Failure from pores in sapphire filaments in high-temperature tensile testing. A) Higher magnification SEM photographs of a filament fracture at 1400°C (0.0024 min.$^{-1}$ strain rate, σ_f = 547 MPa) courtesy of Newcomb (62); published with permission of the Am. Cer. Soc. B) Plot of the ratio of the fracture mirror to that of the pore from which failure initiated. The marked increase in this ratio reflects crack growth around the pores (66); published with permission of J. Mat. Sci. (65).

The limited data on the porosity dependence of hardness at elevated temperatures, i.e., that of Atkins and Tabor (68) and Koester and Moak (69) previously reviewed by Rice (70), is summarized in Fig. 9.6. This indicates similar b values at room and higher temperatures for some materials, e.g., 3–6 for NbC, but higher values at elevated temperatures for others, e.g., 10–20 for TiB$_2$, ~ 20 for TiC, and possibly higher still for ThO$_2$. While the very limited data base and characterization of the materials (e.g., grain size and porosity) make the results uncertain, the similarity to other behavior provides some insight.

Data for effects of porosity on higher temperature compressive strength as previously reviewed by Rice (70), although also limited in extent, indicated two general trends. The first is for the porosity dependence of compressive strength to possibly first decrease as temperature increases, e.g., to several hundred C, then more definitely increase at higher temperatures, e.g., at > 1000–1200°C (Fig. 9.7). Thus, evaluation of grain-size and porosity effects on the compressive strength of Al$_2$O$_3$ indicated b values of 20–25, 20–40, and ~ 60 at 1000, 1600–1700, and 1900°C respectively. Second, this increase in porosity dependence is generally associated with (often substantially) increased intergranular fracture. Again this appears to be primarily due to substantial intergranular porosity (Fig.

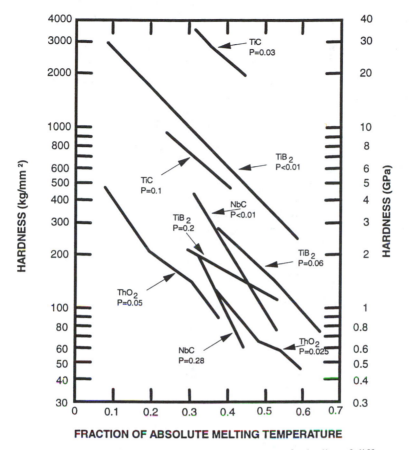

Figure 9.6 Summary of hardness versus temperature for bodies of different porosities after Rice (70). Published with permission of Academic Press.

9.7). No data is known for the effects of porosity on wear or erosion at elevated temperatures (nor for machining, which is not done at elevated temperatures).

9.3.2 Effects of Porosity and Microcracks on Thermal Shock and Stress Resistance

As noted in Chapter 1, the most common technique of testing materials for thermal shock resistance is to heat sets of test specimens in a furnace to set temperatures, then quench them into a liquid bath. Subsequent measurement of flexure strengths typically shows a critical quench-temperature differential (ΔT_C) is reached where strengths at room temperature start to decrease,

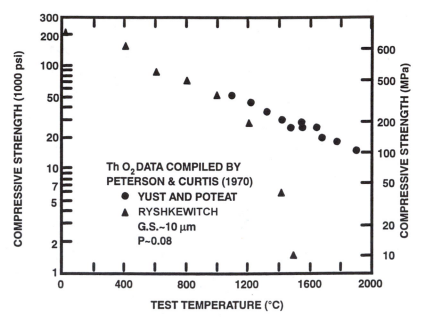

Figure 9.7 Compressive strength of two ThO_2 bodies with differing porosity versus test temperature, after Rice (70). Note the much greater decrease for the body with more porosity at $\geq 1200°C$. While other factors such as purity may play a role, the major effect is attributed to the greater (especially intergranular) porosity. Published with permission of Academic Press.

typically initially precipitously. There have been extensive studies of the effects of porosity on thermal shock resistance, but many are complicated by various combinations of limited porosity ranges, characterization, or both, and especially other variables such as grain size and composition, e.g., in refractories. There are some studies, however, that provide reasonable insight to mechanisms. Thus, Arato et al. (71) showed that different high-purity Al_2O_3 powders sintered to relative densities of 60–97% of the theoretical maximum exhibited the classical strength trends in the water quench test. As shown in Fig. 9.8, as porosity increases, the initial strengths decreases, but the precipitous decrease in strength (at ΔT_C 200–300°C) was reduced, as were further strength decreases as quench temperatures further increased; however, while the relative changes in strength were reduced at higher P, the absolute levels of retained strength were not as expected for porosity effects. This probably reflects the effects of using coarser powders to obtain more porosity at the same sintering temperature (1650°C), and clearly also shows the reduction of strengths with increased grain size.

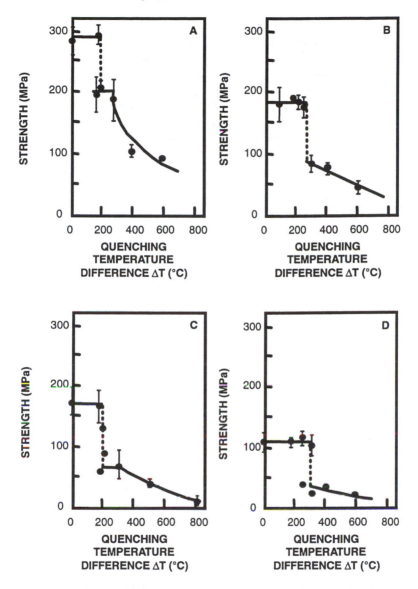

Figure 9.8 Thermal shock resistance (via water quench tests) of high-purity alumina bodies of different porosity (using coarser particles to increase porosity), with some variation in grain size (G). A) P = 0.029 (G = 1.3 μm), B) P = 0.244 (G = 4.4 μm), C) P = 0.26 (G = 5.2 μm), and D) P = 0.26 (G = 6.3 μm). Data after Arato et al. (71), published with permission of J. Cer. Soc. Japan, Intl. Ed.

A similar but somewhat more definitive earlier work is that of Smith et al. (72). They sintered high-purity alumina by itself to ~ 98.5% of theoretical density, or with additions of polystyrene spheres (40–100 μm diameter), wood "fibers" (50 μm long and 12 μm diameter), or organic flakes (40–100 μm maximum dimension) to increase porosities by an additional 1–5%; thus there was little or no variation in grain size. Their results (Fig. 9.9) show substantial reductions in initial strength due to the limited added porosity, little change in the critical ΔT_C [~ 175°C, i.e., similar to but lower than that of Arato et al. (71)], and reduction in the extent of precipitous (and subsequent) strength decreases. They clearly showed that the absolute level of strength retained was generally higher with more porosity, i.e., showing the greater thermal-shock resistance of porous bodies commonly seen in practice. There is also a strong indication that the character of the porosity is important (though the pores from the wood "fiber" and the organic flakes, while different from each other and those from the styrene spheres, were not easily characterizable as might be implied by their source descriptions). Their results for 3% spherical and 5% "cylindrical" pores are very similar to results of Koumoto et al. (73), who used the water quench test on SiC bars sintered with P = 0.25–0.38 and G = 2–8 μm). They showed strengths at the lower porosities (and larger G) decreased gradually from 40 MPa to a limit of ~ 20 MPa, i.e., about half the starting value at 900°C, while those with the highest porosity (and largest G) decreased from ~ 30 MPa to the same value, i.e., about half the decrease.

Besides these two (limited) studies, thermal shock data generated by Larsen et al. (74) for a variety of commercial and experimental Si_3N_4 and SiC bodies (including other property measurements that allowed calculation of the thermal shock resistance per Eq. 9.3) provides useful insight. Comparison of the experimental and calculated values of thermal shock resistance for these materials shows that most follow the trend line for the two values being equal, hence corroborating the typical thermal shock analysis (Fig. 9.10); however, there are distinct exceptions, namely some, but far from all, of the RSSN. This again shows that some porosity levels and characters can give enhanced thermal shock resistance.

The other specific evaluation is that of Arnold et al. (17) who compared calculations based on modeling the porosity dependence of the pertinent properties with data for glass and $CaTiO_3$ (P = 0.01–0.14). Their calculated results showed a small initial increase in the relative thermal shock resistance to a maximum (e.g., ~ 1–2% above the thermal shock resistance at P = 0) at P ~0.01, then a decrease; however, the data is widely scattered over the limited porosity range, thus precluding any definitive support for their model (whose validity is suspect).

Three other studies, while not addressing in detail the porosity dependence of thermal shock, observed greater thermal shock resistance in porous bodies

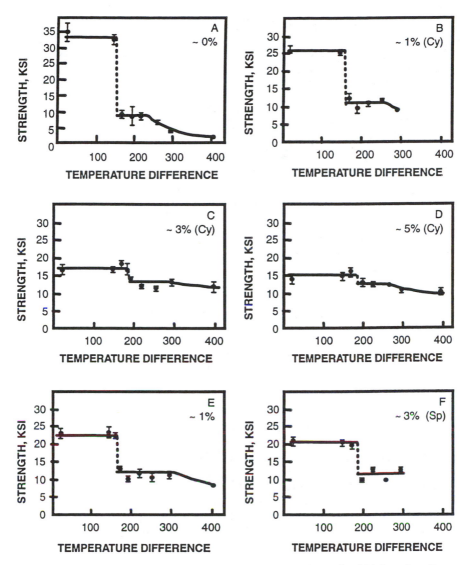

Figure 9.9 Thermal shock resistance (via water quench tests) of high-purity, fine-grain alumina, with ~ 1.5% natural porosity and such bodies with 1–5% added porosity from plastic spheres, "cylindrical" wood fibers, and organic flakes, after Smith et al. (72). A) Natural porosity, B–D) plus 1, 3, and 5% "cylindrical" pores, E) plus 1% "flake" pores, and F) plus 1% spherical pores. Published with permission of the Bull. Am. Cer. Soc.

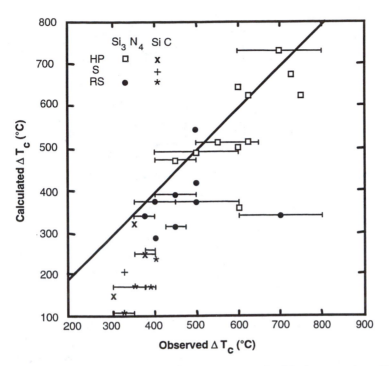

Figure 9.10 Plot of calculated versus measured critical temperature differences for the onset of serious thermal shock damage in typical quench tests of various $Si_3 N_4$ and SiC bodies, after Larsen et al. (74).

versus nearly dense bodies. First, Boch et al. (75) showed that for their hot-pressed AlN, fully dense samples had a critical ΔT_C of 250°C, but samples with $P = 0.12$ gave 300°C (and higher internal friction). They also noted that in the latter case calculated and measured ΔT_C values did not agree. Second, Antsifirov et al. (76) showed, using small samples, that ΔT_C increased from ~ 280 to ~ 350°C as P increased from 0.58 to 0.71 in RSSN. Third, Koumoto et al. (73) showed that sintered SiC with $P = 0.25-0.35$, although starting from lower strengths (30–40 MPa) had much more limited strength loss on quenching (to the limits of test at 900°C), i.e., to ~ 20 MPa, in comparison with a much more precipitous strength decreases to the same or lower levels in dense SiC.

One of two known studies specifically on effects of porosity on thermal stress fracture is that of Coble and Kingery (77), who sintered thicker-wall alumina tubes from two differently sized particles of fused grain (−100 + 150 mesh, and −60 + 100 mesh). Porosities of ~ 10–50% were introduced via inclusion of naphthalene (giving pores ~ 300 μm diameter). By differential heating of the interior versus the exterior, the temperature differentials to cause fracture (mainly axial) of the tubes were determined. This temperature differential decreased in a similar fashion for bodies made from either grain fraction, i.e., from ~ 175°C at P ~ 0 to ~ 70°C at P = 0.2, to a nearly constant value of ~ 55°C for P = 0.35–0.5. The initial decrease (e.g., to P ~0.2) was similar to that expected for tensile strength, i.e., a b value of ~ 5.6, as would be expected. The subsequent approximately steady temperature differential for fracture is consistent with enhanced thermal stress resistance of porous bodies.

As noted earlier, it has been proposed that porous layers at or near the surface(s) exposed to temperature differentials can reduce thermal stresses, and hence thermal stress and shock failure (24). Gruninger et al. (78) demonstrated such effects by forming thin (~ 4.3-μm thick) porous alumina coatings on 96% dense alumina electronic substrate disks (~ 0.64 mm thick). Coatings had high levels of fine pores (~ 7–9 nm diameter), e.g., as reflected in the high surface areas of the coatings (75–110 m/g). Using water quench tests (with biaxial flexure via support on three balls and loaded by a small piston), they showed that the critical ΔT for the uncoated substrates was ~ 150°C, i.e., somewhat lower but similar to other alumina tests. Addition of one layer of porous coating consistently increased ΔT_C by ~ 10°C and a two-coating layer approximately doubled this increase.

Orenstein and Green (79) studied the thermal shock behavior of alumina-silica ceramic open-cell foams by quenching into a water or an oil bath, then measuring the residual compressive and tensile strengths and Young's and shear moduli. No gross degradation was observed to the maximum ΔT of 1500°C, other than some surface spalling in bodies with the finest pores (65/in.); however, in general properties decreased with increasing quench temperatures, especially above 300°C, e.g., to ~ 20–30% of the starting values at 600°C in bodies with P~ 0.83, due to observed strut cracking (and internal friction increased), as expected from the models presented earlier. In particular, there was good correlation of decreases in both tensile and compressive strengths (which were about equal) with that of Young's modulus; however, there were significant deviations from the expected behavior (per Eq. 9.3) in two key aspects. First, all samples showed substantially higher than predicted ΔT values, e.g., by factors of 2–4. Second, the observed respective significant and modest increases in thermal shock resistance increased with cell size and density, as opposed to the decreases predicted by the analysis that yielded Eq. 9.3. These significant differences were attributed to effects of the quench medium in infiltrating the open-cell structure, thus maintaining some macroscopic thermal stress across the foam, not just across individual struts as initially considered in attempting to explain the differences between actual and predicted results.

As noted earlier, microcracked materials are often used for their substantial thermal shock resistance. This arises from the generally significant reduction in bulk (but not lattice) thermal expansion (Fig. 7.10), but a decrease in Young's modulus and an increase in fracture toughness for larger cracks may also be a factor (Section 4.3.3). Such materials typically have modest strengths, but retain most of their strengths through rather extreme thermal shocks. Thus, materials such as Al_2TiO_5-based bodies have been considered for more refractory auto exhaust catalyst supports (section 10.3.2), exhibiting limited strength losses with substantial quench temperatures, e.g., > 1000°C. Similarly, Claussen and colleagues (45,80) showed that their duplex composites of agglomerates of one ZrO_2 + Al_2O_3 composition (mostly monoclinic ZrO_2) dispersed in a matrix of Y_2O_3 that was partially stabilized ZrO_2 +Al_2O_3 resulted in substantial thermal shock resistance with substantial microcracking in the agglomerates. Thus, the ZrO_2+Al_2O_3 matrix alone had a ΔT_C ~ 200°C where strengths decreased catastrophically from 1300–1700 MPa to nearly zero. The addition of increasing agglomerate content and associated increased microcracking [revealed by dye penetrant examination (45)] resulted in no significant change in ΔT_C and a decrease in initial strength (e.g., from 400 to 100 MPa) resulted in strengths retained after quenches to 500°C of 50–100 MPa (with the residual strength generally being inversely related to the starting strength) (79). As noted earlier (Section 4.3.3), however, suitable microcracking in the agglomerates and hence in the composites was dependent not only on the composition and particle sizes, but also on residual porosity in the agglomerates after processing the composite (45). Composites with no porosity, e.g., as a result of HIPing, did not show increased toughness or thermal shock resistance, but bodies having limited porosity in the agglomerates after composite processing did, thus showing a significant interaction of microcracking and porosity. An interaction between pores and microcracks was also shown by thermal expansion of Nb_2O_5 being dependent on porosity (Fig. 7.10A).

9.3.3 Nonmechanical Properties

There is substantial data on the thermal conductivity of porous materials at elevated temperatures due particularly to the use of porous ceramics for refractory insulation; however, there is only modest data for controlled tests of the effects of porosity on thermal conductivity. Data for BeO (Fig.7.9) shows that thermal conductivity versus P at various temperatures to 500°C follow parallel lines, thus showing no change in mechanism as temperature increases. This is consistent with expectations since neither convection nor radiation transfer are expected to become measurable factors since respectively the pores are too fine and the temperatures too modest. Similar trends are shown for die-pressed graphite (81) to substantially higher temperatures (Fig. 9.11). Since graphite is electrically conductive and thus not transparent to infrared radiation, little or no radiative transport would be expected with the expected modest pore sizes; thus, the parallelness of the lines shows no significant changes in mechanism (solid

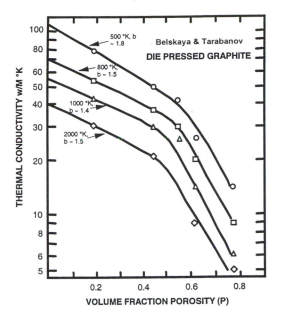

Figure 9.11 Thermal conductivity of die-pressed graphite versus volume fraction porosity (P) at various temperatures after Belskaya and Tarabanov (81). Note the parallelness of the lines, as expected, showing no significant shift in the conduction mechanism; also, while the b values are lower than expected for die pressing they are consistent with indicated P_c values of ~ 0.9.

conductivity). The modest scatter and the limited variation in the b values are consistent, with the latter probably being reduced with more data at lower P. The low b values, ~ 1.6, are lower than expected for typical die-pressed bodies (e.g., ~ 3–4), but probably reflect some orientation of the graphite particles (and resultant pores in pressing) and are reasonably consistent with the indicated P_c values of the order of 0.9.

Data to quantitatively evaluate effects of microcracking on thermal conductivity or diffusivity is difficult to obtain since obtaining bodies with controlled microcrack populations that can be quantitatively characterized is at best difficult. Thermal diffusivities have been measured for some ceramics that often microcrack extensively due to highly anisotropic thermal expansion (depending on the degree of anisotropy, other properties, especially E, and on grain size, as discussed in Chapters 4, 5). Such cracking should be approximately random (and on the scale of the grains), and is known to close up (and probably heal) on heating and open up or reform on cooling, but with hysteresis. Thus studies of the temperature dependence of the thermal diffusivity of a substantially anisotropic material such as Fe_2TiO_5 (82) (Fig. 9.12) qualitatively corroborate

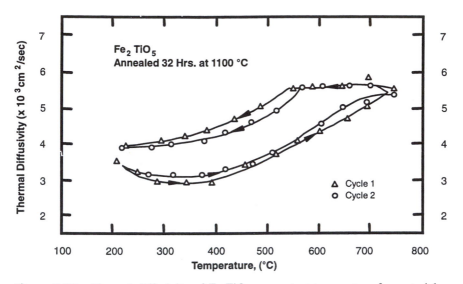

Figure 9.12 Thermal diffusivity of Fe_2TiO_5 versus test temperature for material as hot-pressed, and after annealing at 1100°C for 32 hours. Arrows indicate direction of temperature change. The as-pressed material, being of sufficiently fine grain size, had little or no microcracking, showed the normal decrease in thermal diffusivity as temperature increases or decreases, and thus no hysteresis; however, annealing of samples to increase grain size (from 1 to ~10 μm), thus initiating microcracking, substantially reduced thermal diffusivity at lower temperatures, and resulted in mostly increased or decreased diffusivity as temperature increased or decreased, respectively, due to respective closing and opening of microcracks (and some hysteresis). Data after Siebeneck et al. (82). Published with permission of Dr. J. A. Pask.

the random model of Table 7.2 and Fig. 7.2. There, despite the normal decrease in diffusivity with increasing temperature, the thermal diffusivity increases over most of the range of increasing temperature and decreases over most of the range of decreasing temperature. These trends are counter to the normal temperature dependence and reflect respectively the decrease of microcrack density and its increase. Thus, replotting these changes in diffusivity in terms of crack density, although only qualitatively possible, is consistent with the random crack model, i.e., microcrack density decreasing as temperature increases, and vice versa. Data for less anisotropic $MgTi_2O_5$ and more anisotropic Al_2TiO_5 show similar trends, except that because of the greater anisotropy of the latter it was not possible to get dense bodies of sufficiently fine grain size to have no microcracking. Data for $AlNbO_4$ is similar, showing normal conductivity decreases for finer grain (2 μm) bodies, but with progressively lower conductivity and increased hysteresis as grain size increased to 3 and then to 7 μm (83).

Both graphite and BN bodies commonly show significant effects of micro-cracking on mechanical properties (Section 4.3.3; Figs. 9.2, 9.4), so the temperature dependence of the thermal conductivity of these materials is also of interest in this regard. Comparison of such data for various graphite and BN materials and some other ceramics from manufacturers literature and other reviews (41,84–87) (Fig. 9.13) generally shows more similarities than differences, which is in marked contrast to the significant differences typically shown for mechanical properties. There is varying anisotropy as a function of the measurement direction relative to grain orientation, with high anisotropy associated with a high degree of orientation. Thus, there is high conductivity in the a direction of PBN and especially PG (i.e similar to diamond), versus much less in their c direction, and much less anisotropy in particulate-derived bodies, especially those with modest crystalline anisotropy, i.e., reactor graphite (RG). Except for a substantial increase in conductivity of PBN in the c direction, however, all other graphites and BN bodies show decreases in conductivity as temperature increases (except at low, usually cryogenic temperatures where materials show conductivity maxima, e.g., some in Fig. 9.13 have maxima at ~ − 200°C or less).

There are several complications in sorting out effects of microcracking on these materials. First, since graphite is an electrical conductor, it has (along with lattice conduction) some thermal conduction via electrons, which decreases with increasing temperature, but no radiant transfer through it (but some across pores or microcracks). On the other hand, BN and diamond are insulators so they have only lattice conduction at lower temperatures, then some radiant transfer through them and across pores or cracks at higher temperatures, as with other dielectric ceramics such as Al_2O_3, BeO, and ZrO_2. The lower conductivity versus temperature slopes of PG relative to other ceramics, and the constant and increasing conductivity of PBN in the a and c directions, respectively, may indicate some effects of microcrack closure on conductivity as temperatures increase; however, comparison of thermal conductivity with mechanical effects suggests that much of the microcracking that effects mechanical properties is generated by stressing the body during mechanical tests, especially in conjunction with crack propagation. This shows the need and potential value of comparing different tests (and indicates value in comparing thermal and electrical conductivities of carbon bodies as a function of temperature and pore character).

Litovsky and colleagues (88,89) reviewed and analyzed thermal conductivity of ceramic refractories, which are commonly composite materials, as a function of porosity, microcracks, or both, and atmosphere and temperature. They divide the conductivity behavior into a lower porosity regime, P < 0.3, and a higher porosity regime, P > 0.3. In the former, crystalline materials with primarily intragranular porosity and little or no microcracking have a higher thermal

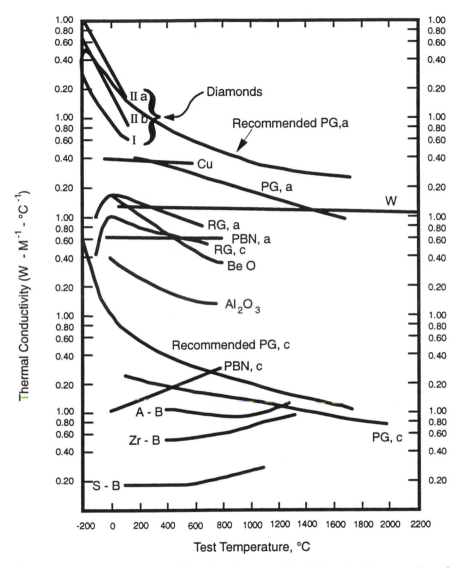

Figure 9.13 Thermal conductivity of various graphite, BN, and other ceramic and metal bodies versus test temperature from manufacturer's literature and other reviews (41,84–87). Note data for pyrolitic graphite (PG), BN (PBN), and a reactor graphite (RG) in both the a and c directions, as well as data for alumina, silica, and zirconia furnace bricks (A-, S-, and Zr-B).

conductivity that depends primarily on the porosity, temperature, and material, with increasing temperature causing substantial progressive decreases in conductivity. As noted earlier, glassy materials have lower conductivity, which decreases with increasing porosity but increases with increasing temperature, which may approach levels to which crystalline materials have decreased at higher temperatures, e.g., $\geq 1200°C$. The combination of crystalline and glassy phases can thus have varying conductive behavior as a function of temperature, depending on the microstructure of the body, with a continuous glassy phase having the greatest effect on conductivity and its temperature dependence.

Introduction of substantial intergranular porosity, microcracking, or both in crystalline materials makes conductivity sensitive to the type and especially pressure of the gas atmosphere on conductive behavior. In such bodies, conductivity is much lower at low gas pressure, i.e., lowest in vacuum, and increases with increasing temperatures. On the other hand, as gas pressure increases, lower temperature conductivities increase substantially, but at higher, e.g., atmospheric, pressures conductivity decreases with increasing temperature. The lower and higher pressure conductivities commonly approach each other at temperatures of the order of 1200°C. Glassy materials with microcracks simply have their conductivity proportionally reduced as gases of lower conductivity or pressure are present. In a given material, bodies with the same level of porosity, i.e., P, can have substantially different conductivities, even in the absence of microcracks, showing again that the nature of the porosity, not just its amount is important (88). Thus, ceramic refractory thermal conductivity behavior over these regimes clearly shows the need to address a variety of phenomena which requires families of models that address the diversity of behavior exhibited.

The diversity of behavior and the need for families of models is also shown by refractories of higher porosity, though variations at high porosity are less diverse (89). This arises from the increasing openness of the pore structure at higher porosity levels and the decreasing impact of the solid phase as it becomes a smaller volume fraction of the body, which commonly also reduces or eliminates microcracking. At low temperatures conductivity of more porous bodies commonly follows a $(1 - P)^{3/2}$ dependence (i.e., Eq. 4 of Table 7.1); however, at higher temperatures radiation effects become increasingly important. This introduces a substantial effect of pore sizes, which increases with increasing temperature; thus, for example, at $\sim 1000°C$ conductivity is approximately five times greater in a refractory silicate concrete with pores ~ 2 mm versus those $\sim < 0.5$ mm diameter, with somewhat greater increases occurring between pore sizes of 0.5–1 mm. An important consequence of the increased radiation contribution as temperature increases is that materials increasingly show higher conductivity minima as a function of porosity. The relative and absolute

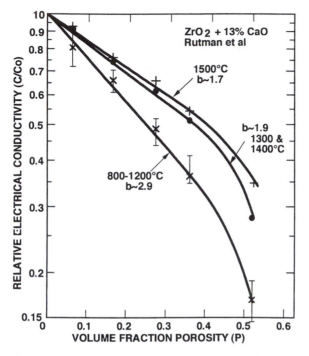

Figure 9.14 Electrical conductivity versus volume fraction porosity (P) at various temperatures for $ZrO_2+13\%$ CaO after Rutman et al. (90). Note that while the overall trends follow those of MSA models, the reduction of b values with increased temperature implies basic changes in the conduction mechanisms.

conductivity minima levels are commonly at P = 0.8–0.95, where there are more pores than material, and increase with temperature).

One known study of electrical conductivity versus P at higher temperature is that of Rutmen et al. (90) for ZrO_2 + 13% CaO (Fig. 9.14). This data clearly follows trends consistent with MSA models (with b values of ~ 1.7–3); however, decreasing b values as temperature increases indicates the conduction mechanism is changing with temperature, or, with more extreme temperature exposures, that the pore structure is changing. This is clearly possible since both the ionic and electronic conductivities change with temperature, so the balance of these is likely to change. It is not clear why, however, this would result in a change in mechanisms, specifically a decreased reduction of conduction with increased porosity as temperature increases. Whether mechanisms such as increased conduction across the pores, e.g., via thermionic emission, are occurring to cause these changes are not known.

9.4 DISCUSSION

Three general topics need to be addressed: 1) separation of grain size, porosity, and microcracking effects, 2) self-consistency; and 3) overall trends. As grain size increases, especially above the typical lower level of ~ 1 μm commonly found in practice, it can affect apparent porosity dependence in two ways. First, many mechanical properties, especially tensile and compressive strengths, decrease with increasing grain size (Chapters 5, 6). Thus, failure to account for effects of increased grain size on such properties can give incorrect porosity dependence. Since grain size typically increases at lower porosity, this will falsely imply reduced porosity dependence. For tensile strength, a single larger grain or larger grain cluster can often be a failure source at lower stress than for the rest of the grain structure, primarily in noncubic materials, accentuating this effect. Second, increased grain size from grain growth is accompanied by varying degrees of transforming some intergranular to intragranular pores (Figs. 1.2–1.4), and typically by an increase in the size and possibly number of intergranular pores as well as often by increased concentration of grain-boundary phases from additives or impurities. As test temperatures increase these grain-related changes can have differing effects on properties, as can intrinsic changes in material properties such as thermal expansion anisotropy (TEA, present in noncubic materials) and elastic anisotropy (EA, present in essentially all crystalline materials) (91); however, two effects stand out. First, increased intragranular porosity commonly reduces the porosity dependence at elevated temperature, as at lower temperatures (see previous chapters), e.g., as indicated by the Al_2O_3 data of Passmore et al. (56) (Table 9.2). Second, in all processes involving fracture as the direct result of the test as well as to some extent in tests where fracture may be a supplemental aspect of the test (e.g., hardness, erosion, and wear), intergranular porosity enhances intergranular fracture. Such fracture is often accompanied by reduced properties, especially strength, particularly with grain-boundary phases. Thus, there will commonly be an increase in the porosity dependence of such properties at higher temperatures, but with the rate and extent of this depending substantially on the temperature and body character.

Separation of porosity and microcracking effects, the latter primarily via TEA or phase transformation, is also often necessary. While in some cases these effects may be approximately independent, as assumed in most modeling, there is usually some coupling, e.g., as indicated in Hasselman's (43) modeling of graphite. Increased porosity also increases the thermal expansion of Nb_2O_5 (Fig. 7.10A), indicating reduced microcracking, which is supported by the elastic moduli being approximately independent of P (Fig. 3.16A). There is also an important interaction of grain size with most such microcracking since the onset and extent (and possibly character) of microcracks is a result of the grain size and key physical properties. While such separation of porosity and grain size effects on microcracking can be important, of greater need is the ability to characterize the nature and spatial distribution of such cracks to better relate results to models and, in turn to improve models. The diversity of thermal conductive behavior of ceramic refractories illustrates in part the complex combination of porosity and microcracking effects.

While the generally limited extent of porosity effect data at elevated temperatures means that there is very limited evaluation or demonstration of self-consistency, this is partially mitigated by the nature of most temperature dependence. Since, many mechanisms do not change until high temperatures are reached, data at intermediate temperatures provide the opportunity for evaluating self-consistency. Thus, the parallelness of curves for elastic moduli (Fig. 9.1) and thermal conductivity (Fig. 9.11), or the same curve for the relative property at various temperatures show reasonable self-consistency for these cases.

Two overall trends are indicated. First, many properties continue to be determined by the same porosity dependence at elevated temperatures as at lower temperatures. The extent of this continuation of mechanisms and departures from this due to changes in the existing (or the introduction of other) mechanism(s) clearly varies with the property, material, and especially the microstructure; however, this continuity provides a valuable basis for evaluation. The first of two key changes that occur in thermal conductivity is effects of gases in bodies with grain-boundary pores, microcracks, or both, mainly at low to intermediate temperatures. The second is the increasing introduction of radiative heat transfer in all bodies, especially of increasing porosity, as temperatures increase to and above moderate levels. There can also be changes in electrical conductivity as well as dielectric breakdown as temperature increases.

A second overall trend for fracture properties is the transition to predominantly intergranular fracture mode (4). This appears to be enhanced by intergranular porosity and significantly accentuates the porosity dependence of such properties. The rate and extent at which this occurs clearly depends on material and microstructural character. Such increased porosity dependence is particularly shown for tensile and compressive strengths, but is also indicated for hardness and elastic moduli. The latter manifestation must be studied further to determine the extent to which it is intrinsic (due to grain-boundary character) or extrinsic (due to the strain rate of the test allowing creep or other nonelastic deformation).

Brittle-creep failure transitions usually occur well below melting, but studies of ice (isostructural to α-Al_2O_3 so properties of both are similar as a function of the fraction of absolute melting temperature, T_m) show brittle fracture extending to > 95% of T_m. Tabata's (92) sea ice studies at $-9°C$ gave b ~ 4 for dynamic E (P ~ 0.12–0.25); nonbrittle to brittle failure transitions with increasing loading rate (~0.2–1 Kg/cm^2·sec), with higher rates for higher P and static E increased (especially at lower P); and b ~ 4 for flexural strengths for testing above this stressing rate range. The two b values are about the same and their absolute values are consistent with brittle failure and the expected larger (~ 1 cm) grains with laminar distributions of saline pockets and air bubbles. Nakaya and Kuroiwa (93) obtained b ~ 4 for Young's modulus (grain size ~ 1/4–1 mm, P ~ 0.35 to a rollover at ~ 0.54 and P_c ~ 0.7) and b ~ 3 for compressive strength of compacted snow (P ~ 0.1 to 0.3, with a probable rollover at P ~ 0.3 and P_c ~ 0.7).

9.5 SUMMARY

Data on porosity dependence of properties at elevated temperatures is limited, indicating the need for more study as well as caution in interpreting results. Complications in separating porosity and grain size effects can also be a challenge, especially for strength and some other mechanical properties, as can separation of microcracking and porosity effects in the more limited bodies in which this occurs. The general continuity of mechanisms as temperatures increase provides substantial guidance as well as a evaluation of the uniformity of the body and the quality of the data. The two major trends observed were: 1) the general continuity of mechanisms as temperature increases, and 2) the increased, typically substantial, porosity dependence of strength properties at high temperatures (typically associated with predominantly intergranular fracture). Both of these trends vary in their occurrence with the material and especially the microstructure involved. While there are obvious uncertainties given the limited data and these observations, the general conclusion is that the models used for room-temperature behavior are still generally appropriate for elevated-temperature behavior, possibly with some modification. Thus, MSA models are still generally useful and thus far superior, especially for predictive purposes. Further evaluation of other models is recommended, but with correlation to MSA models and much better microstructural characterization for all tests.

Important changes in mechanisms of thermal conductivity with increasing temperature can occur due to gas effects in bodies with intergranular porosity, microcracks, or both, and in more porous bodies due to increasing radiation effects at higher temperatures. Changing electrical conductivity, fracture mode, and electrical breakdown can also be important.

REFERENCES

1. O. L. Anderson, "Derivation of Wachtman's Equation for the Temperature Dependence of Elastic Moduli of Oxide Compounds," Phy. Rev., **144**(2), pp. 553–57, 1966.

2. J. B. Wachtman, Jr., W. E. Teft, D. J. Lam, Jr., and C. S. Apstein, "Exponential Temperature Dedendence of Young's modules for Several Oxides," Phy. Rev., **122**, pp. 1754–59, 1961.

3. M. L. Nandanpawar and S. Rajagopalan, "Wachtman's Equation and Temperature Dependence of Bulk Moduli in Solids," J. Appl. Phy., **49**(7), pp. 3976–79, 1978.

4. R. W. Rice, "Ceramic Fracture Mode-Intergranular vs. Transgranular Fracture," Fractography of Glasses and Ceramics III (J. R. Varner and G. Quinn, Eds,), American Ceramics Society, Westerville, OH, pp. 1–53, 1996.

5. A. W. Ruff and S. M. Wiederhorn, "Erosion of Solid Particle Impact," National Bureau of Standards Report, NSBIR 78–1575, 1/ 1979.

6. A. W. Ruff and L. K. Ives, "Measurement of Solid Particle Velocity in Erosive Wear" Wear, **35**, pp. 195–99, 1975.

7. B. A. Boley and J. H. Weiner, "Theory of Thermal Stresses," John Wiley & Sons, Inc., New York, 1960.

8. S. Manson and R. W. Smith, "Theory of Thermal Shock Resistance of Brittle Materials Based on Weibull's Statistical Theory of Strength," J. Am. Cer. Soc., **38**(1), pp. 18–27, 1955.

9. D. P. H. Hasselman, "Unified Theory of Thermal Shock Fracture Initiation and Crack Propagation in Brittle Ceramics," J. Am. Cer. Soc., **52**(11), pp. 600–04, 1969.

10. D. P. H. Hasselman, "Thermal Stress Resistance Parameters for Brittle Refractory Ceramics: A Compendium," Am. Cer. Soc. Bul., **49**(12), pp. 1033–37, 1970.

11. D. P. H. Hasselman, "Figures-of-Merit for the Thermal Stress Resistance of High-Temperature Brittle Materials: A Review," Ceramurgia Intl., **4**(4), pp. 147–50, 1978.

12. D. P. H. Hasselman, "Thermal Stress Crack Stability and Propogation in Severe Thermal Environments," Ceramics in Severe Environments, (W. W. Kriegel and H. Palmour III, Eds,), Plenum Press, New York, pp. 89–103, 1971. Materials Science Research, vol. 5.

13. D. P. H. Hasselman, "Analog Between Maximum-Tensile-Stress and Fracture-Mechanical Thermal Stress Resistance Parameters for Brittle Refractory Ceramics," J. Am. Cer. Soc., **54**(4), p. 219, 1971.

14. T. Oztener, K. Satamurthy, C. E. Knight, J. P. Singh, and D. P. H. Hasselman, "Effect of T- and Spatially Varying Heat Transfer Coefficient on Thermal Stress Resistance of Brittle Ceramics Measured by the Quenching Method," J. Am. Cer. Soc., **66**(1), pp. 53–58, 1983.

15. H. Henecke, J. R. Thomas, Jr., and D. P. H. Hasselman, "Role of Material Properties in the Thermal-Stress Fracture of Brittle Ceramics Subject to Conductive Heat Transfer," J. Am. Cer. Soc., **67**(6), pp. 393–98, 1984.

16. P. F. Becher, "Effect of Water Bath Temperature on the Thermal Shock of Al_2O_3," J. Am. Cer. Soc., **64**(3), pp. C-17–18, 1981.

17. M. Arnold, A. R. Boccaccini, and G. Ondracek, "Theoretical and Experimental Considerations on the Thermal Shock Resistance of Sintered Glasses and Ceramics using Modelled Microstructure-Property Correlations," J. Mat. Sci., **31**, pp. 463–69, 1996.

18. A. G. Evans and E. A. Charles, "Structural Integrity in Severe Thermal Environments," J. Am. Cer. Soc., **60**(1–2), pp. 22–26, 1977.

19. D. P. H. Hasselman, "Analysis of the Strain at Fracture of Brittle Solids with High Densities of Microcracks," J. Am. Cer. Soc., **52**(8), pp. 458–59, 1969.
20. D. P. H. Hasselman and J. P. Singh, "Analysis of Thermal Stress Resistance of Microcracked Brittle Ceramics," Am. Cer. Soc. Bul., **58**(9), pp. 856–60, 1979.
21. O. L. Bowie, "Analysis of an Infinite Plate Containing Radial Cracks Originating at the Boundary of an Internal Circular Hole," J. Math. Phy., **35**(1), pp. 60–71, 1956.
22. D. P. H. Hasselman, "Role of Physical Properties in Post-Thermal Buckling Resistance of Brittle Ceramics," J. Am. Cer. Soc., **61**(3–4), p. 178, 1977.
23. J. P. Singh, K. Niihara, and D. P. H. Hasselman, "Analysis of Thermal Fatigue Behavior of Brittle Structural Materials," J. Mat. Sci., **16**, pp. 2789–97, 1981.
24. D. P. H. Hasselman and G. E. Youngblood, "Enhanced Thermal Stress Resistance of Structural Ceramics with Thermal Conductivity Gradient," J. Am. Cer. Soc., **61**(1–2), pp. 49–52, 1978.
25. K. Satyamurthy, J. P. Singh, M. P. Kamat, and D. P. H. Hasselman, "Thermal Stress Analysis of Brittle Ceramics with Density Gradients Under Conditions of Transient Convective Heat Transfer," Trans. & J. Brit. Cer. Soc., **79**(1), pp. 10–14, 1980.
26. L. J. Gibson and M. F. Ashby, "Cellular Solids, Structure & Properties," Pergamon Press, New York(1988).
27. J. B. Wachtman, Jr. and D. J. Lam, Jr., "Young's modules of Various Refractory Materials as a Function of Temperature," J. Am. Cer. Soc., **42**(5), pp. 254–60, 1959.
28. M. O. Marlowe and D. R. Wilder, "Elasticity and Internal Friction of Polycrystalline Yttrium Oxide," J. Am. Cer. Soc., **48**(5), pp. 227–233 (1965).
29. O. Hunter, Jr., H. J. Korklan, and R. R. Suchomel, "Elastic Properties of Polycrystalline Monoclinic Sm_2O_3," J. Am. Cer. Soc., **57**(6), pp. 267–68 (1974).
30. S. L. Dole, O. Hunter, Jr., and F. W. Calderwood, "Elastic Properties of Polycrystalline Scandium and Thulium Sesquioxides," J. Am. Cer. Soc., **60**(3–4), pp. 167–68 (1977).
31. J. A. Haglund and O. Hunter, Jr., "Elastic Properties of Polycrystalline Monoclinic Gd_2O_3," J. Am. Cer. Soc., **56**(6), pp. 327–330, 1973.
32. R. L. Coble and W. D. Kingery, "Effect of Porosity on Physical Properties of Sintered Alumina," J. Am. Cer. Soc., **39**(11), pp. 377–385, 1956.
33. H. Neuber and A. Wimmer, "Experimental Investigations of the Behavior of Brittle Materials at Various Ranges of Temperature, Technical Report AFML-TR-68–23, Air Force Materials Laboratory, Wright-Patterson Air Force Base, OH (March 1968).
34. B. A. Chandler, E. C. Duderstadt and J. F. White, "Fabrication and Properties of Extruded and Sintered BeO," J. Nuc. Mat., **8**(3), pp. 329–347, (1963).
35. R. E. Fryxell and B. A. Chandler, "Creep, Strength, Expansion, and Elastic Moduli of Sintered BeO as a Function of Grain Size, Porosity, and Grain Orientation," J. Am. Cer. Soc., **47**(6), pp. 283–91 (1964).
36. S. Spinner, F. P. Knudsen, and L. Stone, "Elastic Constant-Porosity Relations for Polycrystalline Thoria," J. Res. Natl. Bu. Stds., **67C**(1), pp. 39–46 (1–3/1963).
37. B. T. Kelly, "Physics of Graphite," Applied Science Pub, London, 1981.
38. M. O. Marlowe, "Elastic Properties of Three Grades of Fine Grained Graphite to 2000°C," General Electric Nuceonics Lab. Report for NASA, CR-66933 for contract NAS1–9852, N70–42099, 6/25, 1970.
39. S. Sato, H. Awaji, and H. Akuzawa, "Fracture Toughness of Reactor Graphite at High Temperatures," Carbon, **16**, pp. 95–102, 1978.

40. J. E. Brocklehurst, "Fracture in Polycrystalline Graphite," Chemistry and Physics of Carbon, vol. 13 (P. L. Walker, Jr. and P. A. Thrower, Eds.), Marcel Dekker, Inc., pp. 145–279, 1977.

41. W. C. Riley, "Graphite," High Temperature Materials and Technology (I. E. Campbell and E. M. Sherwood, Eds.), John Wiley & Sons, Inc. New York, pp. 188–234, 1967.

42. O. D. Slagle, "Thermal Expansion Hysteresis in Polycrystalline Graphite," Carbon, **7**, pp. 337–44, 1969.

43. D. P. H. Hasselman, "Voight and Reuss Moduli of Polycrystalline Solids with Microcracks," J. Cer. Soc., **53**(3), p. 170, 1970.

44. W. R. Manning, "Anomalous Elastic Behavior of Polycrystalline Nb_2O_5," Ph.D. Thesis, Ames Laboratory, USAEC, Iowa State University (June 1972), Ames, IA.

45. E. H. Lutz, N. Claussen, and M. V. Swain, "K^R-Curve Behavior of Duplex Ceramics," J. Am. Cer. Soc., **74**(1), pp. 11–18, 1991.

46. Y. Ohya, Z. E. Nakagawa, and K. Hamano, "Crack Healing and Bending Strength of Aluminum Titanate Ceramics at High Temperatures," J. Am. Cer. Soc., **71**(5), pp. C-232–33, 1988.

47. S. L. Dole, O. Hunter, Jr., F. W. Crawford, and D. J. Bray, "Microcracking of Monoclinic HfO_2," J. Am. Cer. Soc., **61**(11–12), pp. 486–90, 1978.

48. O. Hunter, Jr., R. W. Scheidecker, and S. Tojo, "Characterization of Metastable Tetragonal Hafnia," Ceramurgica Intl., **5**(4), pp. 137–, 1979.

49. R. W. Davidge, "Grain Boundary Cracking in Alumina," Proc. Brit. Cer. Soc., **32**, pp. 199–203, 3/1982.

50. S. W. Freiman and J. J. Mecholsky, Jr., "Effect of Temperature and Environment on Crack Propogation in Graphite," J. Mat. Sci., **13**, pp. 1249–60, 1978.

51. T. Oku, S. Ishiyama, M. Eto, Y. Goto, K. Urashima, and M. Inagaki, "Effects of Notch Sharpness and Specimen Size on Fracture Toughness of Nuclear Graphite," J. Cer. Soc. Jpn., Inter. Ed., **96**, pp. 743–57, 1988.

52. K. Tamagawa and H. Takahashi, "Acoustic Emission Characteristics and Fracture Toughness of Artificial Graphite Electrodes," J. Eur. Cer. Soc., **7**, pp. 49–54, 1991.

53. M. Sakai, J. I. Yoshimura, Y. Goto, and M. Inagaki, "R-Curve Behavior of a Polycrystalline Graphite: Microcracking and Grain Bridging in the Wake Region," J. Am. Cer. Soc., **71**(8), pp. 609–16, 1988.

54. R. Baddi, J. Crampon, and R. Duclos, "Temperature and Porosity Effects on the Fracture of Fine Grain Size MgO," J. Mat. Sci., **21**, pp. 1145–50, 1986.

55. K. C. Goretta, R. Brezny, C. Q. Dam, D. J. Green, A. R. Arellano-Lopez, and A. D. Dominguez-Rodriguez, "High Temperature Mechanical Behavior of Porous Open-Cell Al_2O_3," Mat. Sci. & Eng., **A124**, pp. 151–58, 1990.

56. E. M. Passmore, R. M. Spriggs, and T. Vasilos, "Strength-Grain Size-Porosity Relations in Alumina," J. Am. Cer. Soc., **48**(1), pp. 1–7, 1965.

57. K. P. Knudsen, "Dependence of Mechanical Strength of Brittle Polycrystalline Specimens on Porosity and Grain Size," J. Am. Cer. Soc., **42**(8), pp. 376–88. 1959.

58. F. P. Knudsen, H. S. Parker, and M. D. Burdick, "Flexural Strength of Specimens Prepared from Several Uranium Dioxide; Its Dependence on Porosity and Grain Size and the Influence of Additions of Titania," J. Am. Cer. Soc., **43**(12), pp. 641–47, 1960.

59. M. D. Burdick and H. S. Parker, "Effect of Particle Size on Bulk Density and Strength Properties of Uranium Dioxide Specimens," J. Am. Cer. Soc., **39**(5), pp. 181–87, 1956.

60. E. V. Clougherty. D. Kalish, and E. T. Peters, "Research and Development of Refractory Oxidation Resistant Diborides," Man Labs, Inc. Tech Report AFML-TR-68–190 for Air Force contract No. AF33(615)-3671, July 1968.

61. S. G. Seshadri, M. Srinivasan and K. Y. Chia, "Microstructure and Mechanical Properties of Pressureless Sintered Alpha-SiC," Silicon Carbide (J. D. Cawley & C. E. Semler, Eds.), Am. Cer. Soc., Westerville, OH, pp. 215–25 (1989). Ceramic Transactions, vol. 2.

62. S. A. Newcomb, "Temperature and Time Dependence of the Strength of C-Axis Sapphire from 800–1500°C," M.S. Thesis, Pennylvania State University, University Park, PA, May 1992.

63. S. A. Newcomb and R. E. Tressler, "Slow Crack Growth of Sapphire at 800 to 1500°C," J. Am. Cer. Soc., **76**(10), pp. 2505–12, 1993.

64. S. A. Newcomb and R. E. Tressler, "High Temperature Fracture Toughness of Sapphire," J. Am. Cer. Soc., **77**(11), pp. 3030–32, 1994.

65. R. W. Rice, "Corroboration and Extension of Analysis of C-Axis Sapphire Filament Fractures from Pores," J. Mat. Sci. Let., **16**, pp. 202–6, 1997.

66. R. W. Rice and P. F. Becher, "Comment on 'Creep Deformation of 0 Sapphire,' - J. Am. Cer. Soc., **60**(3–4), pp. 186–88, 1977.

67. R. W. Rice, "Pores as Fracture Origins in Ceramics," J. Mat. Sci., **19**, pp. 895–914, 1984.

68. A. G. Atkins and D. Tabor, "Hardness and Deformation Properties of Solids at Very High temperatures," Proc. Royal Soc., **292**, pp. 441–59, 1966.

69. R. D. Koester and D. P. Moak, "Hot Hardness of Selected Borides, Oxides and Carbides to 1900°C," J. Am. Cer. Soc., **50**(6), pp. 290–96, 1967.

70. R. W. Rice, "Microstructure Dependence of Mechanical Behavior, "Treatise on Materials Science and Technology, **II**, Properties and Microstructure (R. K. McCrone, Ed.), Academic Press, Inc., pp. 199–381, 1977.

71. T. Arato, K. Nakamura, and M. Sobue, "Characteristics of Thermal Shock Durability of Porous Al_2O_3 Ceramics," J. Cer. Soc. Jap., Intl. Ed., **97**, pp. 790–97, 1989.

72. R. D. Smith, H. A. Anderson, and R. E. More, "Influence of Induced Porosity on the Thermal Shock Characteristics of Al_2O_3," Am. Cer. Soc. Bul., **55**(11), pp. 979–82, 1976.

73. K. Koumoto, H. Shimizu, W. S. Seo, C. H. Pai, and H. Yanaqgida, "Thermal Shock Resistance of Porous SiC Ceramics," Trans. & J. Brit. Cer. Soc., **90**(1), pp. 32–33, 1991.

74. D. C. Larsen, J. W. Adams, L. R. Johnson, A. P. S. Teotia, and L. G. Hill, "Ceramic Materials for Advanced Heat Engines: Technical and Economic Evaluation," Noyes Publications, Park Ridge, NJ, 1985.

75. P. Boch, J. C. Glandus, J. Jarrige, J. P. Lecompte, and J. Mexmain, "Sintering, Oxidation and Mechanical Properties of Hot Pressed Aluminum Nitride," Ceramics Intl., **6**(1), pp. 34–40, 1982.

76. V. G. Antsifirov, V. G. Gilev, A. G. Lanin, V. P. Popov, and A. L. Tkatchyov, "Thermal Stress Resistance of a Porous Silicon Nitride," Ceramics Intl., **17**, pp. 181–85, 1991.

77. R. L. Coble and W. D. Kingery, "Effect of Porosity on Thermal Stress Fracture," J. Am. Cer. Soc., **38**(1), pp. 33–37, 1955.
78. M. F. Gruninger, J. B. Wachtman, Jr., and R. A. Haber, "Thermal Shock Protection of Dense Alumina Substrates by Porous Alumina Sol-Gel Coatings," Cer. Eng. Sci. Proc., **8**(7–8), pp. 596–601, 1987.
79. R. M. Orenstein and D. J. Green, "Thermal Shock Behavior of Open-Cell Ceramic Foams," J. Am. Cer. Soc., **75**(7), pp. 1899–1905, 1992.
80. E. H. Lutz, M. V. Swain, and N. Claussen, "Thermal Shock Behavior of Duplex Ceramics," J. Am. Cer. Soc., **74**(1), pp. 19–24, 1991.
81. E. A. Belskaya and A. S. Tarabanov, "Experimental Studies Concerning the Electrical Conductivity of High-Porosity Carbon-Graphitic Materials," Institute of High Temperatures, Academy of Sciences of the USSR, Tr. from Inzhenerno-Fizicheskii Zhurnal, **20**(4), pp. 654–59, 1971.
82. H. J. Siebeneck, J. J. Cleveland, D. P. H. Hasselman, and R. C. Bradt, "Thermal Diffusivity of Microcracked Ceramic Materials," Ceramic Microstructure, 76 With Emphasis on Energy Applications (R. M. Fulrath and J. A. Pask, Eds.), Westview Press, Bolder, CO., pp. 743–62, 1977.
83. W. R. Manning, G. E. Youngblood, and D. P. H. Hasselman, "Effect of Microcracking on the Thermal Diffusivity of Polycrystalline Aluminum Niobate," J. Am. Cer. Soc., **60**(91–10), pp. 469–70, 1977.
84. R. Taylor, K. E. Gilchrist, and L. J. Poston, "Thermal Conductivity of Polycrystalline Graphite," Carbon, **6**, pp. 537–44, 1968.
85. A. Goldsmith, T. E. Waterman, and H. J. Hirschorn, "Handbook of Thermophysical Properties of Solid Materials, vol. 1: Elements, Revised ed.," Macmillan Co., New York, 1961.
86. Y. S. Touloukian, Ed., "Thermophysical Properties of High Temperature Solid Materials, vol. 1: Elements," Macmillan Co., New York, 1967.
87. Y. S. Touloukian, R. W. Powell, C. Y. Ho, and P. G. Klemens, "Thermophysical Properties of Matter, vol. 2: Nonmetallic Solids," IFI/Plenum, New York-Washington, 1970.
88. E. Litovsky and M. Shapiro, "Gas Pressure and Temperature Dependences of Thermal Conductivity of Porous Ceramic Materials: Part 1, Refractories and Ceramics with Porosity Below 30%," J. Am. Cer. Soc., **75**, pp. 3425–29, 1992.
89. E. Litovsky, M. Shapiro, and A. Shavit, "Gas Pressure and Temperature Dependences of Thermal Conductivity of Porous Ceramic Materials: Part 2, Refractories and Ceramics with Porosity Exceeding 30%," J. Am. Cer. Soc., **79**, pp. 1366–76, 1996.
90. D. S. Rutman, A. F. Maurin, G. A. Taksis, and Y. S. Toropov, "Relationship Between Electric Resistance of Zirconia Ceramics and Porosity," Refractories, **5–6**, pp. 371–74, 1970.
91. R. W. Rice, "Possible Effects of Elastic Anisotropy on Mechanical Properties of Ceramics," J. Mat. Sci. Let., **13**, pp. 1261–6, 1994.
92. E. A. Durbin and C. C. Harman, "An Appraisal of the Sintering Behavior and Thermal Expansion of Some Columbate," Battelle Memorial Institute Report BMI-791, 1952.

10

SUMMARY OF POROSITY AND MICROCRACKING EFFECTS, APPLICATIONS, SPECIAL FABRICATION, AND ENGINEERING

KEY CHAPTER GOALS

1. Summarize porosity and microcrack dependences of properties and the need for further improvements.
2. Outline applications for designed porosity.
3. Summarize novel processing methods for obtaining a variety of designed porous bodies, including membranes.

10.1 INTRODUCTION

This chapter has four purposes. First is to summarize the porosity dependence of mechanical and other properties, addressing both data and models, and noting the status and needs of these. The more limited observations on the effects of microcracking on properties and its interaction with other microstructural variables, especially porosity are also summarized. The second purpose is to outline the applications of porous materials, especially ceramics, with particular attention to newer and emerging applications. The third purpose is to outline fabrication/processing techniques to obtain the generally higher level and character of porosity needed for many applications. These include both derivatives of normal processing and novel techniques, both of which are large, specialized, subjects in themselves. The fourth purpose is to outline some of the engineering for porous materials, especially property tradeoffs, and the implications of porosity effects on failure for nondestructive evaluation (NDE).

Before proceeding to the above topics it is appropriate to again consider the hierarchy of the porosity dependence of properties. Thus, as summarized in Table 2.1, the three categories of porosity dependence of properties are those that are: I) not dependent on porosity, II) dependent on only the amount of porosity, and III) dependent on not only the amount of porosity but also on one or more characteristics of the porosity. The latter category in turn can be divided into three subcategories, namely properties determined: A) primarily or exclusively by transmission of stress or flux through the solid phase, B) primarily or exclusively by flux through the pore phase, and C) by combinations of these. Examples of properties in these three categories include respectively: I) theoretical density and thermal expansion, II) density, heat capacity per unit volume, and some dielectric constant behavior, IIIA) mechanical properties and much electrical and thermal conductivity, IIIB) fluid flow, and IIIC) some thermal conductivity (mainly at high temperatures) and some electrical, magnetic and related properties. Some properties may be shifted from one category to another, e.g., dielectric constant and thermal conductivity depending on material, porosity (or other microstructural) and environmental conditions.

There is also a hierarchy for the microcrack dependences of properties with categories parallel to those for porosity with many similarities, as well as some important differences between the two hierarchies. Thus, there are the same basic categories, i.e., I) theoretical density, II) actual density, thermal expansion, and IIIA) mechanical and conductive properties; however, Category IIIB (and hence also IIIC) behavior is far more limited for microcracks than pores because of the typically much smaller volume and little or no interconnection of microcracks. Also, while many properties fall in the same categories for both pores and microcracks, some do not, e.g., thermal expansion.

This book has mainly addressed Category IIIA porosity dependence, with much less discussion of the other categories. Category IIIB dependence is a large and complex subject in itself, and although a major factor in many of the applications discussed later in this chapter, could only be cursorily treated in this book (Sections 7.2.5, 7.2.6, 7.3.5); however, understanding Category IIIA behavior is important to understanding Category IIIB and IIIC behavior since they also depend on IIIA behavior. Considerable attention is also given to microcrack dependence for Category III behavior.

Briefly, consider the porosity parameters necessary for characterizing the porosity dependence of properties. Obviously none are needed for Category I properties and Category II properties only need the volume fraction porosity, P. For Category IIIA properties both P and at least a second parameter are needed, with both pore shape and the minimum solid area (MSA, Figs. 2.16, 2.17) being candidates; however, pore shape is hard to define in even many ideal pore systems, let alone in real ones, and even harder to measure. Furthermore, it may reflect some extraneous information. On the other hand, MSA is very definable and measurable for ideal pores, and probably for many, if not all, real pore systems. It is the average solid contiguity normal to the reference direction, i.e., of the overall stress or flux. For some properties pore size and spatial distribution of larger pores are also important. This is particularly true for tensile strength and electrical breakdown, but can also be important for some other, e.g., magnetic, properties. Furthermore, the heterogeneity of the general porosity, not just larger pores, is a serious issue for all Category III properties since these usually entail nonlinear porosity effects, which typically increase in their severity as P increases. Thus, areas of higher than average porosity out weigh the effects of corresponding areas of less porosity, increasing the porosity dependence; however, even for properties that have a linear dependence on P, heterogeneity of the porosity can result in problems, commonly due to the area of measurement having a different level of porosity than the overall sample itself, e.g., due to porosity gradients from die-wall friction in pressing test disks. Thus, measurements of, for example, dielectric constant, electrical breakdown, and thermal conductivity, are often purposely made to exclude the peripheral area in order to avoid edge effects. They are therefor susceptible to possible porosity gradient effects which often change with processing parameters to achieve different levels of porosity. Issues and recommendations for characterizing porosity were discussed in Section 2.3 and Table 2.3, and briefly summarized in Section 10.5.2.

Microcrack characterization parameters parallel those for porosity, i.e., only microcrack content needed for Category II behavior, and both amount and character, i.e., size, shape, orientation, and stacking, needed for Category III behavior. Heterogeneity of microcrack size, shape, orientation and stacking (and hence spatial distribution) is also expected to be important, but has not been addressed. Microcracks are inherently anisotropic in shape so their preferred orientation can be very important. This is particularly important and more common in highly anisotropic materials with frequent associated anisotropy of grain shape that results in oriented microcracks, e.g., graphite and BN.

10.2 POROSITY AND MICROCRACKING EFFECTS IN CERAMICS AND RELATED MATERIALS

10.2.1 Wave Velocities and Elastic Properties at Low to Moderate Temperature

First, the porosity dependence of wave velocities and elastic properties at lower temperatures is summarized along with observations on the effects of microcracking as in earlier chapters. Then the porosity dependence of other mechanical and nonmechanical properties at such temperatures are summarized, followed by consideration of all properties at higher temperatures. Elastic properties and wave velocities which yield the elastic properties are again considered first since they are important in themselves, as well as for their relevance to most other mechanical properties as discussed in earlier chapters. In principle, they also are generally, most amenable to more rigorous mechanics analysis, but, as discussed in earlier chapters and further here, this has often been somewhat of a dilemma.

Wave velocities and elastic moduli all continuously decrease with increasing volume fraction porosity, P, often nearly linearly over the lower portion of the pertinent P ranges, especially for wave velocities since they inherently have lower P-dependence. Pertinent P values and the extent of linerarity vary with the microstructure and the property. Linear elastic moduli decreases are inherent over most or all of the range to P = 1 for stress parallel with aligned tubular pores; however, linearity becomes only an approximation of increasingly restricted applicability as pore orientation (e.g., for cylindrical pores oriented normal to the stress direction) and especially shape changes, e.g., to spherical and angular pores, and especially pores between packed particles, i.e., the latter to P ≤ 0.1–0.2. Thus, while the occurrence of linerarity has encouraged use of linear plots, extrapolations using them are often tenuous and can be in serious error to higher porosities. The limitations arise since most porosities lead to increasing nonlinearity, then to precipitous decreases of all properties to zero at critical P, i.e., P_C, values that vary from ~ 1 down to ~ 0.2, roughly in the order of porosities listed above (Figs. 2.11, 3.1; Tables 2.5, 3.6). Wave velocities intrinsically have less P-dependence, e.g., having exponents of ½ (b or n)–1 where b or n are the exponents of the P dependences of the elastic moduli over the approximately first half of their P ranges (Section 3.2.2).

Both the substantial property decreases and especially the above nonlinear decreases lead to use of log-log and two types of semilog plots, each with their advantages and disadvantages. Log-log plots, which have been fairly widely used, give the greatest extent of linear plots, but this may be only an apparent linerarity. Thus, important shifts in the porosity dependence may yield only modest deviations on log-log plots that may be missed, especially if the shifts are only partly covered by the P range of measurement. Log-log plots seriously obscure P_C values and are often not efficient for displaying a range of data. There are indications of correlations of b and n values (respectively slopes on semilog and log-log plots) at low and moderate P levels based on correlations of b values and pore character that indicate extension to n values; however, evaluation of log-log plots over larger P ranges has not demonstrated a clear systematic variation with pore character (Fig. 2.14B).

Semilog plots using linear property and log-P scales have been used in some case. The plots appear to be effective in indicating extrapolated P_C values, but are inefficient in displaying large changes in properties, i.e., over a large P range. The main and extensive use of semilog plots is via a log scale of properties versus a linear P scale. This is an efficient and effective way to display a wide range of data, although at high fractions of P_C (e.g., 85–100%) where property decreases are greatest expanded scales are quite useful. Such semilog plots are particularly effective in showing: 1) major aspects of changes in P-dependence for a given porosity (including P_C values), and 2) major differences in the P-dependence for different pore characteristics and related MSA models. These allow approximations for combinations of different porosities or changes in a given pore character, as discussed earlier and below. The typical character of such a semilog plot is initially an ~ linear decrease (at lower porosities relative to the limit of the porosity achievable in a body of a particular microstructural character), in the absolute or relative modulus (i.e., the value at some P divided by the value at P = 0). Much of the data does not extend to or beyond the P range (i.e., P = 0.10–0.5) where the semilog linerarity is typical. There is, however, sufficient data to show that the linear decrease of the moduli transitions to a progressively greater rate of property decrease, i.e., rollover, with increasing P to ultimately fall precipitously to zero moduli values at varying specific porosities designated as P_C. This transition to progressively greater decreases in moduli values toward their going to zero is required by basic geometrical considerations, namely that any mix of pores and solid must reach a percolation limit of the solid. At this limit the body reaches the point where it is just barely a coherent solid, i.e., where there is still just barely contiguity of the solid phase so the body can sustain a tensile stress (or, if an electrical conductor, sustain an electrical current in the absence of some conduction across the pores). Any further increase in porosity of bodies at this point eliminates this coherence. This limiting, i.e., critical, volume fraction of porosity, P_C, varies from ~ 0.2 to 1, depending on both pore shape and stacking, and can also be dependent on the nature of the stress and specimen to microstructure scales as discussed earlier.

There are two key factors to note about this elastic moduli versus porosity relations. First is the very important but widely neglected fact that the specifics of the above characteristic relations depend on the character of the porosity. The character that impacts the moduli versus P relations is the pore geometry, which includes not only key aspects of the shape but also the stacking of the pores and their orientation relative to each other (and hence to some reference axis). As discussed earlier and further below, the key measure of these geometrical characteristics is the minimum solid area (MSA), which is the mean solid contiguity normal to the reference axis (typically the stress of flux direction). Furthermore, while there is at best limited characterization of the porosity, there are typical, approximate correlations between fabrication/processing methods and parameters and pore character that provide some general guidance. Thus, extrusion commonly produces some aligned tubular pore character, the extent depending on extrusion parameters such as pressure and body parameters such as particle packing and shape, and higher binder content. Similarly, die pressing at higher

pressures and with more die-wall friction can result in laminar pores and especially laminar arrays of pores. Otherwise, higher die (and hot-pressing) pressures increases particle packing density (i.e., C_n), hence changing pore shape
and stacking as well as reducing P_C values. In most processes, increasing the
platelet character of the particulate material increases opportunities for aligned
platelet-type pores.

The second key factor is that there is a basic interrelationship between the
three resulting characteristics of the above trends, namely the initial approximately linear semilog slope (b value), the rollover from the linerarity toward
zero moduli at P_C, and P_C itself. Of these the slope is the most definitive since it
is unique to a single pore character. P_C is the next but less definitive characteristic since it is not unique to a single given pore character, and the rollover still
less so since it occurs over varying ranges rather than at a fixed value. The
relation of these three characteristics is very logical, namely that the lower the
linear slope, the higher the P range for both the onset and the occurrence of the
rollover toward zero properties at P_C, and the greater the P_C value. Thus, tubular
or platelet pores aligned parallel with the stress or flux have the lowest b values
(~ 1–2), high, long rollovers (P ~ 0.6–0.8), and a high P_C ~1 (Tables 2.5, 3.2,
3.3, Figs. 2.11 and 3.1). Such pores stressed normal to their aligned axes as well
as spherical pores (from foaming, bubbles, or burnout of spherical particles) or
various cubic pores (reflecting burnout of more angular particulates and some
intragranular pores) give slopes, b values, of b ~ 2–4, e.g., depending on
orientation and stacking, rollover ranges of P ~ 0.6–0.75 and ~ 0.5–0.9,
respectively, and P_C values of ~ 0.8–1. For pores between packed uniform
particles, b values range from ~ 5–9 (random packing and simple cubic packing
being ~ 5, and progressively higher values for progressively denser packing)
with respective rollover ranges of P~ 0.35–0.45 and 0.2–0.25, and P_C values of
~ 0.45 to 0.25, respectively.

The similarity of shape and systematic changes in the above semilog plots of
absolute or relative moduli versus P provides an important advantage over other
plots, namely interpolation between curves. This commonly allows a reasonable
estimate for the trends for bodies of mixed or changing porosity. Handling such
porosities is important since many processes yield varying mixes of pores (e.g.,
of pores between packed particles and tubular or laminar pores for extrusion and
die pressing, as noted above). Furthermore, pore character changes, sometimes
substantially, with densification (Figs. 1.2–1.4), e.g., shifting of some pores
from inter- to intragranular locations (with accompanying shape changes) at
lower porosity with grain growth (Figs. 1.2–1.4). Usually a simple, linear rule of
mixtures, interpolation is a good approximation, but more work is needed to
refine this since pores may act in series or parallel, the former deviating from
linear interpolation. The extreme of such deviations are hierarchial pore structures, which can be particularly advantageous for good property/mass ratios (1).

The similar and systematically changing shapes of the moduli versus P relations on semilog plots also have the advantage that they very effectively use the whole graph area in comparison to generally less effective use of such area for linear or log-log plots.

While stacking of the particles that define the pores, or stacking of the pores themselves, is important, this has been modeled only for pores between uniform spherical particles and aligned cylindrical pores. More analysis beyond simple cubic stacking for other pores is needed. Mixed stackings can be reasonably addressed by combining models for the different stackings involved; however, modeling based on the coordination number, C_n, of stacked particles is a promising alternative or complementary approach that deserves further attention, especially if it can be extended to other pores, e.g., spherical and cylindrical ones.

Three summary observations applicable not only to wave velocities and elastic properties but also to all mechanical (and some other) properties should be noted. First, pore character is important. While some different pore types give similar behavior for at least some and often many properties, other pore types result in a diversity of behavior. Thus, cylindrical pores normal to the flux or stress axis and spherical pores give similar property dependences, which are overlapped by the range of property dependences for cubically shaped pores of differing orientations. All of these fall in about the middle of the range of porosity dependences, and thus form a most common or frequent range, i.e., a "main stream," of porosity dependence; however, property dependences of cylindrical pores parallel with the stress or flux axis are definitely lower than for the main stream (again and more clearly illustrating the importance of pore orientation), and the very important family of porosities between packed particles are all higher than the main stream (i.e., lower and higher b values respectively). Furthermore, the important variable of pore stacking further broadens the range for each type of porosity and thus also for the whole range of porosity. Thus, while the fallacy of neglecting pore character will not be obvious in some cases, it is critical for any comprehensive understanding of porosity effects.

The second summary observation is that it is necessary to understand effects of mixes of different porosities in a given body. This is essential since most fabrication methods produce varying mixes of pore types, with the mix changing with both the porosity level and processing parameters, and grain growth. Such methods also often result in varying spatial distributions of the pores, including heterogeneities, that must also be addressed. While such pore mixtures increase the occurrence of pore behavior in the main stream, it also adds to the breadth of that main stream and the frequency of property dependence beyond the main stream. Thus, handling these factors requires handling pore mixtures, which can only be effectively done by understanding the behavior of the constituent porosities. This and the impact of the character of different pore types make it essentially impossible for a single model to cover much, let alone all, of the porosity range, i.e., of both type and amount of porosity. Thus, universal pore-property models are not viable. Instead a family of rationally related models based on different basic pore types is needed, possibly one that is primarily for specified ranges of P that can be combined to address real porous bodies. The MSA models are thus far the only set of models meeting a fair portion of the requirements.

10.2.2 Elastic Property Specifics and Models

Consider now specific differences in the relations of bulk (B), Young's (E), and shear (G) moduli as well as Poisson's ratio (ν). Evaluation of self-consistency (usually totally neglected, see Section 3.4.1) shows that even for an isotropic material the porosity dependence of these (and hence other moduli) cannot be the same. Data supports this, showing the initial semilog slope, i.e., the b value, for B is typically > that for E (e.g., by ~ 5–10%), that the value for E is typically > that for G (also typically by ~5–10%), and that the b value for ν is approximately the difference in the b values for B and G. Thus, since over a range of porosities b averages ~ 4 for E, it averages about 4.5 for B, about 3.5 for G, and about 0.5–1 for ν. The latter is contradictory to claims that ν could increase or be constant as a function of P, and is consistent with the fact that such claims are at best uncertain in their validity. There are, however, theoretical uncertainties of the P-dependence of ν as P approaches 1 (Section 3.4).

Note that there have been some higher values of moduli measured in compression versus those in tension, e.g., for bodies of sintered zirconia beads (Fig. 3.9A) and of pyrolytic graphite (Fig. 9.2). There appears to be a theoretical basis for this in the specifics of stress-microstructure interactions in maintaining or enhancing stress (or flux) transmission, which also ultimately defines P_C values. As discussed earlier (Section 2.4.4) some ideal (and potentially some real) pore structures break into strands through the length of the specimen as they approach the upper porosity limits of their solid character. In such cases the strands may still transmit some tensile stress (and conductive flux) parallel with their length, but possibly not parallel compressive stresses due to buckling, depending on the stress (or flux) coupling to the body. In some of these cases, Poission's ratio effects may play a role, i.e., enhancing tensile and limiting compressive stress transmission. Thus, the P_C limits may be higher for tension and lower for compression in such cases, with corresponding directions but diminishing levels of changes occurring as P decreases from the limiting P_C values back to or into the rollover range. In other cases, however, e.g., in bodies of partially bonded beads or with microcracks, compressive stresses may maintain or enhance stress transmission and tension reduce it. Clearly hydrostatic compression increases stress transmission of bodies, e.g., rocks, with microcracks (and possibly some macrocracks) and hydrostatic tension decreases it. The situations under biaxial and uniaxial compression are more complicated since, while there may be a drive toward greater coherency parallel to the applied stress directions, Poisson's expansion in the remaining normal direction(s) may cause behavior in that direction to be similar to that in tension. Much of the resultant response depends on details of the microstructure, especially of orientation and anisotropy; however, differing stress effects can be real and are another of the many areas for further research.

Consider now modeling of the porosity dependence of elastic properties, which can often be classified as one of three approaches: 1) purely empirical, 2) based on load-bearing concepts applicable over the full, feasible P range pertinent to each of a diverse array of pore geometries, and 3) more rigorous mechanistic models of limited, isolated pores, usually of spheroidal (mainly spherical) shape. Purely empirical models have not been considered in any detail since there is currently no way to relate them to pore character (which is usually not defined) and hence to judge their validity for other than data already fit by them, i.e., for application to other porosity ranges or character (known or unknown).

While the more rigorous mechanistic models are, in principle, more desirable, their utility has been limited. The reason for this basically stems from modeling of porosity effects being a diverse and complex problem with uncertainties as to what are reasonable simplifications and approximations and which are not. These models typically assume an ideal pore shape, as do most other models, but, unlike the load-bearing models, mechanistic models are not yet amenable to addressing pore stacking and are still constrained to a limited, usually low (undefined) range of pore content and character. While such models now provide a variety of two-dimensional (tubular) pore solutions, they are still restricted to spheroidal pores for three-dimensional pores. As noted earlier, a family of models is needed to cover the range of amounts and character of porosities, so there are still critical gaps, e.g., especially of pores between packed particles of various stacking and degrees of bonding. These and other limitations also typically arise for models derived for composite materials, then applied to porous materials by setting the properties of the second phase = 0, usually without addressing the issue of how the greater difference in pore versus matrix properties has on the validity of the assumptions of the original composite model. Such composite or dilute concentration approaches typically mean that there is no indication of P_C values and related changes in the porosity dependence, including the fact that P_C values can be as low as ~ 0.25, i.e., in the range that some may consider (questionably) in the dilute concentration range. These complications and limitations are illustrated by there being a variety of (mainly earlier) models for similar or identical porosity, with many not agreeing with each other. The problem is further illustrated and compounded by these models typically being modified in a totally ad hoc fashion, often in several stages, and sometimes with questionable rational. Such modifications clearly reduce claims for such models being more rigorous, even if to a more limited range of applicability. Furthermore, these models have typically been applied in an arbitrary fashion, i.e., to any (usually unknown) pore character and to any porosity level i.e., to or beyond the limit of the ad hoc introduced P_C values.

The substantial advances in mechanistic modeling, especially of two-dimensional (tubular) pores, in recent years show excellent promise. Thus, besides a variety of tubular pores, mixes of them and the effects of specific shape aspects have been addressed. The latter include cylindrical pores being stiffer than prismatic pores, square tubular pores being transversely very anisotropic, and greater proportional reductions in elastic moduli in the order of rounded, sharp, and cupped edges of tubular pores. These results, along with spherical pores being stiffer, are all consistent with MSA models for both two- and three-dimensional pores; however, mechanistic models go beyond MSA models by probably being capable of analytically predicting the porosity dependences of Poisson's ratio and the differences in the porosity dependences of B, E, and G. Furthermore, they have provided useful solutions for microcracks as well as pore-microcrack combinations and transitions (i.e., for "inflated" microcracks). Thus, further development of mechanistic models is needed to verify, modify, or replace simpler MSA models; however, it is likely that MSA models will have continued use in aiding improvement and extension of mechanistic or combined models as well as providing overall conceptual guidance for understanding and designing porosity structures.

Originally, the geometrical load-bearing (specifically the MSA) models were also generally arbitrarily applied to any porosity; however, more recently the application of these models individually to pores of the same or similar character as specified in the original model is at least approximately valid, including their inherent identification of P_C values (2,3). Furthermore, the collection of these models provides a fairly comprehensive covering of real porosity character as well as a reasonable approximation for combinations and changes in porosity, as discussed earlier. They also have the advantages of being applicable to several other important mechanical and nonmechanical properties, being simple to apply, and being consistent with extensive past empirical characterization of porosity effects via the exponential approximation to the initial approximately semilog linearity. Furthermore, these models are applicable over the complete range of porosity valid for the particular pore geometry, thus giving guidance in the changes in properties with major changes in pore character, e.g., from open to closed porosity, and in transitioning from more specialized models, e.g., for foams (which have also thus far been basically MSA-type models). MSA approaches have not been successful for microcracks. Thus, MSA models are the only ones currently having some real capability for an important task for models, namely true prediction of properties given only

Table 10.1 Comparison of Models and Experimental Porosity Dependences for Foam Materials[a]

Porous material	Young's modulus (E)		Shear modulus (G)		Fracture toughness (K_{IC})		Tensile strength (σ_t)		Compression strength (σ_c)	
	A	n	A	n	A	n	A	n	A	n
Bonded balloons									0.77	2
									0.24	~2[d]
Closed-cell foams	1	3			1	2				2
					0.65	2				
Open-cell foams[b]	1	1								
		1.2								
Open-cell foams[c]	1	2	0.4	2	1	1.5	1	1.5	0.65	1.5
	0.3		0.14	2	0.65	1.5			0.3	2.2

[a] Comparison of predicted (top) and experimentally (bottom) values of A and n in the expression $A(1 - P)^n$.

[b] Foam structure of Ghent and Thomas (Fig. 3.3A).

[c] Foam structure of Gibson and Ashby (Fig. 3.3B).

[d] Exponent is only ~ 2 since at higher strengths a bilinear trend is found (Eq. 6.5).

the amount and character of the porosity; however, while these models are consistent with the experimentally observed differences in the porosity dependences of B, E, and G, and that of Poisson's ratio, they have not directly given such differences. Thus, they also need improvement for these and other reasons, as discussed above and earlier (Chapters 3–6) and in Section 10.5.2.

Consider briefly now data for foams and specific models for them. While the basic MSA models are applicable to foams, specialized foam models have been derived (Table 10.1). These are basically specialized MSA models, whose more limited modeling of different pore (foam cell) stackings (Figs. 3.3A,B) reinforces the important but often neglected role of pore stacking. Overall the limited evaluation to date shows the regular MSA models covering the full P range and those derived for foams giving similar agreement with data, i.e., consistent with overall trends but typically needing improvement in absolute agreement. Experimental data for elastic moduli (and other mechanical properties) show some deviations in the exponent n and typically substantial deviation of the A values from those predicted for the foam-specific models, i.e., in the $A(1 - P)^n$ dependence. Thus, A values range from 1/3 to 2/3 of those predicted, which probably reflects statistical variations in strut and wall dimensions and their quality, hence load-bearing ability, that need to be addressed. Such variations are also important to more general MSA models since the many fewer struts and walls of foams versus denser bodies are again important to properties, and also to P_C values, i.e., lowering them some from theoretical values for ideal stackings.

10.2.3 Fracture Energy/Toughness and Tensile Strength at Low to Moderate Temperature

Other mechanical properties have both theoretical and experimentally demonstrated correlations with elastic properties, especially Young's modulus, and specifically with their corresponding porosity dependences, which is inherent in MSA models. Thus all of the general comments and trends as well as many of the specific ones for the porosity dependence of elastic properties are applicable to other mechanical properties. The focus is thus on examining the modifications or shifts that occur in the porosity dependence of these other properties from that of the elastic properties. Such shifts may be intrinsic to the particular property or may be due to interactions of the property in one or both of two different fashions with the grain structure. The first is a real porosity effect due to changes in pore character with grain growth (and hence grain size) that commonly accompanies reduction in porosity. Most common and typically significant is the shifting of some pores from inter- to intragranular porosity, which is a real porosity effect that also impacts elastic properties. Some intergranular pores becoming intragranular pores reduces property decreases from the pores, hence increasing the apparent P-dependence; however, increased size (and hence grain-boundary coverage) of cusp- (e.g., lenticular-) shaped intergranular pores results in reduced property levels, hence an apparent reduction in the P-dependence. The second is effects of the grain size changes themselves on the properties of interest (which typically do not affect elastic properties). Direct effects of grain size, while not a true effect of porosity, frequently occur as a result of grain growth while reducing porosity, which often substantially effects other mechanical properties; however, pore changes with grain growth can also change the property-porosity dependence from that of the elastic properties since other mechanical properties are often more sensitive to inter- versus intragranular pore location. Other mechanical properties may deviate from the porosity dependence due to both less direct correlation with elastic properties and greater effects of other porosity parameters, as shown in the next section. Most mechanical properties are more sensitive to pore size, shape, orientation, and especially heterogeneity of porosity than elastic properties, with tensile strength being amongst the most susceptible, as discussed in earlier chapters and summarized below.

The overall trend for fracture energy, and especially for fracture toughness, versus P is similar to that of E as theoretically expected for simple linear elastic fracture; however, there are frequent and often significant deviations to both higher and lower P-dependence that arise due to grain size, porosity, and test factors singly, or in various combinations, as summarized in Table 4.4. These deviations, which have greater effect on fracture energy and less (although still often substantial) effect on fracture toughness due to the mitigating effects of the direct porosity dependence of K_{IC} on that of E, can be for more or less porosity dependence than that of E over part or much of the pertinent P range. Thus, increasing grain size as P decreases can give an apparent increase or decrease in fracture energy or toughness depending on the material, grain size range, test method, and test parameters, that is mainly due to the size of the crack relative to the scale of the key microstructural features impacting crack propagation. These effects can be compounded by effects due to changing pore location from inter- to intragranular as grain growth occurs. The first and most significant of two key effects of porosity itself ranges from a temporarily reduced decrease (or ever an increase) in fracture toughness (and especially energy) with increased P (mainly at low-intermediate P), which is attributed to crack branching, bridging, and related phenomena. These effects, which have been almost totally neglected, occur primarily at larger crack sizes, but appear to be a key factor in the significant increases in thermal shock that occur in some porous bodies. The second effect of porosity is that of porosity heterogeneity or anisotropy (especially when unrecognized), which can be interactive with any of the preceding factors as well as giving data scatter and enhancing the nonlinearity of the P-dependence of these properties. The net effect of these variations of the P-dependence is to give average dependences of fracture energy and toughness less than those for E, i.e., lower average b values, e.g., by ~ 1.

Finally, a critical issue in the use of fracture energy and toughness values in evaluating tensile strength failure is whether the crack scales in the crack propagation and strength evaluation reflect comparable microstructural interactions. This is an important issue in the impact of effects attributed to crack branching and related phenomena as noted above. This is also important in the many cases in failure initiating from pores or pore clusters. In many such cases the critical stages of failure-causing crack propagation is through denser material and not through such pores; this often substantially impacts typical fracture energy and toughness values, thus making the use of such typical values for such cases suspect. Although the theoretical versus experimental deviations of K_{IC} for foams are of uncertain statistical significance as compared to those for E and strengths (Table 10.1), they appear more extreme, as for lower porosity materials.

Turning now to the porosity dependence of tensile (hence also flexure) strength, this also shows an overall similarity to the porosity dependence of E as expected theoretically. Overall, the deviations of the tensile strength-P from E-P relations are substantially less than those for fracture energy, due in part to their also being less for fracture toughness which is directly proportional to strength; however, the porosity dependence of strength usually does not reflect the reduced porosity dependence often seen for fracture toughness, and instead can show somewhat overall increased porosity dependence. The first of two key reasons for these differences is again the larger crack scales commonly used for fracture toughness versus those typically controlling strength. Branching, bridging, and related effects of larger cracks on fracture toughness are reduced or are not present for typically smaller cracks controlling strength. The second key reason for strength differing from the porosity dependence of fracture toughness, especially for somewhat greater P-dependence of strength relative to fracture toughness and to a lesser extent relative to E, is effects of porosity heterogeneity. The elongation of flaws (e.g., from machining perpendicular to the tensile axis) is often progressively reduced with increasing porosity that is heterogeneous. Such elongation reduction has greater effect on strength since strength is determined by weak links, e.g., extremes of more porous areas on the scale of strength-controlling flaws, rather than other, e.g., elastic, effects of the heterogeneous porosity. The result in such cases is thus a reduction in the rate of decrease in strength with increasing P (Fig. 5.10); however, in the absence of such flaw elongation changes, heterogeneities of porosity increase the porosity dependence. They may be the flaws themselves or simply reduce E and K_{IC} in the immediate proximity of other failure-causing flaws. On the other hand, pores or porous areas acting as fracture origins at lower porosity (a frequent occurrence) can disproportionately reduce strengths and give an apparent reduction in the P-dependence there. The same effect results from failure associated with larger grains which also is a common occurrence at lower P; however, this effect is often different from that of larger grains on K_{IC}-P relations since, as noted earlier, larger grains can increase or decrease K_{IC}, while they do not increase strength (and generally decrease it).

An important corollary to the effect of porosity on tensile strengths is its effect on the statistics of such failure, as measured by the Weibull modulus. There is limited data to suggest that introducing a population of larger isolated pores in a body such that the pores act as the failure origins can increase the Weibull modulus, at least modestly. The demonstration of this was in otherwise relatively dense bodies, but could conceptually apply to more porous bodies where the main porosity was fine pores with isolated larger pores or pore clusters. A key question is the broader dependence of the Weibull modulus on porosity over a range of porosity levels, i.e., of P values. Limited data shows the Weibull moduli for porous bodies from sintering packed powders to various degrees decreased with increasing P, e.g., from ~10–14 to a possible limit in the range of 2–6 (Fig. 5.7); the latter range is also seen for foams. There may, however, be a possible limited increase in Weibull moduli due to larger, isolated (e.g., spherical) pores at modest porosity levels.

10.2.4 Hardness, Compressive Strength, and Related Behavior at Low to Moderate Temperature

Consider now the porosity dependence of hardness, which is commonly similar to that of related properties of compressive strength, wear, erosion, and machining since they commonly correlate with hardness. Furthermore, all of these properties entail some local tensile stress, but substantial compressive stress, which also correlates with elastic moduli, though not as explicitly as tensile properties summarized in the previous section. Thus the trends with porosity are similar and sometimes the same as for E, but also with tendencies toward greater porosity dependence, i.e., somewhat higher b values.

Indentation hardness has similar and sometimes possibly somewhat greater porosity dependence as E. Again, grain-size effects can change the apparent P-dependence of H, but this is both more limited and more complicated since increasing grain-size decreases H over most of the grain-size range, but often increases H at large grain sizes. Also, changes from inter- to intragranular pores with grain growth may reduce the actual P-dependence due to intergranular pores enhancing spalling often accompanying indentation; however, heterogeneous porosity in terms of either spatial or size distribution or both, or of uniform larger pores, on the scale of the indent can have complex effects, since they often give "bad" indents which are discarded and not even taken as a clue to heterogeneity of the porosity. In scratch and possibly also indentation hardness, crushing of porous material becomes an important factor at higher porosity. Also in scratch hardness, significant effects can occur due to enhanced penetration and plowing resulting from more porous areas or larger pores. Both of the latter can significantly increase the porosity dependence, often above a load-, material-, and especially pore structure-dependent threshold.

Consider next effects that may limit the porosity dependence of compressive strength. Increasing grain size with decreasing porosity can give an apparent decrease in the porosity dependence of compressive strengths, as for tensile strengths. There may also be some effects similar to those noted earlier giving higher E values in compression than tension, i.e., reduced P-dependence, but these effects are probably limited to modest stresses and do not occur at the high stresses and extensive cracking associated with compressive failure of relatively dense bodies. On the other hand, compressive strength is less susceptible to strength control by a single or a few isolated or clustered pores (or grains) since compressive failure is typically one of cumulative damage, i.e., cracking, rather than catastrophic failure from the single weakest flaw as in tensile strength. Thus, such pores or clusters often have less effect in compression; however, the high stresses and progressive failure character of compressive strength may provide more opportunity for effects of pore shape and related stress concentrations to come into play via progressive cracking as load increases, possibly increasing the P-dependence of compressive strength somewhat. An important factor in much compressive failure of highly porous bodies, i.e., foams, is the collapse of cells, which may have greatest effect over an intermediate range of foam porosities, i.e., not beginning until sufficient P is reached and then diminishing at high P, with both the onset and termination probably being foam structure dependent. Again, there is variation of theoretically predicted and experimentally obtained values for foams, not only for A, but also for n (Table 10.1).

Finally, consider wear, erosion and to some extent machining, which are more complex since they often involve a varying variety of phenomena; however, overall they show many similarities to the porosity dependence of hardness, including crushing and plowing (especially for some wear). As with scratch hardness, both result in a significant increase in wear (and probably erosion) due to crushing above load-, material-, and microstructural-dependent levels. Also, at least some wear applications of commercial aluminas indicate that shifting some of the pores from inter- to intragranular locations with grain growth may reduce wear (Section 6.3.3).

10.2.5 Microcracks

Microcracking affects a variety of materials and their properties, including some monolithic ceramics (especially ones with very anisotropic thermal expansion), some common traditional ceramics, a number of ceramic composites, and some rocks, mostly due to the multiphase character of most of these. Detailed quantitative documentation of their effects as a function of the extent and character of the microcracks is often uncertain and complex due to both the difficulties of characterizing the microcracks and their changing with test and other conditions. The latter include varying (often unknown) degrees of preexisting microcracks versus those formed during stressing or increases and decreases with test temperature; however, considerable advance has occurred due to more qualitative observations, modeling, and comparison of the two.

Microcracking commonly occurs due to stresses generated on a microstructural scale from mismatch strains from phase changes or different local thermal expansion from different phases or crystalline anisotropy of grains. Their formation occurs at or above a critical grain or particle size given approximately as (4):

$$D_S \sim 9\gamma[E(\Delta\epsilon)^2]^{-1} \tag{10.1}$$

where D_S is the critical grain or particle diameter (for an equivalent volume sphere as the grain or particle), γ = the fracture energy involved (e.g., to fracture 1 or a few grain-boundary facets or for transgranular fracture), E = the local Young's modules, and $\Delta\epsilon$ = the local mismatch strain. Microcracking thus commences at larger grain or particle sizes meeting the combinations of factors needed, and proceeds to finer grains or particles as strain mismatches increase, e.g., due to decreasing temperatures. Hence microcrack sizes are generally on the scale of the grains or particles and form on cooling and often close and heal on heating, usually with hysteresis that increases with the extent of anisotropy, e.g., Figs. 9.3, 9.12.

Consider first the effects of microcracking on elastic moduli, which is of importance in itself, and also provides guidance for microcrack dependence of other properties, as well as some guidance to effects of porosity. MSA concepts do not appear suitable for modeling microcracking, but mechanistic models have been developed for both dilute, i.e., noninteracting, microcracks (e.g., Table 3.5) as well as for higher contents with some microcrack interaction. The former show linear decreases for the limited microcrack contents, but with significant dependence on crack shape, stacking (i.e., spatial distribution), and orientation. The primary changes due to microcrack interaction are nonlinear property decreases with increasing microcrack contents. Thus, these show microcrack effects on properties with basic similarity to those of porosity effects on properties.

Quantitative validation of models is limited by the difficulties of characterizing microcrack contents and possible dependences of these on stress level, character, and history, as well as temperature level and possibly history (Section 3.3.4); however, cross-correlation of elastic properties that are dependent on microcracking support the models (Fig. 3.16B) as do evaluations of the temperature dependence of elastic properties of bodies that are known or expected to microcrack (Figs. 9.2, 9.3). Similar but less extensive modeling and experimental evaluations show similar effects on thermal conductivity (Fig. 7.2, Table 7.2).

Thus, microcracks make varying, often significant, reductions in elastic moduli, with many similarities to porosity dependence; however, these reductions are not nearly as catastrophic as application of simple extrapolation of pore-crack stress considerations would suggest (e.g., Figs 2.6, 2.7, 3.14 and the following section). Furthermore, limited but important results for HfO_2 raise a more serious and basic complication. As grain size was increased the temperature range and breadth of the moduli hysteresis as a function of temperature increased as expected; however, on reaching a grain size of ~ 17 µm, some microcracks did not close or heal on heating, so temperature cycling resulted in extensive progressive decreases of properties (Section 3.3.4). Whether this reflects general effects of the absolute scale of the microcracks or their relation to grain size (e.g., encompassing more than one grain or grain-boundary facet), or of the phase transformation of HfO_2 is unknown. In all of these cases some increasing reduction of registry of the two crack faces could occur to progressively limit closure, but this is not verified, and what parameters may be important is not known.

Less modeling and data exists for fracture energy and toughness, but both clearly show effects that are often significant. Thus, in anisotropic materials fracture energy and toughness, at least for most values measured with larger cracks, first begin to increase above some grain size level (e.g., at ~ 40% of D_S per Eq. 10.1). Accelerating increases reach a maximum (at ~ D_S), then decrease substantially with increasing grain size, with the absolute level of these properties and their changes generally decreasing as the TEA increases (Fig. 4.7). The changes of at least the less anisotropic materials have been attributed to microcrack generation in conjunction with the propagation of the macrocrack used for test, much of this based on subsequently developed models.

Originally, microcrack generation was postulated to be mainly ahead of and above and below the main crack tip, but recent models have focused on microcrack generation primarily or exclusively in the wake of the crack, i.e., in or near the plane of the macrocrack, and behind its tip. Limited experimental data supports a mainly wake association of macrocrack-generated microcracks, but there is also some evidence for some generation and related effects such as crack deflection and branching at or closely ahead of the macrocrack. The general decrease in levels and changes of toughness as anisotropy increases probably reflects more preexisting microcracks; however, many uncertainties remain in these trends and from other observations, e.g., that while microcracks may also increase the fracture energies or toughnesses of rocks, those without microcracks had still higher values (Section 8.3.2). Furthermore, limited observations in composites comparing preexisting versus macrocrack-generated microcracks in very similar bodies indicates some but often limited and not necessarily fully understood differences between effects of the two types of microcracks on fracture roughness (Section 8.3.7). Other issues exist as discussed at the end of this section and immediately below (for strength).

There is limited data on the microcrack dependence of strength, but all that there is clearly shows strength being significantly decreased by microcracks (Sections 5.3.1, 8.3.7), which is basically inconsistent with the trends of much fracture toughness data. Thus, where data is available as a function of temperature of very anisotropic materials such as Al_2TiO_5 (Section 4.3.3) and graphite (Fig. 9.4), and hence indirectly their dependences on microcrack contents, show trends for both strength and Young's modulus closely following each another in their increase as temperature increases to reduce microcracks. (Unfortunately there have apparently been no comparable fracture toughness measurements.) Microcracking gives materials high thermal shock resistance, but only modest strength, the latter consistent with the above strength observations. Their increasing thermal shock resistance is due mainly to their reduction in thermal expansion (Fig. 7.10) and secondarily to reductions in Young's modulus.

There is almost no information on effects of microcracks on other mechanical properties, except some primarily on effects of cracking from indentations as outlined in Section 6.3.1. Spall cracking occurs mainly along grain boundaries in conjunction with Knoop and especially Vickers indents when the indent and grain sizes are similar. Possible effects of grain-boundary phases and stresses from thermal expansion and elastic anisotropy have been suggested. More extensive is the reduction and often elimination of Hertzian cracks associated with spherical indentations in ceramics with substantial indent-induced microcracking. These include plasma-sprayed alumina coatings and bulk bodies of partially stabilized zirconia, some ceramics with some larger grains such as "self-reinforced" Si_3N_4, and some ceramic composites such as from crystallization of glasses forming substantial amounts and sizes of micatious grains. Generation of (initial) microcracks from pores in compressive stressing of bodies was also shown (Fig. 6.1). There has been little or no information reported on effects of microcracks on other related properties such as erosion and wear; however, while microcracking should reduce the formation of Hertzian cracks from individual particle impact, effects on resultant erosion are more uncertain since microcracking may aid erosion from multiple impacts. One of the reason why there is little published data on microcrack effects on wear is that limited tests indicate that weak microstructural boundaries, and thus presumably the associated microckacking, is often catastrophic for wear resistance.

Microcracks also reduce thermal (Fig. 7.2, Table 7.2) and electrical conductivities and increase optical scattering and electrical breakdown. They also play various but usually poorly studied roles in other, e.g., of some magnetic, piezoelectric, and ferroelectric, properties.

In summary, two sets of observations should be noted based on both modeling and data. First is the general similarities of the effects of microcracks and pores on properties, i.e., that the character (shape, orientation, and stacking) of both pores and microcracks are important in their effects in addition to the levels of their occurrence in bodies. Similarities of effects are seen over the range of most properties, including crack propagation effects, although these are more complex as discussed below. In both cases models are important, but experimental limitations on adequate characterization of microcrack contents make their modeling more important, with again a family of models being needed; however, one important modeling difference is that MSA approaches have thus far not worked for microcracks, but have been more successful for pores, while approximately the reverse is true of mechanistic models. This in turn is potentially encouraging for improvement of mechanistic models for pores, their marriage with MSA models, or both, since the mechanistic models for cracks nicely parallels the MSA results for pores. Finally, it would ultimately be desirable to develop correlation of pore and microcrack characterization so crack and pore effects could be directly compared, and their interaction or transition from cracks to pores better understood.

The second set of summary observations is that there are many uncertainties left in understanding microcrack effects. These include better defining the relative character, roles, and mechanisms of preexisting microcracks versus those generated in conjunction with propagation of a stressed macrocrack, and effects of the crack size, shape, and velocity, and the opposite trends of many toughness and strength measurements. Generation of microcracks in the bulk of at least some materials (e.g., at least some graphites) by some stressing, as implied by effects on elastic properties, and the apparent effects of different stress states in measuring elastic properties are other issues to be better clarified. The issue of why microcrack closing and probably healing ceases at larger grain sizes of HfO_2, and what the causes and resultant ramifications of this are, is also important. Similarly, while effects of pores and microcracks are commonly assumed to be independent, this is not always the case and raises further questions. Thus, some probable interaction of porosity and microcracks in sandstone (via fracture energy/toughness and R-curve effects, Fig. 4.5), clear interaction in Nb_2O_5 (via thermal expansion, Fig. 7.10A), and critical effect on the formation and resultant effects of microcracks in duplex composites (Sections 9.3.2) were shown. Finally, the generally assumed microcrack-grain size ratio of ~ 1, while qualitatively observed (and quantitatively in TiO_2), has been reported to substantially increase progressively as grain sizes decreased (e.g., below ~ 200 μm) where microcracks were generated during strength and static (but not dynamic) E tests of Al_2O_3 (Sections 4.3.2, 4.4, and 4.5).

10.2.6 Other Materials and High Temperature

More broadly, the porosity dependence of other materials is generally similar to that of ceramics, but with some common deviations. Thus traditional ceramics, e.g., porcelains and whitewares that are composites of glass and crystalline phases, show the same overall trends as modern ceramic composites, metals, rocks, and cements and plasters (Chapter 8). This includes similar b values, rollovers, and probably P_C values for comparable porosity (in the limited cases where there is a reasonable indication that the pore character and measurements were carried out to sufficiently high porosity levels); however, there is often a trend for higher b values, especially in rocks and cements, where heterogeneity of porosity is a factor in both, especially in rocks. In cements and to a lesser extent in ceramic composites, variation of which phase the pores are most associated with is also a probable factor. This appears to be particularly important in ceramic fiber composites where different fabrication gives very different results (Fig. 8.15). Also, in cements as well as some other materials made with mass added in the process of densifying the material, modeling of the properties, e.g., via modified MSA models, should improve understanding and accuracy of predictions. Such models for these materials may show increased b values. Note that the close similarity of the porosity dependences of ceramics and metals strongly argues against an important role of stress concentrations from pores on mechanical properties. If such concentrations were important in metals, their limitation by local ductility should result in substantially different porosity dependence in them, which is contrary to extensive observations.

So long as the mechanisms controlling the porosity dependence remain constant, or nearly so, with temperature increases, the porosity dependence remains the same. This is shown for mechanical properties, especially elastic moduli, and thermal conductivity (Figs. 7.9, 9.1, 9.11); however, mechanisms can change with temperature, e.g., as frequently happens with thermal conductivity due to changing contributions of convective and radiative transport (both at higher P and larger pores, and radiative transport in bodies of higher transmission in the IR wavelengths). Changes in electrical conductivity may also occur (Fig. 9.14). Greater decreases in elastic moduli at higher temperatures pose problems of whether they represent intrinsic relaxations or creep effects manifesting themselves simply due to the use of too low a strain rate.

10.3 APPLICATIONS FOR POROUS MATERIALS

10.3.1 Physical Functions

There are a variety of existing, developing, and potential applications for porous materials, very often ceramics, that individually or as groups represent substantial technological and economic opportunity. While all of these applications involve varying degrees of physical, chemical, and sometimes biological functions, it is useful to classify them by which of these functions is most essential or unique to the application. Thus, this section outlines the nature of applications based mainly on physical functions, and the following section addresses those based substantially on chemical or biological functions. This is followed in turn by some of the process technology that is used to obtain the necessary or desired porosities. Then broader issues are noted, in particular some of the engineering necessary to balance the often partially to substantially conflicting requirements, as well as status, potential, and competition. The latter addresses some of the issues determining whether a ceramic or other material is used for these applications and if the application may be economically viable. Most applications are for free-standing, porous (including some fibrous) bodies, but porous coatings can also be important.

Thermal insulation is a major application of porous materials, especially ceramics, and is a long-standing technology and business. Thus, both rigid porous ceramic refractories, especially bricks, and fibrous (usually felt, rigidized or not) insulations are used. While much of such insulation is made of oxide materials, there are some nonxide refractories, e.g., Si_3N_4, and carbon felt is widely used for nonoxidizing applications (the latter having extensively replaced the use of carbon black for many applications). There are newer possible applications for porous ceramic insulators, e.g., as exhaust port liners (in limited use in at least one high performance car), and interest has been expressed in complete manifold liners for diesel engines. The latter might be as coatings, but most consideration has been given to monolithic parts. Such parts pose challenges in fabrication that may be addressed by processing and fabrication technologies discussed in the next section. Substantial use has also been made of porous ceramic, especially ZrO_2-based, coatings on metal components, i.e., thermal barrier coatings in engines, for which there is substantial technology, e.g., plasma spraying. Much of the advance in this latter field has come from better process control, but other process technology may find use. While such porous ceramic coatings have been typically used on metals, they can have applicability on ceramics (e.g., Section 9.3.2). In all cases varying tradeoffs between the insulating capability, thermal stress/shock resistance, and other factors such as particle erosion, corrosion, or combinations must be addressed in order to engineer suitable or, preferably, superior solutions. In some cases issues such as installation and servicing of the porous component are also factors.

Consider now newer or lesser known applications, as well as well-established applications, especially those for which there is or may be substantial growth, which are outlined in Table 10.2. A lesser known but established application with substantial potential growth is battery and fuel cell separators. While chemical durability is important in these applications, the fundamental requirement is electrical insulation. Thus, in recent fuel cells, fibrous, electrically insulating mats of BN or AlN have been used. One of the important challenges, as in almost all applications, is cost.

Other applications for electrical functions are based on reducing the dielectric constant of materials of modest dielectric constant (typically < 10) with no porosity. Two potential applications are in substrate and packaging materials for electronics and for electromagnetic (radar, microwave, etc.) windows. In the former case, lower dielectric constants increase the speed of electrical signals traveling in conductors on top of or buried in the dielectric. Key issues are closed pores so that the conductor material does not penetrate into the dielectric on a microstructural scale and the tradeoff with porosity reductions of other pertinent properties such as thermal conductivity, stiffness, and strength. Lower dielectric constant windows allow a frequently desired broader range of microwave wavelength operations since windows are designed for optimal operation at a given frequency based on dielectric constant-window–thickness relations; however, increased porosity also reduces strength and erosion resistance that are often important in many, especially aerospace, applications where an increased bandwidth of operation may be desired. Another important but very different area for the design of porous materials is as filters, which represents a large and diverse array of long-established, emerging, and potential applications (5–11). Some of the former are air and fuel filters for engines, e.g., for autos, but many of these filters are organic materials. Ceramics have been used for fuel filters and ceramic filters to trap (and usually subsequently allow burning of) particulates in diesel engine exhaust have been extensively investigated. Much larger ceramic "candle" filters are under development for removal of particulate material from hot coal-combustion gases to replace gas combustion to drive gas turbines for electric power generation. Ceramic filters

Table 10.2   Summary of Evolving or Emerging Applications of Porous Ceramics

Application	Basic function	Technical needs/issues	Status	Potential	Competition
Battery/ fuel cell separators	Electrical insulation chemical stability and permeability	Chemical durability, electrical insulation	Existing to development, emerging	Moderate	Plastics, composites
Biological media	Effective bacterial growth	Biocompatibility/limit nutrient, impurity fluctuation impacts	Limited commercialization	Substantial	Natural, plastic, carbon materials
Burners	Reduce NOX emissions, uniform heating	Permeability, thermal shock resistance/ durability	Some different commercial applications	Moderate to substantial	Metal screens, felts
Catalyst supports	Chemical and physical compatibility	High surface area, permeability, mechanical integrity tradeoffs	Largest material volume and dollar application	Including existing, new applications (membrane reactors)	Metals, new technology
Filters	Remove particals, bacteria, and other matter from fluids	Uniform porosity, permeability/ stability	Some particulate (juice, molten metal, especially Al), and bact. (e.g., beer) filters	Moderate to substantial increase	Polymers, metals (low-mod. temperatures) powder, pebble beds
Gas sensors	Measurable changes with atmosphere change	Response time, durability	Research and development	Several applications, small components in electrical circuits	Other materials, designs, technologies
Sonar transducers	Higher sensitivity under hydrostatic pressure	Reduce effects of hydrostatic pressure/ mechanical reliability	Development halted	Hydrophones, submarines	New polymers, designs

were extensively used in the gas diffusion process to separate uranium isotopes because of their thermal and chemical stability, and a variety of other gas and liquid filtering applications have been considered. The main factor that determines whether organic, metallic, or ceramic materials are used is durability, i.e., ceramics are favored for more extreme chemical and thermal requirements. Although a fair amount of research has been conducted on other ceramic filter applications, there has been only modest success in lower temperature filtrations of liquids. Thus, there has been some filtration of bacteria, e.g., from beer and drinking water, and of pulp from fruit juice, but many issues remain limiting the growth of these and related application. More immediate issues are cleaning, where in principle the stability of ceramics (and often also of metals) to temperature and solvents may be an important advantage, e.g., for sterilization; however, the logistics and demands of such cleaning can be serious issues, e.g., the time, resistance, and incompleteness of back flushing and the thermal stresses in sterilization. These issues are related to other system issues, such as the size, manifolding, pressure control, and so on, of the filters for an effective system, which often entail issues of mechanical reliability, or exacerbate those from cleaning, e.g., thermal stress and back pressures. These issues, along with cost, have been limitations on ceramics and other materials finding broader application in what could be an extensive array of uses.

The area of ceramic filters that has seen somewhat greater success is in filtering molten metals, especially aluminum. A substantial fraction of cast aluminum, especially billets for foil making, is filtered to remove inclusions, from impurities, insulation, and furnace debris. These inclusions are a major source of foil breakage during its manufacture, and restringing a foil line after the foil has broken is a substantial cost. Thus there is an important incentive to minimize such failures and hence to filter out the inclusions that are a major cause of such failures. Some other castings of aluminum, iron, and nickel alloys are filtered, but the latter two metals represent substantially greater challenges because of the higher temperatures and greater chemical reactivity. The largest scale molten-metal filtering in the aluminum industry is through pebble beds of alumina particles, but considerable filtration is through manufactured filters, typically ceramic reticulated foams (e.g., Fig. 1.13B). In all of these applications cost is an important issue, e.g., since the filter must be removed and it is typically discarded along with the metal in it. Of course the cost constraints scale with the value of the product, so casting of expensive turbine blades is more cost tolerant. Cost issues can also be related to mold design and casting procedure. Thus, if a filter can be inserted to filter the stream feeding a mold of several parts, with modest cost to place it there, this can be more cost effective. Obviously such filtration places severe thermal shock requirements on filters, and more refractory metals also impose serious chemical reactivity issues as noted above, i.e., the filter must not contaminate the cast part.

Other specific applications considered for ceramic filters are summarized as follows. Ceramic foam filters over commercial kitchen stoves remove fat/grease better than metal filters, but do not allow effective recovery/removal of the fat and grease (12); however, coating the ceramic struts with glass (to seal the pores in the struts) gave competitive recovery, and so such coated filters were overall superior to metal filters. Other applications considered include various separations from liquids, e.g., (13,14). Porous alumina molds for slip-casting ceramics have been considered to avoid contamination from conventional gypsum molds as well as providing higher penetration pressures and casting rates (15). Microporous gels are useful desiccants for some applications (16). A possible novel application is the use of porous ceramics "pipes" for water-conserving irrigation systems for concentrated, rapid growth of food plants, e.g., lettuce (17). Porous ceramics also give finer bubbles in aeration than other porous materials due to differences in wetting (18). Another novel application considered that may suggest a variety of other uses of porous structures and their interaction with electrical or magnetic fields is the separation of different gases based on their magnetic properties via use of a porous superconducting ceramic filter (19,20). This takes advantage of the fact that a superimposed magnetic flux, being excluded by the superconductor and hence concentrated in the pores, will highly interact with the magnetic characteristics of gases, e.g., allowing potential separation of oxygen (paramagnetic) and argon (diamagnetic).

Another very different application which (although current development has been suspended) may have important future potential, is improving sonar transducer performance under hydrostatic pressure, i.e., for hydrophones and submarines. This may also be an example of other opportunities for newer applications of using designed porosity to give improved electrical (and possibly magnetic) performance. As noted in Sections 7.2.4 and 7.3.3, d_h can decrease or increase with increasing porosity depending on the effects of the porosity on the components of d_h ($= d_{31} + d_{32} + d_{33}$), which depends on the nature of the porosity. Nagata et al. (21) were apparently the first to explicitly use porosity to increase d_h, by partially sintering spray-dried agglomerates of PZT; however, as in most applications, there are trade offs that must be considered, which in this case (and in many others) is mechanical integrity. Rice and colleagues (22,23) recognized that pores between particles, e.g., spherical particles, caused the greatest decreases in mechanical properties (e.g., Figs. 2.11, 3.1), i.e., such bodies are the least effective use of remaining mass to sustain mechanical loads. They first considered spherical pores for much greater mass efficiency, but chose aligned platelet pores for somewhat greater mass efficiency, as well as potential processing practicality. They initially made bodies by tape-casting PZT sheets with a thin layer of pure binder on one side. Thus, when the sheet was dried and then broken into fragments that were die pressed and sintered, the burnout of the binder layer left platelet pores extensively aligned normal to the pressing direction. Such bodies were promising. Kahn and colleagues (24,25) recognized that greater uniformity and control of porosity and processing could be obtained by printing burn-out material for the pores on PZT tapes that were stacked and fired to produce desired pore structures.

Bodies made using metal or ceramic balloons may have application for various electrical or magnetic functions. An example of this is the use of PZT balloons to make ultrasonic transducers (26).

10.3.2 Chemical or Biological Functions

Ceramic and metal burners have found use in several applications, with potential for a number of other, possibly much larger volume (e.g., home hot water heater), applications. There are important physical requirements for such burners, mainly maintenance of integrity, e.g., from thermal shock, and in some cases emittance for greater efficiency; however, there are important environmental improvements that such burners can offer that have been an important factor in their development and application. These improvements can be appreciated by recalling the basic nature of a gas flame. The cross-section of the typical elongated, pointed flames ranges outward from a hot (blue) center core to progressively cooler layers of increasing redder hues. The high temperatures of the inner core of such flames generates the great bulk of the undesired nitrogen oxide products of such flames while their, elongated, pointed shape means their heat is not particularly uniform over the area to be heated. Insertion of a suitably porous body in the flame that can sustain the transient and steady state thermal stresses and distribute the gas stream (with limited back pressure) and combustion heat can substantially reduce all of these problems. Thus, with such porous bodies in a flame, there is a far more uniform temperature distribution in the flame, primarily at the expense of the hot core, which can therefore dramatically reduce undesired nitrogen oxide (NOX) emissions from such burners. Thus, increasingly stringent emissions limitations on gas flames has been an important factor in the development of such burners. Such burners also can be more efficient in the heat delivered, and can deliver it more uniformly due to their extensively shaping the flame (even over substantial dimensions, e.g., of meters) and providing more, often much more, heat energy from radiation.

Both metal and ceramic burners are in use, with metals (typically as felts, with coating to improve emissivity) having the advantage of being somewhat more robust, but typically at about the upper limits of their use temperatures. Metals can be used for many of these applications since the flame temperatures at which undesired gas emissions are adequately low are within the upper limits of their operating ranges. Use of ceramics that are catalysts, or are compatible with catalysts, that reduce emissions at higher temperatures could extend the use

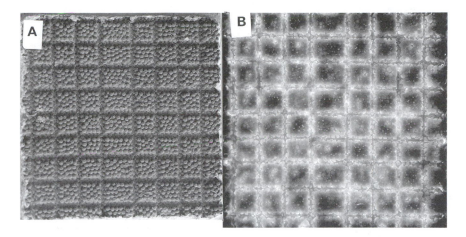

Figure 10.1 Ceramic burner fabricated by bonding ceramic balloons together at the contact points. Photographs of 5.5 square in. test burner plates. A) As fabricated, and B) in operation as a burner. Brighter and darker areas reflect, respectively, areas of greater and lesser combustion. Photos courtesy of Dr. J. Cochran of Ceramic Fillers Inc.

of ceramic burners; however, ceramics already have considerable application, mainly in two forms. The first is fibrous (industrial) burners where they can be as individual burners a meter or more long or clusters of several to many smaller or larger individual burners. Another form of ceramic burners (e.g., for fast food deep fat fryers) consist of ceramic plates with holes formed in them. Various porous, e.g., foams or foam-like, bodies are important candidates (27) (Fig. 10.1). The key needs are sufficient uniformity, robustness (including during both installation and cyclical use, often in less than clean atmospheres), and low cost.

A long-standing, large, diverse, important and growing class of applications of porous ceramics that clearly often have physical and chemical functions, with the latter commonly driving the application is for catalyst systems. The largest and oldest of these has been catalyst supports, which can be roughly divided into industrial and vehicular supports. Initially both ceramic beads and cellular ceramic monoliths were considered for vehicular application. Candidates for the latter included foams and fiber systems which were abandoned for honeycombs made by either rolling up calendered (e.g., corrugated) ceramic tapes or by extrusion (earlier with 400 cells/in.^{2} with wall thicknesses of 0.006 in., and now with up to 600 cells/in.^{2} and with 0.004 in. thick walls). Both the (porous) beads and extruded cordierite honeycomb made it to market, with the extruded material coming to dominate the market, especially for autos. The area of the extruded cordierite honeycomb cells combined with the substantial open porosity in the cell walls provide the surface area necessary for the desired catalytic activity. This activity is obtained from a very fine dispersion of platinum on top of an alumina wash coat that is deposited via a slurry technique on the cordierite honeycomb. The alumina wash coat, which is used to increase surface area and provide greater compatibility for the platinum catalyst, must provide good adherence and stability. The typical auto exhaust catalyst support consists of two extruded cordierite "biscuits" that commonly have a flattened oval shape about 2 in. thick, 4 in. wide and 4 in. long (see Fig. 1.14A for a partial cross-section). Thermal stress and shock resistance, while not much of a factor for beads (which are subject to fracture and attrition from shock and vibration) was a major factor in the selection of cordierite (also favored by cost) for the honey-comb. Packaging the honeycomb biscuits to avoid mechanical shock and vibra-tion from driving was an important task accomplished by the design of a compliant blanket based on expanded (i.e., porous) vermiculite.

Additional demands are being placed on auto catalyst supports that may change the technology. Thus, stability of the high surface area is an increasing problem as temperatures increase, which can also pose problems for the more modest temperature capabilities of cordierite. Composites based on aluminum titanate (Al_2TiO_5) have been extensively explored as alternatives because its microcracking enhances thermal shock resistance (28,29); however, reducing emissions that pass through the catalytic converter before it gets heated to its effective operating temperature, i.e., on starting the vehicle, have become a major factor in further reducing automotive air pollution. Two approaches considered to address this problem have been the use of preheated monolith catalyst supports and of highly porous, active absorbers. Metal monoliths (often made of rolled-up corrugated metal foils) which can be resistively heated (and have also been considered as an alternate to ceramic supports for other reasons) are a major candidate for preheated monoliths. The absorber concept is to absorb the initial emissions in a porous body, from which they are then released by heating from the catalytic converter as it heats up, then passed through it and destroyed by the now fully functioning converter. This latter, simpler concept is thus another use for designed porous materials.

The broad area of industrial catalysts can only be briefly outlined here because of its scope and diversity. There is some use of ceramic as well as metal honeycombs, mainly in the removal of emissions from industrial power plant exhausts. Thus, for example, large, porous TiO_2 extruded monoliths (e.g about 5 in. square and 2–3 ft. long with cells widths of ~1/4 in.) have been used; however, the great bulk of industrial catalysts consist of beds of beads of tailored porosity and composition, again mainly as supports for the actual catalyst but occasionally as the catalyst themselves. Besides the key factor of the pore character, typically a high surface area (hence open porosity) and pores within specific size ranges are required, as well as abrasion and crushing resis-tance and low cost. The former are needed for operation of the catalyst bed, while the latter is needed because of the volumes required. A potentially very significant extension of the use of ceramics in catalysis applications is in membrane reactors (30,31). These will require ceramic bodies to be a substantial component of the reactor itself rather than as items placed in the reactor.

Three other applications included under the broad heading of chemical applications are sensors, chromatography media, and media for biological growth, e.g., for bioremediation. Extensive research has shown the very porous, small bodies of ceramic semiconductors such as SnO_2 and TiO_2 can be used as sensors for gases via changes in the resistivity of the very porous body due to differing interactions with gasses on the high surface area (32–37). Presumably, these changes result from the effects of adsorbed species on the surface conductivity being a factor in the overall conductivity because of the very high surface area and hence measurable surface to bulk effect. Sensors consisting of a porous layer on a section of an optical fiber have also been studied (38). Such applications are clearly of technological interest, but are rather specialized because of their use of small quantities of materials and their typically being a component of an electronic or fiber optic circuit.

Media for chromatography and biological growth (39–41) reflect respectively existing and potentially much greater volumes of materials than sensors, but are generally modest in comparison to catalysts. Both applications use beads or granular particles, e.g., of chopped extruded material for some biological growth. In both cases the pore structure is important along with various chemical factors, some of which are related to the choice of material, and some of which are met by treatment of the porous beads or particles. Thus, for example, in bioremediation—the use of bacteria to remove toxic materials in liquid streams—it is important that the bacteria and nutrients for them have reasonable access to the interior of the beads. This is not only for the effective use of volume, but also often the need to survive serious transient reductions of nutrients or the presence of poisons. Thus, having bacterial growth through most or all of the beads and with reasonable but not excessive time for diffusion of liquid ingredients in and out of the bead interiors (as well as possible buffering agents there) can be important. Low cost for most biological media is particularly important, so in addition to low-cost manufactured materials natural materials such as diatamatious earth have been used and the use of pumice (42,43) has been proposed.

Another biological-type application for porous ceramics is for prothesis. This is primarily for materials which are compatible with bone intergrowth, i.e., apatite materials (44–46). Here the porosity is a compromise between maximal initial properties versus greater opportunity for growth of bone into the prothesis. Porosity at, near, and connected to the surface, with sizes amenable to intergrowth of bone is often important.

10.4 FABRICATION AND PROCESSING

10.4.1 Beads and Balloons (and Tubes)

Beads and balloons have a variety of applications by themselves (e.g., as catalyst supports and in chromatographic, and biological applications) in addition to important applications for making bodies. There are a variety of processes for fabricating ceramic (and often also metal and plastic) beads or balloons, with the former possibly having a variety of internal pore structures. The first set of such processes is from the melt. Molten droplets can be made by spinning off of a rotating, melting tip, or by various methods of atomizing molten streams; although these are used more broadly for making metal particles, they have also been used for making ceramic particles or beads (47) (Fig. 10.2). There is some degree of control of the size range of the droplets via material and especially process parameters such as temperature, rotation speed and so on. Such solidified droplets typically have a fairly spherical shape and differing surface

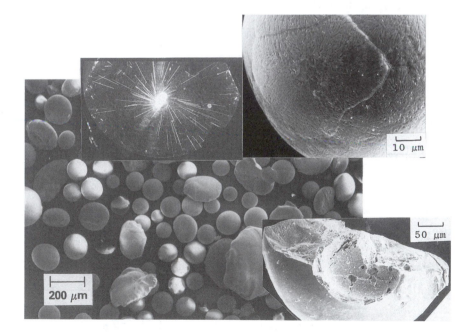

Figure 10.2 Melt-derived ceramic beads from molten PSZ droplets. Inserts: droplets being spun off of a rotating PSZ rod tip heated with a laser beam, and higher magnifications of the outer surface and the internal microstructure of resultant beads. Note that much but not all of the substantial porosity is due to one large pore in about the bead center.

Table 10.3 Bead and Balloon Fabrication Methods and Results

Forming technology	● or ○[a]	Sphericity[b]	Diameter[c]	~ Size range	Material application
Die pressing	●	M–H[d]	VN	mm to cm	Wide
Agglomeration	●	P-M	M	≤ 100 μm to 1 cm	Wide
Spray dry	●	P-M	M	≤ 10 μm to 0.5 mm	Wide
Sol drop	●	VH	VN	μm to mm	Selective[e]
Droplet stream	●, ○	H- VH	H	300 μm to 5 mm	Wide
Emulsion	●	H	M	≤ 1 μm to 1 mm	Wide
Plastic bead replic.	○	H	N	≤ 10 μm to 5 mm	Wide
Molten droplets	●	M–H	W–M	≤ 10 μm to mm	Fairly wide
Agglomerate Surface Melt	○	P–H	W–M	≤ 10 μm to mm	Selective[f]

[a] Beads (●) or balloons (○).

[b] P = poor, M = medium, H = high, V = very.

[c] V = very, N = narrow, M = medium, W = wide.

[d] Tailorable, e.g., equatorial, belts produced.

[e] Demonstrated with sols, i.e., for oxides, but potentially applicable to other polymeric precursor, e.g., for nonoxides.

[f] Applicable only to materials that melt with limited vapor pressure, especially certain silicate glasses.

textures. Glass surfaces are commonly very smooth (e.g., Fig. 10.5), but while solidified polycrystalline surfaces are overall fairly smooth, there are typically varying degrees of grain faceting and grain-boundary grooving (Fig. 10.2). Internally, solidified polycrystalline droplets may have equiaxed or elongated grains, and almost always have varying amounts of internal porosity that may be distributed in various fashions, but much of this near the center (Fig. 10.2). Some of this porosity is intrinsic from the typically substantial liquid-to-solid volume decrease (i.e., density increase), which for crystalline ceramics is commonly of the order of 10% and can approach or exceed ~ 30% [e.g., ~ 20% for various alumina compositions (48)]. Additional porosity may frequently be generated on solidification due to exsolution of gases dissolved in the melt. Some tailoring of internal pore structures in melt-solidified droplets is feasible via use of phase separating, eutectic, compositions, and postsolidification leaching of one phase, but does not appear to have been applied (although it clearly could be for Rainey nickel and phase-separating glasses).

Several techniques available for making beads (mainly via sintering of particulate compacts derived by one of several forming methods) are summarized in Table 10.3. For larger beads (e.g., down to ~ 0.25 in. dia.), die pressing, as widely used for milling media, is feasible (Fig. 10.3A,B). Many sand milling media are made by various powder agglomeration techniques, e.g., by feeding powder and binder onto a rotating table. For smaller beads (e.g., ~ 10 to a few

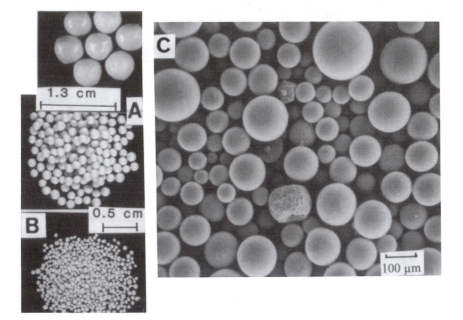

Figure 10.3   Ceramic beads made by die pressing or agglomeration and firing (oxides, A, B), and rigidization and firing of emulsion droplets (glassy carbon, C).

hundred microns) the primary method is spray drying; however, there are at least three other methods based on liquid processing that yield beads, often having one or more of the following attributes: more uniform sphericity, size, and pore structure. Developed for nuclear reactor fuel and in long use for some catalyst beads is the dripping of sol droplets into a fluid bath in which they gel while settling to the bottom of the fluid container.

The first of two newer bead making techniques is via emulsions (49)· which can commonly produce beads from tens to hundreds of microns in diameter. Emulsion techniques essentially involve three requirements: 1) having a liquid precursor of the product ceramic or other material, 2) another liquid in which the fluid precursor can be emulsified (usually with the aid of a surfactants), and 3) a technique to rigidize the precursor after it has been emulsified. Thus, by forming an emulsion to form spherical droplets of a fluid ceramic precursor in another fluid and then having the ceramic-producing droplets rigidize, they can be filtered out and fired to produce the product beads. The two main classes of liquid precursors for ceramics are particulate slurries (i.e., slips) and polymeric precursors (preceramic polymers and sols). The latter are rigidized by polymerization, e.g., via polymerization catalysts (such as for polyfurferal alcohol for glassy carbon beads, Fig. 10.3C, or water for sols) or thermally via heating of the emulsion, or both. Beads from slurries are rigidized by polymerization of the binder phase, which may be organic binders or some ceramic precursor. Emulsion techniques produce a range of bead sizes whose mean and range can be adjusted some by selecting or modifying the emulsification fluids or the surfactants used, and the parameters such as degree of mixing shear. In some cases flowing the emulsion fluid mix through a screen can help control the droplet diameter and its range.

The second and newer technique for making beads is to generate a controlled stream of a liquid precursor to the ceramic or other product desired via proprietary nozzles so this stream breaks into individual, very uniform droplets due to inherent surface tension effects (50–52). By again having a method of rigidizing the droplets before the end of their flight, beads are obtained, usually of quite uniform sphericity and size (the same as for balloons, Fig. 1.14C). The latter can typically be varied from several hundred to a few thousand microns in diameter by adjusting material and operating parameters. While use of other precursors is feasible, the ones extensively used are slurries, either organic- or water-based, with beads of either of the latter being rigidized by drying during free fall in a vertical tube with an upward flow of warm air.

An important extension of bead technology is the generation and control of bead internal microstructure, especially pore structure. This can be done to varying extents in any bead sintering operation by control of the particle size, burn out of fugitive (typically organic) phases, and the degree of sintering; however, the use of more sophisticated techniques offer greater variation and control, with precursor phase separation (discussed above and below), emulsions or foaming within beads made by either of the newer, fluid-based bead methods outlined above. Thus, making beads from mixes that undergo binder phase separation can give internal bead pore structures as in Fig. 10.8 by either emulsion or fluid-stream methods. Beads can also be made from an emulsion in which the fluid forming the beads is in turn an emulsion in which one phase is removable chemically or thermally. A result of such a double-emulsion technique is shown in Fig. 10.4A. While foaming within beads is a more demanding and limited process, it can be done with interesting results (Fig. 10.4B).

The extreme of pore volume in beads is to make them hollow, i.e., balloons. Three methods available for making balloons of ceramics and other materials are also summarized in Table 10.3. The first is to coat a precursor, commonly a slip, onto an expendable organic bead such as polystyrene. The second is via the fluid-stream method since proprietary nozzles are readily made that yield hollow, thin, fluid streams that form uniform, thin-wall green balloons that are readily fired to ceramic (Fig. 1.14C) or metal balloons. Metal balloons (and beads) can also be made directly from molten metal streams, especially for

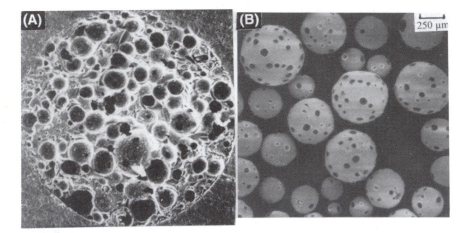

Figure 10.4 Examples of beads with natural plus added induced internal porosity. A) Cross-section of a single glassy carbon bead made by the double emulsion technique. B) Foamed alumina beads (note the large foam pores intersecting the bead surfaces). Photographs courtesy of Dr. R. Brezny of W. R. Grace.

lower melting metals and alloys. The third technique of making ceramic (and potentially also some metal) balloons is via melt processes. Thus, many tons of differing glass balloons are made for a variety of applications by feeding either droplets of a liquid (slip) precursor or an already solid, agglomerated particulate precursor, primarily from spray drying, into an upward pointed flame. The key to making good balloons is to have the proper balance of outgassing from the resultant agglomerated particulate precursor (e.g., from binders) and the formation of glass melt on its surface. Forming enough glass melt on the surface to effectively seal it by the time the remaining outgassing from the rest of the particulate agglomerate is just sufficient to blow the forming balloon to the desired size range is the key to forming glass balloons. Careful control of (low-cost) precursor chemistry, agglomerate size and structure, binder chemistry and content (as the primary source of gas, i.e., blowing agents), and flame conditions are keys to making good balloons. Such balloons inherently cover a range of sizes, wall thicknesses, etc. (Fig. 10.5), which are somewhat adjustable. Elimination of poor balloons, shot, fragments, and so on, is done in part by the upward direction of the flame since much of this is much less buoyant in the hot gas stream, and hence falls back down, while balloons are collected above. Subsequent mixing of collected balloons and a fluid, usually water, further removes many poor balloons due to imperfections, i.e., separating "sinkers" from "floaters." Sizing, e.g., via sieving, and posttreatment such as surface strengthening, coating, and so on, can also often be done. Larger, more irregular

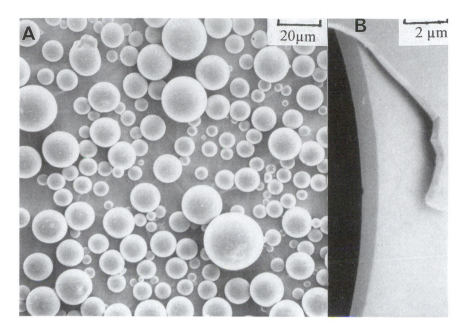

Figure 10.5 Examples of glass balloons. A) Lower magnification photograph showing the generally good sphericity, smooth surfaces, and typical range of balloon sizes. B) Higher magnification showing the wall cross-sections of two nesting balloon fragments. Photograph courtesy of Dr. J. Block of W. R. Grace.

balloons of more refractory, polycrystalline ceramics, mainly from alumina and zirconia, are available commercially, apparently from an extension of the glass balloon process, but using plasma torches for blowing the more refractory materials.

These are illustrative of more advanced (but generally not mature) developments that are promising. Other processes exist or have been shown in laboratory tests that can extend the scope of such processing. A few examples of other demonstrations and studies are described in refs. 53–56.

There has been much less development of hollow rods, i.e., tubes, which are more restricted in their use, but may have specialized uses if sufficiently further developed. A few examples are given to indicate the range and state of development. Some possible tubular character was indicated in making porous Cr_3C_2 from Cr_2O_3 (57). Small-scale laboratory demonstrations gave zirconium phosphate tubes a few hundred microns in diameter with a central hole ~ 100 μm in diameter by wet chemical means (58). Converting aragonite ($CaCO_3$) whiskers to tubes of polycrystalline CaF_2 with aqueous HF solutions and leaching the remaining $CaCO_3$ with HCl gave tubes ~ 2 μm in diameter with central holes

~ 1 μm in diameter (59). ZrO_2 tubes ~ 100 μm in diameter and with ~ 3 μm wall thickness were made by CVD onto a surface-oxidized Ni wire (60). The former two examples illustrate more specialized chemical techniques and the latter CVD process a more general technique of making fine tubes. Carbon nanotubes (61) and related materials, by themselves or as templets for various depositions, are in principle another source of very fine tubular pores.

10.4.2 Special Bulk Porous Bodies from Beads and Balloons

While the above beads and balloons have a variety of uses as individual beads or balloons as discussed earlier, they also have important applications in making monolithic bodies. Beads or especially balloons may be added to a body mix to contribute pores to a monolith, or in actually forming most of the monolith itself, with both offering important processing advantages. Thus, beads or balloons can be used as fillers in a matrix to contribute their internal and in some cases some external porosity (i.e., as pores between themselves or the matrix) in the body. In such cases they can be added as a single size range or as a mix of or graded sizes in a uniform or graded fashion across one or more of the body dimensions, thus giving substantial design flexibility. Usually such use of beads or balloons is designed to preserve their chemical and physical identity by choosing their and the matrix character accordingly; however, in some cases the beads or balloons may purposely be selected to lose their chemical but not their physical identity via dissolution or reaction. This may be done for various reasons, e.g., to accommodate differential shrinkage between previously sintered beads or balloons and the sintering matrix. This may be done by use of incompletely sintered beads or balloons, their being friable, dissolving in, or reacting with the matrix, or all three. An example of this is the use of glass balloons in an appropriate matrix, e.g., alumina, where the glass dissolves in the matrix, aiding sintering, accommodating differential sintering, and resulting in higher refractoriness than with the glass beads or balloons retaining their identity. Other reaction processes, e.g., for Si_3N_4 and SiC bodies, are also possible. These extend the versatility and use of processing with addition of ceramic or metal beads or balloons.

A large extension of the use of beads and especially balloons is as the primary constituent in making bodies, which can be done with either green or partially or fully sintered beads or balloons (52,62). Clearly fragility can be an issue or limitation in some forming operations (increasing in the order listed), particularly with beads with extensive internal pore structure and especially with balloons; however, even in cases of considerable friability, bodies can be fabricated from beads or balloons, e.g., via various liquid precursor infiltrations, with the use of slips often being particularly practical. There are both various fabrication methods and important advantages to making bodies by bonding prefired beads or balloons together. The advantages are again extensive

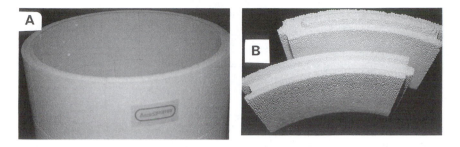

Figure 10.6 Photographs illustrating both the size and shape capability of making bodies by bonding ceramic beads or balloons together. A) Photograph of the top portion of a cylindrical insulation liner (27 in. OD, 24 in. ID, 40 in. tall) for an induction-heated furnace fabricated from 3-mm diameter mullite balloons. Cylinders were fabricated both as a single piece and assembled from smaller tongue-and-groove sections (as shown here and in B). B) Photograph of tongue-and-groove sections used to make the cylinder in (A), which gave greater thermal shock resistance than a single monolithic cylinder. Photos courtesy of Dr. J. Cochran of CFI.

tailorability of body character and microstructures and in nearly zero shrinkage. The latter results from the sintered beads or balloons being in contact so there is essentially no net shrinkage of the body they form. This means that large or complex shaped bodies can very readily be made, and typically with low-cost tooling, e.g., plastic molds (e.g., Fig. 10.6). The tailorability includes not only the selection of the chemical and microstructural characters of the beads or balloons and their distribution, e.g., gradation of their character, in the body along with gradation of the amount and character of the bonding phase, as well as of possible fiber reinforcement. There are various bonding techniques (some proprietary), but a practical and widely used one is simply to use a slip to bond beads or balloons. This can commonly be confined mainly near the contact points, hence minimizing differential sintering shrinkage between the already sintered beads or balloons and the particulate material from the slip. There are various other general and speciality bonding techniques that relax or eliminate differential shrinkage issues and that can be varied to give nearly fully dense filling of the interstices between beads or balloons, and dense, smooth surfaces. Further weight savings and increases in insulating and other capabilities can often be gained by incorporation smaller beads and especially balloons in the interstices between larger beads or balloons.

10.4.3 Direct Fabrication and Processing of Special Bulk Porous Bodies

There are some emerging, very novel methods of fabricating unusual pore structures, especially of very fine (commonly nm scale), uniform pores that can only be noted here. A traditional one with important newer developments is fabrication of zeolites and related materials in other compositions (e.g., phosphates) and different pore size ranges. While zeolite particles can be fabricated into bulk bodies, doing so while maintaining extensive connection of the fine crystallographic pores poses problems; however, small-scale laboratory demonstrations have been made of forming arrays of fine channel pores in other materials. Two other but similar, partially related methods with potential for fabricating/processing novel porous materials are: 1) development of methods to mimic natural processes of making porous inorganic-based skeletal and shell materials (some having a substantial single-crystal character), and 2) developing molecular self-assembly/replication techniques using metal or ceramic precursors. While these may hold large future potential, much of it beyond existing technology, the latter provides substantial existing and developing opportunities, as summarized below.

A common way of producing bodies for application of their porosity is to adapt existing processing and fabrication technology to the need. Thus one of the simplest ways is to incompletely sinter compacts of particulates of the material selected. The engineering is then to select the combination of particle source (hence also often character), size (and often size distribution), consolidation method and parameters, and the extent of sintering to produce the desired function. This mainly empirical approach reflects the lack of ability to adequately characterize porosity and relate its character to specific functions such as needed for most if not all of the applications outlined in the preceding section. This approach, although typically the most inefficient use of mass, i.e., there is the greatest loss of properties while achieving the least porosity, is commonly used and is functional. Its use for manufacture of ceramic filter tubes, discussed in the next section, is a good example.

The other traditional but more versatile way of making more porous bodies is via incorporation of material that is removed during or after partial or complete fabrication of the component. The most common manifestation of this is inclusion of (usually organic) material to be burned out (or volatilized) from the body. Much of the engineering is then in choosing the material (especially for suitable burnout), its size and size distribution, concentration and degree and uniformity of mixing in the green body, and firing schedule. This is a quite versatile process, but can be difficult to control precisely. It gives bodies ranging from closed to open pores (Table 10.4).

Another manifestation of the above process is to make bodies where one of the phases can subsequently be removed, primarily by liquid leaching, mainly as a result of phase separation on cooling (63–66). Phase separation and leaching, which inherently produces open pores, i.e., in the foam regime, is done mainly with glasses, including for making filters, at least experimentally. Phase separation and leaching is generally not applicable to polycrystalline bodies because most or all of the second phase forms along the grain boundaries rather than in

Table 10.4   Ceramic Foam Fabrication Techniques

Foam type	Fabrication/processing technique		
Closed cell	Bonding beads/balloons	Direct blowing	Burnout
Open cell	Phase separate/leach	Direct blowing	Replicate

the grains; however, this apparently can have some applicability to some single crystals, e.g., apparently being the process by which highly porous sapphire crystals have been produced on a laboratory scale. Some single crystals with open porosity and considerable surface area can be obtained by growth conditions that give unstable crystal growth (Fig. 10.7).

The primary method of making reticulated ceramic foams is by replicating polyurethane foams; this represents an advanced, readily available technology (67–69) (Table 10.4). This is done by coating the urethane foam with a ceramic slip, drying and firing to remove the urethane, and then sintering the ceramic. This works fairly well but can present problems of cracking of the resultant ceramic struts due to shrinkage problems in drying the slip, or more commonly from differential expansion of the urethane and the dried ceramic during removal of the urethane. This also leaves the ceramic struts hollow, which may have some advantages, but also complicates understanding the properties of these materials (67–69) (Sections 3.3.1, 4.3.1, 5.3.5, and 6.3.2).

Figure 10.7   Example of $KTaO_3$ crystal surface from unstable crystal growth conditions giving substantial open porosity and resultant surface area. Photograph courtesy of Dr. L. Boatner of ORNL.

Ceramic foams can also be produced directly, with the applicable processes sometimes being material dependent. A basic method is blowing of foams such as in making polymer foams. The basic requirements are to generate sufficient and uniformly distributed gas during a sufficiently fluid stage of the processing, then to have the body rigidize in order to preserve the foam structure (and not to fall as in the improper baking of bread or cake). Glassy carbon (reticulated) foam is made by foaming the polyferferall precursor, rigidizing by polymerazation, and then pyrolysis. Such reticulated glassy carbon foams have also been used as substrates to deposit metal or ceramic coatings by chemical vapor deposition (70); other vapor deposition, e.g., PVD, of both metals and ceramics as well as electrochemical deposition of metals should be feasible. Similar potential is indicated for direct foaming of some other polymeric precursors to ceramics, e.g., some preceramic polymers for SiC and Si_3N_4. The challenge in such cases can be adequate release of pyrolysis gas in the proper stages of viscosity of the precursor.

Addition of a separate blowing agent may be feasible in some cases, as is commonly the case for glass foams (e.g., where $CaCO_3$ is a common blowing agent). Similar processing is indicted for some sol-gel processed materials, although their large mass losses on conversion to ceramics leads to large amounts of fine porosity and resultant surface areas (71–73). The extremes of such processing are zero- and especially aerogels, but the use of a blowing agents, then rigidizing by gelation is also feasible (74–77). Another way of making some foams is by whipping a slurry of another liquid or the ingredients with a surfactant to control surface tension and hence the degree of foaming, then rigidizing the foam via a setting agent such as a cement or plaster-type material. The residue from the latter must be suitably compatible with the desired product; for example, using plaster of paris in rigidizing ZrO_2 foams is appropriate since the CaO from the plaster decomposition can be the stabilizer for the ZrO_2. A more general way of making a substantial range of foams is from porous or hollow beads or balloons as discussed below.

There are at least two other novel techniques of making ceramic foams, especially finer structure ones. First, some, mainly finer structure, foams can also be made by using phase separation in the binder phase. Thus, Brezny and Spotnitz (78) showed that mixing some ceramic powders in a suitable mixture of polyethylene and the proper mineral oil at temperatures where the two organics are mutually soluble can lead to quite uniform ceramic foam structures (Fig. 10.8). This occurs in systems where the polyethylene and mineral oil phases separate on cooling. Apparently the ceramic particles stay primarily with the polyethylene because of its substantially higher viscosity. Thus, upon removal of the oil (chemically or thermally) and firing, a uniform foam on the scale of the phase separation is obtained. Second, fine foams (e.g., with pores in the 0.1– 1 μm range or less) of a number of ceramics have been reported by Yamamoto

Figure 10.8 Example of an alumina foam made by phase separation of the binder phase (76). Photograph courtesy of Dr. R. Brezney of W. R. Grace.

and Masuda (79). Their process consisted of using conventional CVD to produce submicron powders (of Si_3N_4 in their demonstration of the process); however, the reactor was modified such that the particles generated in the gas phase were electrically charged, and the charged particles were deposited by a DC field on an electrode. The particulate deposit, apparently of loose chains of submicron ball-particles, results in a quite uniform, fine foam-like body on firing. The process appears applicable to a wide range of ceramics, i.e., those that can be produced by gas-phase nucleation in CVD.

10.4.4 Porous Membranes

Finally, consider other techniques to form filter structures, especially the membranes on them, or both. As noted earlier a common method for making filters is via partial sintering of consolidated powders, usually done in stages to produce progressively finer pore structures via the use of progressively finer powders (6). Thus, the first step is to make a body with a coarser grain-pore structure that

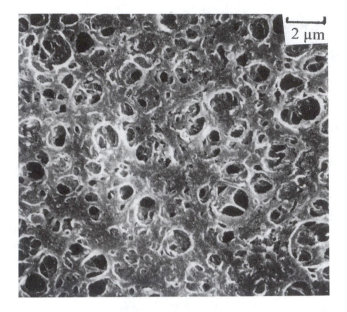

Figure 10.9   Example of ceramic (alumina) membrane formed on an alumina substrate by sol-gel processing, in this case apparently involving a phase separation of the sol and some of the fluid phase (81). Photograph courtesy of Dr. J. Block of W. R. Grace.

not only forms the support structure for the membrane, but also provides the structural integrity. Since tubes are the dominant form of filters, the coarse grain structure is typically extruded, commonly as a multibore tube. Then, usually after firing of the support body, progressively finer structure layers are added as needed on one surface to form the actual filter membrane, usually with each layer being fired before the next layer is added. Most layers are applied via slips, but in some cases the actual filter layer is made by sol-gel processing (80,81). The latter is typically conventional processing, but a promising newer method of forming a membrane layer, apparently involving phase separation between organic ingredients and a sol has been demonstrated (81). This results in fine, quite uniform open pores as shown in Fig. 10.9. (Other methods such as CVD might also be used, e.g., feeding reactive gases from opposite sides of the filter.) This layering allows the tailoring of the filter, makes it aysmetrical in structure, and aids both through flow and reverse flow for possible cleaning via back flushing.

This filter membrane fabrication yields random porosity which has substantial inefficiency of fluid flow as a function of porosity. There are several fabrication methods that can substantially improve on this. One of these is foaming, which allows two important advantages: 1) foams can be made in substantial sizes and shapes with nominal tooling costs, and 2) foams made in a mold have a skin, which can be the membrane or an important component of it, Thus the very open character of open-cell foams can allow much better fluid flow to or away from (for cleaning) the membrane. The skin-membrane can be reinforced or refined in pore structure by adding slip or other fluid cast layers as in conventional filter fabrication.

There are also at least three fabrication concepts (and various degrees of demonstrating the novel processing that they are based on) that can yield essentially straight tubular, channel pores. These are typically or nearly normal to the filter surface and may or may not taper in their cross-section. The first uses burn out of organic fibers (20–200 μm) that can be reasonably oriented and arrayed (e.g., with spacings of (70–400 μm) (82). An extension of this is a conceptual process of growing whiskers normal to a surface (e.g., on a plate) or the inside or outside of a tube, rod, or fiber, then to cast the membrane-filter material against this surface. Selecting a removable whisker material, e.g., graphite, that can be burned out or metal or salt whiskers that can be leached out in processing, leaves through channels, typically with some taper. Whiskers reduce the resultant channel hole diameters to a few microns or less.

The second concept is based on anodization of aluminum films or foils (e.g., 2–250 μm thick) with sulfuric acid (83,84). This has produced nearly circular cylindrical holes normal to the foil/film plane that have fairly uniform diameters in the range of 2–20 nm, with similar spacings between pores (i.e., $P \geq 0.5$). The pore size and spacing can be controlled a fair amount by process parameters, with smaller pores being obtained in thinner sheets; obviously, the resultant foils are fragile and require support. Similar electrochemical generation of pores may be feasible in a few other materials. More recently laboratory demonstration of fabricating through holes in an alumina tube have been reported by this technique. Kobayashi et al. (85) report converting an aluminum tube 50 mm long and 2 mm outer diameter (with a 0.2-mm wall thickness) to an alumina tube with radial holes ~ 150 nm in diameter. Tubes offer greater mechanical integrity, e.g., their tubes could with stand external pressures of 1 MPa.

The third method, for which there has been promising fabrication demonstrations of forming dielectric ceramics with tubular channels that taper substantially, is by a simple, versatile adaptation of electrophoretic deposition (86) (Fig. 10.10). This is very much like electroplating, but instead of depositing atoms from ions in a conducting bath under an applied electrical field, colloidal particles are deposited from a nearly nonconducting bath due to the inherent but adjustable charges on their surfaces. This is done with very simple equipment and can produce substantial size and versatile shape components, as shown by other applications, e.g., demonstration of beta alumina battery tubes (87). The generation of tubular pores takes advantage of a basic limitation that this

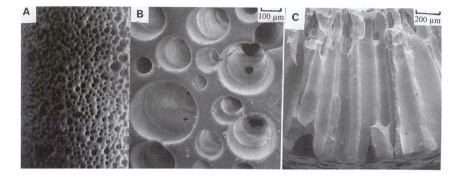

Figure 10.10 Example of radial pores in an alumina tube formed by electrophoretic deposition, using bubbles formed by electrolysis of the water-based slurry to form the U- or V-shaped radial pores (86). Note the large opening on the outer surface and the fine opening (a few microns) at the inside and essentially no porosity elsewhere (making the material quite strong for the level of porosity). Photograph courtesy of Dr. A. Kerkar of W. R. Grace.

process often experiences, especially when used with (typically lower cost and safer) aqueous-based colloidal suspensions. In such cases deposition rates are normally limited by some electrolysis (i.e., electrical breakdown) of the water and resultant bubbling at the electrodes. Taking advantage of this results in pores having a length-wise U or V cross-section, which is attributed to bubbles nucleating at the electrode surface then growing in diameter and moving their centers outward from the electrode as ceramic particles deposit around them. The large pore diameter depends on the wall thickness, but is typically substantial, while that at the base of the U or V is of the order of a few microns. Again finer membrane layers can be added to the filter surface and, as with foam methods, the process is versatile in shape, including possibilities of including some aspects of manifolding in the fabrication of filters (or burners). In a much earlier stage of development are efforts indicating the feasibility of combining two technologies noted earlier, namely forming a single layer of phosphate-based zeolite-type crystals or an anodized aluminum substrate having parallel holes (83).

10.5 DISCUSSION AND CONCLUSIONS

10.5.1 Engineering Trade-offs, Costs, Nondestructive Evaluation, and Systems Aspects

Engineering to make application of porous (in fact any) materials successful require that the components offer suitable tradeoffs between different properties and fabrication/performance versus cost, while also providing suitable levels of reliability and system compatibility. These are each substantial topics themselves and so can only be noted or outlined here. Consider first property tradeoffs, which depend directly on the porosity dependence of the different properties.

Whatever the primary function of the porosity, e.g., for lighter weight, lower thermal conductivity, lower dielectric constant, higher thermal shock resistance, there are invariably other physical or chemical requirements effected by porosity that the components must meet. Since the additional requirements are also usually dependent on porosity, meeting them typically requires some tradeoff between the porosity dependence of the different properties involved, which requires understanding the similarities and differences in the porosity dependences of different properties. This was briefly outlined for porous sonar transducers in Section 10.4.1; important examples of such additional requirements are outlined in Table 10.5. Thus, all high-temperature applications impose serious thermal shock requirements from startup, temperature variations, and shut downs (emergency and accidental). Other challenges to maintenance of component integrity arise from various factors such as thermal/chemical attack, erosion, and mechanical stressing. The latter arises when, for example, molten metal filters must support a substantial initial head of molten metal, and industrial exhaust stack catalysts must support a substantial load of their own mass.

Meeting the additional requirements for applications requires different tradeoffs that bear directly on the amount, type, or both of the porosity selected (along with material selection and fabrication). If all properties involved have the same or very similar porosity dependences, then the challenge is to select the type and level of porosity. This will be dictated by the property requirement necessitating the lowest level of porosity, which commonly depends on the type of porosity (hence also commonly on the fabrication/processing) selection. Such

Table 10.5 Examples of Other Requirements in Application of Porous Materials

Application[a]	Basic requirement[a]	Other requirements[a,b]
Auto exhaust catalyst	High, stable surface area	TS, T/C, erosion, melting
Industrial exhaust stack catalyst	High, stable surface area	TS, load support, T/C, erosion
Lower temperature filters	Particulate removal	TS, system stresses, erosion
Coal combustion gas filter	Particulate removal	TS, slag attack/stresses
Molten metal filters	Particulate removal	TS, T/C, molten head load
Microwave windows, domes	Low dielectric constant	TS, erosion

[a] See Table 10.1.
[b] TS = thermal shock, T/C = thermal/chemical. Listed in approximate order of decreasing importance in material selection and porosity requirements.

selection may or may not be adequate for the other requirements; if not, design or other tradeoffs must be successfully made if the application is to be realized.

The other, more common, possibility is where the different requirements impose differing porosity dependences. There are many examples of this. Thus, per Table 10.5, thermal shock resistance is often required, which often improves with increasing porosity, at least over certain ranges and characters of porosity (that are not well understood, Sections 4.4, 9.3.2); however, other accompanying requirements commonly require varying resistance to component degradation, e.g., from erosion, thermal/chemical attack, and mechanical failure. The latter, while typically being dominated by tensile strength, may also be impacted by other factors such as extensive crack growth or elastic buckling of members. All of these entail various types of porosity dependences, but usually with accelerating property decreases as the level of porosity increases. Similarly, many catalyst and related applications require high surface areas as well as high permeability; however, greater surface area is obtained from pores between fine, packed particles and less from larger pores providing channels into the body, while the reverse is true for permeability. Thus, a balance of the two types of porosity is needed, along with balances needed to meet other performance requirements such as maintaining mechanical integrity. While this is more complicated, it also allows for more creative engineering and invention. It is essential to recognize that variation in the porosity dependence with pore character is an essential factor in such engineering. Thus, modeling and testing which neglect pore character contribute little to the needs, except confusion.

A key requirement for applications of porous ceramic or other material components, as for most dense ones, is that they meet the system requirements. Thus, for example, filters not only must remove the particulates or bacteria desired, they must do this at the required input and output stream pressures, flow rates, and their fluctuations. This typically requires much larger filters than typically made in the laboratory and series, and especially parallel, connections that place important manifolding requirements that can significantly impact the manufacture and use of the filters. Similarly, cleaning by back flushing, sterilization, or both can add significant demands of the filter pore structure and resultant component integrity.

Another key engineering tradeoff is that of cost, which is impacted by the material(s) and the amount and character of porosity selected since these impact the fabrication/processing selected. For larger volume applications, e.g., of the applications in Table 10.2, the costs of the ceramics for most of these must be less than a few dollars per pound. This often favors or dictates use of natural materials, either as the product or as key constituents in it. Pumice has been considered as a natural material for a host material for biological growth (42,43), and diatamatious earth has been used for filtering applications. Similarly, the selection and fabrication of cordierite as the support for automobile exhaust catalysts in part reflected the use of clay as a major ingredient that kept raw materials costs down. This also had the unexpected benefit of resulting in preferred orientation of the clay particles in the extruded honeycomb. This preferred orientation carried over to the reacted, sintered body enhancing its thermal shock resistance because of more favorable orientation of the anisotropic thermal expansion of cordierite (88).

An important aspect of applications is reliability of the system and hence of the components, which becomes more challenging as the level of property demand increases relative to the levels commonly achieved by available fabrication/process technology. With regard to porous ceramic components, it was shown earlier that the limited data on Weibull moduli of test samples gave lower moduli, e.g., 3–9, than denser bodies, e.g., 9–12 (Fig. 5.7). Such moduli will typically be lower in components, especially as their size, surface area and shape complexity increases. Thus, mechanical reliability may be challenging for at least some of the applications. Proof testing may be helpful, but has typically been restricted to simple stress states, mainly those obtained in rotating parts (e.g., for grinding wheels and turbo charger rotors). Nondestructive evaluation (NDE), may also be of useful, but conventional approaches of identifying probable sources of failure in components presents three challenges. The first is in identification of probable sources of failure. Large isolated pores or pore clusters that are frequent sources of failure in denser bodies (Sections 5.2 and 5.3) are often detectable; however, in more porous bodies the challenge is typically to find areas of combinations of: 1) much less extreme statistical differences in the amount and sizes of pores from the matrix, and 2) other, mainly surface defects. The latter are typically machining flaws if the component has been machined, impact or other handling flaws, or processing defects, and mainly surface irregularities, in as-fired or as-processed components. These present challenges that can often be met by practical, conventional, or innovative engineering.

There is substantial opportunity to improve the effectiveness of porous bodies for various applications driven by the functions their porosity serve. A key to this is better understanding of how the amount and character of porosity affect different properties. A longer term potential opportunity to improve the design of porous bodies is to develop fabrication technology for and conduct evaluations of proposed benefits from hierarchial pore structures (1). These are structures with two or more scales of pores such that the walls or struts around the largest scale of pores in turn have similar pores, but of a much finer scale dictated by the scale of the larger pores and surrounding walls or struts. This hierarchial scale of porosity is expected to improve the effectiveness of the porosity in giving lightweight structures with the addition of each pore scale.

The relative property curves such as Figures 2.11 and 3.1 are basic guides for the weight, i.e., mass, efficiency for the many properties they are pertinant to, whether for hierarchial porosity development or just weight efficiency for a simple pore structure. Thus, note that there are basically two groups of pore types: 1) tubular and spherical pores as well as other pores larger than the particle or grain structure, and 2) pores between packed and bonded particles. Pores of the first group reduce component properties much more slowly with increasing porosity than do those of the second group. Since component mass is simply the volume fraction solid (= 1 − P) times the component volume and material density, the two pore groups give respectively much greater and much lower effective use of their mass for a given mass (and hence P level) for property retention. Thus selection of pores for efficient use of mass in preserving as much of the properties as possible is based primarily on selection of the first group of pores. Further selection is based on first meeting other requirements of the application and then refining the mass efficiency by selecting the particular pore geometry (or combinations) that give the best property retention. The latter includes more detailed analysis of the property dependence for a given pore structure, which includes effects of pore stacking. Other requirements that may often be dominant are the total porosity allowed, the degree to which: 1) anisotropic properties are allowed, and 2) open or closed porosity may be needed. Again, as noted in conjunction with surface area effects, combinations of different pore types may often be most beneficial, by raise challenges to both design and processing.

10.5.2 Measurement and Modeling Needs

A critical need in further study and development of porous materials is for measurements and modeling to be carried out in a more comprehensive fashion with a broader perspective on pore structures. This requires addressing three key issues, the first of which is more and better characterization of the pores themselves, i.e., clearly beyond only the volume fraction porosity typically given, as discussed in Section 2.3, Table 2.3. Thus, some evaluation of pore shape, stacking [e.g., as indicated by pore or particle (or both) coordination numbers], location, orientation, and size are needed. While addressing heterogeneity is particularly necessary in some natural materials such as rocks, it is important in all materials. Since pore character is important, direct (even qualitative) observations for anisotropy of pore shape and degree of preferred orientation as well as direct tests to ascertain isotropy or anisotropy of resultant properties are important to quantify. This task can be aided by the interrelation that typically exists between these factors and their dependence on fabrication/processing methods and parameters. Thus, pore association with other microstructural features, e.g., within grains, at grain boundaries, and so on, can be important, the extent of this should be sought, along with changes of this and other pore characteristics as P changes. Similarly, in composites, the location of the pores and their associated character in the different phases can be as or more important, especially in fiber composites as discussed in Sections 8.2.1, 8.3.6 (Figs. 8.1, 8.15).

The second key issues in pore characterization is identifying and quantifying anisotropy of pore shape, or shape of pore cluster, both locally and on a larger scale. The latter results in anisotropy of properties, so checking for this (and if indicated then quantifying the anisotropy) becomes a key factor in evaluating pore-shape orientation effects. Furthermore, since almost all pore-property anisotropy results directly from fabrication/ processing factors, reporting and evaluating their effects becomes a major factor in pore characterization.

The third key aspect of pore characterization is the spatial distribution of both the amount and character of the pores, which entails several aspects. One of these is the location of pores at grain or phase boundaries, i.e., intergranular, versus intragranular pores. Intergranular pores were shown in earlier chapters to play an important role across a wide variety of properties, with particularly pronounced effects indicated in fiber composites. While pores typically start in intergranular locations, grain growth often significantly changes this, and thus becomes an important indicator of changes in pore location (and associated character) change. The other, often more prevalent and serious aspects of pore distribution, are the broader issues of the more global spatial distribution of the amount or character of pores. This entails both gradients and heterogeneities of the pore distributions, which if identified can be handled separately, but both are treatable by their effect on data scatter. Both of these, but especially gradients, are again related to fabrication/processing methods and parameters which again need to be used as clues to such effects. Laminar arrays of pores are an important example.

More statistical evaluation of data is needed to address the above issues, but is particularly important for addressing effects of varying spatial distribution. While, standard statistical techniques are an important aspect of this, it has been suggested in this book that use of the Weibull distribution, widely used for strength (especially tensile strength) of ceramics, be extended to other properties as well as to the porosity measurements themselves. This is illustrated by the summary of such application in Table 10.6. Thus, for example, the Weibull modulus (m) for P values themselves vary widely, with high values indicating narrower distributions and hence less heterogeneity of the distribution of the amount of porosity. Good correlation of Weibull moduli with P and properties should indicate important effects of the distribution of the amount of porosity, while poorer correlation indicates other factors such as the distribution of the character of the pores.

Weibull moduli for various properties also show wide variations, often with expected trends. Thus, higher m values for sound velocities versus for elastic moduli is consistent with the latter depending on the square of the former (Eqs. 1.1, 1.2), as are lower m values for properties that are dependent on fracture.

Table 10.6 Summary of Weibull Moduli (m) of Porous Ceramics and Rocks[a]

Material[b]	m(P)	m(v_l)	m(E)	m(G)	m(B)	m(v)	m(σ_t)[c]	m(σ_c)[c]	Investigator
Al$_2$O$_3$	600	500–1000	150–300	200–300	80–190	200–300			Lang (89)
Al$_2$O$_3$	40–80		~150				7–26		Neuber and Wimmer (90)
MgAl$_2$O$_4$	30		100				18		
ZrO$_2$	50		45				10		
ThO$_2$	16–140		80–430	170–470					Spinner et al. (91)
FXG	24		6				7		Pernot et al. (92)
Rock		10–50	3		4	3	2	2	Price (93)

[a] Weibull moduli for: P = volume fraction porosity, v_l = sound velocity, E = Young's modulus, G = shear modulus, v = Poisson's ratio, σ_t = tensile (or flexure) strength, and σ_c = compressive strength.

[b] All compounds shown are for bodies from compacted powders sintered to various densities; FXG = foamed, crystallized glass; Rock refers to Yucca Mt. tuff.

[c] These and other fracture properties also generally depend on grain size, in which case its variation should also be addressed in evaluating property variations. The Al$_2$O$_3$ data of Neuber and Wimmer allows some evaluation of this, giving m values for it of 36–80 (with no clear correlation with the m values for σ_t). Note that while substantial data is not available, m values for fracture toughness are commonly of the order of 10–20.

Although there is some trend for lower m values for elastic properties at higher P, this is far from universal, showing substantial variations in heterogeneity of porosity at all P levels. The very low m values for various properties of rocks from more comprehensive studies is consistent with heterogeneities that are more likely in such materials. Similarly, low Weibull moduli for ceramic foams (e.g., 3–8) are consistent with variations expected from properties being dependant on fewer microstructural elements and variations in these. The latter are indicated by estimated m values of struts of two ceramic foams being in the range of 3–6.

Improved characterization is thus one of the most critical needs for improved documentation and understanding of porosity effects on properties. Understanding the fabrication/processing techniques used to produce the bodies being studied and the effects of their parameter changes on the resulting porosity can be an important tool that should be much more widely used and reported. Similarly, measurement of properties beyond the immediate range of porosity of interest, as well as of other properties on the same bodies can be very valuable in better characterizing and understanding porosity effects on properties. It is important to recall that many of these advances will be iterative in nature. Thus measurements of porosity effects improves the characterization of porosity,

Table 10.7 Comparison of MSA and Mechanistic Models for the Porosity Dependence of Mechanical Properties

Model type	Nontu-bular pore type range	Pore stacking range	Typical P range [a]	P_C values identi-fied [b]	Combine pore effects	Rigorous mechanics	Prediction of indivi-dual v-, B-, E-, and G-P depen-dences	Demon-strated model-data agreement
MSA	Wide	Wide	0–1	Yes	Yes	No	No [c]	Substantial
Mechanics	Narrower, mainly spherical	None	P to ~ 0.1–0.4, but not defined	No	Some	Yes (for B and some-times G)	Yes	Limited

[a] P = volume fraction porosity.

[b] P value at which the solid falls below its percolation limit so that the body is no longer a coherent solid.

[c] MSA models do not predict the difference in P dependencies, but are not inconsistant with them.

which in turn improves understanding of pore effects and hence also further improves measurement, and so on.

Improved modeling is also an important need. While MSA models are widely applicable to many properties, i.e., most mechanical and conductive properties (Table 10.7), improvements are needed. This includes not only modeling other stackings of idealized pore geometries, but also modeling of bodies with mixed pore sizes, shapes, and stackings, and broader spatial heterogeneity. Similarly, for foam models, modeling of other cell and stacking geometries as well as of tapered struts and walls and statistical aspects of their failure are important. Another basic need is to improve the accuracy/validity of the underlying mechanics, which might be done by marrying MSA modeling with mechanistic or computer modeling, or both. A critical need is to develop MSA models that are not based on conservation of mass as for sintering, and hence would be more applicable to reaction-processed materials, e.g., cements, as well as CVI materials.

Other models, especially those based on more rigorous mechanical analysis of elastic behavior, have substantial potential, but have been limited, for example, to isolated (i.e., dilute contents) spheroidal (mainly spherical) pores. Both some of their potential as well as many of their limitations thus far are summarized in Table 10.7. One advantage of mechanistic models not shown there is their success in modeling elastic effects of microcracks and the promise of handling mixes of microcracks and pores. (MSA models have not been successfully applied to microcracks, and can only be used for combined pore and microcrack effects by interpolation between pore and microcrack models.) Some improvements in other models may be via correlation and guidance from MSA models, e.g., considering pore stackings based on solutions of arrays of holes, and possibly a marriage of mechanistic and MSA models should be useful. An overall need is to better identify the bounds of their limited applicability. A particular need is to identify the extent to which initial quantification (94) of the aspect ratio and degree of preferred orientation of spheroidal pores can be done in real bodies and compared on a substantial basis to resultant modeling. A major improvement would be modeling one or more other basic pore shapes, with pores between spheroidal particles being particularly important. This, with improved spheroidal pore modeling, would give a more comprehensive treatment of pores. Evaluation of the correlation and possible marriage of these models with MSA models should also be quite useful, but must be done in the context of the realities of porosity. Thus, for example, all porosities have some P_C value, and commonly associated with this are increasing nonlinearities in property dependences in approaching this value. The location, character, heterogeneities of the porosity and changes of these as the processing and porosity change must also be considered. The recent modeling of Roberts and Knackstedt (95) that essentially generates microstructures appears to offer new modeling possibilities, e.g., a possible marriage with MSA models.

An important step in improving all models is to better identify the specific contributions of different factors. This has been done to a substantial extent on a quantitative basis for effects of the amount and character of the porosity via MSA models. It has also been done to some extent, more on a qualitative basis, for other factors such as for heterogeneity of the size, stacking, shape, and especially spatial distribution of porosity; however, this needs to be done in a more quantitative and more systematic fashion. This may often be done by adjusting key exponential factors such as b and n, but other approaches should also be considered, e.g., multiplier factors of the main P-dependence function. Since contributions of these other factors are also generally functions of P and other microstructural parameters, these must be identified. Computer modeling appears to be a particularly valuable tool for this, but to be particularly effective this must be done with a much broader understanding of pore and related microstructural parameters than has commonly been the case.

REFERENCES

1. R. Lakes, "Materials with Structural Hierarchy," Nature, **361**, pp. 511–15, 2/11/1993.
2. R. W. Rice, "Evaluation and Extension of Physical Property-Porosity Models Based on Minimum Solid Area," J. Mat. Sci., **31**, pp. 102–8, 1996.

3. R. W. Rice, "Comparison of Physical Property-Porosity Behavior with Minimum Solid Area Models," J. Mat. Sci., **31**, pp. 1509–28, 1996.
4. R. W. Rice and R. C. Pohanka, "Grain-Size Dependence of Spontaneous Cracking in Ceramics," J. Am. Cer. Soc. **62**(11–12), pp. 559–63, 1979.
5. A. S. Michaels, "New Vistas for Membrane Technology," Chemtech, pp. 162–72, 3/1989.
6. J. Charpin, A. J. Burggraaf, and L. Cot, "A Survey of Ceramic Membranes for Separations in Liquid and Gaseous Media," Ind. Cer., **11**(2), pp. 84–89, 1991.
7. J. Haggin, "Ceramic, Metallic Devices Extend Membrane Separation Technology," C & E News, pp. 34–35, 7/31/1989.
8. K. K. Chan and A. M. Brownstein, "Ceramic Membranes-Growth Prospects and Opportunities," Am. Cer. Soc. Bul., **70**(4), pp. 703–7, 1991.
9. J. F. Zievers, P. Eggersted, and E. C. Zievers, "Porous Ceramics for Gas Filtration," Am. Cer. Soc. Bul., **70**(1), pp. 108–11, 1991.
10. M. A. Alvin, T. E. Lippert, and J. E. Lane, "Assessment of Porous Ceramic Materials for Hot Gas Filtration Applications," Am. Cer. Soc. Bul., **70**(9), pp. 1491–97, 1991.
11. H. E. Johnson and B. L. Schulman, "Assessment of the Potential for Refinery Applications of Inorganic Membrane Technology—An Identification and Screening Analysis," Final report for U.S. Department of Energy contract No. DE-ACO1–88FE61680 (Task 23), 5/1993.
12. T. Masuda, K. Tomita, and T. Iwata, "Application of Ceramic Foam," Porous Materials (K. Ishizaki, L. Shepard, S. Okada, T. Hamasaki, and B. Huybrechts, Eds.), American Ceramics Society, Westerville, OH, pp. 285–93, 1993. Ceramic Transactions, vol. 31.
13. F. Takahashi and Y. Sakai, "Amylose Separation from Starch and Glucose Separation from Starch Saccharified Solution by Filtration Through a Ceramic Membrane," Porous Materials (K. Ishizaki, L. Shepard, S. Okada, T. Hamasaki, and B. Huybrechts, Eds.), American Ceramics Society, Westerville, OH, pp. 401–10, 1993. Ceramic Transactions, vol. 31.
14. M. Asaeda, K. Okazaki, and A. Nakatani, "Preparation of Thin Porous Silica Membranes for Separation of Non-Aqueous Organic Solvent Mixtures by Pervaporation," Porous Materials (K. Ishizaki, L. Shepard, S. Okada, T. Hamasaki, and B. Huybrechts, Eds.), Ceramic Transactions, vol. 31, American Ceramics Society, Westerville, OH, pp. 411–20, 1993.
15. Y. Kondo, Y. Hashizuka, S. Okada, and M. Shibayama, "Application of Porous Alumina Ceramics for Casting Molds," Porous Materials (K. Ishizaki, L. Shepard, S. Okada, T. Hamasaki, and B. Huybrechts, Eds.), American Ceramics Society, Westerville, OH, pp. 325–33, 1993. Ceramic Transactions, vol. 31.
16. S. Komarneni and P. B. Malla, "Microporous Gels As Dessicants," Porous Materials (K. Ishizaki, L. Shepard, S. Okada, T. Hamasaki, and B. Huybrechts, Eds.), American Ceramics Society, Westerville, OH, pp. 295–303, 1993. Ceramic Transactions, vol. 31.
17. M. Kubota and T. Kojima, "Development of Porous Ceramics for a Negative-Pressure Difference Irrigation System," Porous Materials (K. Ishizaki, L. Shepard, S. Okada, T. Hamasaki, and B. Huybrechts, Eds.), American Ceramics Society, Westerville, OH, pp. 353–60, 1993. Ceramic Transactions, vol. 31.
18. N. Ueno, "Effect of Porous Materials on the Generation and Growth of Bubbles in Aeration," Porous Materials (K. Ishizaki, L. Shepard, S. Okada, T. Hamasaki, and

B. Huybrechts, Eds.), American Ceramics Society, Westerville, OH, pp. 343–52, 1993. Ceramic Transactions, vol. 31.

19. Y. Sawai, K. Ishizaki, S. Nigata, and R. Jain, "Prospects for Obtaining a Superconducting Filter to Purify Oxygen from Argon," Porous Materials (K. Ishizaki, L. Shepard, S. Okada, T. Hamasaki, and B. Huybrechts, Eds.), American Ceramics Society, Westerville, OH, pp. 335–41, 1993. Ceramic Transactions, vol. 31.

20. S. Reich and I. Cabasso, "Separation of Paramagnetic and Diamagnetic Molecules Using High-TC Superconducting Ceramics," Nature, **338**, pp. 330–32, 3/23/1989.

21. K. Nagata, H. Igarashi, K. Okazaki, and R. C. Bradt, "Properties of an Interconnected Porous Pb(Zr, Ti)O$_3$ Ceramics," Jap. J. Appl. Phy., **19**(1), pp. L37–40, 1980.

22. M. Kahn, R. W. Rice, and D. E. Shadwell, "Preparation and Piezoelectric Response of PZT Ceramics with Anisotropic Pores," Adv. Cer. Mat., **1**(1), pp. 55–60, 1986.

23. R. W. Rice, M. Kahn, and D. E. Shadwell, "Ceramic Body with Ordered Pores," U.S. Patent No. 4,683,161, 7/28/1987.

24. M. Kahn, "Acoustic and Elastic Properties of PZT Ceramics with Anisotropic Pores," J. Am. Cer. Soc., **68**(11), pp. 623–28, 1985.

25. M. Kahn, A. Dalzell, and B. Kovel, "PZT Ceramic-Air Composites for Hydrostatic Sensing," Adv. Cer. Mat., **2**(4), pp. 836–40, 1987.

26. R. Meyer, H. Weitzing, Q. Xu, Q. Zhang, R. E. Newnham, and J. K. Cochran, "PZT Hollow-Sphere Transducers," J. Am. Cer. Soc., **77**(6), pp. 1669–1772, 1994.

27. J. E. McEntyre, J. K. Cochran, and K. J. Lee, "Permeability and Porosity of Bonded Hollow Sphere Foams for Use in Radiant Burners," Spheres and Microspheres: Synthesis and Applications (D. L. Wilcox, M. Berg, T. Bernat, J. K. Cochran, and D. Kellerman, Eds.), Materials Research Society, Pittsburg, PA, pp. 165–71, 1995. Materials Research Society Proceedings, vol. 372.

28. F. J. Parker, "Low Thermal Expansion ZrTiO$_4$–Al$_2$TiO$_5$–ZrO$_2$ Compositions," U.S. Patent 4,758,542, 7/1988.

29. F. J. Parker, "Al$_2$TiO$_5$–ZrTiO$_4$–ZrO$_2$ Composites: A New Family of Low-Thermal-Expansion Ceramics," J. Am. Cer. Soc., **73**(4), pp. 929–32, 1990.

30. H. P. Hsieh, "Inorganic Membrane Reactors," Catal. Rev. Sci. Eng., **32**(1&2), pp. 1–70, 1991.

31. T. Kokugan and G. Keitoh, "Applications of Microporous Materials to Membrane Reactors," (K. Ishizaki, L. Shepard, S. Okada, T.Hamasaki, and B. Huybrechts, Eds.), American Ceramics Society, Westerville, OH, pp. 421–31, 1993. Ceramic Transactions, vol. 31.

32. D. C. Hill and H. L. Tuller, "Ceramic Sensors: Theory and Practice," Ceramic Materials for Electronics (R. C. Buchanan, Ed.), Marcel Dekker, Inc., New York, pp. 265–374, 1986.

33. N. Ichinose, "Electronic Ceramics for Sensors," Am. Cer. Soc. Bul., **64**(5), pp. 1581–85, 1985.

34. G. Fisher, "Ceramic Sensors: Providing Control Through Chemical Reactions," Am. Cer. Soc. Bul., **65**(4), pp. 622–29, 1986.

35. M. Uo, M. Numata, I. Karube, and A. Makishima, "Mercuric Ion Sensor with FIA System Using Immobilized Mercuric Reductase on Porous Glass," Porous Materials (K. Ishizaki, L. Shepard, S. Okada, T.Hamasaki, and B. Huybrechts, Eds.), American Ceramics Society, Westerville, OH, pp. 361–70, 1993. Ceramic Transactions, vol. 31.

36. K. S. Patel and H. T. Sun, "Ceramic Semiconductors for Gas Detection," Porous Ceramic Materials: Fabrication, Characterization, Application, Key Engineering

Materials, vol. 115 (D. M. Liu, Ed.), Trans Tech Publications, Lebanon, NH, pp. 181–90, 1996.

37. H. T. Sun, C. Cantalini, and M. Pelino, "Porosification Effect on Electroceramic Properties, Porous Ceramic Materials: Fabrication, Characterization, Application, Key Engineering Materials, vol. 115 (D. M. Liu, Ed.), Trans Tech Publications, Lebanon, NH, pp. 167–80.

38. J. Y. Ding, M. R. Shahriari, and G. H. Sigel, Jr., "Fiber Optic Moisture Sensors for High Temperatures," Am. Cer. Soc. Bul., **70**(9), pp. 1513–18, 1991.

39. H. Abe, H. Seki, A. Fukanga, and M. Egashira, "Bimodal Porous Coordierite Ceramics for Yeast Cell Immobilization," Porous Materials (K. Ishizaki, L. Shepard, S. Okada, T. Hamasaki, and B. Huybrechts, Eds.), American Ceramics Society, Westerville, OH, pp. 371–80, 1993. Ceramic Transactions, vol. 31.

40. H. Horitsu, "A New Approach That Uses Bioreactors with Inorganic Carriers (Ceramic) in the Production of Fermented Foods and Beverages," (K. Ishizaki, L. Shepard, S. Okada, T.Hamasaki, and B. Huybrechts, Eds.), American Ceramics Society, Westerville, OH, pp. 381–89, 1993. Ceramic Transactions, vol. 31.

41. M. Kawase, Y. Kamiya, and M. Kaneno, "Porous Ceramic Carrier for Bioreactor," (K. Ishizaki, L. Shepard, S. Okada, T.Hamasaki, and B. Huybrechts, Eds.), American Ceramics Society, Westerville, OH, pp. 391–400, 1993. Ceramic Transactions, vol. 31.

42. F. J. Parker, A. W. Kerkar, and R. Brezny, "Low Density Glassy Materials for Bioremediation Supports," U.S. Patent No. 5,397,755, 3/14/1995.

43. F. J. Parker, A. W. Kerkar, and R. Brezny, "Low Density Glassy Materials for Bioremediation Supports," U.S. Patent No. 5,519,910, 5/21/1996.

44. F. H. Lin, C. C. Lin, H. C. Liu, and C. Y. Wang, "In-Vivo Evaluation of Porous $Ca_2P_2O_7$ with Sodium Phosphate Addition in Orthopaedics," Porous Ceramic Materials: Fabrication, Characterization, Application, Key Engineering Materials, vol. 115 (D. M. Liu, Ed.), Trans Tech Publications, Lebanon, NH, pp. 191–208, 1996.

45. D. Mo. Liu, "Porous Hydroxyapatite Bioceramics," Porous Ceramic Materials: Fabrication, Characterization, Application" Porous Ceramic Materials: Fabrication, Characterization, Application, Key Engineering Materials, vol. 115 (D. M. Liu, Ed.), Trans Tech Publications, Lebanon, NH, pp. 209–32, 1996.

46. W. Q. Chen, S. X. Qu, Z. J. Yang, X. D. Zang, and M. Q. Yuan, "The Histological Observation of the Early Osteogenesis Induced in Porous Calcium Phosphate Ceramics in Muscular Tissue of the Dogs," Porous Ceramic Materials: Fabrication, Characterization, Application, Key Engineering Materials, vol. 115 (D. M. Liu, Ed.), Trans Tech Publications, Lebanon, NH, pp. 233–36, 1996.

47. J. R. Spann, R. W. Rice, W. S. Coblenz, and W. J. McDonough, "Laser Processing of Ceramics," Emergent Process Methods for High-Technology Ceramics, (R. F. Davis, H. Palmour III, and R. L. Porter, Eds.), Plenum Publishing Corp, New York, pp. 473–503, 1984.

48. J. J. Rasmussen, "Surface Tension, Density, and Volume Change on Melting of Al_2O_3 Systems, Cr_2O_3, and Sm_2O_3," J. Am. Cer. Soc., **55**(6), p. 326, 1972.

49. R. Brezny, "Porous Ceramic Beads," U.S. Patent No. 5,322,821, 7/21/1994.

50. P. R. Chu and J. K. Cochran, "Coaxial Nozzle Formation of Hollow Spheres from Liquids I: Model," J. Am. Cer. Soc., **78**, pp, 1995.

51. J. H. Chung, J. K. Cochran, and K. J. Lee, "Compressive Mechanical Behavior of Hollow Ceramic Spheres," Spheres and Microspheres: Synthesis and Applications

(D. L. Wilcox, M. Berg, T. Bernat, J. K. Cochran, and D. Kellerman, Eds.), Materials Research Society, Pittsburg, PA, pp. 179–86, 1995. Materials Research Society Proceedings, vol. 372

52. R. B. Clancy, J. K. Cochran, and T. H. Sanders, "Fabrication and Properties of Hollow Sphere Nickel Foams," Spheres and Microspheres: Synthesis and Applications (D. L. Wilcox, M. Berg, T. Bernat, J. K. Cochran, and D. Kellerman, Eds.), Materials Research Society, Pittsburg, PA, pp. 155–63, 1995. Materials Research Society Proceedings, vol. 372.

53. X. Yang and T. K. Chaki, "Millimetre-Sized Hollow Spheres of Lead Zirconate Titanate by a Sol-Gel Method," J. Mat. Sci., **31**, pp. 2563–7, 1996.

54. A. K. Bhattacharya and A. Hartridge, "The Preparation of Spinel Ferrites with Spherical morphology," J. Mat. Sci. Let., **15**, pp. 1842–43, 1996.

55. Z. Zhang, Y. Tanigami, and R. Terai, "A Novel Comparison of the Growth of SiO_2 Spheres," J. Mat. Sci. Let., **15**, pp. 1902–4, 1996.

56. G. Pravdic and M. S. J. Gani, "The Formation of Hollow Spherical Ceramic Particles in a D. C. Plasma," J. Mat. Sci., **31**, pp. 3487–95, 1996.

57. S. Hashimoto and A. Yamaguchi, "Growth of Hollow Cr_3C_2 Polycrystals with Cr_2O_3" J. Am. Cer. Soc., **78**(7), pp. 1985–88, 1995.

58. R. Kataoka and Y. Mizutani, "Formation of Fibrous Zirconium Phosphate," J. Mat. Sci. Let., **15**, pp. 174–5, 1996.

59. Y. Ota, S. Inui, T. Iwashita, and Y. Abe, "Preparation of CaF_2 Fibers," J. Am. Cer. Soc., **79**(11), pp. 2986–88, 1996.

60. A. Mineshige, M. Inaba, Z. Ogumi, T. Takshashi, T. Kawagoe, A. Tasaka, and K. Kikuchi, "Preparation of Yttria-Stabalized Zirconia Microtube by Electrochemical Vapor Deposition," J. Am. Cer. Soc., **78**(11), pp. 3157–59, 1995.

61. S. Iijima, "Carbon Nanotubes," Materials Research Society Bul., pp. 43–49, 11/1994.

62. R. W. Rice and R. Brezny, "Method for Making Sintered Bodies," U.S. Patent No. 5,167,885, 12/1/1992.

63. T. Yazawa, "Present Status and Future Potential of Preparation of Porous Glass and its Application," Porous Ceramic Materials: Fabrication, Characterization, Application Key Engineering Materials, vol. 115 (D. M. Liu, Ed.), Trans Tech Publications, Lebanon, NH, pp. 124–46, 1996.

64. S. Morimoto, "Porous Glass: Preparation and Properties," Porous Ceramic Materials: Fabrication, Characterization, Application Key Engineering Materials, vol. 115 (D. M. Liu, Ed.), Trans Tech Publications, Lebanon, NH, pp. 147–58, 1996.

65. H. Abe, H. Tsuzuki, A. Fukungua, H. Tateyama, and M. Egashira, "Preparation of Microporous Material from Coordierite by Acid Treatment," Porous Ceramic Materials: Fabrication, Characterization, Application Key Engineering Materials, vol. 115, (D. M. Liu, Ed.), Trans Tech Publications, Lebanon, NH, pp. 159–66, 1996.

66. D. A. Hirschfeld, T. K. Li, and D. M. Liu, "Processing of Porous Oxide Ceramics," Porous Ceramic Materials: Fabrication, Characterization, Application Key Engineering Materials, vol. 115, (D. M. Liu, Ed.), Trans Tech Publications, Lebanon, NH, pp. 65–80, 1996.

67. H. Hagiwara and D. J. Green, "Elastic Behavior of Open-Cell Alumina," J. Am. Cer. Soc., **70**(11), pp. 811–15, 1987.

68. R. Brezny and D. J. Green, "Fracture Behavior of Open-Cell Ceramics," J. Am. Cer. Soc., **72**(7), pp. 1145–52, 1989.

69. R. Brezny, D. J. Green, and C. Q. Dam, "Evaluation of Strut Strength in Open-Cell Ceramics," Cer. Eng. & Sci. Proc., **9**(7–8), pp. 649–52, 1988.

70. A. J. Sherman, B. E. Williams, M. J. Delarosa, and R. Laferla, "Characterization of Porous Cellular Materials Fabricated by Chemical Vapor Deposition," Mechanical Properties of Porous and Cellular Materials," (K. Sieradzki, D. J. Green, and L. J. Gibson, Eds.), Mat. Res. Soc. Proc., **207**, pp. 141–49, 1991.

71. R. B. Bagwell and G. L. Messing, "Critical Factors in the Production of Sol-Gel Derived Porous Alumina," Mechanical Properties of Porous and Cellular Materials, (K. Sieradzki, D. J. Green, and L. J. Gibson, Eds.), Materials Research Society, Pittsburgh, PA, pp. 45–64, 1991. Materials Research Society Proceedings, vol. 207.

72. T. C. Huang and H. I. Chen, "Synthesis and Characterization of Gas Permselective Alumina Membranes," Mechanical Properties of Porous and Cellular Materials (K. Sieradzki, D. J. Green, and L. J. Gibson, Eds.), Materials Research Society, Pittsburgh, PA, Mat. Res. Soc. Proc., **207**, pp. 81–92, 1991. Materials Research Society Proceedings, vol. 207.

73. L. C. Klein and R. H. Woodman, "Porous Silica by the Sol-Gel Process," Mechanical Properties of Porous and Cellular Materials (K. Sieradzki, D. J. Green, and L. J. Gibson, Eds.), Materials Research Society, Pittsburgh, PA, pp. 109–24, 1991. Materials Research Society Proceedings, vol. 207.

74. S. J. Teichner, "Aerogels, Why They are in Vogue," Chem. Tech., pp. 372–77, 6/1991.

75. J. Fricke and A. Emmerling, "Aerogels," J. Am. Cer. Soc., **75**(8), pp. 2027–36, 1992.

76. D. A. Lindquist, et al., "Formation and Pore Structure of Boron Nitride Aerogels," J. Am. Cer. Soc., **73**(3), pp. 757–60, 1990.

77. F. Cluzel, G. Larnac, and J. Phalippou, "Structure and Thermal Evolution of Mullite Aerogels," J. Mat. Sci., **26**, pp. 5979–84, 1991.

78. R. Brezny and R. M. Spotnitz, "Method of Making Microcellular Ceramic Bodies," U.S. Patent No. 5,427,721, 6/27/1995.

79. H. Yamamoto and S. Masuda, "Electrostatic Formation of Ceramic Membrane, Porous Materials," (K. Ishizaki, L. Shepard, S. Okada, T.Hamasaki, and B. Huybrechts, Eds.), American Ceramics Society, Westerville, OH, pp. 305–14, 1993. Ceramic Transactions, vol. 31.

80. M. A. Anderson, M. J. Gieselmann, and Q. Xu, "Titania and Alumina Ceramic Membranes," J. Membrane Sci., **39**, pp. 243–58, 1988.

81. J. Block, "Inorganic Membrane," U.S. Patent No. 4,980,062, 12/5/1990.

82. K. Uchida, S. Isami, M. Kityotsuka, and K. Iwasaki, "A Ceramic Filter with Uniformly Distributed Cylindrical Holes," Ceramics Today-Tomorrow's Ceramics (P. Vincenzini, Ed.), Elsevier Sci. Pub., **66D**, pp. 2639–45, 1991.

83. K. N. Rai and E. Ruckenstein, "Alumina Substrates with Cylindrical Parallel Pores," J. Cat., **40**, pp. 117–23, 1975.

84. Y. F. Chu and E. Ruckenstein, "Design of Pores in Alumina," J. Cat., **41**, pp. 384–96, 1976.

85. Y. Kobayashi, K. Iwasaki, T. Kyotani, and A. Tomita, "Preparation of Tubular Alumina Membrane with Uniform Straight Channels by Anodic Oxidation Process," J. Mat. Sci., **31**, pp. 6185–87, 1996.

86. A. V. Kerkar, "Manufacture of Conical Pore Ceramics by Electrophoretic Deposition," U.S. Patent No. 5,340,779, 8/23/1994.

87. M. A. Andrews, A. H. Collins, D. C. Cornish, and J. Dracass, "The Forming of Ceramic Bodies by Electrophoretic Deposition, Fabrication Science 2," Proc. Brit. Cer. Soc., **12**, pp. 211–29, 3/1969.

88. I. M. Lachman, R. B. Bagley, and R. M. Lewis, "Thermal Expansion of Extruded Coordierite Cramics," Am. Cer. Soc., Bul., **60**(2), pp. 202–05, 1981.

89. S. M. Lang, "Properties of High-Temperature Ceramics and Cermets, Elasticity and Density at Room Temperature," National Bureau of Standards Monograph 6, U.S. Govt. Printing Office, Washington, D.C. (March 1960).

90. H. Neuber and A. Wimmer, "Experimental Investigations of the Behavior of Brittle Materials at Various Ranges of Temperature, Technical Report AFML-TR-68–23, Air Force Materials Laboratory, Wright-Patterson Air Force Base, OH (March 1968).

91. S. Spinner, F. P. Knudsen, and L. Stone, "Elastic Constant-Porosity Relations for Polycrystalline Thoria," J. Res., Natl. Bu. Stds., **67C**(1), pp. 39–46 (Jan-Mar 1963).

92. F. Pernot, J. Zarzycki, F. Bonnel, P. Rabischong, and P. Baldet, "New Glass-Ceramic Materials for Prosthetic Applications," J. Mat. Sci., **14**, pp. 1694–1706, 1979

93. R. H. Price, "Analysis of the Elastic and Strength Properties of Yucca Mountain Tuff, Nevada," Res. and Eng. Appl. of Rock Masses, **1**, pp. 89–96, 1985. Proceedings of the 26th U.S. Symposium on Rock Mechanics (E. Ashworth, Ed.,).

94. A. R. Boccaccini and G. Ondracek, On the Porosity Dependence of the Fracture Strength of Ceramics, Third Euro-Ceramics, **3** Engineering Ceramics (P. Duran and J. F. Fernandez, Eds.), pp. 895–900, 1993.

95. A. P. Roberts and M. A. Knackstedt, "Mechanical and Transport Properties of Model Foamed Solids," J. Mat. Sci. Let., **14**, pp. 1357–59, 1995.

INDEX